QUANTICS

Rudiments of
Quantum Physics

QUANTICS
Rudiments of
Quantum Physics

Jean-Marc Lévy-Leblond
and
Françoise Balibar

(*English translation by*
S. Twareque Ali)

1990

ELSEVIER
AMSTERDAM - LAUSANNE - NEW YORK - OXFORD - SHANNON - TOKYO

Published by:

North-Holland

Elsevier Science Publishers B.V.
P.O. Box 211
1000 AE Amsterdam
The Netherlands

ISBN: 0 444 87424 0 (hardbound)
         0 444 88120 4 (paperback)

First printing: 1990
Second edition: 1996

Library of Congress Cataloging in Publication Data

Lévy-Leblond, Jean Marc.
     [Quantique. English]
     Quantics: rudiments of quantum physics / Jean-Marc Lévy-Leblond and Françoise
Balibar ; English translation by S. Twareque Ali.
         p.   cm.
     Translation of: Quantique.
     ISBN 0-444-87424-0. — ISBN 0-444-88120-4 (pbk.)
     1. Quantum theory.   I. Balibar, Françoise. II. Title.
QC174.12.L4813   1989
530.1′2—dc20   89-23450
                                                                                            CIP

Printed in The Netherlands

Foreword

No other area of science has possibly witnessed such a proliferation of textbooks as quantum physics. To write yet another, surely requires an amount of presumption. The present text originated in our desire to do, not necessarily *better* than our predecessors, but at least something *different*.

Quantum physics stands out because of its conceptual difficulty and experimental complexity. Instead of hiding behind, as is often the case, an undoubtedly efficient and well polished mathematical machinery, we have decided to face these difficulties head-on. To the dull exploration of the solutions of a highly particular differential equation (the Schrödinger equation), we have preferred the adventurous discovery of a vast domain of the physics of our times. In other words, our intention is to open up full access to the ideas of quantum theory, beyond the techniques of manipulations to which the subject is so often reduced.

This reduction, which for fifty years has dominated the way in which quantum physics has been expounded and taught, has obvious historical roots. Quantum theory, after its debut and its early successes of the period 1920–1930, was helped by the clarity and rigour of its formalism, to attain wide dissemination, in spite of the difficulties and obscurities of its concepts. The early epistemological debate over its foundations and the interpretation of these concepts, which was far from being closed, has slowly given way to a grudging consensus, resigned to the efficiency of its methods. Hence followed this long line of books, which always devote a few pages, for the sake of form, to the philosophical problems, present a resumé of the orthodox vulgate ('the Copenhagen interpretation!'), and then just as quickly forget these major problems in order to pass on to serious matters: separation of variables in the Schrödinger equation, coupling of angular momenta, perturbation techniques and other formal gymnastics.

By now quantum physics has been maturing for half a century, and far from becoming an esoteric and marginal branch of physics, it underlies very many of its most active domains. Among these are molecular, atomic, nuclear, elementary particle, solid state, polymer and macromolecular physics (the latter of interest in biology). It also lies at the very heart of numerous experimental methods: quantum and neutron optics, electron microscopy,.... Some instrumentation techniques, partly quantum in nature, such as those which rely on lasers or electron microscopy, have extended beyond the precincts of university laboratories and are currently being used in industrial production or medicine.

As a consequence, the teaching of quantum physics at universities, which thirty years ago was relegated to graduate school, as initiation to research, now-a-days begins in the early undergraduate years, as part of the basic training. This transformation has not, to our knowledge, been accompanied by a parallel change in the didactical methods – a change that would have taken into account, the present day situation in quantum physics, as it is done and understood. Such a change had, of course, been initiated, in the sixties, by the pioneering work of R. Feynman (Volume 3 of the Lectures on Physics) and E. Wichmann (Volume 4 of the Berkeley Lectures on Physics), but they have, up to now, remained without sequel. It is not that these works have gone unnoticed – on the contrary, they have universally been acclaimed as being revolutionary, but only to adopt a more reformist approach. Nobody really dares to base an entire course in the spirit of these textbooks, and often, they are only used to breathe an extra bit of spirit (in some physical sense, let us say) into the traditional abstract and scholastic way of teaching. The teaching method of Feynman and Wichmann is not, after all, taken seriously. It is right in the wake of these predecessors that we would like to place our book – to the extent that it is meant to provide the reader with an introduction to quantum physics, starting at the undergraduate level, with the emphasis more on ideas than on formulae, and on demonstrating the concrete aspects of the concepts introduced.

Briefly, what we hope is the innovative character of this book, rests upon: *(1) A systematic and an early introduction of the qualitative and heuristic methods of physics,* such as are currently used in laboratory, but very rarely revealed in the classroom. Thus, in the first chapter, we attempt to familiarize the reader with the orders of magnitude of the quantum domain, using the full force of dimensional analysis. We do this by showing that theoretical work in physics cannot be reduced to just a formal manipulation of mathematical symbols. It also means teaching physics in the very idiom of its practitioners. Research in the quantum domain is never limited to (and often does not even include) solving a Schrödinger equation. Far from engaging in computations only, the researcher must *think* quantically, in terms of quantum angular momenta, life-times related to the widths of energy spectra, interferences between probabilities, etc. That electrons have a quantic behaviour *like* photons, is evident to someone working with an electron microscope – and it is based on such evidences that experiments on quantum systems are conceived, planned and executed. The heuristic point of view also allows us to quickly introduce certain ideas, which usually require the development of a sophisticated mathematical arsenal. Thus, for example, angular momentum, with its quantic specificities, appears as early as the second chapter, and the quantum theory of scattering is extensively treated in the fifth chapter.

(2) The constant appeal to modern experiments in quantum physics, in all the diverse domains in which they are carried out. Rather than being content with citing the usual historic experiments (Davisson and Germer, Stern and Gerlach, etc.), we have included a large number of examples of fundamental quantum phenomena, borrowed from nuclear physics, as well as from atomic physics, optical coherence, physics of solids and neutron physics, chosen from professional journals of the past few years. In this way, we have tried to demonstrate the broad sweep and fertility of quantum ideas. In particular, a number of these phenomena, sometimes of great importance, have been included in the exercises following each chapter. These exercises must often be looked upon as further development of the text, more than as mere academic applications.

(3) The adoption of modern theoretical points of view on quantum physics, as opposed to the very customary chronological approach. Rather than forcing the student to follow the long and arduous historical route, leading to the elaboration of the theory, we have chosen to build on the more recent, but more profound, foundations – cleared and polished by the past half-century of clarification and recasting. Consequently, we only require a rudimentary knowledge of classical mechanics, from the reader, and no familiarity with its more sophisticated aspects (Hamiltonian formalism, Poisson brackets, etc.) On the other hand, in keeping with the contemporary approach, we attach great importance to invariance principles and to notions of symmetry, which are extremely powerful in quantum theory. But we do not assume any sophisticated previous acquaintance with group theory, and these principles are applied on the most elementary level.

(4) A conscious effort to shed light on the conceptual difficulties, without concealing them behind technical complexities, and thus confronting them directly. At the same time, we have attempted a delicate operation of renewing and trimming the terminology. After all, quantum theory is littered with a whimsical or unfounded vocabulary, which often only serves to obscure the problems. A notable example is the term 'observable', pertaining to a very specific philosophical point of view, and for which we have preferred to use the simpler and clearer term 'physical magnitude' – even though that means having to underscore the quantum specificity of this notion. Similarly, we have deliberately rejected the 'uncertainties' attributed to Heisenberg – the source of so much pedagogical and philosophical misunderstanding. It is in this same spirit that we propose to do away with the term 'quantum *mechanics*', which is at once too restrictive (it does not encompass quantum field theory), and too strongly reminiscent of the very physical (mechanistic)

concept which it supersedes. The way in this case has been shown to us by our students who readily talk about their 'quantics examination', 'quantics course'*, etc. We have decided to follow this route, which after all is rather commonplace, since in this way we simply place quantics side by side with mechanics, and of course, thermodynamics, electronics, acoustics – not to mention … physics itself.

The present text, the fruit of many years of actual teaching, has gone through many successive versions, and has been tried out by numerous colleagues with many groups of students. Above all, it is therefore, a teaching manual. But, we would wish that it might also be used in the laboratory, in particular, by the experimentalists from whom we have borrowed so freely. One often hears research workers expressing the desire to widen their professional culture, to deepen or rejuvenate their primary education. Such an aspiration does not come from an abstract desire to become generally cultured. Rather, it reflects the desire to increase their ability to picture, interpret and understand physics – *their* physics. To satisfy this need, these researchers all too often only have at their disposal daunting and sophisticated treatises, which they find intimidating, since they have the impression that they would only find abstract answers to their very concrete questions. Perhaps, the present book would live up to their expectations better.

Our project consists of three separate volumes:

(A) Rudiments. This first volume is an introduction, essentially at the early undergraduate level. The qualitative, heuristic and experimental aspects have been given precedence. The idea is to impart an active comprehension of quantics and a global view of its applications, not necessarily demanding a subsequent in-depth understanding. In other words, this book is aimed at all those who wish to, or have to, learn what quantics is all about, even though it may not be their preoccupation later as specialized students or professional practitioners. Similarly, future chemists, biologists and engineers should find this book an adequate baggage, free from the all to numerous technicalities, which are so often imposed upon them. Indeed, we would wish that this volume might serve the purposes of affording a scientific education to interested non-scientific lay people, be they philosophers, journalists or amateurs of science.

---

* Translator's footnote: The original French terms are 'examens de quantique' and 'cours de quantique'. The translated English terms, suggested here, are not *yet* popular among Anglophonic students!

(B) Elements. The second volume will put into action, in a rigorous way, the general ideas introduced in the *Rudiments.* Giving an overall mathematical structure which would be more modern than found in many existing texts, it would still remain at a level of technical difficulty which does not go appreciably beyond that of the early undergraduate years. It would, however, go beyond that framework in abundance of material. This book would offer, therefore, both the means to extend the undergraduate teaching of quantics to, for example, future specialists in physics, and to make the bridge to the exigencies of graduate work. The material contained in this book would, however, remain restricted to the fundamentals of quantics.

(C) Complements. This third volume would consist of a development of the theory, which would allow it to be applied effectively in most of the complex situations encountered in practice, for example, the various commonly used approximation methods. It would also sketch the way to extend elementary quantics to field theory, the N-body problem as well as its applications to various particular but important phenomena (such as superconductivity, or the Mössbauer effect).

J.-M.L.-L., F.B.
Nice-Paris 1974–1984.

Acknowledgements

We would like, first of all, to thank Chantal Djankoff and Patricia Meralli, who, with patience and competence, have typed the text in its successive versions and numerous modifications. Equally, we would like to thank Marcel Desban, who produced the illustrations in this book, Marianne Maury, who drew fig. 2.18, as well as Julie Brumberg-Chaumont and Alice Lévy-Leblond for their help in producing fig. 2.21.

We have benefitted immensely from numerous suggestions and corrections, proposed by our colleagues Odile Betbeder, Cécile Malgrange, Alain Laverne and Michel Hulin, who have been such attentive and helpful readers, as well as, for certain passages, Daniel Boutet, Claude Guthmann and Bernard Roulet.

A host of other people have helped us enrich the documentation. We are deeply indebted for our information to Bernard Cagnac (multi-photonic processes, ch. 2), Ferenc Mezei and Michel Schlenker (neutron physics, chs. 2, 4 and 6), Luc Valentin (Coulomb scattering, ch. 7), Franck Laloë (quantum fluids, ch. 7), Madeleine Gandais (electron microscopy, ch. 4). Exercise 6.4 is an adaptation of a problem made up by J. L. Basdevant and Y. Quéré. The introduction of the bosonic factorial (ch. 7, sect. 5.3) is based on an idea of Gérard Vallée, Jean-Pierre Provost and Jean-Louis Meunier.

For the corrections incorporated into the second edition, we are greatly indebted to many careful readers, in particular, Jean-Louis Meunier and Guy Plaut.

Finally, we must thank the successive generations of students who served as test groups, during the development of this course.

Translator's apology

The laws of physics, like all other laws, can be translated. The mathematics of physics transcends the idiosyncrasies and nuances of language. It does not need to be translated. But an author's perception of the subject, and the manner in which this understanding is conveyed to the reader draws, in a book, heavily upon the idiom of the language in which the book is written. That can never be translated. Rendering from French, a language which prides itself in its explicit subtleties, into English, in which subtleties are expressed mostly by being implied, is almost like telephoning a grimace. That is not to say, however, that I have not enjoyed, enormously, translating this sumptuous book. I wish that as an undergraduate, I had learnt quantum mechanics, or rather quantics, from this book. Needless to say, residual obscurities in the translated version are attributable to my own incomplete mastery over both languages – and of course, over the subject matter itself.

More than anything else, I feel proud and honoured to have made two good friends in the two authors. To the extent possible, I have tried to incorporate all their suggestions for clearing up errors and mistranslations. The result, I hope, is an English version which, at least in its 'quantic content', reflects the French original accurately.

I would like to express my gratitude to Françoise and Jean-Marc for their unending patience.

There are several other people, thanking whom is a pleasure: Cécile DeWitt-Morette, who suggested that I translate the book, Mark A. J. R. Eligh, Acquisition Editor for the North-Holland Physics programme of Elsevier Science Publishers B.V., in whose patience I indulged beyond limit, my son Nabeel, whose language skills are infinitely superior to mine, Brigeen Badour, Tezeta Taye and Carole Vachon, who took turns through the thankless chore of typing and retyping, and Julia Patterson, who read, re-read and and proofread.

S.T.A.
Montréal
Jan. 1989.

xiii

Contents

1

The quantum regime

1.1. Quantics today

The science of quantics saw the light of the day, at the turn of the present century, as physicists attempted unsuccessfully to understand certain phenomena, such as the photoelectric effect, the radiation from atoms and their constitution, using the existing concepts of mechanics – henceforth to be called 'classical mechanics'. This mechanics, developed since the seventeenth century, had witnessed, particularly during the preceding hundred years, a success as remarkable in the study of the movements of heavenly bodies as in its applications to the fields of technology. One might have been tempted to believe that the concepts of classical mechanics were universal and even final. Indeed, notions such as those of a point particle, its trajectory, its instantaneous velocity, or that of a force, seemed so natural that one could nearly forget the tortuous process which led to the emergence of these ideas, rooted as they are in our limited material experience. It was certainly not without reluctance that these ideas were abandoned, as it was decisively revealed that, in the new domains in which nature was being examined, it refused to conform to the traditional conceptual framework of the physicist.

Ever since its emergence, however, quantum physics has continually consolidated its hold by deepening its conceptual basis. Although new physical phenomena have continued to be discovered for half a century now, none of these has endangered as yet the validity of the basic quantum concepts. One obviously has to admit that these concepts will some day be rendered inadequate, and be replaced by newer ideas, in much the same manner in which the concepts of quantum mechanics have themselves replaced those of classical mechanics. For the time being, however, they do hold well. It is even more remarkable, that although quantum theory found its origin in investigations within the atomic domain, which is characterized by distances of the order of an angström, and by energies of the order of a few electron volts, its concepts have actually been applied to the study of elementary particles at distances a million times smaller and at energies a billion times higher.

One is thus led to affirm the universality of present-day quantum theory, to the extent that it offers one a general conceptual framework encompassing physical theories, such as, e.g., those of electromagnetism or of nuclear forces, all of which describe a specific set of physical phenomena. It is customary, therefore, to use as we have done, the term 'quantum theory' to generically describe this universal framework, while reserving the term 'quantum mechanics' to describe the simplest of its realizations. It is specifically this latter which will be the subject of a detailed study in this book. Table 1.1 summarizes the different types of quantum theories in use today. It is probably worthwhile to underscore the fact that since the adopted terminology reflects a certain historical development of the subject, it is often inadequate from a modern point of view. Thus, terms such as 'quantum mechanics' and 'quantum field theory', reflect in their very wording a classical connection, a framework from which they have largely broken away, and this mainly is the reason why we propose the use of the simpler and more compact term 'quantics'.

The universality of quantum theory, however, is basically one of principle. Indeed it is clear that there are numerous situations where the concepts of

Table 1.1. *Summary classification of various quantum theories*

It is possible to build a quantum theory in accordance with a specific notion of space–time. Within the Galilean framework, the number of particles in the system under study is usually assumed to be fixed – a restriction which allows one to construct theories where this number could be very small (one or two in the simplest of cases). By contrast, in the Einsteinian framework, neither the mass of a particle, nor, in general, the particle number is a conserved quantity, and right away the theory has to admit situations where this number might well vary.

	Space–time frame		
	Galilean		Einsteinian
Number of particles in the system	Small and constant	Large (infinite or variable)	Variable (infinite)
Relevant quantum theory	Quantum mechanics	Quantum field theory	
		'Non-relativistic' ('*N*-body problem')	'Relativistic'
Typical domain of application	Atomic and molecular physics	Condensed matter physics (solids and liquids)	Elementary particle physics (quantum electrodynamics, quantum chromo-dynamics).
	Nuclear physics		

classical mechanics retain their validity – thus, neither the flipper (pin-ball) fan, nor the cosmonaut, for example, has the need in his trade, to abandon the notion of a force or of a trajectory. However, classical theories (of both mechanics and fields) appear as approximations to the quantum theory: approximations whose validity is excellent over a wide range of different physical conditions. Yet one should guard against trying too hastily, either to identify the domain of validity of classical theories with the entire macroscopic world, or to limit the use of quantum theory exclusively to the microscopic domain. The importance of quantum theory is not restricted merely to the description of certain exotic phenomena, far removed from one's daily world of experience. Today, there not only exist products and instruments, such as lasers, transistors and superconductors, which quantics alone has rendered possible, but even the existence of matter itself, as we know, is incomprehensible within the context of classical physics (fig. 1.1). For example, classical physics is able neither to explain the stability of matter, nor to describe the structure of atoms at either the collective or at the individual level. We begin, therefore, by a sharper delineation of those areas of the physical world where a recourse to quantum theory is imperative.

a) b)

Fig. 1.1. *Quantum objects*
(a) The 'microchip', the basic constituent of modern micro-electronic technology, owes its existence basically to the development of the quantum theory of solids. Document Group Bull.
(b) But even 'ordinary' matter at our level, in its most natural manifestations, cannot be properly understood except in quantum terms. Ph. Jeanbor © Archives Bordas.

1.2. The quantum constant

1.2.1. Fundamental constants and quantum physics

Every physical theory establishes relationships ('laws') between some of its fundamental concepts. Thus, it introduces 'fundamental constants' which depend upon an arbitrary choice of units of measurement, and which, upon a subsequent new choice of units, consistent with the laws of the theory, may *apparently* be eliminated. To illustrate the situation by a simple example, let us suppose that a unit of length (called the *foot*) and a unit of area (called the *palm*) have been independently defined. A fundamental law of geometry, considered here as constituting the physics of space, establishes the relationship $S = Ka^2$ between the length a of the side of a square, expressed in feet, and its surface area S, expressed in palms, where K is a 'fundamental constant', having the approximate value $K \simeq 8.78$ palms/foot squared. Naturally, on account of the geometrical law so established, one would be interested in changing the system of units by replacing, for example, the unit of area by a foot squared (i.e., by a square foot*). One would then have, in this new system, the relationship $S = a^2$, and the 'fundamental constant' K, having been reduced to unity, would seem to have disappeared. 'Classical' mechanics, because it is classical (i.e., having been dove-tailed over a long period of time) does not possess explicit fundamental constants any longer – they all lie effectively buried within the systems of units which are adopted to express its laws. In electromagnetism, the situation is less simple. There, the rationalized MKSA (metre–kilogram–second–ampere) system of units introduces into the expression $F = (1/4\pi\varepsilon_0)(qq'/r^2)$, for the Coulomb force, a fundamental constant $1/4\pi\varepsilon_0$, resulting from separate choices for systems of mechanical and electrical units. Statistical mechanics may be characterized by the Boltzmann constant k (appearing in the relation $E = kT$, cf. exercise 1.1) and the Einsteinian relativistic mechanics by c, the velocity of light (appearing, e.g., in the relation $E = mc^2$).

The application of a physical theory to a particular process must, therefore, lead to an expression of the magnitudes being studied as functions of the relevant physical parameters and the fundamental constants of the theory at hand. Very often, dimensional analysis alone enables one to write down the expressions being sought, up to a simple numerical factor, which a detailed application of the theory then allows one to evaluate exactly. For example, the oscillations of a simple pendulum of length l and mass m, in a field of

---

* Parentheses added by translator.

gravitational acceleration g, must necessarily have the period $T = k(l/g)^{1/2}$. The laws of mechanics then demonstrate that $k = 2\pi$. Conversely, if some physical magnitude of interest cannot be expressed, even dimensionally, with the help of the characteristic parameters of the system being studied and the fundamental constants of the theories being used, then one has by the same token, the proof of the inadequacy of these particular theories. It was in this manner, that at the turn of the century, the classical theories (of mechanics, electromagnetism and thermodynamics) revealed themselves as being incapable of explaining various important phenomena, such as the photoelectric effect, black-body radiation or even the stability of the atoms which constitute matter.

The photoelectric effect, for example, had had classical physics decisively checkmated. Towards the end of the nineteenth century, it was discovered that ultraviolet electromagnetic radiation was, in general, capable of expelling electrons from metals. However, contrary to what might have reasonably been expected, the energy of the expelled electrons did not depend upon the intensity of the incident radiation (only the number of electrons emitted was proportional to it). Using monochromatic radiation, experimentalists were able to demonstrate that the kinetic energy E of the electrons depended linearly on the angular frequency (pulsation) ω of the radiation, i.e., $E = \hbar\omega - W$, where the constant $\hbar$ (called '$\hbar$-bar': we shall discover later the reason for this bizarre nomenclature) depended neither on the intensity of the radiation nor on the particular metal being used, while, by contrast, W, the 'threshold energy' was a characteristic of the metal (fig. 1.2). Had this constant to be explained classically, it would only have been possible to do so after a detailed analysis of the electromagnetic interaction between the radiation field and the electrons of the metal. In that case, it ought to have been possible to express it in terms of the quantities characterizing the electron, namely, its mass and charge, and the fundamental constants of electromagnetic theory. All attempts in this direction ran aground, and it soon became clear that $\hbar$ was a new fundamental constant of physics (cf. exercises 1.2 and 1.3). As is clear, the constant $\hbar$ sets up a connection between two physical concepts which so far had been disparate – the energy and the pulsation* (or the frequency). We shall return to this point later in ch. 2. Here we content ourselves by stating that $\hbar$ is *the* fundamental constant which

---

* The word 'pulsation' is used here, and later, in preference to the more commonly used term, 'angular frequency'. This choice has the advantage of being more descriptive, as well as being symmetric to the term 'undulation', introduced in the following to denote the wave number. (Note added by translator.)

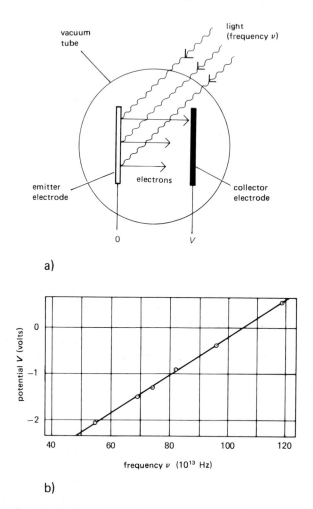

Fig. 1.2. *The photoelectric effect*

(a) A beam of light of frequency v impinges upon an electrode. If the energy hv of a photon is greater than the work W needed to release an electron, the latter is emitted with a kinetic energy $E = hv - W$, measured by the negative (repulsive for electrons) potential $V = E/q_e$ necessary to neutralize the photoelectric current. If on the other hand, $hv < W$, the electron would be short by the amount of energy $E = W - hv$, in order to be emitted from the metal; a photoelectric current would only flow if a negative (attractive for electrons) potential $V = E/q_e$ is applied. In both cases, the threshold potential necessary for the passage of a current is given by $V = (hv - W)/q_e$.

(b) This linear relationship between the potential and the threshold frequency is perfectly confirmed by experiment. This graph shows the 'historic' results of R. A. Millikan [Nobel Lectures in Physics (1922–1941); Elsevier, Amsterdam 1965, p. 63] for sodium.

characterizes quantum theory. We shall call it 'the quantum constant'. Its numerical value in the rationalized MKSA system of units is

$$\hbar = (1.054592 \pm 0.000006) \times 10^{-34} \text{ MKSA units.} \tag{1.2.1}$$

It is particularly easy to remember the approximation,

$$\hbar \simeq 10^{-34} \text{ MKSA units.} \tag{1.2.2}$$

Historically, it was not $\hbar$, but rather the quantity

$$h = 2\pi\hbar, \tag{1.2.3}$$

having the numerical value

$$h = 6.625 \cdots \times 10^{-34} \text{ MKSA units,} \tag{1.2.4}$$

that had been introduced by Planck (in 1900) in the study of black-body radiation, marking thereby the birth of quantum physics, or rather its conception (for indeed, its period of gestation was to last some two decades). In fact, if one were to reason in terms of the frequency, rather than the pulsation, one would write for the 'quantum' of the energy of radiation,

$$\varepsilon = \hbar\omega = 2\pi\hbar v = hv. \tag{1.2.5}$$

The 'Planck's constant' h was, thus, born in association with the concept of frequency. It is, however, the concept of pulsation which is preferable by far, for the needs of the theory. (Do we not prefer to write $\sin(\omega t)$ rather than $\sin(2\pi v t)$ for harmonic evolution?) The fact is, that it is ω^{-1}, and not the period v^{-1}, which describes the characteristic time evolution of a harmonic phenomenon better (fig. 1.3). Using the pulsation ω and the constant $\hbar$, the superfluous factor of 2π is eliminated from the theoretical expressions. It is $\hbar$, and not h, which is the true constant of quantics – the one which minimizes the importance (i.e., their deviation from unity) of the unknown numerical coefficients which figure in the various theoretical expressions. Its use confers maximal validity to what we ought to call the Zeroth Principle of Physics: '*All dimensionless numerical coefficients must approximately be equal to one*'. We sharpen this immediately by saying that this principle only holds if one adds to it the proviso, '*... once one has correctly chosen the physical magnitudes which are considered fundamental to the problem being studied*'. This choice is a

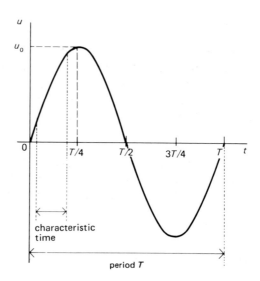

Fig. 1.3. *Frequency or pulsation (angular frequency)?*
Consider a sinusoidal phenomenon $u = u_0 \sin(2\pi v\tau)$, of frequency v and hence of period $T = v^{-1}$.
The characteristic time of the phenomenon is the time over which there is a 'notable' change, i.e.,
the time needed to pass from a small value (close to zero) to a large one (close to the maximum
u_0). One could, for example, take it to be a quarter of the period, $\frac{1}{4}T$, i.e., the time necessary for
the phase to change by $\frac{1}{2}\pi$. It is still more natural to take it to be the time necessary for the phase
to change by one 'natural' unit of angle (the radian), i.e., $\omega = T/2\pi$. Thus the pulsation ω is a
more natural measure of the rate of temporal evolution of the phase than the frequency v, and its
use in computations involving dimensional analysis eliminates cumbersome factors of 2π. We
note, however, that factors of 2π reappear at the interface between theory and experiment:
experimentalists measure periods or entire cycles (not radians) and hence frequencies (not
angular frequencies).

practical matter – the famous 'physical intuition' – and it is to a better
development of this 'feeling' that the present chapter is devoted (cf. sect. 1.3.2,
and most of the exercises, in particular, the counterexamples in exercises 1.3
and 1.5). Thus, order of magnitude estimates, using dimensional analysis,
turn out to be remarkably good if one uses $\hbar$, while the use of h may lead to
serious errors (a factor of $2\pi \simeq 6$, squared or cubed, is not to be neglected).
Going back to the example of the classical pendulum, a dimensional
evaluation of the angular frequency yields $\omega = k(g/l)^{1/2}$, with an unknown
numerical factor which the theory proves to be equal to one – and not to
0.159... as would have been the case with frequency.

1.2.2. Dimensional analysis

From the law governing the photoelectric effect, $E = \hbar\omega - W$, we can extract the dimensional equation for the constant $\hbar$. In a system which has as basic magnitudes, the mass (with dimension $\mathcal{M}$), the length ($\mathcal{L}$) and time ($\mathcal{T}$), one has the usual dimensional equations, $[\omega] = \mathcal{T}^{-1}$ and $[E] = \mathcal{M}\mathcal{L}^2\mathcal{T}^{-2}$, from which the dimension of $\hbar$ appears as

$$[\hbar] = \mathcal{M}\mathcal{L}^2\mathcal{T}^{-1}. \tag{1.2.6}$$

But then, what is the significance of this expression? It is worthwhile to rewrite this in terms of some of the usual physical magnitudes. We know already, since this has been the basis of our calculations, that

$$[\hbar] = [\text{energy}] \times [\text{time}]. \tag{1.2.7}$$

From the relation $\mathcal{M}\mathcal{L}^2\mathcal{T}^{-1} = \mathcal{M}\mathcal{L}\mathcal{T}^{-1} \times \mathcal{L}$, one obtains

$$[\hbar] = [\text{momentum}] \times [\text{length}]. \tag{1.2.8}$$

One observes, between eq. (1.2.7) and eq. (1.2.8), the remarkable spatio-temporal correspondence which exists, respectively, between the pairs energy–time and momentum–position: energy is linked to time (more precisely, its conservation is associated to invariance under time trans-lations), just as momentum is linked to space. This remark already renders plausible the compatibility of the structure of quantum physics with the Galilean theory of relativity (or the Einsteinian theory, for that matter), i.e., the possibility of adapting to quantum theories, without major alterations, the notions of space–time germane to the particular theory of relativity being considered. By invoking the notion of the angular momentum, related to angles (its conservation being associated to the invariance under rotations), one can immediately conjecture a third relation, analogous to eqs. (1.2.7) and (1.2.8), and write

$$[\hbar] = [\text{angular momentum}] \times [\text{angle}]. \tag{1.2.9}$$

Since one works in a system of units, such as the MKSA system, where angles are dimensionless quantities (the radian is a ratio of two lengths), one has, in fact,

$$[\hbar] = [\text{angular momentum}], \tag{1.2.10}$$

a relation which, in view of eq. (1.2.8), is rather obvious. The constant $\hbar$ appears as the natural quantum unit of angular momentum.

It will be useful in the following to have at our disposal, a generic term for any physical magnitude which has the dimension of $\hbar$. The Lagrangian formalism of classical mechanics defines under the name 'action', either an integral of the energy over time, or an integral of the momentum over space, which has, therefore, this dimension. Furthermore, this concept played an important role in the course of the pre-history of quantics, during which time Planck's constant had received the appellation 'quantum of action'. Quite apart from the concept itself, we retain here the terminology, and christen as 'action' any magnitude which has the dimension $\mathcal{M}\mathcal{L}^2\mathcal{T}^{-1}$, denoting it, in general, by A. It will be equally useful for our purposes to give a name to the unit of action in the MKSA system of units. We christen this unit 'the Lagrange' and abbreviate it as L – which in addition, is eminently suitable as a unit of angular momentum (often denoted by L or l). Thus one has

$$1 \text{ L} = 1 \text{ kg m}^2 \text{ s}^{-1} = 1 \text{ J s} \simeq 10^{34}\hbar. \tag{1.2.11}$$

As has been amply demonstrated in the dimensional equations [eqs. (1.2.7)–(1.2.10)], one can evidently express an action with the help of many different physical magnitudes. A frequently used relation is the relation which expresses the action in terms of a mass, a length and an energy – physical magnitudes which are used particularly often. These three magnitudes are independent, and there must necessarily exist a unique monomial combination which yields the dimension of an action. One verifies easily, upon writing $(\mathcal{M}\mathcal{L}^2\mathcal{T}^{-1})^2 = (\mathcal{M}\mathcal{L}^2\mathcal{T}^{-2})(\mathcal{M})(\mathcal{L}^2)$, that this relation is

$$[\text{action}]^2 = [\text{energy}] \times [\text{mass}] \times [\text{length}]^2. \tag{1.2.12}$$

1.3. Limits of validity of classical physics and the need for a quantum physics

1.3.1. The quantum criterion

The Planck's constant characterizes, as we have seen, quantum physics. It is the natural standard for expressing any 'action' which is characteristic of a physical system. In this capacity, its primary role is to delimit the domain of validity of classical theories. These theories do not constitute valid approximations to the quantum theory, and hence are not trustworthy, unless in the given situation, all physical magnitudes of the type of an 'action' are very large compared to $\hbar$. As soon as an action is comparable to $\hbar$, it becomes impossible to neglect quantum effects and to content oneself with the classical approximation. Conversely, before proceeding to analyze a physical phenomenon with the entire arsenal of a quantum theory, it is worth the trouble to assure oneself that the classical theory would not suffice. The recipe for this is simple: using the parameters characterizing the physical problem, form quantities of the type of an 'action' and compare them to $\hbar$.

> Action of the order of $\hbar$ = Quantum physics

Thus, whenever the characteristic action of the phenomenon is of the order of the quantum constant, a recourse to quantum theory is necessary and reciprocally, if the action is much larger than $\hbar$, the approximation furnished by the classical theory is sufficient. It is appropriate at this point to realize, however, that the line of demarcation between the domains of the quantum and the classical theories does not coincide with the one separating the macroscopic from the microscopic world. Indeed, a characteristic action can be computed using many different physical magnitudes, some of which may well assume values that are quite noticeably macroscopic in scale, with the action thus calculated still being of the order of the quantum constant. In other words, there exist macroscopic physical systems, some of whose properties are unexplainable classically – requiring for this purpose, a quantum theory. This is the case for the radiation from a black body (cf. exercise 1.7), for superconductors and superfluids [cf. example (f) below and exercise 1.6], for lasers, and above all, for ordinary matter [cf. example (e) below and exercises 1.11, 1.12]. Conversely, it is not excluded, and in fact is rather frequent, that in certain respects microscopic phenomena can be correctly described by the classical theory.

Finally, one ought to add that by virtue of quantum theory itself, it is not possible for a physical phenomenon to display a characteristic action much smaller than $\hbar$. If a combination of physical variables yields such an action, then it must be devoid of any physical significance*; there do not exist actual phenomena which are characterized by these variables (cf. example (f) and exercise 1.5).

---

* We can understand the role of the constant $\hbar$ by reconsidering, for example, the constant K which appeared in the geometry of the plane, for the expression of the area as a function of the length, $S = Ka^2$, for a square of side a. To prove the necessity of a theory of areas, we form, as before, characteristic quantities of dimension (length)^2/(area) and compare them to $1/K$ (which

1.3.2. Examples

(a) An ordinary mechanical *watch* has mobile parts (it is these rather than the whole watch itself that need to be considered here) which typically are of size $d \simeq 10^{-4}$ m, mass $m \simeq 10^{-4}$ kg and have a typical time $\tau \simeq 1$ s (obvious!). The characteristic action is, therefore, $A \simeq md^2\tau^{-1} \simeq 10^{-12}$ L $\simeq 10^{22}\hbar$. Watch makers do not possibly need to study quantum theory (even with quartz watches, having microprocessors).

(b) *A radio antenna* having a power output $P = 1$ kW, emits at a frequency $v = 1$ MHz. As we have said before, it would be better to use here the angular frequency $\omega = 2\pi v = 6 \times 10^6$ s^{-1}. Since $[P] = \mathcal{M}\mathcal{L}^2\mathcal{T}^{-3}$ and $[\omega] = \mathcal{T}^{-1}$, it is easy to obtain the characteristic action, $A = P\omega^{-2} \simeq 3 \times 10^{-11}$ L $\simeq 3 \times 10^{23}\hbar$. Later, we will see (ch. 2, sect. 2.2.1) that the number 3×10^{23} is, up to a factor of 2π (once again!), the number of photons emitted by the antenna during one cycle of its oscillation.

(c) *An oscillating electrical circuit* consisting, e.g., of a capacitance $C = 10^{-10}$ F and an inductance $L = 10^{-4}$ H is traversed by a current $I =$

---

plays here the role of $\hbar$). As an example, we consider a highly elongated rectangle, with sides a and b with $a \gg b$, and having an area S, this latter being assumed to be measured directly by some non-geometrical method (for example, by measuring the weight of a lamina of known surface density and cut to the size of the surface being measured). Suppose that we are unaware of the relation $S = Kab$, given by the theory of areas, and we ask ourselves if S could be calculated by means of a theory involving the constant K. We can form the three quantities ab/S, a^2/S and b^2/S. The first combination is evidently of the order of $1/K$, from which we conclude that a theory of plane surface areas characterized by K has a good chance of permitting us to calculate S in terms of a and b. The second combination is much larger than $1/K$ which indicates that the rectangle, considered here in terms of the single dimension a – thereby, degenerating into a kind of a strip – depends on a simpler theory, a one-dimensional geometry in which the constant K does not feature (in this theory, the area S is simply proportional to a, up to a certain empirical constant: the area per unit length). Finally if $b^2/S \ll 1/K$, then it simply means that b alone is not sufficient for characterizing the surface and evaluating its area; in plane geometry there do not exist figures having a length of the order of b, as a single characteristic dimension and an area S satisfying such an inequality – one always has $S \geqslant Kd^2$, where d is the minimal characteristic dimension of a surface of area S. At the very least, one may affirm that if one were to find a surface obeying the inequality $d^2/S \ll 1/K$, then such a surface could not obey the laws of plane geometry, i.e., its very existence would be testimony to their inadequacy – the surface could, for example, belong to a curved Riemannian space. Similarly, if one were to discover one day a physical phenomenon possessing a characteristic action, which is definitely pertinent to it and whose value is much smaller than that of the quantum constant, then this phenomenon would have escaped the scope of quantum mechanics, bringing to evidence the limits of its validity.

10^{-3} A. We leave to the reader the task of showing that $A = L^{3/2} C^{1/2} I^2$ (it is not necessary to calculate $[L]$ and $[C]$ – suffice it to recall that LI^2 is an energy and $(LC)^{1/2}$ a time!), hence,

$$A \simeq 10^{-17} \, L \simeq 10^{17} \hbar.$$

(d) *The atom.* We consider the example of hydrogen. Quite independently of any theoretical knowledge of the atomic structure, the experimental results obtained at the beginning of this century indicate that the ionization energy of a hydrogen atom is $E = 13.6 \, \text{eV} \simeq 2 \times 10^{-18}$ J. Furthermore, its spectrum is characterized by a minimal wavelength $\lambda \simeq 10^3$ Å (in the ultraviolet), implying a maximal angular frequency $\omega \simeq 2 \times 10^{16} \, \text{s}^{-1}$. One deduces from this the characteristic action $A = E/\omega \simeq 10^{-34} \, L \simeq \hbar$, and concludes, therefore, that there is no possibility of understanding the hydrogen atom without recourse to quantics.

(e) *Crystalline structure.* A typical solid crystalline substance, such as, e.g., ordinary table salt (NaCl), has its atoms arranged regularly in a cubic lattice (fig. 1.4). The distance between an atom of sodium and a neighbouring atom of chlorine can be measured by means of X-ray diffraction experiments, in which the crystal plays the role of a diffraction grating. One finds in this manner a distance $a = 2.81$ Å as a characteristic of the crystalline structure.

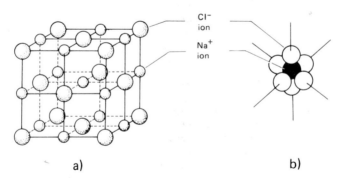

a) b)

Fig. 1.4. *The salt crystal*
(a) In an ordinary salt crystal NaCl, the ions Na^+ and Cl^- alternately occupy the vertices of a cubic lattice. For clarity of illustration, the distances between the ions have been shown considerably magnified compared to the sizes of the ions.
(b) A somewhat more realistic view of a Na^+ ion, surrounded by its six nearest-neighbour Cl^- ions (at the vertices of an octahedron).

Furthermore, the binding energy per 'molecule' of NaCl – deduced essentially from energy balance considerations in chemical reactions involving dissociation of the crystal (e.g., dissolution in water) – is measured by classical methods. One finds that $E = 183$ kcal/mol $= 8.0$ eV/molecule. Finally, taking a typical intermediate mass between that of sodium (20 atomic units or amu – recall that 1 g $= N$ amu, $N = 6 \times 10^{23}$ being the Avogadro number) and that of chlorine (35 atomic units), namely, $M \simeq 30$ amu $= 5 \times 10^{-26}$ kg, one can form a characteristic action $A = (ME)^{1/2}a$ [cf. eq. (1.2.12)]. Numerically, $A \simeq 7 \times 10^{-32}$ L $\gg \hbar$. However, it could be the case here that the correct characteristic mass to consider is that of the electron, if after all it is true that the bonding of crystals is due to the dynamics of the valence electrons. In the absence of a detailed theory, we may at the very least pose such a question: one would have then a characteristic action $A' = (ME)^{1/2}a \simeq 2.9 \times 10^{-34}$ L $\simeq 2.9\hbar$. Thus, if the structure of the crystal is determined by the behaviour of the electrons, then it has a quantum nature. This example demonstrates the ambiguity of a purely formal reasoning and the necessity of adding a little bit of 'physical intuition' to dimensional analysis.

(f) *Superfluidity.* At ordinary pressures, helium liquefies at the extremely low temperature of $T_l \simeq 4.2$ K; it undergoes, at an even lower temperature, a phase transition which endows it with curious properties. Below the 'lambda point', at a temperature $T_\lambda = 2.18$ K, helium (called 'helium-II' in this phase) becomes *superfluid*, i.e., it flows with zero viscosity. As a result, it can, by spontaneous capillary action, flow out of a container in which one tries to hold it. It also possesses interesting thermal properties, in particular an extraordinarily high thermal conductivity (fig. 1.5). Does this phenomenon have a quantum nature? The transition temperature T_λ furnishes us with an energy kT_λ; in addition, we may use the density of helium $\rho = 1.46 \times 10^2$ kg m^{-3} and its atomic mass, $M \simeq 4$ amu $\simeq 6.7 \times 10^{-27}$ kg, to obtain the characteristic action, $A = M^{5/6}(kT_\lambda)^{1/2}\rho^{-1/3}$. The reader, surprised by these fractional exponents, should note that $a = (M/\rho)^{1/3}$ is the average distance between neighbouring helium atoms, in terms of which the calculated action may be rewritten as $A = M^{1/2}(kT_\lambda)^{1/2}a$, in conformity with eq. (1.2.12). In any case, numerically $A \simeq 1.7 \times 10^{-34}$ L $\simeq 1.7\hbar$, from which one concludes the necessity of a quantum explanation of superfluidity. Contrary to the preceding case, here it is not the mass of the electron which intervenes, but rather that of the atom; the phenomenon is, therefore, not specifically electronic, and seems more to depend on the helium atom as a whole. Indeed, had we utilized the mass of the electron to calculate the characteristic action, we would have obtained $A' \simeq 2.5 \times 10^{-4}\hbar$. This would have indicated that

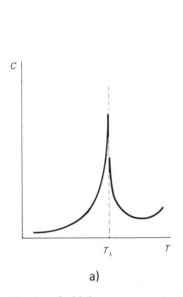

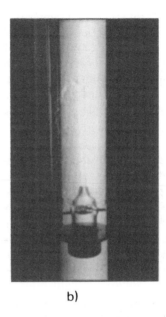

a)   b)

Fig. 1.5. *Superfluid helium*
(a) Liquid helium undergoes, at low temperatures, a phase transition which is manifested, in particular, on a curve representing the specific heat as a function of the temperature. The shape of this curve at the discontinuity is responsible for the name 'lambda point'. Below $T_\lambda = 2.18$ K, helium (called He-II in this phase) becomes superfluid, i.e., it flows without any viscosity. The behaviour is specifically of a quantum nature.
(b) The 'fountain effect' is a spectacular manifestation of superfluidity. Heating by means of a resistance, the lower (larger) portion of a tube, partially immersed in helium-II, gives rise to an upward jet of helium. The principle of this curious pump depends upon the transition between superfluid and normal helium caused by the raising of the temperature (photograph communicated by courtesy of J. F. Allen, St. Andrew's University, Scotland).

the quantities used were not pertinent to the phenomenon being studied: the mass of the electron is irrelevant to the nature of superfluidity.

(g) *Nuclei.* Nuclear energies are measured in million electron volts (1 MeV = 10^6 eV). This is the unit of energy which is best suited to the measurement of energy levels, binding energies, etc. For example, the binding energy per nucleon (proton or neutron) in an ordinary nucleus, i.e., the energy which would have to be supplied in order to extract a nucleon, is typically $E \simeq 8$ MeV $\simeq 1.3 \times 10^{-12}$ J. Furthermore, the dimensions of nuclei, deduced, for example, from electron scattering experiments, are measured in 'fermis'

(1 F $= 10^{-15}$ m): one often defines for a nucleus of mass number A, an average radius $r_A \simeq A^{1/3} r_0$, with $r_0 \simeq 1.2$ F. With the mass of the nucleon, $M = 1.6 \times 10^{-27}$ kg, one may form a characteristic action $A = (ME)^{1/2} r_0 \simeq 5 \times 10^{-35}$ L $\simeq 0.5\hbar$, for the atomic nucleus. Hence, nuclear physics is necessarily of a quantum nature. It is a remarkable fact that quantum theory, which was initially founded upon atomic physics, remains valid in nuclear physics as well, on a scale of distances a hundred thousand times smaller.

1.4. Quantum orders of magnitude

Once we have assured ourselves that a quantum treatment is actually necessary for studying a given physical phenomenon, by having compared $\hbar$ to the characteristic actions of the phenomenon, we may go a step further. The above comparison is, in effect, made by using the quantities which are most directly measurable and accessible to *experiment*. It is then possible to explore, with the aid of dimensional analysis, the relations which a quantum *theory* would eventually establish between these physical quantities and the fundamental parameters of the problem. An evaluation of the orders of magnitude of the quantities so estimated would indicate exactly whether the choice of the fundamental parameters, and hence of the points of departure of the theory being sought, has been judicious or not. Such a procedure is common practice among physicists in the course of their research work. However, it is only rarely made explicit in their publications and even less so in the teaching of the discipline. It is, nevertheless, an essential stage in the work, through which one may observe the difference between research activity – the theoretical output of physics – as it is actually carried out, and the purely formal presentations – the mathematical crystallizations – which it engenders. *A physical theory cannot be reduced to pure formalism*; one must erase the stigma from all heuristic activity.

1.4.1. Atomic and molecular physics

The simplest case is that of the hydrogen atom, composed of one proton and one electron. We would like to build a theory based upon the idea that these are bound together to form the atom by means of electromagnetic forces, resulting from the interaction of the charge of the proton (q_e) with that of the electron $(-q_e)$, even though we know nothing as yet about the quantum dynamics of these forces.

An evaluation, using dimensional analysis, of the quantities of interest to us (energy, size of the atoms) should, therefore, be made in terms of these charges, having dimension $[q_e] = \mathcal{2}$, and the fundamental constant of electrostatics, $1/4\pi\varepsilon_0$. From the classical expression for the Coulomb force, $F = (1/4\pi\varepsilon_0)(qq'/r^2)$, one extracts the dimensional equations $[1/4\pi\varepsilon_0] = \mathcal{M}\mathcal{L}^3\mathcal{T}^{-2}\mathcal{2}^{-2}$. The combination*

$$e^2 \triangleq q_e^2/4\pi\varepsilon_0 \qquad\qquad (1.4.1)$$

has dimensions,

$$[e^2] = \mathcal{M}\mathcal{L}^3\mathcal{T}^3, \qquad\qquad (1.4.2)$$

which no longer involve any electrical quantities, and, therefore, it is this constant which would necessarily enter into a calculation of any mechanical quantity of the atom. One can say that e^2 measures the elementary electrical charge, or rather its square – i.e. the intensity of interaction between two elementary charges – in purely mechanical (non-electrical) units. Its value in the MKSA system of units is

$$e^2 = 2.30 \times 10^{-28} \text{ kg m}^3 \text{ s}^{-2}. \qquad\qquad (1.4.3)$$

The expressions obtained by using e^2 are formally analogous to those which would arise in a system of units having three fundamental magnitudes $\mathcal{M}$, $\mathcal{L}$ and $\mathcal{T}$, with the unit of electrical charge having been derived, for example, from Coulomb's law, as in the CGS electrostatic system.

Besides e^2, which characterizes the intensity of the electrical forces, and h which expresses the quantum nature of the expressions sought, we ought to note that the theory must also involve the masses of the particles present. But then, we have two of these at our disposal: the mass of the electron (m) and that of the proton (M). A derivation based exclusively upon dimensional analysis would, therefore, seem to be doomed to failure, since every correct expression could generate an infinity of others, upon multiplication by a function of the (purely numerical) variable m/M. It is, therefore, necessary to use some additional argument, and for this to reflect upon the physical role played in this problem by these masses. As in the classical case, they express the inertia of the particles in relative motion. We may think, therefore, that this motion is essentially that of the electron 'around' the much more massive

---

* In the expression which follows, and generally in this book, the symbol $\triangleq$ signals a *definition* of the term appearing on the left.

and practically immobile (cf., however, exercise 1.14) proton. The mass of this latter should, therefore, not intervene, to a first approximation. This idea is confirmed by the experimental fact that the isotopes of hydrogen (deuterium and tritium) have atomic properties (energies, dimensions, spectra) which are practically identical to its own, in spite of their having double and triple its nuclear mass. Hence, to a first approximation, it is the mass of the electron alone which ought to enter the calculation. One could have used a similar argument to justify the use of the electronic mass in the analysis of the crystalline structure above [sect. 1.3, example (e)]. To describe the hydrogen atom, we have at our disposal the three quantities m, e^2 and $\hbar$, which have dimensions $[m] = \mathcal{M}$, $[e^2] = \mathcal{M}\mathcal{L}^3\mathcal{T}^{-2}$ and $[\hbar] = \mathcal{M}\mathcal{L}^2\mathcal{T}^{-1}$, respectively. With their aid, one may form, in a unique manner, an energy, a length, etc., which happen to be the appropriate standards for the atomic domain. Reasoning in this way, one calls the quantity

$$E_H \triangleq \frac{me^4}{2\hbar^2}, \quad E_H = 13.6 \, \text{eV} = 1 \, \text{Ry}, \tag{1.4.4}$$

the Rydberg energy or 'rydberg' (Ry) (where the factor of 2, in the denominator, has been introduced for later convenience), and the length

$$a_0 \triangleq \frac{\hbar^2}{me^2}, \quad a_0 = 0.53 \, \text{Å}, \tag{1.4.5}$$

the Bohr radius. Thus, a dozen or so electron-volts and an angström are the typical energies and dimensions of the hydrogen atom. Although the notion of a length in the quantum domain is for us still undefined, eq. (1.4.5) shows us that the region of space in which the movement of the electron of a hydrogen atom takes place, has dimensions of the order of an angström. Finally, one can obtain a typical velocity for the electron inside the atom,

$$v_A \simeq \frac{e^2}{\hbar}, \quad v_A = 2 \times 10^6 \, \text{m s}^{-1}. \tag{1.4.6}$$

This velocity is about ten times less than c, the velocity of light*. Thus, the motion of the electron is non-Einsteinian (in other words, it can be

---

* The velocity c is a fundamental constant characterizing the structure of space–time. It turns out, provided photons have vanishing mass, that this is also the velocity of light. However, that is a different story altogether.

understood within the framework of a Galilean space–time), which justifies a posteriori our 'overlooking' c amongst the fundamental parameters of atomic physics. In conclusion, we have therefore, acquired the conviction that a quantum theory *could* explain the properties of the hydrogen atom.

Since the orders of magnitude, which are characteristic to the hydrogen atom, depend only upon the mass (m) and charge (e) of the electron, within the framework of a quantum theory ($\hbar$), these same orders of magnitude characterize, therefore, all quantum phenomena related to the electron. This is the case for the whole of atomic and molecular physics, solid or liquid (cf. exercises 1.8 to 1.19). The physics of molecules, and, of course, chemistry, both of which study the same objects – albeit with their own specific methods – are based entirely upon the behaviour of electrons, which ensures the stability of the molecules by binding the atoms, as well as the transformations which they undergo through their mutual interactions. Molecular combinations or reactions are secondary compared to phenomena which are specifically atomic: electrons first bind themselves to nuclei forming atoms, and it is only by virtue of residual effects that, at a subsequent stage, the formation of molecules and their reactions is ensured. One can expect, therefore, that the characteristic energies of molecular physics would be lower than those of atomic physics. In fact, molecular binding energies are in general of the order of 1 electron-volt (1 eV) rather than a rydberg (1 Ry $\simeq$ 10 eV). It might seem curious that the electron-volt, defined as it is, with the help of the macroscopic unit of potential – the volt – would be so well adapted to the uses of molecular physics. However, this is not just a mere coincidence. Indeed, the volt is initially defined in terms of the electromotive force of ordinary electric cells. Now, these latter operate through chemical reactions, which furnish the electrons with energies of the order of ... just 1 electron-volt! Thus, from the microscopic point of view the volt appears as the ratio of two specifically electronic quantities,

$$\text{volt} = \frac{\text{electron volt (molecular binding energy)}}{\text{charge of the electron}}.$$

If one uses expression (1.4.4) for the rydberg, and takes into account eq. (1.4.1), one sees that the 'macroscopic' unit of potential may be written as,

$$1 \text{ volt} \simeq \frac{1}{25}\left(\frac{1}{4\pi\varepsilon_0}\right)^2 \frac{mq_e^3}{\hbar^2} \simeq \frac{1}{25}\frac{1}{4\pi\varepsilon_0}\frac{q_e}{a_0}. \tag{1.4.7}$$

It is related to the characteristic quantities (m, q_e) of the electron as well as to the fundamental constants of electromagnetism ($1/4\pi\varepsilon_0$) and quantum theory ($\hbar$). Here, again, is an example of the manner in which the ordinary world, at our level, is indebted to its deeper quantum nature.

1.4.2. *Quantum electrodynamics and the fine structure constant*

Since the combination $e^2/\hbar$ is a velocity, its ratio with respect to the velocity of light is a dimensionless number. Its importance greatly transcends the significance that we have just discovered for it. This number is denoted by

$$\alpha \triangleq e^2/\hbar c, \tag{1.4.8}$$

and conventionally, its value is given in terms of its inverse,

$$1/\alpha = 137.037 \dots . \tag{1.4.9}$$

(independently of any system of units!) It carries the unfortunate name, 'the fine structure constant', since (historically)* it first appeared in connection with the fine structure of atomic spectra. Actually, it ought to be considered as the square of the elementary charge, evaluated in terms of the universal standards $\hbar$ and c, or, better still, as an absolute measure of the strength of the electromagnetic interaction. It characterizes, in an intrinsic way, the force of coupling between the elementary electric charge and the electromagnetic field; thus, in more modern terminology, one also calls it 'the electromagnetic coupling constant'! It must be expected to show up in all situations which reveal quantum ($\hbar$) and Einsteinian relativistic (c) properties of electrically charged (e^2) particles.

The coupling constant α measures the modifications brought about to the properties of charged particles because of their interaction with the electro-magnetic field. As an illustration, let us consider the particles which go by the name of 'pions'. They show up in three states of charge: π^+ (with positive charge, q_e), π^0 (neutral, with zero charge) and π^- (with negative charge, $-q_e$). They have identical properties, except for their charges. The mass of the π^+ (and of the π^-) would have been equal to that of the π^0, had the charge which it carries not given rise to an electromagnetic field, which is linked to the particle and which is the seat of a self-energy, contributing to the mass. The ratio of this mass difference to the mass of the neutral pion – which is most complicated to calculate in detail – is of the order of the coupling

---

* Parentheses added by translator.

strength between the charge and the field. As a matter of fact, experimentally one obtains $m_{\pi^0} \simeq 2.41 \times 10^{-28}$ kg and $m_{\pi^\pm} \simeq 2.49 \times 10^{-28}$ kg, so that $\Delta m/m \simeq 8/250 \simeq 4\alpha$, which is comparable to α in order of magnitude*.

More generally, in all quantum and Einsteinian relativistic theories of interactions between charged particles and the electromagnetic field, the quantities studied appear as functions of the constant $\hbar$ and the natural standards of the theory. For example, electrodynamics (the theory of single electrons, considered coupled to the electromagnetic field) has, as characteristic units, the mass–energy of the electron,

$$mc^2 \simeq 0.51 \text{ MeV}, \tag{1.4.10}$$

and the length (called the 'Compton length' of the electron)

$$\lambda_C \triangleq \frac{\hbar}{mc} = 3.8 \times 10^{-13} \text{ m}, \tag{1.4.11}$$

while the binding energy of the electron in the hydrogen atom may be written, using eqs. (1.4.4) and (1.4.10), as

$$E_H = \tfrac{1}{2}\alpha^2 mc^2. \tag{1.4.12}$$

One can easily understand this expression if one notes that the binding of the electron to the proton is really an electromagnetic phenomenon, brought about by the coupling of the Coulomb field to the electron on the one hand and to the proton on the other (one may say, if one so wishes, that the proton 'creates' the field which 'acts' on the electron). It is this double coupling which explains the presence of the constant α raised to the second power, in eq. (1.4.12). The coupling constant measures, therefore, the effect of the interaction: if it were zero, the electrons and the electromagnetic field would have been 'free' or non-interacting. It is a windfall for quantum electrodynamics that its coupling constant α is so small numerically [cf. eq. (1.4.9)]. It permits one to embark upon a computational procedure by which the quantities of interest (functions of α) are expanded as series, in powers of α, wherein one recognizes the zeroth order terms as evidently corresponding to a theory without interactions. The smallness of α allows one to hope that these

---

* The complexity of the phenomenon is demonstrated by the example of the K mesons, in which the charged particles (K^+, K^-) are *lighter* than the neutral particles. However, the relative mass difference is still of the order of α.

expansions would converge, at least in some sense. Sophisticated computational techniques have been employed to evaluate these expansions, which make quantum electrodynamics – despite certain conceptual obscurities – into the most precise of all physical theories. As an example, it is enough to mention the 'Lamb shift': it concerns the frequency difference between two lines in the spectrum of the hydrogen atom, for which the theory predicts the value $\Delta\nu^{th} = 1057.911 \pm 0.012$ MHz, while experiment gives $\Delta\nu^{exp} = 1057.90 \pm 0.06$ MHz. One notes that this represents a shift which is 10^6 times weaker than the principal levels (of the order of 10^{15} Hz) in the spectrum of hydrogen! In point of fact, these are corrections of the order of α^3, compared to the principal structure of the spectrum.

1.4.3. Particle physics. Fundamental interactions

The physics of 'elementary' particles, aims, in particular, to explore the notion of elementariness itself, and one of its results has been to demonstrate the limits of this notion – hence the quotation marks. It studies the intrinsic properties of these particles under conditions in which Einsteinian relativity and quantum theory cannot be ignored. The two standards c and $\hbar$ are, therefore, available, and hence allow us to associate to each particle of mass m,

an energy $= mc^2$,

a length $\ = \hbar/mc$,

a time $\ \ \ = \hbar/mc^2$,

etc.

These standards depend only on the typical mass and characterize those physical aspects of the particle which are independent of the nature of the 'forces' exerted between them – aspects which may be called 'kinematic'.

These 'forces' or interactions as they are called today, may according to our present knowledge be classified into several categories, as displayed synthetically in table 1.2.

(a) *The gravitational interaction* which we cite for record only, since at the level of particle physics its strength is weak (exercises 1.21 and 1.22). The quantum theory of gravitational interaction encounters serious conceptual

Table 1.2
Fundamental interactions.

	Interaction	Range	Strength (coupling constant)	Examples (in microphysics)
G r a n d **u n i f i c a t i o n ?**	**S t r o n g** Chromic	Very short ($\sim 10^{-2}$ F)	$\gg 1$	Constitution of hadrons by binding between quarks
	S t r o n g Hadronic	Short (~ 1 F)	A few units	Binding between hadrons (nuclear forces)
	E l e c t r o — w e a k Electromagnetic	Long ($1/r$)	10^{-2}	Constitution of atoms and molecules (electron–nuclear binding)
	E l e c t r o — w e a k Van der Waals	Medium ($1/r^6$)	–	Intermolecular forces
	E l e c t r o — w e a k Weak	Very short ($\ll 1$ F)	10^{-6}	Radioactive beta decay
	Gravitational	Long ($1/r$)	10^{-38}	?

and technical difficulties, which are still unresolved, and in any case for all experimental effects so far studied in particle physics, gravitational forces have been negligible compared to the other forces. It is because of its cumulative character and its long range that gravity becomes important, indeed dominant, at macroscopic scales.

(b) *The electromagnetic interaction.* This encompasses the electric and magnetic interactions, already well-known classically, between electrically charged particles (regardless of their nature or their properties), mediated by the electromagnetic field whose quantum elements are the photons. This interaction has a long range, which is responsible for the slow fall-off of the usual Coulomb field ($1/r$ dependence of the potential) with distance.

The absolute strength of this interaction is characterized by the coupling constant $\alpha \simeq 10^{-2}$ already introduced. In a heuristic sense one can understand a coupling constant as a measure of the 'effectiveness' of the corresponding interaction – i.e., the proportion by which physical quantities are modified as a result of the interaction. This is what we have seen in the case of the mass of the pion, where the electric charge modified the mass of the charged pion, as a fraction of that of the neutral pion, to order α. Similar considerations hold for the other quantities such as, e.g., the interaction times. Thus, for the electromagnetic interaction, in view of the weakness of its strength, the time necessary for the interaction to manifest itself is larger than the kinematical time which is typical of the phenomenon being studied. As an example, the η meson decays under the influence of the electromagnetic interaction into two photons: $\eta \rightarrow \gamma + \gamma$. One expects from this that its life time would be longer, by the factor $1/\alpha^2 \simeq 2 \times 10^4$ (the square arising from the double electromagnetic coupling), than the characteristic kinematical time of 3×10^{-24} s, and hence that it would be of the order of 10^{-19} s. It is, in fact, in the neighbourhood of 10^{-20} s which, given the coarseness of our approximation, is quite satisfactory. It ought to be added that the electromagnetic interaction can have subtle effects in systems which are even slightly complex. Thus, between electrically neutral atoms or molecules there still exist forces which are generally attractive. These arise from the combined interplay of Coulomb-type attractions and repulsions which, because of the phenomenon of polarization (Van der Waals forces), in particular, never quite compensate for each other.

(c) *The hadronic interaction.* This is the strong interaction which governs, e.g., the cohesiveness of nuclei. It has a short range (we shall see why, in ch. 3) and

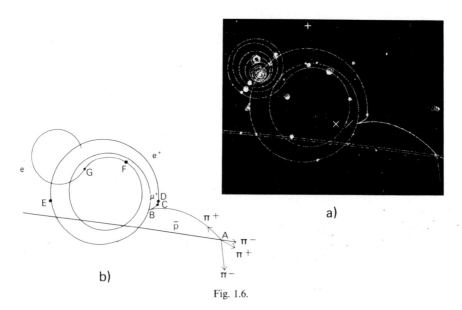

Fig. 1.6.

cannot be felt beyond a distance of about a fermi. Inside a nucleus, a nucleon is held by a binding energy of the order of 10 MeV. The ratio of this binding energy, typical of nuclear forces, to the characteristic mass of the π meson furnishes us with an estimate of the coupling constant g (or rather its square) of the strong interaction [just as α^2 allowed us to estimate (experimentally) the coupling constant of the electromagnetic interaction, cf. eq. (1.4.12)]. Hence, it follows that $g^2 \simeq 10/140$, so that $g \simeq 0.3$. But this value is high enough to render the power-series technique, restricted even to the lowest orders, to be of doubtful applicability, and unlike in the case of electrodynamics, the method ceases to be adequate. The ratio of the binding energy to the mass of the pion is a complicated function of the coupling constant,

---

Fig. 1.6. *Interactions*
This photograph was taken inside a 'bubble chamber' in which the boiling of a superheated liquid is triggered by the passage of charged particles, the trajectories of which are thus materialized in the forms of strings of tiny bubbles. A magnetic field bends the trajectories of these particles in a sense which depends upon the sign of their charges. The smaller the momentum of the particle, the more pronounced is the bending.
In (a) one sees two antiprotons $\bar{p}$, entering from the left, following parallel trajectories. One of them, which alone is shown in the diagram (b) interacts at A with a proton of the liquid which fills the chamber. There occurs an 'annihilation' at A and the production of pions (the neutral ones are invisible)

A: $\bar{p} + p \rightarrow \pi^+ + \pi^+ + \pi^- + \pi^- + n(?)\pi^0$ *strong interaction.*

One of the positive pions, hurled backwards, decays at B into a muon and a neutrino (invisible).

B: $\pi^+ \rightarrow \mu^+ + \nu_\mu$ *weak interaction.*

The muon, in its turn, decays at C emitting a positron and invisible neutrinos

C: $\mu^+ \rightarrow e^+ + \nu_e + \bar{\nu}_\mu$ *weak interaction.*

The positron now spirals around the magnetic field, losing its energy little by little through collisions with the electrons of the liquid, as seen at D, E, F and particularly at G, where it transfers practically all its energy to an electron:

D, E, F, G: $e^+ + e^- \rightarrow e^+ + e^-$ *electromagnetic interaction.*

The positron, stopped at G, rapidly annihilates itself against an electron, emitting photons (invisible)

G: $e^+ + e^- \rightarrow \gamma + \gamma$ *electromagnetic interaction.*

(Photo: Nuclear Physics Laboratory, College de France.)

which cannot be approximated simply by the lowest-order term in g^2. Furthermore, the existence of different masses (those of the pion and the nucleon) still complicates the problem. We can only conclude that the coupling constant is of the order of unity. Thus, the interaction modifies the properties of particles by amounts comparable in value to what they would have been in the absence of the interaction. It is for this reason that this interaction is called 'strong'. Another method for estimating g, is to consider an unstable particle, such as the ρ meson, which decays into two π mesons under the effect of the strong interaction. Now, the mass of the ρ meson is 770 MeV, while that of its two constituents together is only about 280 MeV. The relative mass difference is of the order of unity and so also is, therefore, the coupling constant g, whatever might be the (unknown) expression of the mass difference as a function of g. The characteristic dynamical time of the strong interaction is related to the interaction energies which, as we shall see later, are themselves of the order of a hundred MeV. This time, like the kinematical time, is therefore, of the order of 10^{-24} s – the typical time for the strong interaction to manifest itself. In fact, this is the decay time of unstable particles such as the ρ meson.

Today, one knows that the strong interaction is only a manifestation of a more fundamental interaction between 'more elementary' particles – the quarks – of which the ordinary hadrons (nucleons, mesons, etc.) are composed. A new species of charge, arbitrarily named 'colour' is considered to be the source of these forces, just as the electric charge is the source of the electric force. By analogy with electrodynamics, the name 'chromodynamics' is, therefore, given to the theory of these interactions which we shall call *'chromic'* interactions. The hadronic forces would thus be an effect of the chromic forces within the quark conglomerates (the hadrons), just as the Van der Waals forces are a residual effect of the electromagnetic forces within the electron–nucleon conglomerates (the atoms). This tells us that the underlying chromic interactions are very strong (since their residue, the hadronic interaction, is still strong) and must be characterized by a coupling constant which is notably greater than unity – assuming that the notion of a coupling constant itself still retains a meaning. Finally, just as the electromagnetic interaction between electric charges is mediated by the photons of the electromagnetic field, there exists a 'chromic field', whose particles are called 'gluons' and which mediates the chromic interaction between quarks. Nevertheless, chromodynamics is very considerably more complex than electrodynamics; in particular, quarks most probably cannot be separated from one another (they are said to be 'confined'). Thus, hadrons would have to be composite and at the same time indecomposable ('atomic' in an etymological sense).

(d) *The 'weak' interaction*, which manifests itself mostly in certain particle decays, e.g., in radioactive β-decay, is an interaction with a very short range and weak strength, having a coupling constant of the order of 10^{-6}. The characteristic time of this interaction, necessary for its full manifestation, is very much longer than those of the strong or even the electromagnetic interactions. It is, therefore, often masked by these latter. The weak interaction is similarly mediated by specific particles, currently known as 'weak bosons' – a terminology which is rather unsatisfactory and hopefully provisional. (The word boson will be introduced in ch. 7.) The weak interaction involves not only the hadrons already mentioned, but also the entire family of leptons (electrons, neutrinos, muons, etc.). Today there exists strong evidence that the weak interaction will be unified with the electromagnetic interaction, just as this latter itself had formerly resulted from a unification of the magnetic (weak) interaction with the electric interaction. Within this perspective one speaks of an *'electro-weak' interaction* and considers the 'weak bosons' as the cousins of the photon.

Table 1.3 indicates the present classification of the fundamental particles according to the interactions in which they participate.

Table 1.3
Classification of particles by their interactions.

Interaction	Particle			
	Quarks, gluons (and hadrons)	Leptons (electrons, muons, neutrinos, ...) and weak bosons	Photon	Graviton?
Strong	Yes	No	No	No
Electromagnetic	Yes	Yes (if charged)	Yes	No
Weak	Yes	Yes	No	No
(Gravitational)	Yes	Yes	Yes	Yes

1.5. Some open problems

1.5.1. Limits of validity of quantum physics

There is no reason for quantum theory to be different from other physical theories: its domain of validity is certainly limited and it must one day relinquish its place to a deeper theory, to which, of course, it shall then constitute a good approximation over a certain domain. However, there are no indications at the present time which would allow one to foresee either that the limits of this domain are being approached, or even the direction in which these may lie. One would think that the exploration of smaller and smaller spatial regions could have led one to encounter these limits. However, quantum theory, which was developed at the scale of atomic physics (10^{-10} m), has remained valid at the scale of nuclear physics (10^{-15} m), where it seems to be holding its ground at the even further reduced scale (presently 10^{-17} to 10^{-18} m) of particle physics. Certainly, there are difficulties aplenty, as well as complications arising essentially from the need for it to obey the constraints imposed by the (Einsteinian) theory of relativity. But these seem more to be problems for the theoreticians than for the theory, in the sense that its fundamental concepts have not as yet been questioned. It is possible that the (future) criterion for the validity of quantum physics would have to be formulated not in space–time terms, but in terms of other physical quantities, which would play on the quantum–post-quantum frontier the role played by the 'action' on the classical–quantum frontier.

1.5.2. Conditions for the validity of classical physics

We know today that the theories of classical physics (mechanics, field theory, etc.) are merely approximations to an underlying quantum theory. Although we have a criterion for the empirical validity of these theories (through the estimation of the characteristic actions), it is remarkable that there does not exist any coherent and satisfying way to construct them, starting from their quantum foundations. The formal limiting procedure, whereby one 'lets $\hbar$ tend to zero', allows one to study some of the quantum-to-classical structural transitions. But $\hbar$ being what it is (having the value that it has), it is clear that the physical limit corresponds, in fact, to an indefinite increase in the number of particles. But then how does an N-particle system, interacting through a certain type of force and obeying the laws of quantum physics, end up *seeming* to obey the laws of classical physics as N becomes large? That this problem is both real and complex becomes clear if one thinks of the existence of macroscopic systems – lasers, superfluids, superconductors – which exhibit quantum effects that are qualitatively incomprehensible (and in any case unpredictable) in classical terms. The macroscopic limit is not automatically classical, as we have said before, and there do exist specific conditions for the validity of the classical approximation. Some progress has recently been accomplished in studying the conditions for the validity of classical thermodynamics. It has been shown, e.g. – and this is a highly nontrivial result – that its validity depends on some typically quantum collective properties of the particles in question (see ch. 7), which appear to be crucial for the ordinary macroscopic behaviour of matter. The nature of the forces (Coulombic in the case of ordinary matter) plays an equally determinative role here.

1.5.3. The coupling constants

The question naturally arises as to how one could explain the numerical values of the coupling constants. The problem has a definite meaning only for those constants (or their combinations) which are dimensionless. It is the electromagnetic coupling α which has attracted the greatest amount of attention in this regard. This 'pure number' has genuinely fascinated numerous physicists, who have indulged in some amusing, although dubious, neo-Pythagorean tendencies. One might cite here the unfortunate case of the famous astrophysicist Eddington whose theory purported to 'explain' the value of α^{-1}, at a time when experiment seemed to show it to be compatible with the integral number 136, by finding it to be $4^2(4^2 + 1)/2$! This theory could possibly have been adapted to yield the value 137 for α^{-1} following more precise experimental measurements, but it would have had to stumble in front of the decimals. Some more recent attempts have hardly met better success. It is probably necessary to develop a general theory for the coupling constants of the various fundamental interactions, explaining in particular the remarkable spread in their orders of magnitude (table 1.2). This would call for the unification already discussed of these different theories of interactions. At the present time, much hope is based upon such a unification within the framework of the so-called 'gauge' theories – quantum field theories which generalize certain particularities of electrodynamics.

A second problem concerning the coupling constants is their very constancy. Temporal variations, capable of explaining certain intriguing numerical coincidences (cf. exercise 1.22), had been proposed for the gravitational constant (Dirac) and later on for that of electromagnetism (Gamow). Heavy with cosmological and astrophysical consequences, these ideas seem to have been relegated to the background – at least in their initial forms – simply because of the sheer weight of their ultimate predictions invalidated by observation.

Exercises

1.1. The *Boltzmann constant* has the value $k = 1.38 \times 10^{-23}$ J K^{-1}.

(a) Express it in eV K^{-1}.

(b) What are the values, in eV, of the typical thermal agitation energies ε at the temperature T, given that $\varepsilon = kT$, for $T = 3$, 30, 300 and 3000 K?

1.2.   From the experimental results in fig. 1.2, obtain an estimate for the *Planck constant.*

1.3.   One can ask if it might not be possible to set up a classical theory for the *photoelectric effect*, and hence to compute the quantum constant h, using the physical parameters of the electron (its mass m, charge $-q_e$) and the fundamental constants of classical electromagnetism. Show that, using these magnitudes, one can in effect write down a quantity h' having the dimensions of the quantum constant h. Compare the numerical values of h and h'. Comment on the result.

1.4.   A beam of electrons having kinetic energy $E = 20$ eV (the mass of the electron is $m = 0.9 \times 10^{-30}$ kg) is sent through a *metallic crystal* whose neighbouring atoms are situated at a distance $a = 2$ Å. Calculate the characteristic action. Is it necessary to use quantum theory to describe the interaction of the electrons with the crystal? Answer the same question for electrons having kinetic energies $E = 20$ keV and $E = 2 \times 10^{-2}$ eV, respectively.

1.5.   One could attempt to evaluate the characteristic action for a *hydrogen atom* given the following: its mass $M = 1.6 \times 10^{-27}$ kg, binding energy $E_H = 13.6$ eV, minimal wavelength in its spectrum $\lambda_m = 10^3$ Å. Express this action as a function of M, E_H and λ_m, and compute it numerically.
Why is it so much larger than h, is devoid of any physical significance and hence should not be taken into consideration?

1.6.   *Superfluid helium* behaves in an unusual way under rotation: instead of rotating *en masse* together with the container in which it is placed, and dragged along by its viscosity (as would be the case for an ordinary liquid), it becomes the seat of multiple local rotational movements (vortices), about axes which are parallel to the general axis of rotation. These rotational movements are detected by observing tiny suspended particles of solid hydrogen, which are carried along by the liquid helium and which can be seen as describing some sort of loops, typically with diameters of the order of $d \simeq 0.1$ mm and times $\tau \simeq 0.1$ s. Show that in spite of these macroscopic magnitudes, the action associated with *one* atom of helium (of mass $M = 6.4 \times 10^{-27}$ kg), executing these motions, is clearly quantum. (It can be proved that the curl or circulation of the velocity of rotation within a vortex is 'quantized': $\int \boldsymbol{v} \cdot d\boldsymbol{r} = nh/M$, where n is an integer. The rotation of superfluid helium is a specifically quantum phenomenon, manifested macroscopically.)

1.7.   *Black-body radiation* is the name given to electromagnetic radiation in thermal equilibrium (at a temperature T) with a material which is a perfect emitter and absorber. Let $u(\omega, T)$ be the spectral density (per unit volume) of the energy of this radiation; in other words, the electromagnetic energy per unit volume within the band of angular frequencies between ω and $\omega + d\omega$ is

$$dE = u(\omega, T)\, d\omega.$$

(a) Show that the dimensional equation for u is

$$[u] = \mathcal{M}\mathcal{L}^{-1}\mathcal{T}^{-1}.$$

(b) Recall that in classical thermodynamics, to each degree of freedom of a system at the temperature T, there corresponds an average amount of energy of the order of kT, where k is the Boltzmann constant. Show that (in the classical theory) u is necessarily of the form: $u_{cl} = A(c)^x(\omega)^y(kT)^z$ ('Rayleigh–Jeans' formula); c is the velocity of light (what is it doing here?) and A is an unknown numerical constant. Calculate the exponents x, y and z.

(c) Show that the preceding expression is physically absurd, by evaluating the total energy per unit volume (integrated over the frequencies of radiation)

$$E_{cl}(T) = \int_0^\infty u_{cl}(\omega, T)\, d\omega.$$

Does the difficulty arise from the high or from the low frequencies?

(d) It is quantics which allows one to resolve this paradox. Show that the energy density is actually of the form

$$u(\omega, T) = u_{cl}(\omega, T)\, f(\hbar\omega/kT).$$

Are quantum concepts necessary for $\hbar\omega \gg kT$ or for $\hbar\omega \ll kT$? Hence deduce the limits of the function f at the origin and at infinity [taking into account that it has the role of eliminating the difficulty which appeared in (c)].

(e) Show that the energy density at the temperature T has a maximum at an angular frequency $\omega_m(T)$ such that $\omega_m(T) = CT$ (Wien's empirical law), and express the 'Wien's constant' C in terms of $\hbar$ and k up to a numerical constant. Give a numerical estimate for C (recall that $k = 1.4 \times 10^{-23}$ MKSA units). Compare it with the exact (experimental and theoretical) value: $C \simeq 13.7 \times 10^{11}$ s^{-1} K^{-1}. Calculate the angular frequency and the wavelength for the maximum energy density at a temperature of 1000 K. In which region of the spectrum does this maximum lie? How does it shift as the temperature rises? Show how this paragraph could aptly have been termed, 'In the red hot iron, in the blushing embers – It is of quantics that one remembers.'

(f) Hence, deduce that the total energy per unit volume

$$E(T) = \int_0^\infty u(\omega, T)\, d\omega,$$

is given by

$$E(T) = \sigma T^4$$

(Stefan's empirical law) and express the 'Stefan's constant' σ in terms of k, $\hbar$ and c, to within a numerical coefficient. From this, obtain a numerical estimate for σ. Compare this to the exact value, $\sigma = 1.90 \times 10^{-16}$ J m^{-3} K^{-4}.

(g) For high frequencies, an approximate formula of empirical origin

$$u_w(\omega, T) \propto \omega^3 \exp(-a\omega/T),$$

where a is a certain constant, was proposed by Wien. This implies (prove!) that for $s \to \infty$, $f(s) \simeq se^{-s}$. Taking into account the behaviour of $f(s)$ for $s \to 0$, suggest a plausible interpolation – i.e., an expression for $f(s)$ obeying all the required conditions. (It was basically in this manner that Planck discovered the formula which bears his name, before 'proving' it.) How would you verify if it is correct?

1.8. The *hydrogen atom* is characterized by the mass of the electron $m = 0.9 \times 10^{-30}$ kg, a dimension $a_0 = 0.5$ Å and a binding energy $E_H = 13.6$ eV. Evaluate dimensionally in terms of m, a_0 and E_H, and then numerically, the following physical magnitudes, characteristic of the state of the electron in the hydrogen atom:

(a) the mass per unit volume (density);

(b) the pressure;

Compare with the corresponding orders of magnitude for states of matter on the macroscopic scale.

1.9. Repeat the above problem for the nucleons inside a *nucleus of deuterium* (the simplest one being a proton and a neutron), of size $a = 4$ F and binding energy $E_d = 2.2$ MeV. The mass of the nucleon is $M = 1.6 \times 10^{-27}$ kg.

1.10. The binding of two (or more) atoms to form a molecule is a quantum phenomenon of an electronic nature.

(a) From this deduce the order of magnitude of the distance between the atoms in the molecule – usually called the 'size' of the molecule.

(b) If a molecule is sufficiently asymmetric, such as, for example, the diatomic molecules HCl or SiO, it possesses an *electric dipole moment* D. Estimate theoretically, then numerically, the molecular dipole moment. Compare with the typical experimental values: $D = 0.35 \times 10^{-29}$ C m for HCl, $D = 1.0 \times 10^{-29}$ C m for SiO, $D = 0.60 \times 10^{-29}$ C m for H_2O and $D = 2.9 \times 10^{-29}$ C m for NaCl.

1.11. Consider certain physical properties of *water at the macroscopic level*. These are to be evaluated in a system of units appropriate to the atomic scale, in which the basic magnitudes and their units are energy (1 eV), length (1 Å) and action ($1 L \simeq 10^{-34}$ MKSA units). For the thermodynamic magnitudes, one similarly takes the Boltzmann constant as the unit and thereby measures temperatures in units of energy (300 K $= \frac{1}{40}$ eV). Calculate and discuss the values of:

(a) the compressibility $\chi = 45 \times 10^{-12}$ cm^2 dyne^{-1};

(b) the surface tension $\gamma = 73$ dyne cm^{-1};

(c) the viscosity $\eta = 10^{-2}$ dyne s cm^{-2};

(d) the heat capacity $Q = 4.18$ J g^{-1} K^{-1};

(e) the coefficient of thermal expansion $\beta = 0.21 \times 10^{-3}$ K^{-1};

(f) the thermal conductivity $\lambda = 6 \times 10^{-3}$ W cm^{-1} K^{-1};

[in (a), (b), (c), (d), (e) and (f) values are given at 20°C].

(g) the parameters of the critical point: temperature $T_c = 647$ K, pressure $P_c = 218$ atm and density $\rho_c = 0.33$ g cm^{-3}.

1.12. Consider some physical properties of *iron at the macroscopic scale*.

(a) The density of iron is $\rho = 7.9$ g cm^{-3}. Express this in terms of the atomic mass of iron (56 amu) per Å^3.

(b) The Young's modulus of iron, which measures its elasticity, is $E = 1.2 \times 10^{12}$ dyne cm^{-2}. Express this in eV Å^{-3}.

(c) The heat of vaporization of iron is $\Gamma_V = 84$ kcal/mol and the temperature of vaporization $T_V = 3160$ K. Express Γ_V in eV/molecule and kT_V in eV (where k is the Boltzmann constant).

1.13. In a *hypothetical world* where the Planck's constant is twice as large and electrons are three times lighter (everything else being the same), what would the following magnitudes become as ratios of their usual values:

(a) the ionization energy of the hydrogen atom;
(b) the minimal wavelength in the spectrum of hydrogen;
(c) the size of the hydrogen molecule;
(d) the density of ordinary table salt;
(e) the latent heat of vaporization of one litre of water;
(f) the temperature of fusion of ice;
(g) the energy density of black-body radiation, for example, at 1000 K (see exercise 1.7);
(h) the height of the mountains on a planet with the same radius as the earth;
(i) the velocity of sound in the atmosphere of this planet, supposing that the mass and the temperature of its atmosphere are identical to those of the terrestrial atmosphere?

1.14. The nucleus of the hydrogen atom, the proton, is not infinitely heavy, although its mass M is considerably larger than that of the electron m; $M/m \simeq 2 \times 10^3$, approximately. Thus, the proton also moves inside the atom.

(a) in the classical theory, both the proton and the electron move around their centre of mass. As a function of the ratio M/m, what would be the ratios of
– the radii of their orbits (assumed circular);
– their angular velocities;
– their linear velocities?

(b) Supposing that these results remain correct in the quantum theory, estimate the orders of magnitude of
– the domain of spatial localization;
– the velocity;
– the kinetic energy;
of the proton in quantum motion.

1.15. A *muonic atom* is an atom with one of the electrons replaced by a muon, which is some sort of heavy electron of mass $m' = 1.9 \times 10^{-28}$ kg $\simeq 200m$, where m is the mass of the electron.

(a) If the atomic number (the number of protons in the nucleus) is Z, what is the order of magnitude of the distance of the muon from the nucleus? Compare this to that of the distance of the electrons from the nucleus and show that the presence of the electrons does not influence the behaviour of the muon.

(b) The nucleus of an element of mass number A (number of protons and neutrons) has a 'radius' of the order of $r = A^{1/3}r_0$, where $r_0 \simeq 1.2$ F $= 1.2 \times 10^{-15}$ m. Compare this radius to the nucleus–muon distance. (Recall that for ordinary nuclei, $A \simeq 2Z$.) From what values of A would the nuclear structure begin to influence the behaviour of the muon?

(c) What is the order of magnitude of the velocity of the muon? Are there values of Z for which relativistic effects become important?

(d) What is the order of magnitude of the energy of the muon? In which region of the electromagnetic spectrum should the spectroscopy of muonic atoms be done?

1.16. Consider a classical model of the atom, or more generally, an arbitrary system of N *classical charged particles* interacting via Coulomb's law *only*.

(a) Consider the effects on the distances, velocities, accelerations and forces of the system, of a spatial dilation by a factor λ and a temporal dilation by a factor μ ($r' = \lambda r$ and $t' = \mu t$).

(b) Show that the equations of motion are invariant if $\lambda^3 = \mu^2$. Hence deduce that for each system, $r_k = f_k(t)$ ($k = 1, 2, ..., N$), of possible trajectories, $r_k = \sigma^2 f_k(\sigma^3)$ is an equally possible system, where σ is an arbitrary number. Hence, find a way to obtain Kepler's third law of planetary orbits.

(c) How does the energy of the system vary under the spatio–temporal change of scale by the factor σ described above? Hence, deduce that for a classical atom, the energy is not bounded below, so that there does not exist a stable state.

(d) How does the action vary under a change of scale by the factor σ? Hence deduce that the existence of a standard unit of action – the quantum constant $\hbar$ – enables one to avoid the preceding conclusions.

1.17. For a *hydrogen atom*, estimate dimensionally the electric field $\mathscr{E}_{at}$, due to the proton and acting on the electron, as a function of the mass m of the electron, its charge q_e, e^2 and $\hbar$. Evaluate its order of magnitude. Placed in an external electric field $\mathscr{E}_{ex}$, the atom is not stable unless $\mathscr{E}_{ex} \ll \mathscr{E}_{at}$, for otherwise the external field dominates that of the proton, thereby pulling out the electron and ionizing the atom. Knowing that a laser beam having power $P = 10^{12}$ W can be concentrated onto a surface of area 10^{-5} cm^2, evaluate the electric field so produced and compare it to $\mathscr{E}_{at}$ to find out if this laser beam would be capable of ionizing hydrogen.

1.18. In classical physics, an electrically charged object, endowed with an intrinsic angular momentum, carries a certain *magnetic moment* as well. A similar situation prevails in quantics.

(a) Establish the dimensional equation for a magnetic moment.

(b) Usual quantum particles have intrinsic angular momenta ('spins') of the order of $\hbar$. Hence deduce, for such a particle of mass m and charge q, a way to evaluate its magnetic moment.

(c) Calculate numerically the order of magnitude of the magnetic moments of the electron and the proton. What would be the order of magnitude of atomic magnetic moments (due to the electron and the nucleus)?

(d) What is the order of magnitude of the frequency of electromagnetic waves capable of modifying the state of the magnetic energy of an atom in a static magnetic field $\mathscr{B}$ (take, e.g., $\mathscr{B} \simeq 1$ Tesla)?

1.19. Considering the electron as a small charged sphere, it is possible to calculate its radius by attributing all its mass–energy mc^2, to the energy of its electrostatic field.

(a) Hence deduce the '*classical radius*' of the electron, $r_e = e^2/mc^2$ up to a numerical factor, depending upon the details of the charge distribution).

(b) Atomic physics prompts one to endow the electron with a spin (intrinsic angular momentum) of the order of $\hbar$. Show that in a model in which the electron is considered as a rotating object of dimension r_e, one can deduce from this a way to evaluate the (linear) velocity of the rotation of its peripheral elements.

(c) What is the order of magnitude of this peripheral velocity, if one adopts the 'classical radius' as the dimension of the electron? Compare it to c. Show that it is impossible to represent the electron as being localized to a domain of dimensions smaller than its 'Compton length' $\lambda_C = \hbar/mc$.

1.20. Consider the *Coulomb scattering* of a nucleus of mass m and charge $Z_1 q_e$, by a second nucleus of charge $Z_2 q_e$, assumed very much heavier than the first and hence immobile. Let v be the velocity of the incident nucleus and $E = \frac{1}{2} m v^2$ its kinetic energy.

(a) Assuming the motion to be governed by classical mechanics, calculate numerically the minimal distance of approach d of the two nuclei, for an α-particle ($Z_1 = 2$) of energy $E = 10$ MeV scattered by a copper nucleus ($Z_2 = 60$) and compare d to the radius of the nucleus of copper, $R \simeq 5$ F.

(b) What is the characteristic action of the system? Show that a classical description is not valid a priori, unless $v \ll Z_1 Z_2 \alpha c$. Up to what kinetic energies can one describe the scattering of α-particles by copper nuclei classically, under the hypothesis of a pure Coulomb interaction? Up to what energy is this hypothesis, which neglects the nuclear interaction, tenable?

1.21. Suppose that the classical expression for the *attractive gravitational* potential for two masses M and M' at a distance d from each other

$$U = -\frac{GMM'}{d}$$

remains correct (at least in order of magnitude) in the quantum theory. Recall that $G = 6.7 \times 10^{-11}$ MKSA units.

(a) For two protons ($M = M' = m_p$), at what characteristic distance d_G does the interaction energy become comparable to the mass–energy $m_p c^2$?

(b) Compare d_G to the quantum characteristic distance (Compton length) $\lambda_p = \hbar/m_p c$. What does one conclude from this? Convert G into a dimensionless coupling constant Γ by means of a monomial combination involving the mass m_p and the universal constants $\hbar$ and c. Find Γ explicity, compute its numerical value and compare it to the coupling constants of the other interactions.

1.22. The *Newton constant*, or the gravitational coupling constant, has a value of $G = 6.7 \times 10^{-11}$ MKSA units.

(a) What is the dimension of G?

(b) Show that the constants G, $\hbar$ and c define a system of mechanical units. Write in terms of G, $\hbar$ and c and evaluate numerically, the units of mass, length and time in this system. To what type of a physical phenomenon would these units be appropriate?

(c) Let m be the mass of the π meson, a typical fundamental particle: $mc^2 \simeq 140$ MeV. Show that the ratio e^2/Gm^2 is a dimensionless number. Calculate this number and discuss its physical significance.

(d) The 'age of the universe' (i.e., the time that has elapsed since a certain explosive event, the so-called 'big bang', which gave birth to the universe as we know it) is estimated to be $T \simeq 10^{10}$ year. Estimate this as a function of the unit of time $\tau = \hbar/mc^2$, associated to the mass of the π meson. Calculate τ and T/τ. Compare the ratio T/τ to the ratio e^2/Gm^2, previously calculated. If the relative proximity of these two large numbers is not just to be a simple coincidence, what physical hypotheses could suggest this closeness?

2

Quantons

2.1. The concepts of classical physics

In 'classical', i.e., pre-quantum physics, every physical system can be understood in terms of two fundamental notions: either that of a wave (or field) or that of a particle (or corpuscle). These are theoretical notions whose characteristics have been developed, based upon our experience of certain objects from our daily macroscopic world – from the rolling waves of the sea and the dust particles dancing in a ray of sunlight, right up to the sun itself, along with its planetary system and its luminous radiation. It is remarkable that the same basic concepts allow us to make theoretical analyses of such diverse phenomena and at such dissimilar scales. There seems, however, to be no hint which might make us suspect that their adequacy in describing the objects of the real world might be universal. It could hardly be surprising, therefore, when at the beginning of this century, experimental physics – aided by technical progress – conquered new domains, surpassing the terrain of classical theory. The notions of a wave and a particle were consequently found to be out of line with a quantum theory. Before introducing their substitute (yes! there is only one) we shall briefly recall the essential characteristics of the classical concepts. Quantum ideas mark, in effect, a complete rupture with those of classical physics, and yet at the same time remain largely indebted to them – even if only on the level of the vocabulary that is employed. To appreciate fully this break, as well as this continuity, and to have quantum theory appear in its proper perspective, it is appropriate to sketch the classical background from which it broke away.

2.1.1. Classical particles

The concept of a *particle* is presented as an idealization of a certain class of real objects, whose prototype is the flipper (pin-ball) or the tennis ball. Like any other physical concept, the concept of a particle is constructed through a process of abstraction and pruning, in terms of which certain characteristic properties alone are retained.

The notion of a particle retains, first of all, the *discrete* character of a pin-ball. In other words, one can individualize the particles and denumerate or count them, one by one.

A tennis ball, like any real object, occupies a finite region of space. However, in a number of circumstances, the spatial extension of these objects may be considered negligible. This would be the case for the dimensions of a planet compared to those of the solar system: it is a matter of scale. The concept of a point particle describes situations where one may disregard the dimensions of the object and, by an idealization, reduce its spatial extension to a geometric point. One speaks, therefore, of *the position* of the particle in space.

This position varies, in general, with time: the particle (which one now designates as 'a moving body') describes a line in space, called a 'trajectory'. The motion of a particle is, therefore, *cursive*, which the usual terminology clearly underscores: one says that the particle 'describes' its trajectory, leaving eventually a 'trace' (like the trail of condensation of an aeroplane in the sky).

At each instant, the position of the particle is well-defined; if one refers it to a certain system of axes, each of the three spatial coordinates assumes one and only one value. To each particle is associated a curve in space–time (three coordinates of space + one of time) which represents the complete history of the particle. This is the 'world line' – some sort of a visualization of its individual destiny – and its trajectory in geometric space is nothing other than the projection of this world line on the hyperplane defined by the three coordinates of space (fig. 2.1). The very fact that one can associate to each particle *one* world line, clearly indicates the individual and discrete character of the spatio–temporal properties attached to the concept of a particle. But although the notion of a particle does not retain the spatial extension of real objects, it does however consider them as physical (and not just mathematical) objects with material properties: to each particle is associated a *mass* which indicates the quantity of matter present in the (point) particle – whence the name 'material (or mass) point' which is sometimes used as a synonym for a particle.

The notion of a particle acquires its full (classical) meaning only with reference to that of an *interaction*. The physics of the classical period (that of Newton) entertained the ambition of explaining the world through the combined interplay of the reciprocal actions between particles – the inter-actions. It is for this reason that the notion of a particle includes, besides its spatio–temporal properties, its *dynamical* properties – i.e., properties related to the movement of the particle under the effect of the interactions.

To each particle, therefore, is associated a set of magnitudes which describe

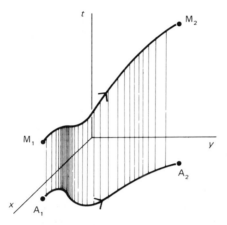

Fig. 2.1. *Trajectories and world lines*
The trajectory of the particle in a space of two dimensions (x, y) is given by the curve $A_1 A_2$ and is the projection of its world line $M_1 M_2$ in the space–time of two + one dimensions (x, y, t).

its state of motion and vary with the latter, governed by the interactions. Among these magnitudes, a very particular importance is attached to those which are conserved. By this, it is meant that the evolution takes place in such a manner, that when one considers the set of particles in interaction, there is a compensation of the individual variations and constancy, on the collective level, of the magnitude in question. These magnitudes, namely *energy*, *momentum* and *angular momentum*, thus derive their importance from the fact that if for an isolated system, composed of different interacting particles, the energy, momentum and angular momentum of each particle varies in the course of the motion, their total energy, momentum and angular momentum – sum of the individual values (variables) for each particle – remains constant in time (fig. 2.2). These conservation laws are, in the final analysis, related to the structure of space–time. In effect, this space–time is isotropic and homogeneous: time always flows in the same fashion; there is no preferred point in space, and all spatial directions are equivalent. Generally one refers to the points in space–time by means of their coordinates with respect to a system of reference (or reference frame). Now, starting with one reference frame, it is possible to define an infinity of others by displacing, e.g., the origin of coordinates, or by changing the orientation of the axes. To say that space–time is homogeneous and isotropic is the same as saying that reference frames which are obtained from the same frame, whether by translations in time, translations in space or rotations in space, are equivalent: the laws of

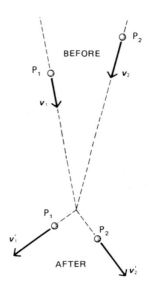

Fig. 2.2.   *Collisions*

In a collision between two particles P_1 and P_2, the state of motion of each (e.g., its velocity) changes. But there exist some conserved magnitudes, whose total values are the same both before and after the collision:

energy: $E_1 + E_2 = E'_1 + E'_2$;

momentum: $p_1 + p_2 = p'_1 + p'_2$;

angular momentum: $J_1 + J_2 = J'_1 + J'_2$.

physics assume the same form in them all. Another formulation consists in stating that the laws of physics are invariant, or symmetric, under all these changes of the reference frame. It is proved in books on classical mechanics (and we shall not demonstrate it here) that for an isolated system one can associate to each symmetry a conserved magnitude – i.e., a magnitude which keeps the same numerical value in the course of time. To the symmetry under time translations is associated the energy, to that under space translations the momentum, and to that under spatial rotations the angular momentum (see table 2.1).

The notion of a particle must also take into account the specific character of physical objects in addition to these general geometrical (trajectory) and dynamical (conserved magnitudes) characteristics. Hence, to each particle is attributed a certain number of intrinsic properties which specify its nature. Each particle is endowed with specific *charges*, which correspond to some

Table 2.1
Symmetries and conserved properties.

Spatio–temporal symmetry	Conserved dynamical magnitude
Translations in time	Energy E
Translations in space	Momentum p
Rotations in space	Angular momentum J

general properties of matter, and which indicate the values of these properties appearing in that portion of matter which constitutes this particle. Among these 'labels', enabling one to recognize the particle, one finds therefore, the electric charge, the gravitational mass, etc. Hence, these charges are related to the particular interactions (electromagnetic, gravitational, etc.) and their values indicate the strengths with which the particle experiences and exerts these interactions. If these interactions are described through the intermediary of a field (see below), the charge of a single particle characterizes the 'coupling' of this particle with the corresponding field, and one comes upon the notion of the 'coupling constant' introduced earlier (see sect. 1.4).

2.1.2. Classical fields

The other key notion of classical physics is that of a *wave* (or a *field*). This notion, like that of a particle, idealizes certain actual phenomena: one cites, in this context, the waves* on the surface of water, or the successive jolts which reverberate through the wagons of a train when a locomotive backs up to it. In both cases, the abstract notion of a wave is visualized as a displacement (transverse in the case of water waves and longitudinal for the train) propagating through successive and contiguous steps within a given medium. But in classical physics there exist as well 'true' fields, which are impossible to visualize in terms of displacements and a medium for propagation, and which ought to be thought of as existing by themselves. The electromagnetic field in vacuum is a case in point.

The notion of a field is in contrast to that of a particle, in that it connotes an essentially *continuous* character. A wave is an entity defined at each point of space–time. It could be a scalar or a vector (or an even more complicated)

* In the original text, in conformity with French usage, the term 'vague' is used to mean a wave in the non-technical sense (e.g., waves in water), while the word 'onde' is used to connote, more technically, its idealization in physics. In this context, the author points out the absence in English of separate words to differentiate between these nuances. (Footnote added by translator.)

quantity, which one denotes by $\Phi(r, t)$, the value at the point r at the instant of time t. The magnitude Φ is called the *amplitude* of the wave. A wave describes an essentially non-localized phenomenon, whose spatial extension is, in principle, infinite. Often, however, the field assumes an appreciable value, at a given instant, only over a limited region of space–time, depending upon the situation considered so that the extension of the phenomenon is physically finite. This spatio–temporal extension distinguishes the concept of a wave from that of a particle. It is in this sense that one often speaks of a 'field' rather than of a 'wave', the usage (somewhat fuzzy, however) being to reserve the term 'wave' for the case where the emphasis is upon the propagation of the phenomenon.

The evolution, in the course of time, of an undulatory phenomenon also differs radically from the cursive motion of a particle: a field *propagates*; i.e., the locus of all those points at which the field $\Phi(r, t)$ simultaneously assumes the same value Φ_0 – the name 'wave front' being given to this locus – is displaced in the course of time. The 'motion' of a wave is of the frontal type. It is governed by certain propagation equations, that is to say, by a set of conditions specifying the spatio–temporal evolution of its amplitude $\Phi(r, t)$. Thus, the propagation of the electromagnetic field is determined by the Maxwell equations.

A particle only occupies one point of space–time and it occupies it completely: two particles cannot find themselves simultaneously at the same point without colliding with each other; i.e., they interact by 'rebounding' against each other, suffering thereby a modification in their subsequent motions. The situation is quite different in the case of fields, precisely because of the continuous character of the notion of a field. What does in fact happen when two fields 'meet' each other? Let $\Phi_1(r, t)$ and $\Phi_2(r, t)$ be two fields, each occupying a region of space and propagating towards each other. There comes a time when their regions of influence overlap. At this point one can no longer speak of *two* fields, since one cannot count fields (as one would count particles). The most one can say is that there prevails in this region a field which is a combination of two others. To boot, it seems that the statement, 'when two fields meet each other', is incorrect: before their combination, as during and after its occurrence, there exists in fact only one field defined throughout the whole of space. This field manifests itself in two spatial regions of localization, which merge into a single one at the moment of their combination and later separate again (fig. 2.3).

Fig. 2.3.   *Superposition of fields*

The 'meeting' of two localized waves: in the region of overlap the validity of the principle of superposition is assumed. The waves combine by simple addition and separate without deformation.

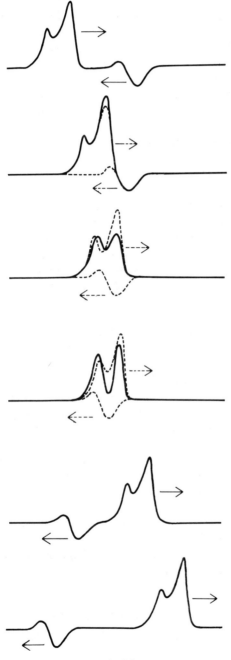

Fig. 2.3.

It remains to define the field amplitude at each point of their region of overlap, or in other words, the nature of the 'combination' of the two fields. The functions $\Phi_1(r, t)$ and $\Phi_2(r, t)$ are solutions of the equation of propagation of the field being studied. In most cases of physical interest this equation is linear. This, however, is no accident, but rather a matter of practical necessity: since linear situations are so much easier to treat mathematically than the others, one always begins by 'linearizing' the theory – i.e., by postulating linear laws. This hypothesis is later abandoned if it appears that it fails to account for the situation completely, considering it in this case only as a first approximation, once the limits of its validity have been explored. Under the hypothesis of linearity, therefore, the function defined at each point by the relation

$$\Phi(r, t) = \Phi_1(r, t) + \Phi_2(r, t), \tag{2.1.1}$$

is also a solution of the equation of propagation of the field. Thus, at each point of the region of overlap, the field is simply the algebraic sum of the amplitudes $\Phi_1(r, t)$ and $\Phi_2(r, t)$: 'combination' reduces to addition – the two fields simply superpose. One says that the field being studied obeys the *principle of superposition* – a principle which appears as one of the constitutive properties of the notion of a field in a linear theory. According to this, 'two' fields which meet, later separate, without their evolution being affected by their 'mixing'.

2.1.3. *Fourier analysis. Plane waves*

When the superposition principle prevails – i.e., when one is dealing with a linear (or a linearized!) field theory – one can decompose an arbitrary field, whose spatio–temporal dependence could be complex, into a sum of previously defined 'simpler' fields.

It is natural to consider as particularly 'simple', a field which is 'everywhere and always the same'. This is not to mean that its amplitude is the same at each point of space–time, reducing thereby to an uninteresting constant. It is intended rather to indicate that the field amplitude, as a function of space and time, has the 'same form' at each point. Considering for a moment the functional dependence of the field on time alone, one demands that this dependence be the same starting at any instant of time taken as the origin. In different terms, the function $\Phi(t)$ should have the same dependence on the time t as the function $\Phi(t + \tau)$, where the time is now counted, starting at the instant $t_0 = -\tau$, instead of $t_0 = 0$. More precisely, one requires that the two functions be the same, up to a factor, i.e.,

$$\Phi(t + \tau) = K \Phi(t), \tag{2.1.2}$$

implying that the graphs of the two functions have indeed the 'same form'. It is not difficult to show that the solutions to eq. (2.1.2) are necessarily exponential functions of the type $\Phi(t) \propto \exp(\alpha t + \beta)$ where α and β are a priori arbitrary complex numbers (exercise 2.1). However, if the real part of α does not vanish, the exponential function grows indefinitely – a fact which is not too well in accord with the idea that a field be 'everywhere the same'! One is thus led to keeping only exponentials with purely imaginary arguments,

$$\Phi(t) = a \, \exp \, i(\omega t + \varphi_0), \tag{2.1.3}$$

whose modulii remain constant.

These functions are called 'harmonic', by virtue of the harmony – if one so pleases – of their form, whose invariance under time translations had lead to the definition. The function

$$\varphi(t) = \omega t + \varphi_0 \tag{2.1.4}$$

is called the *phase*; the number φ_0, which is the value of the phase at the origin is often called the phase, for short, when the temporal dependence of the phase function $\varphi(t)$ is implicitly taken for granted.

The quantity a in eq. (2.1.3) is a 'number' whose mathematical nature has not been made precise, since it can vary widely, depending upon the situation that is being envisaged – a scalar, vector or tensor field, assuming real or complex values, etc. The quantity a is often called an amplitude, which is somewhat ambiguous, since properly speaking, it is the function $\Phi(t)$ which is the amplitude of the field.

It is remarkable that a harmonic function is *periodic*: it assumes the same value each time the phase increases by 2π, i.e., when the variable t increases by $T \triangleq 2\pi/\omega$; T is the period. The elementary trigonometric sine and cosine functions are nothing other than the real and imaginary parts of the harmonic function given by eq. (2.1.3). Their appearance in the analysis of vibrational phenomena, in particular in acoustics, and hence their role in musical theory, is additional justification for the adjective 'harmonic'.

A fundamental theorem in mathematical analysis, first formulated by Fourier, confirms that harmonic functions are well suited for the role for which we have chosen them. According to this theorem, any sufficiently regular function (and mathematical textbooks exist to explain this restriction) may be effectively analyzed as a sum of harmonic functions. This sum is discrete (Fourier series) if the function being analyzed is itself periodic; otherwise it is continuous (Fourier integral).

Applied to the study of physical fields, the theorem of Fourier allows one to decompose an arbitrary field, represented by a function $\Phi(r, t)$ into a superposition of harmonic functions of the spatio–temporal variables, whose phase is, therefore, a linear function of the four variables x, y, z and t, separately

$$\varphi(r, t) = \varphi(x, y, z, t) = \omega t - k_x x - k_y y - k_z z + \varphi_0, \qquad (2.1.5)$$

where ω represents the rate of temporal evolution of the phase, and is called the *angular frequency* (or *pulsation*), and k_x, k_y and k_z represent the rate of spatial progression. By convention, ω assumes positive values only, while the three spatial rates of evolution may be either positive or negative. The choice of the $(-)$ sign which precedes them in eq. (2.1.5) is made for reasons of convenience. The three quantities k_x, k_y and k_z, defined along the three coordinate directions of a system of axes, may be considered to be the components of a single vector, called the *wave vector* and denoted in general by k. Introducing this vector into eq. (2.1.5), one gets

$$\varphi(r, t) = \omega t - k \cdot r + \varphi_0. \qquad (2.1.6)$$

At a given instant, all the points of a plane perpendicular to k ($k \cdot r = $ const.) have the same phase, and hence the same amplitude $\Phi(r, t)$. These, therefore, are the wave surfaces of a harmonic field. Here they are plane, parallel to each other and perpendicular to the wave vector; it is in this sense that a harmonic field is called a 'plane wave'. The direction of propagation of such a wave is that of its plane wave surfaces, i.e., the direction of the wave vector k. With the sign convention adopted in eq. (2.1.5) for the phase, a plane wave propagating in the positive sense X'OX (respectively Y'OY, Z'OZ) corresponds to a positive value of k_x (respectively k_y, k_z). Along the direction k (and along this direction only) the phase, eq. (2.1.5), is written

$$\varphi(r, t) = \omega t - kr + \varphi_0, \qquad (2.1.7)$$

upon introducing the quantity $k = |k|$. One sees that the phase (and hence the amplitude) remains constant in time if $r = (\omega/k)t + \varphi_0/k$. Thus, plane waves move along the direction k with velocity $v_{ph} = \omega/k$; this velocity is called the 'phase velocity', since it is the velocity of those points which are 'in phase'. As regards the quantity k, the modulus of the wave vector, we shall agree in this book to name it 'undulation' by analogy with pulsation. Indeed, by eq. (2.1.7), the undulation k is to the spatial dependence, along the direction of propagation, what the pulsation is to the temporal dependence: just as on any fixed wave front (plane) $(\mathbf{k} \cdot \mathbf{r} = \text{const.})$ the phase varies from 0 to 2π, in an interval of time equal to the period $T = 2\pi/\omega$, so also at a given instant of time $(\omega t = \text{const.})$ the phase varies from 0 to 2π as one passes from one wave front to another, situated at a distance $\lambda = 2\pi/k$ (fig. 2.4). The spatial periodicity λ, along the direction of propagation, is called the *wavelength*. Although this length has been defined using the wave vector, it ought to be noted that there does not exist any 'vector $\boldsymbol{\lambda}$' (cf. exercise 2.2) of which λ is the modulus (albeit that the undulation itself is the modulus of the wave *vector*).

The Fourier decomposition, while mathematically always possible, only acquires its full physical meaning, however, when the harmonic functions are effectively physical *fields* – that is to say, are solutions of the equations of propagation – which is the case when one is interested in waves having the whole of space–time for their domains. It is the homogeneity of space–time which gives a privileged position to the harmonic functions, characterized as they are by their translational properties. In the case where the undulatory system is confined, such as, for example, the electromagnetic field inside a wave guide, or sound waves within a resonator, a spatial Fourier decomposition loses its physical pertinence. However, it can be replaced by an analogous decomposition whose 'simple' constituents are no longer plane waves, but rather are functions which conform to the characteristic constraints of the

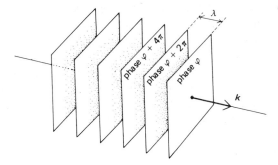

Fig. 2.4. *A harmonic wave*
For a harmonic wave (also called a plane wave), the phase at a given instant of time is constant, all over any one of the wave planes (fronts); λ is the distance, along the direction of propagation, which separates two wave planes whose phases differ by 2π.

situation and represent its proper 'modes' – the name traditionally given to them. On the other hand, since the flow of time is conceived as being inexorable and unconfinable, the harmonic functions are always appropriate for the analysis of temporal evolution.

2.1.4. Fields and particles. Interactions

Experimental practice, developed over three centuries of classical physics, clearly expresses in its very principles the characteristic notions of a field and a particle.

Thus, the discrete character of particles – in other words, the possibility of defining their trajectories – determines the type of experiments which are conducted in the physics of particles. Of particular importance in this context are the *collisions* between particles, where the laws of conservation of energy and of momentum permit one to study the characteristics (e.g., the mass) of one of the particles when those of the others are known. Thus, the physics of particles essentially calls for the determinations of trajectories and the establishment of the balance of conserved quantities – putting thereby its specific concepts (of discreteness) into bold relief. This, moreover, is just as true in celestial mechanics, where the 'particles' might be the planets, the stars, or even the galaxies, as in the motion of electrons – to the extent that the behaviour of electrons in a cathode ray tube or a particle accelerator might be described in classical terms. (We shall soon see that this is possible only approximately, and under certain definite conditions.)

The physics of waves, on the other hand, essentially brings into play the concept of *propagation*. It deals, therefore, with the study of how a field (or a wave), generated by a certain source, progressively affects different domains of space–time. In non-linear theories, such as fluid dynamics (acoustics, hydrodynamics) or the relativistic theory of gravitation (the so-called 'general relativity'), these problems become more complex. Fortunately, a good many theories are either linear (such as electromagnetism), or else can be approximately linearized (such as fluid dynamics, if one neglects viscosity). Thus, the principle of superposition reigns supreme, which permits one to place the emphasis upon phenomena, such as that of *interference* – an essential phenomenon, inasmuch as it underlies a large proportion of experimental work in the physics of waves.

Let us consider the simplest case of two harmonic waves in space–time. Let

$$\Phi_1(r, t) = a \exp \mathrm{i}(\omega_1 t - k_1 \cdot r + \varphi_1),$$

$$\Phi_2(r, t) = a \exp \mathrm{i}(\omega_2 t - k_2 \cdot r + \varphi_2), \tag{2.1.8}$$

where we have further specialized the situation by supposing that the amplitudes are equal. The superposition of these two waves is again a physically realizable field. It is

$$\Phi(r, t) = \Phi_1(r, t) + \Phi_2(r, t)$$
$$= 2a \cos\left[\tfrac{1}{2}(\omega_1 - \omega_2)t - \tfrac{1}{2}(k_1 - k_2)\cdot r + \tfrac{1}{2}(\varphi_1 - \varphi_2)\right]$$
$$\times \exp i\left[\tfrac{1}{2}(\omega_1 + \omega_2)t - \tfrac{1}{2}(k_1 + k_2)\cdot r + \tfrac{1}{2}(\varphi_1 + \varphi_2)\right]. \qquad (2.1.9)$$

If the angular frequencies and the wave vectors are not too different, one has $\omega_1 \simeq \omega_2$ and $k_1 \simeq k_2$, and the total field may be considered as being approximately harmonic, with the average angular frequency

$$\tilde{\omega} = \tfrac{1}{2}(\omega_1 + \omega_2), \qquad (2.1.10)$$

the average wave vector

$$\tilde{k} = \tfrac{1}{2}(k_1 + k_2), \qquad (2.1.11)$$

and the amplitude

$$A(r, t) = 2a \cos\left[\tfrac{1}{2}(\omega_1 - \omega_2)t - \tfrac{1}{2}(k_1 - k_2)\cdot r + \tfrac{1}{2}(\varphi_1 - \varphi_2)\right]. \qquad (2.1.12)$$

The spatio–temporal variations of this amplitude are slow compared to those of the average harmonic function, so that it may be considered to be locally constant (here 'locally' means on the scale of the average period $\tilde{T} = 2\pi/\tilde{\omega}$ and the average length $\tilde{\lambda} = 2\pi/\tilde{k}$). Thus, the result is a harmonic field, whose amplitude is 'modulated'. Generally, one speaks of *interference* (fig. 2.5) when it is a case of spatial modulation ($\omega_1 = \omega_2$, but $k_1 \neq k_2$), or of *diffraction* if the sources, instead of being discrete, form a continuous set. An infinite set of discrete sources – e.g., a grating – furnishes us with an intermediate situation. Temporal modulations ($\omega_1 \neq \omega_2$; $k_1 = k_2$) are in general called *beats*. In all cases, the superposition of waves manifests itself as an alternation of maxima and minima (zeroes in the example above) of the modulus $A(r, t)$ of the amplitude of the total field. The appearance of such 'fringes' in the experimental study of a phenomenon, is indicative of the wave nature of that phenomenon.

It is through the interplay of the two concepts of wave and particle – concepts upon which it is ultimately based – that classical (pre-quantum) physics describes natural phenomena.

In particular, it is through this interplay that the *interaction at 'a distance'*

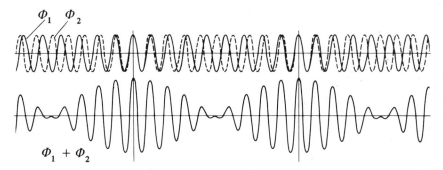

Fig. 2.5. *Interference*
The superposition of two harmonic waves produces, by interference, an alternating sequence of
high and low amplitude zones.

between particles – so eminently difficult to comprehend – is described theoretically. Instead of the effect proceeding directly, without any intermediary, one conceives of it as being mediated by a field. Produced by the first particle, the field propagates towards, and then acts on, the second. Naturally, the second particle also acts on the first via the same mechanism, so that the second particle is a 'source' of the field as well. The laws describing the phenomenon, instead of being reducible to an expression of the force prevailing between two particles, consist of the equations governing the creation and propagation of the field (e.g., the Maxwell equations), on the one hand, and of the expression of the forces exerted by the field on a particle (e.g., the Lorentz forces), on the other. These laws are all local, depending only upon physical magnitudes which are defined at the same point of space. It is in this manner that the difficulties associated with interaction at a distance are circumvented (fig. 2.6).

The notion of a field has, therefore, a double origin. On the one hand, it arose as a purely calculational artifice (thus, originally the gravitational field was merely a formal way to rewrite Newton's laws), while on the other hand, as a description of the motion of an underlying fluid (as in waves in water or sound waves in the air), it has ultimately acquired a reality of its own. A classical field is a full fledged physical entity, divested of the necessity of any material support (such as had at times been invented – as in the case of the ether – in a mere ad hoc fashion). It is a concept which is hardly more abstract than that of the 'point particle' of mechanics.

In classical terms, therefore, matter ought to be conceived as consisting simultaneously of both fields and particles. A complex object consists of a certain number of particles (e.g., 'material points' having weight) and fields

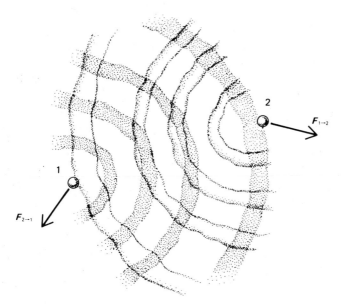

Fig. 2.6. *Field and interaction*
The interaction between two particles is brought about by the intermediary of the field, produced by them, which propagates between and exerts a force on each one of them.

(e.g., gravitational) which they produce and which in turn binds them. From this point of view, the elementary physical entities are, quite unambiguously, either particles or fields.

2.2. The new entities

We shall now describe situations, quite at variance with the classical viewpoint, where the particle–field dichotomy loses its validity. These could be wave phenomena (or at least phenomena which are recognized to be so in classical physics) which manifest discrete properties, or objects traditionally labelled as 'particles', which display a behaviour that is incomprehensible within the framework of classical mechanics.

Since our purpose is not to describe the historic experiments which led to the new concepts that we are about to introduce, but rather to deal with present-day experiments into which these concepts have been incorporated, some of our examples are taken from experimental work coming after 1920.

To look upon them today as being contradictory, it is clearly necessary to adopt an artificially retrospective point of view.

2.2.1. The diffraction of light

To begin with, let us consider an experiment on the diffraction of a light wave by an aperture. In classical physics, the phenomenon of diffraction – a generalization of that of interference – is a characteristic of systems having a wavelike nature.

The set up for such an experiment (fig. 2.7) consists of:
- a source S emitting light, assumed monochromatic for simplicity, of wavelength λ;
- an opaque screen pierced by a circular aperture of adjustable radius R, similar to the diaphragm of a photographic camera;
- an observation screen, at a distance D from the other side of the diaphragm, or a photographic plate, to record the diffraction pattern.

As long as R remains sufficiently large, one observes the projection of the source across the diaphragm as just a circular luminous spot on the screen, bounded by the shadow cast upon it, in accordance with geometrical optics. As one gradually closes the diaphragm, however, one observes the appearance on the edges of this spot, of alternating bright and dark flutings. These fringes, which are the visible manifestation of the principle of superposition, literally invade the entire figure as R becomes of the order of magnitude of the wavelength λ. The calculation of the intensity distribution on the plane of observation is analogous to the (simpler) computation of the intensity distribution due to a slit which is rectangular, rather than circular (cf. exercise 2.3). Instead of the rectilinear fringes, due to such a rectangular slit, one has now – because of the symmetry of the problem – circular or annular fringes. In both cases, the central fringe has dimensions of the order of $\lambda D/R$, and it is here that the brightness is maximal, while that of the others diminishes rapidly as one moves away from the centre of the pattern (actually in practice, one only sees a few rings). The luminosity of the bright fringes is greater the longer the time of exposure of the photographic plate.

Up to this point, everything is completely classical. Let us now reduce the power of the source. We notice, in the beginning, that the fringes do not disappear, which proves that the phenomenon still depends on the principle of superposition. At the same time, the brightness of the pattern diminishes, which again is hardly surprising, until one examines carefully the photographs that are obtained, and discovers that the seemingly continuous pattern is in fact formed by discrete luminous points. Besides, the pattern is

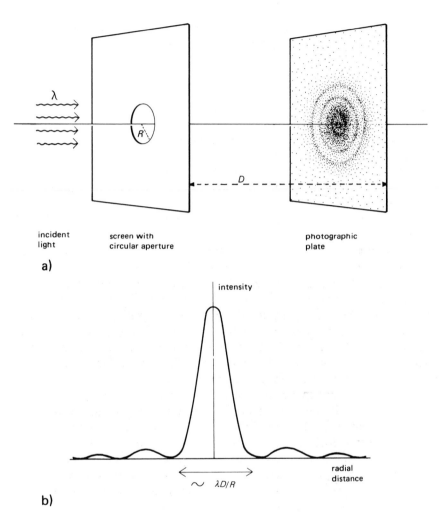

incident   screen with   photographic
light   circular aperture   plate

a)

Fig. 2.7. *Diffraction of a light wave*
(a) Experimental set up.
(b) Radial distribution of the diffracted intensity in the plane of observation.

formed not globally all at once, but rather point by point – its appearance is gradual, and it is only after some time has elapsed, that the density of the luminous points is sufficiently high to convey the impression of a continuous distribution. The intensity of the pattern at any point is proportional to the density of luminous points there. Stated otherwise, the luminous energy arrives at the screen grain by grain, in a random manner. Since it arrives more frequently at certain places, than at others, the respective light and dark fringes appear bit by bit, and after a certain period of time one again obtains the usual system of contrasting rings (fig. 2.8). Nonetheless, as the power of the source is reduced sufficiently, the phenomenon loses its continuous appearance, and the classical explanation in terms of the wave concept loses its validity.

But let us not act more naive than we actually are! We know very well the 'luminous grains' which arrive one after the other at the photographic plate – at least we do know what they are called! They are the *photons* (of the photoelectric effect mentioned in ch. 1). In the experiment which we are considering here, the source sends out N photons per unit time across the opening, which makes it $N\tau$ photons for the duration τ of the experiment, and the points of impact of these $N\tau$ photons give rise to as many luminous points on the screen. For the pattern to appear continuous, and not as a mere cluster

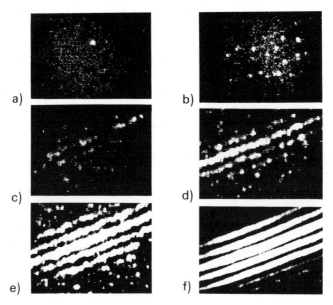

Fig. 2.8. *Interference and discrete impact*
The progressive formation of interference fringes, having a continuous appearance, through discrete impacts. [P. G. Merli, G. F. Missirole and G. Pozzi, Am. J. Phys. 44 (1976) 306, reproduced with the cordial permission of the authors. A (16 mm video) film, illustrating this experiment and its results, is obtainable from the Reparto di Cinematografia Scientifica, Istituto. Lamel, Via dell'Inferno, 40126-Bologna, Italy.]

of discrete, luminous points, it is only necessary for the total number of luminous points (hence of the photons), which make up this image, to be very large, i.e.,

$$N\tau \gg 1. \tag{2.2.1}$$

It is only because the apparatus is traversed by a sufficiently large number of photons that the continuous and wavelike aspect of this particular luminous phenomenon appears. Hence, it is important for the interpretation of this experiment to establish a distinction between the case where only a few photons enter the picture and the case where a large number of photons is involved, which alone – in an approximative way – calls upon the wave concept.

Now, condition (2.2.1), which marks the limit of validity of the wave concept, coincides precisely with the quantum/classical criterion stated in ch. 1. Indeed, the characteristic action for this experiment has to be calculated in terms of the available (classical) parameters, namely, the power P of the source, the size R of the aperture, the wavelength λ of the light, the duration τ of the experiment and the fundamental constant c which characterizes electromagnetic phenomena. The beam of photons transfers during the time τ, an energy $P\tau$, and hence a momentum $P\tau/c$. Multiplying this latter quantity by the size R of the diaphragm, one obtains a quantity which has the dimensions of an action. But the diffraction of the light by the circular aperture becomes apparent only when R is of the order of magnitude of λ, that is when $kR \simeq 1$, where k is the undulation ($k = 2\pi/\lambda$). Thus, the characteristic action of this (diffraction) experiment is

$$A = \frac{P\tau R}{c} \simeq \frac{P\tau}{ck}. \tag{2.2.2}$$

It is this quantity which ought to be compared with $\hbar$. Expressing the power of the source as the product of the number of photons N, emitted per unit time, and the energy $\hbar\omega$ of a single photon, the criterion for the validity of the classical theory assumes the form

$$\frac{N\hbar\omega\tau}{ck} \gg \hbar, \tag{2.2.3}$$

so that, since $\omega = kc$,

$$N\tau \gg 1, \tag{2.2.4}$$

which is just condition (2.2.1) for the validity of a classical description in terms of waves. Thus, the notion of a wave loses its explanatory value as soon as the experimental conditions dictate a recourse to quantum theory. The wave notion cannot form a part of the conceptual arsenal of quantum theory.

2.2.2. Collisions and effective cross-sections

We consider now, in a completely symmetric manner, experiments in which the classical concept of a particle turns out to be inadequate. This involves experiments in nuclear physics, namely, the 'collisions' or the 'scattering' of particles. It is a method which is currently in use for exploring sub-atomic matter: a nuclear target is bombarded using well-known particles (nuclei, protons, electrons, ...), and by studying the deflections of the incident particles by the target particles, one arrives at a characterization of their interaction. One can also, by this same procedure, discover new particles emitted during the course of the interaction. These studies, carried out at the large 'accelerator' sites, constitute the essential part of the work on the 'physics of particles'. The importance of this type of physics is enormous, both from an economic and from a theoretical point of view, and we shall often refer to it in the course of this book. It is, therefore, worthwhile to detail right here the various experimental features and to introduce the notion of the (*effective*) *cross-section*, used by experimentalists in reporting their results.

(a) *Experimental definition:* The incident particles are produced by an accelerator which imparts to them a certain kinetic energy. They form thereby a nearly parallel and mono-energetic beam, characterized by a flux $\mathscr{F}$ – i.e., by the number of particles crossing per unit time, a surface of unit area perpendicular to the direction of the beam. The target, whose dimensions are assumed to be small compared to the other distances characterizing the experimental set-up, consists of particles assumed to be identical. It is characterized by the total number N of particles contained in it, which are apt to cause deflections of the incident particles. After crossing the target, the deflected particles are collected inside a detector, whose position is defined by its directions $(\theta, \varphi) = \Omega$ (fig. 2.9). The detector has a small opening, of solid angle $d\Omega = d(\cos \theta)\, d\varphi$, and for each position Ω, one counts the number dn of deflected particles which enter the detector per unit time. This number is evidently proportional to the opening angle $d\Omega$. One makes the additional assumption that the target is sufficiently thin, so that each incident particle interacts at most once, and with only one particle in the target. Under these conditions, dn is also jointly proportional to $\mathscr{F}$ and N. The coefficient of

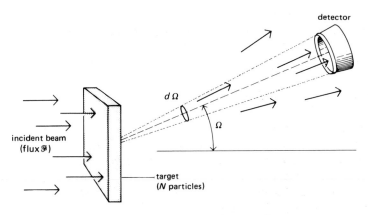

Fig. 2.9. A collision (or scattering) experiment.

proportionality depends upon the position of the detector, i.e., on the angle through which the particles have been deflected,

$$dn = \chi(\Omega) \, N\mathscr{F} \, d\Omega. \tag{2.2.5}$$

The number of particles deflected out of the incident beam – one also says 'scattered' – is

$$n = N\mathscr{F} \int \chi(\Omega) \, d\Omega. \tag{2.2.6}$$

One sets

$$\sigma = \int \chi(\Omega) \, d\Omega. \tag{2.2.7}$$

A dimensional analysis of eq. (2.2.6) shows that σ has the dimensions of an area; this is why it is called the effective cross-section of the scattering. The term 'effective' is justified as follows: imagine that the target consists of objects, each of which presents a geometric cross-section s to the incident flux (fig. 2.10) and the incident particles react only upon direct contact with these objects. Thus, each of these objects intercepts a number $\mathscr{F}s$ of particles, and the total number of particles deflected becomes $n = N\mathscr{F}s$. Thus, from eqs. (2.2.6) and (2.2.7), one gets $\sigma = s$. However, objects making up the target can

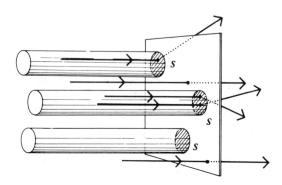

Fig. 2.10. *The notion of the effective cross-section*
An object having a cross-section s, perpendicular to the flux of incident particles, intercepts $\mathscr{F}s$ particles; if all these particles are deflected (complete effectiveness), the effective cross-section σ is equal to the geometric cross-section. Otherwise, the ratio σ/s measures the effectiveness of the target in the scattering process.

deflect the incident particles 'from a distance' through the action of various forces. Additionally, these 'particles' are not necessarily particles in the classical sense of the term: they do not follow definite trajectories. Nevertheless, the quantity

$$\sigma \triangleq \frac{n}{\mathscr{F}N}, \tag{2.2.8}$$

can be written in terms of the experimental quantities of the problem. Everything proceeds 'as though' the target were made up of N objects, each having a cross-section σ. Thus, the area σ characterizes the *effectiveness* of the target in deflecting the beam. Similarly, the magnitude $\chi(\Omega)$, which also has the dimensions of an area, measures the effectiveness of the target in deviating the beam into a given direction Ω. In view of eq. (2.2.7), one often adopts the notation

$$\chi(\Omega) = \frac{d\sigma}{d\Omega}, \tag{2.2.9}$$

and calls this quantity the 'differential effective cross-section'. The nature of σ, the integral of $\chi(\Omega)$ over all directions Ω, is made more explicit by calling it the '*total* effective cross-section'.

In summary, an experiment on the scattering of particles has, as its outcome, a measurement of the effective cross-section – differential, if one is interested in the directional dependence of the process and total, if one only considers its global outcome.

(b) *Classical theory:* Since the notion of an effective cross-section is purely experimental, it is valid regardless of the nature of the interaction producing the deflection, or of the theory which describes it. Thus, the notion makes sense, just as well in the classical as in the quantum theory; but, of course, a theoretical computation of the effective cross-section involves very different ideas and magnitudes, depending upon the theory being employed. In the classical case, where the notion of a trajectory is valid, the principle of this calculation is extremely simple. Outside the region of interaction, the initial trajectory of a particle is a straight line defined by its 'impact parameter' b, i.e., by its distance from an axis passing through the centre of the target which, for simplicity, we assume to be spherically symmetric (fig. 2.11). This

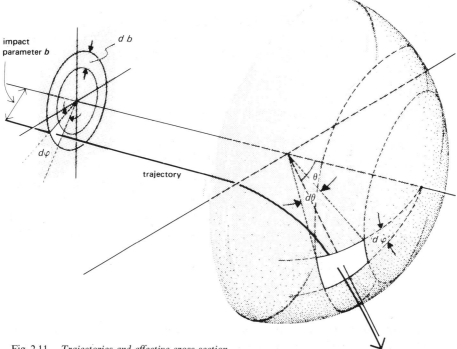

Fig. 2.11. *Trajectories and effective cross-section*
The classical computation of an effective cross-section is done using the relationship between the impact parameter b of a trajectory and the scattering angle θ.

trajectory is bent by the interaction through an angle θ, the 'scattering angle', which depends upon the impact parameter. (For example, the particle is deviated more or less depending upon whether it tends to pass close to or far from the target.) If one now follows the trajectories 'backwards', one notes that the particles registered by the counter, within the opening angle $d\Omega = d\cos\theta \, d\varphi$, are cut out of a sector of the initial beam corresponding to an opening angle $(\varphi, \varphi + d\varphi)$ of the circular annulus having width $(b, b + db)$, where b is the impact parameter corresponding to the scattering angle θ (and $b + db$ for $\theta + d\theta$). The area of this sector being $ds = b \, db \, d\varphi$, a number of particles

$$dn = b \, db \, d\varphi \, \mathcal{F}, \tag{2.2.10}$$

crosses it per unit time, where $\mathcal{F}$ is the incident flux. Comparing with definition (2.2.5), where now $N = 1$ (a single-particle target), and using the notation of eq. (2.2.9), one obtains the effective differential cross-section in the form

$$\chi(\Omega) = \frac{b \, db \, d\varphi}{d\Omega} = \frac{b \, db}{d\cos\theta} \tag{2.2.11}$$

(showing clearly that it has the form of an area). It is, therefore, sufficient to know the relation $b(\theta)$ – i.e., the geometry of the trajectories – to calculate an effective scattering cross-section in classical mechanics (exercise 2.4, fig. 2.12).

(c) *Example: Coulomb scattering.* A special case of great importance is that of scattering under the influence of Coulomb forces. Consider, for this case, a projectile particle of charge $q = Z_1 q_e$ and energy E, and a fixed target of charge $Q = Z_2 q_e$, so that an electrostatic force $Z_1 Z_2 e^2 / r^2$ exists between them. It is well-known, as a traditional exercise in mechanics, that the particle describes a hyperbolic trajectory (fig. 2.13a). The impact parameter b and the angle of deflection θ are related by the formula

$$b = \frac{Z_1 Z_2 e^2}{2E} \cotan(\tfrac{1}{2}\theta). \tag{2.2.12}$$

From eq. (2.2.11) one thus obtains the differential effective cross-section,

$$\chi_c(\Omega) = \left(\frac{Z_1 Z_2 e^2}{4E}\right)^2 \frac{1}{\sin^4(\tfrac{1}{2}\theta)} \tag{2.2.13}$$

illustrated in fig. 2.13b.

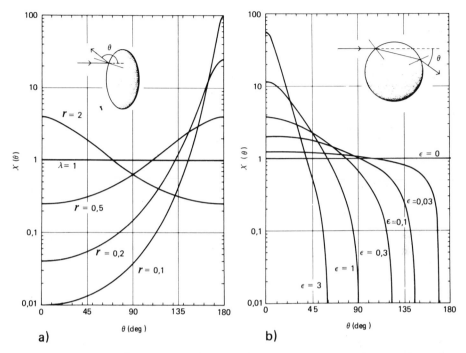

a)   **b)**

Fig. 2.12. *Classical effective cross-sections*
(a) The target is an ellipsoid of revolution, having transverse radius R and horizontal radius R', from which the projectile particle rebounds elastically. The curves show the differential effective cross-section $\chi(\theta)$ in units of $\frac{1}{4}R^2$ for different values of $r = R'/R$ (cf. exercise 2.4a).
(b) The target is a sphere, of radius R, at a potential $-V_0$, so that a particle of energy E moves in a straight line, obeying at the surface the classical laws of refraction in a medium of refractive index $n = (1 + V_0/E)^{1/2}$. The curves show the differential effective cross-section $\chi(\theta)$ in units of $\frac{1}{4}R^2$ for different values of $\varepsilon = E/V_0$ (cf. exercise 2.4b).

It ought to be remarked that the corresponding total effective cross-section is not defined in this case, since the angular integral, eq. (2.2.7), diverges for the function given by eq. (2.2.13). That the total effective cross-section should thus be infinite, results from the long-range character (slow decrease with distance) of the Coulomb forces: no matter how large the impact parameter, the particle always feels the deflective action of the target. The Coulombic differential effective cross-section, eq. (2.2.13), is called the 'Rutherford' cross-section, because of the major role played by this formula in Rutherford's analysis of his famous experiments on the scattering of α-particles, at the turn of the century. It was in this manner that he was able to display the effect of point-like scattering centres and to demonstrate the existence of atomic

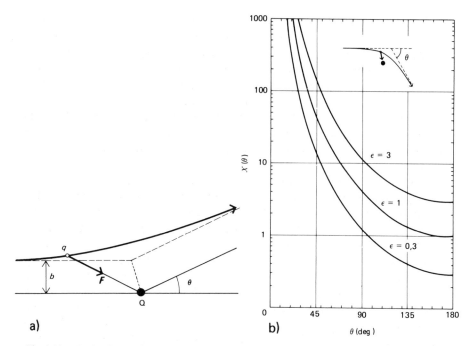

Fig. 2.13.   *Coulomb scattering*
(a) The trajectory of a charged particle deflected by a charged target is hyperbolic.
(b) The 'Rutherford effective cross-section' for different values of the energy. (The dimensionless parameter ε depends on the energy through $\varepsilon = (Z_1 Z_2 e^2/4E)^2 A^{-1}$, where A is the arbitrary unit of area used to measure χ.)

nuclei. It is a most fortunate historic accident that the effective cross-section, eq. (2.2.13), calculated in the classical theory, retains the same expression in the quantum theory: but for this remarkable coincidence (in particular, for the Coulomb forces), the interpretation of these experiments – which we now know to be indicative of the existence of quantum entities, and requiring a concomitant theory – would have taken much longer to emerge (cf. exercise 2.5).

2.2.3. *Scattering of α-particles*

We now envisage an experiment involving the bombardment of atomic nuclei of a certain species, such as aluminium or copper, e.g., by α-particles ($^4_2$He nuclei).

Such an experiment aims to establish the size, form and nature of the target nuclei, to the extent that it would allow one to decide between alternative models of the nucleus. Thus, for example, if one takes the nucleus to be a ball of nuclear material, one can expect, among other possibilities, one in which the particles rebound from this ball or cross it, suffering changes in their trajectories. Whichever model is considered, one discovers that the differential effective cross-section depends, in a 'simple' regular way, upon the angle of deviation θ – as confirmed by calculations using crude models (with the incident particles either rebounding or being deflected while crossing) (cf. exercise 2.4 and fig. 2.12).

If one performs the experiment, one discovers that for certain values of the energy of the incident α-particle, the effective cross-section does not resemble the simple models expected (fig. 2.14). The curve showing the variation of the differential effective cross-section $d\sigma/d\Omega$, as a function of θ displays a series of peaks of decreasing heights, as one gradually moves away from the $\theta = 0$ axis. This alternation of maxima and minima in the angular dependence, definitely recalls the classical diffraction pattern formed by a light wave crossing a circular aperture, discussed in the section preceding fig. 2.7. Similarly, one discovers (fig. 2.15) that as the energy of the incident α-particles increases, the peaks become sharper and narrower, just as in a diffraction pattern, when the aperture is illuminated with light of a shorter wavelength. Thus, the effect in question becomes more prominent the larger the value of E. It appears to be extremely difficult to reconcile these experimental results with the classical theory. In order that formula (2.2.11), giving the classical differential effective cross-section, might be able to account for these results, it would be necessary for the relation $b(\theta)$ between the scattering angle and the impact parameter to display a strange oscillatory behaviour, requiring very unusual trajectories and extremely special interaction potentials. On the other hand, analogous results, leading to diffractive type effective cross-sections are observed with practically every possible type of projectile (protons, neutrons) and with widely different nuclei (fig. 2.16). Moreover, these phenomena go beyond the framework of nuclear physics and are observed in extremely diverse experimental situations. Thus, what we see here is not just a peculiarity of nuclear interactions, but rather a manifestation of a very general property. Indeed, the notion of a trajectory, or the concept of a particle itself, loses its descriptive validity here.

One should not be surprised to discover, therefore, that the situation eludes the jurisdiction of the classical laws of physics and demands a recourse to quantum theory. Indeed, let us evaluate the characteristic action for the scattering of an α-particle, of energy E and mass M, by a nucleus of radius R

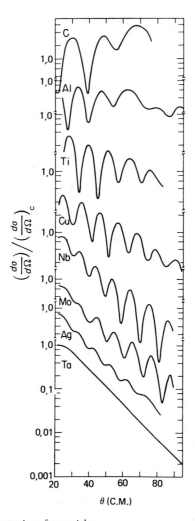

Fig. 2.14.   *Nuclear scattering of α-particles*
Effective cross-sections for the elastic scattering of α-particles by various nuclei (silver, molybdenum, niobium, copper, titanium, aluminium, tantalum and carbon). The incident α-particles have an energy of 40 MeV. The ratio of the measured effective cross-section to the classical Rutherford cross-section $(d\sigma/d\Omega)_c$, has been taken in order to exhibit the nuclear effects only. [G. Igo, H. E. Wegner and R. M. Eisberg, Phys. Rev. 101 (1956) 1508.]

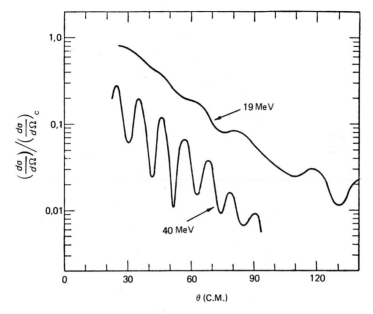

Fig. 2.15. *Nuclear scattering of α-particles*
Effective cross-sections for the scattering of α-particles by copper nuclei for two incident energies, $E = 40$ MeV and $E = 19$ MeV. The distance between successive maxima of the effective scattering cross-section diminishes as the energy of the incident particles increases, everything else remaining the same. [Same reference as for fig. 2.14.]

(and very much heavier than it, so that its mass does not intervene). Using relation (1.2.12), one may write

$$A = E^{1/2} M^{1/2} R. \tag{2.2.14}$$

For the scattering of 40 MeV α-particles, with $M \simeq 6.7 \times 10^{-27}$ kg, on nuclei of, let us say, copper for which $R \simeq 4.8$ F, this becomes

$$A \simeq 10^{-33} L = 10\hbar, \tag{2.2.15}$$

which is a typically quantum order of magnitude.

It follows from relation (2.2.14), that the quantum character of the reactions being studied would become more prominent at lower energies, since the characteristic action diminishes with energy. It ought to be remarked, however, that for very low energies, the α-particles do not interact

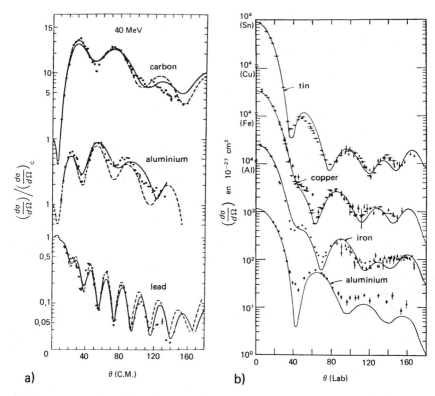

Fig. 2.16.   *Nuclear scattering of protons and neutrons*
(a) Effective cross-sections of the elastic scattering of *protons* of 40 MeV energy from various nuclei. [N. Hintz, Bull. Am. Phys. Soc. 2 (1957) 14.]
(b) Effective cross-sections of the elastic scattering of *neutrons* of 14 MeV energy from various nuclei. [N. Fernbach, Rev. Mod. Phys. 30 (1958) 415.]
The experimental results are indicated by the dots; the broken and continuous curves correspond to different theoretical models.

directly at all with the nucleus, since the Coulomb repulsion prevents a close approach. For a direct 'contact' it is necessary that the energy E of the incident α-particles be such as to enable them to overcome the Coulomb potential barrier. This potential, which is $V(r) = 2Ze^2/r$, for a nucleus of charge Zq_e and an α-particle of charge $2q_e$, has its maximum around $r \simeq R$, the nuclear radius (fig. 2.17). One must have, therefore,

$$E > \frac{2Ze^2}{R},$$

(2.2.16)

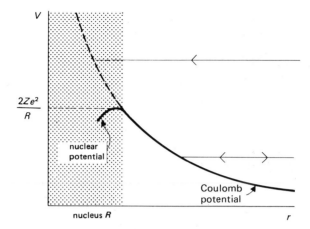

Fig. 2.17. *The Coulomb barrier*
The variation, as a function of the distance r, of the repulsive potential between a particle and a nucleus of charge Zq_e. To effectively reach the nucleus, an α-particle must have an energy greater than $2Ze^2/R$ (R = radius of the nucleus).

which in the case of copper considered above ($Z = 29$), becomes

$$E \gtrsim 20 \text{ MeV} \qquad\qquad (2.2.17)$$

(cf. also exercise 2.6).

At low energies, the α-particles only 'feel' the Coulomb field of the nucleus, which scatters them with the perfectly regular Rutherford effective cross-section described above, and at which quantum effects do not show. Moreover, at the energies considered here, the Coulomb repulsion is always present, and its effects combine with those coming from the nuclear interactions. It is for this reason that the experimental curves in figs. 2.14, 2.15 and 2.16a display the experimental effective cross-sections as ratios of the Rutherford cross-section in a manner which exhibits only the nuclear effects. It is possible to give a more detailed analysis of these experimental results, but we always discover that the classical notion of a particle is unable to explain these phenomena which, as we have seen, are specifically quantum in nature. Like the notion of a wave, that of a particle is alien to quantum theory.

We remark, however, that as long as the α-particles are sufficiently far from the nucleus (in particular, while they are inside the detector or within the source), the characteristic action of the nucleus–α-particle system remains very appreciably larger than $\hbar$, since in that situation the radius of the nucleus in eq. (2.2.14), for the characteristic action, is replaced by the distance of the nucleus from the α-particle. Classical physics is valid and there is no problem in speaking of particles. It is only when the particles reach the nuclei, that the system becomes sufficiently localized and its action sufficiently low for a quantum theory to be necessary. This also justifies the name '*particle physics*' for this branch of physics, which has precisely as its objective a detailed study of what happens exactly when the concept of a particle loses its significance.

2.3. The new concepts

2.3.1. Quantons

The two experiments which we have just analyzed demonstrate the impotence of the two fundamental concepts, of a wave and a particle, of classical physics to account for reality, when by modifying the experimental parameters, one passes from a classical situation (characterized by an action $\gg \hbar$) to a quantum situation (where the characteristic action is of the order of $\hbar$). Thus, the notion of a wave encounters its limit when the experimental set-up is no longer traversed by a large number of photons, without implying, however, that these photons have to be considered as being particles in a

Fig. 2.18. *Quanton, wave and particle*
Quantum theory recognizes only one sort of objects, which in the classical approximation ($A \gg \hbar$) are described by either the notion of a wave or of a particle.

classical sense (since the diffraction rings never disappear, or equivalently since one has to wait for a while to see them appear). Similarly, the notion of a particle loses its descriptive value when one manages to have two particles localized within a sufficiently narrow domain; the experimental curves for the variation of the effective differential cross-section, as a function of the deviation, manifest in this case the 'fringes' characteristic of a wave phenomenon without again implying that one could really speak of waves here, since what one effectively receives and counts in the detector are very definitely individual discrete entities.

We must, therefore, abandon the idea that every physical object is either a wave or a particle. Neither is it possible to say, as is sometimes done that particles 'become' waves in the quantum domain and conversely, that waves are 'transformed' into particles. Nor should it be said that quantum objects have a dual nature, which is simultaneously wavelike and corpuscular (something which is logically absurd, since the two concepts are mutually exclusive).

It is, therefore, necessary to acknowledge that we have here a different kind of an entity, one that is specifically quantum. For this reason we name them *quantons*, even though this nomenclature is not yet universally adopted. These quantons behave in a very specific manner, and it is to an elucidation of this behaviour that this book is devoted.

The essential aspect of the notion of a quanton is its uniqueness. From the point of view of quantum theory, there only exists one sort of objects – the quantons. If in a certain scattering experiment, the α-'particles' begin to manifest wave-like properties while, in a low-intensity diffraction experiment, the light 'waves' manifest particle-like properties, it is due to the fact that helium nuclei and photons are both objects of the same species, namely, quantons (even though they are also different in many respects). Under certain conditions, namely for values of the characteristic action which are much larger than $\hbar$, these quantons can display one or the other of the two particular aspects and be approximately described either as particles or as waves (fig. 2.18). Paraphrasing Feynman (in *The Character of the Physical Law*) and adopting the viewpoint of a classical physicist, who would consider himself 'normal', we might say, α-particles behave in exactly the same way as photons; they are both screwy, but in exactly the same way!

The difference in behaviour between classical and quantum objects ought necessarily to be reflected in the concepts which describe them. Nothing proves, a priori, that the concepts of energy, momentum and angular momentum, which enable one to give a theoretical treatment of the α-particles (in the classical sense) remain adequate for a description of the α-quantons. Similarly, would the concepts of pulsation and undulation enable one to describe the diffraction of photons at low intensities? It does not seem probable. It is reasonable to assume that the classical concepts must, in the quantum domain, relinquish their roles to the newer quantum concepts.

However, just as the behaviour of quantum objects simultaneously manifests characteristics, which in classical terms are specifically those of either particles or of fields (thus, the effective cross-section for the collision of quantons resembles the intensity distribution of a wave diffracted by an aperture), the new, properly quantum concepts that we are trying to construct must be related simultaneously to two different classical counterparts: corpuscular and undulatory.

2.3.2. The Planck–Einstein relation

We already possess, in fact, the bases for such a construction. Everyone knows today – since it is taught in elementary physics courses and is often repeated in popular expositions – that light energy is transported by means of photons. For light of a given pulsation ω, each photon is the carrier of a definite amount of energy

$$E = \hbar\omega. \tag{2.3.1}$$

This is the Planck–Einstein hypothesis (1900–1905); among other things, it allows one to understand the photoelectric effect, which the classical electromagnetic theory was unable to explain (cf. exercise 1.3). But what exactly is the significance of this relation?

It establishes a connection between a classical corpuscular concept, the energy E, and a classical wave concept, the pulsation ω. If one considers that in classical physics waves and particles constitute two distinct classes of objects, one realizes how much boldness and intellectual authority was necessary to claim this relation. Now, the link established here is effected through the intermediary of the quantum constant $\hbar$ – which actually modifies the interpretation of the two concepts being related. The Planck–Einstein relation is a purely quantum relation: it allows one to characterize a quanton, i.e., an object which is *neither* a wave *nor* a particle, by importing from the classical theory into the quantum theory the concepts of energy and pulsation, and joining them by a quantum knot – the constant $\hbar$.

In this manner, a new concept is forged, which is properly quantum, is *neither* wave-like *nor* particle-like, and which one might call a 'quantum-energy-pulsation'. This new concept differs as much from the classical concepts from which it originated, as a quanton differs both from a wave and a particle. It is only in the classical limit (where the characteristic action of the system is appreciably larger than $\hbar$) that the quantum-energy-pulsation may be identified with one or the other of the two classical concepts of energy *or* pulsation.

Thus, under conditions of the classical approximation – namely, when a large number of photons is present – the energy-pulsation of the photons has to be identified with the pulsation (or the frequency) of the electromagnetic wave which these photons collectively constitute. It is in this sense that table 2.2, giving the energy–frequency correspondence for the entire electromagnetic spectrum, has to be understood: the energy appearing opposite to

Table 2.2. *The electromagnetic spectrum*
Corresponding to each region of the spectrum, its usual origin – molecular, atomic or nuclear – has been indicated (the shaded 'balls' schematize the atoms).

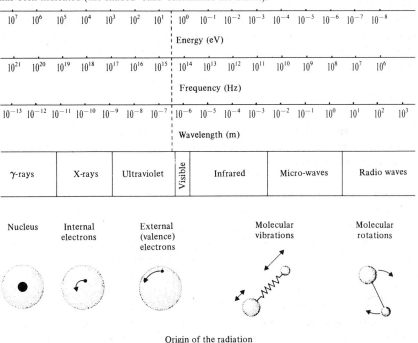

| 10^7 | 10^6 | 10^5 | 10^4 | 10^3 | 10^2 | 10^1 | 10^0 | 10^{-1} | 10^{-2} | 10^{-3} | 10^{-4} | 10^{-5} | 10^{-6} | 10^{-7} | 10^{-8} |

Energy (eV)

| 10^{21} | 10^{20} | 10^{19} | 10^{18} | 10^{17} | 10^{16} | 10^{15} | 10^{14} | 10^{13} | 10^{12} | 10^{11} | 10^{10} | 10^9 | 10^8 | 10^7 | 10^6 |

Frequency (Hz)

| 10^{-13} | 10^{-12} | 10^{-11} | 10^{-10} | 10^{-9} | 10^{-8} | 10^{-7} | 10^{-6} | 10^{-5} | 10^{-4} | 10^{-3} | 10^{-2} | 10^{-1} | 10^0 | 10^1 | 10^2 | 10^3 |

Wavelength (m)

| γ-rays | X-rays | Ultraviolet | Visible | Infrared | Micro-waves | Radio waves |

| Nucleus | Internal electrons | External (valence) electrons | | Molecular vibrations | | Molecular rotations |

Origin of the radiation

each frequency is not that of the electromagnetic wave, but that of each one of the photons which together make up this wave.

By contrast, in the scattering experiment of α-quantons from nuclei, described in sect. 2.2.3, the energy-pulsation of these quantons ought to be identified –.when the classical approximation becomes applicable, i.e., at the level of the source or the detector – with a mechanical energy. Its identification with a pulsation does not have a meaning here, since even in large numbers, the objects in question do not form a wave, but only a discrete collection of individual particles.

Under these conditions, one should, therefore, not speak of the energy (or of the pulsation) of a quanton, but rather of its quantum energy-pulsation, or else one ought to coin a new term for it. The fact that this is not done is only a matter of convenience. The situation is somewhat similar to that in the 18th century, when people in France decided to call a potato – a hitherto unknown vegetable, imported from the Americas – a 'pomme de terre' (or 'ground apple'), since it had the shape of an apple and grew in the ground. The term 'pomme de terre' hardly makes more sense than the expression 'the energy of a quanton', for just as 'pomme de terre' is not an apple (since it does not grow above on an apple-tree), so also the quanton does not possess an energy (in the classical sense) since it is not a particle. However, today no one thinks of the 'pomme de terre' as some kind of an apple and neither do physicists, in the course of their reasonings or their computations, confuse the energy of a quanton with that of a billiard ball. To talk about the energy of a quanton amounts inevitably to an abuse of language, which physicists commit all the time. It is harmless, as long as it is committed with complete knowledge of its cause (and effect). Even worse, physicists currently use the term 'particle' to designate a quanton (so that one speaks of an α-particle, even as it undergoes a quantum scattering from a nucleus) in the same way as the owner of a French restaurant might put 'entrecôte, pommes frites' (sirloin steak with fries) on his menu. Only an uninitiated ignoramus might think that it has something to do with apples. Nevertheless, it is necessary to be on guard, so as not to risk serious confusions. How, for example, would one interpret the entry 'boudin aux pommes' (blood sausage with apples) in the same menu?

Although they are applied to two distinct classes of classical objects, the (classical) concepts of energy and pulsation are nevertheless related (for otherwise, their fusion into a single quantum concept would not have made sense). As a matter of fact, they are both intimately related to the *temporal evolution of physical systems*.

Indeed, it is in this very manner that the concept of energy is introduced into classical mechanics (see sect. 2.1.1): One associates to the invariance of physical laws under time translations a constant of motion (i.e., a magnitude which preserves its numerical value during the motion), the energy. The concept of pulsation, in its turn, is associated in classical wave physics to the notion of harmonicity in time. We have seen (sect. 2.1.2) that a harmonic wave is defined by the notion that a translation (here in time) leads to a simple change in its amplitude, which in view of the relation

$$\exp i\omega(t + \tau) = \exp i\omega\tau \exp i\omega t, \tag{2.3.2}$$

preserves its temporal dependence in the course of this translation. This is a type of invariance which is characteristic of harmonic functions, and a harmonic wave in time is defined by its pulsation.

Against this perspective, the Planck–Einstein relation appears as being completely dominated by the principle of invariance of the laws of physics under time translations. Under the aegis of this principle, it brings about a quantum synthesis of the (undulatory) notion of harmonicity with the (corpuscular) notion of the constant of motion. The relation $E = \hbar\omega$ indicates that symmetry under the operation of time translations is also valid in the quantum theory and hence that the unified concept of the quantum-energy-pulsation is also directly associated to the invariance properties under time translations. We have here a kind of 'correspondence principle' between classical and quantum physics, which must necessarily apply to the other fundamental symmetries – under spatial translations and rotations – of the laws of physics, as well.

2.3.3. The de Broglie relation

In classical mechanics, the momentum p is the conserved magnitude which is associated to the invariance under spatial translations. From the previous reasoning it follows that in the quantum theory it must merge with the magnitude describing the harmonicity (periodicity) of a wave in space, namely, with the wave vector k. A priori, there exists therefore, a relation, built upon the model of the Planck–Einstein relation, which by means of the quantum link $\hbar$ brings about a synthesis of the magnitudes p and k:

$$p = \hbar k. \tag{2.3.3}$$

This is the de Broglie relation.

This relation was originally written by de Broglie, in 1924, in a different form. Recall that the undulation $|k| = k$ is related to the wavelength λ through

$$\lambda = \frac{2\pi}{k} = \frac{2\pi\hbar}{p}, \quad \text{where } p = |p|. \tag{2.3.4}$$

Thus,

$$\lambda = \frac{h}{p}.$$
(2.3.5)

The 'de Broglie wavelength', thus introduced characterizes the spatial harmonicity of a quanton having momentum p. Naturally, the terms 'momentum' and 'wavelength' are employed here by a deliberate misuse of language, for they designate notions which are different from their classical homologues – as demonstrated precisely by their specifically quantum (since it involves the constant $h = 2\pi\hbar$) relationship, eq. (2.3.5). Here it is particularly necessary to be aware of the name 'wavelength', since a quanton is not a wave (see exercise 2.23), and one ought to consider the term as a whole, without attaching any meaning to its individual parts.*

Going back to the example in sect. 2.2.3, we say that an α-quanton of kinetic energy E, and hence having momentum $p = (2mE)^{1/2}$, has the wavelength $\lambda = h/p$. This length necessarily plays a role in any situation where the quanton finds itself in interaction with an object of comparable dimension, and this is exactly the situation considered in the example of sect. 2.2.3. For the scattering of α-particles, of energy $E = 40$ MeV, from copper nuclei, $\lambda = h/(2mE)^{1/2}$ has the value 3 F. This is comparable to the radius of a copper nucleus, which is 4.8 F. In this situation, 'edge effects' are displayed, similar to those which show up when a wave is forced to pass through an aperture of a size comparable to the wavelength (see sect. 2.2.1). It is in this sense that one refers to the α-particle as a 'wave', and says that it has been 'diffracted by the nucleus', whose size is of the same order of magnitude as the wavelength of the quanton. For macroscopic objects the wavelength is so small that no wave-like effects are apparent (see exercise 2.7).

Just as the phenomenon of the diffraction of a wave allows us to 'see' objects whose sizes are of the order of magnitude of the wavelength, so also the 'diffraction' of material particles may be used as a means for 'observing' matter. The resolving power of this technique is fixed by the quantum wavelength, eq. (2.3.5), of the particles. For example, what energy should be imparted to electrons or neutrons in order to 'perceive' the atoms inside a crystal? Here, the characteristic length of the quanton in question has to be of the order of magnitude of the interatomic distances within the crystal, which

---

* In the original French text, the authors use the hyphenated word 'longeur-d'onde' to connote a 'wavelength' in the above sense, in order to differentiate it from the usual (unhyphenated) term 'longeur d'onde'. (Translator's note.)

is approximately 10^{-10} m. Using the formula $\lambda = h/(2mE)^{1/2}$, and taking into account the respective masses of the electrons and the neutrons, we find from this an energy of 150 eV for the electrons (these are also called 'slow' electrons), and an energy of $\frac{1}{13}$ eV for the neutrons (also called 'thermal' neutrons, since this energy is of the order of the energy of thermal agitation at ordinary temperatures). These two numbers have to be comparable to the energies of the photons which would have to be used to 'see' the atoms. A wavelength of the order of 10^{-10} m corresponds to the hard X-ray region of the electromagnetic spectrum, and hence, from table 2.1, to an energy of 10^4 eV (see exercise 2.8).

Conversely, electrons having energies appreciably smaller than about a hundred electron-volts, have wavelengths which are appreciably larger than the interatomic distances within a crystal, and hence do not 'feel' the atoms individually. This, in particular, is the case for the conduction electrons in a metal (see exercise 2.11): thus, one cannot explain the conductivity (and resistivity) of a metal using a classical model of collisions, in which each electron would behave like a little ball in a game of nine-pins, with respect to the atoms. An explanation of the electrical properties of a metal requires a quantum theory (see chs. 6 and 7).

Let us recall the manner in which the magnitude p of the momentum depends upon the kinetic energy E of the quantons (since, in general, it is this parameter which is fixed experimentally) and its mass m. We have the relation,

$$p^2 c^2 = E_{\text{tot}}^2 - m^2 c^4 = E(2mc^2 + E), \tag{2.3.6}$$

given by Einsteinian relativity, where $E_{\text{tot}} = mc^2 + E$ is the total energy (mass energy + kinetic energy). There are two interesting special cases:

(1) For a 'non-relativistic' quanton (more precisely, one that is described by Galilean relativity), for which the condition $E \ll mc^2$ (or $p \ll mc$) holds, one has the limiting relation,

$$p = (2mE)^{1/2}, \tag{2.3.7}$$

which follows directly from the Galilean relation

$$E = \frac{p^2}{2m}, \tag{2.3.8}$$

(2) For an 'ultra-relativistic' quanton, for which the condition $E \gg mc^2$ (or $p \gg mc$) holds, one has the approximate relation

$$p = \frac{E}{c}. \tag{2.3.9}$$

The last relation is, in particular, rigorous for photons, whose mass is zero ($m = 0$). But, in general, it is necessary to bear in mind that quantons which have the same energy may have different wavelengths, depending on their masses. Numerical computations are often facilitated by the use of purely numerical (dimensionless) magnitudes. Thus, it is convenient to express the wavelength of a quanton of mass m, in units of a natural standard furnished by its 'Compton wavelength'

$$\lambda_C \triangleq \frac{h}{mc}, \tag{2.3.10}$$

as a function of the ratio of its kinetic energy E to its mass energy mc^2. Thus, rearranging eqs. (2.3.5) and (2.3.6)–(8–9), we obtain the useful expressions:

general formula,

$$\lambda = \frac{h}{mc} \frac{E}{mc^2} \left(2 + \frac{E}{mc^2} \right)^{-1/2}, \tag{2.3.11}$$

'non-relativistic', Galilean formula,

$$\lambda = \frac{h}{mc} \left(\frac{2E}{mc^2} \right)^{-1/2}, \tag{2.3.12}$$

'ultra-relativistic' formula,

$$\lambda = \frac{h}{mc} \left(\frac{E}{mc^2} \right)^{-1}. \tag{2.3.13}$$

These relations become particularly easy to work with if one knows the mass of the quanton in terms of its energy equivalent, usually expressed in electron-volts (see exercises 2.8 to 2.12). Figure 2.19 displays the function $\lambda(E)$ for different quantons.

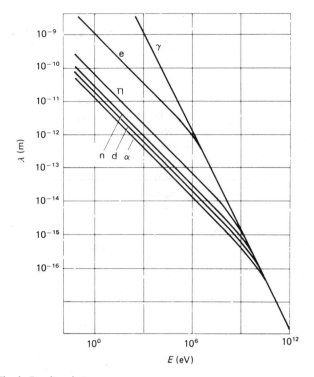

Fig. 2.19. *The de Broglie relation*
The relation between the kinetic energy and the wavelength for various quantons: photon (γ), electron (e), pion (π), neutron (n), α-particle, deuteron (d).

2.3.4. *Quantization of angular momentum*

(a) *Components.* The angular momentum is the magnitude associated in classical mechanics to the invariance under spatial rotations. For a particle placed in a situation of invariance under a rotation about an axis Oz, the component J_z of its angular momentum is a constant of the motion. In quantics, this magnitude should merge with its (classical) undulatory homologue, namely, the magnitude which characterizes the harmonicity of a classical wave, vis à vis a rotation about Oz, or in other words, its functional dependence on the azimuthal angle with respect to the Oz-axis. The identification of this magnitude is not quite as immediate as was the case with the other two spatio–temporal symmetries. Nevertheless, an electromagnetic wave, inside a cylindrical wave guide, or a sound wave inside a cylindrical pipe, furnishes us with relatively simple examples of waves which are harmonic under rotation about an axis (fig. 2.20). These waves are periodic in the angular coordinate φ, about the Oz-axis and during a rotation through an arbitrary angle around Oz, their amplitudes get multiplied by a simple phase factor, with no change in magnitude. Let α be the angular periodicity of such a harmonic wave. Corresponding to it, there is an angular rate of variation of the phase, or an 'angular wave number',

$$m \triangleq \frac{2\pi}{\alpha}, \qquad\qquad (2.3.14)$$

defined through α, just as ω and k are defined through the period T and

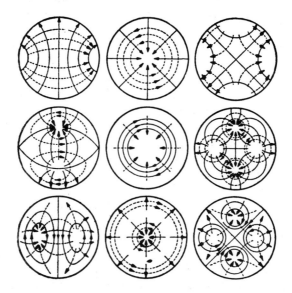

Fig. 2.20. *Angular harmonic waves*
Inside a cylindrical wave guide, viewed in cross-section, the lines of force of the electric (continuous lines) and magnetic (broken lines) fields, for various modes of oscillation, illustrate the harmonic angular dependence of undulatory phenomena. Here they correspond to stationary waves of the type $F(r, \varphi, t) = f(r) \sin(m\varphi) \, e^{i\omega t}$, which may be considered as being the sum of two progressive waves 'turning' in the two opposing directions:

$$F(r, \varphi, t) = f(r)[e^{i(\omega t - m\varphi)} + e^{i(\omega t + m\varphi)}].$$

[From A. Sommerfeld, Electrodynamics, 1964.]

wavelength λ, respectively:

$$\omega = \frac{2\pi}{T}, \qquad k = \frac{2\pi}{\lambda}. \tag{2.3.15}$$

Hence, m is rigorously analagous to the pulsation ω and the undulation k (one might even call it 'angulation'!).

One thus deduces that the relation we are looking for, which is analogous to the Planck–Einstein relation, now has the form

$$J_z = \hbar m. \tag{2.3.16}$$

Of course, the same result would hold for a component of J about an arbitrary axis. Nevertheless, we shall see in the following chapter that a simultaneous consideration of several components gives rise to new problems.

Besides allowing us to attach a quantum significance (through the use of $\hbar$) to the concept of angular momentum, relation (2.3.16) brings out a new kind of property: the components of the quantum angular momentum are *quantized* magnitudes. Indeed, since a rotation through 2π, around Oz must leave the physical system unchanged, the angular periodicity must necessarily be a sub-multiple of 2π, so that $\alpha = 2\pi/(\text{integer})$. Thus, in view of eq. (2.3.14), the rate of variation of the phase can only assume integral values and the component J_z of the angular momentum is quantized by integral multiples of $\hbar$ (see exercises 2.13 and 2.14).

(b) *Double turn and half-integral spin.* But in fact, and completely contrary to normal intuition, it is the *double* turn (4π) which is the basic cyclic unit. At the end of a rotation through an odd number of turns, an object is not in exactly the same relation with respect to its surroundings as after an even number of turns. The difference in nature between a simple and a double turn (or more generally, between an odd and an even number of turns) is a mathematical characteristic of the group of rotations in space. While it plays an essential role in quantum theory, it practically has no manifestation at the macroscopic level. One can, however, demonstrate this, even at our level, by horizontally suspending an oriented object, such as a fork, at the end of a ribbon which is secured at its upper end, and by subjecting the fork first to one complete turn, and then making a double turn (fig. 2.21a). The twisting of the tape reflects the spatial relation of the fork with respect to its surroundings. One can see now that while, in the case of the double turn, it is possible

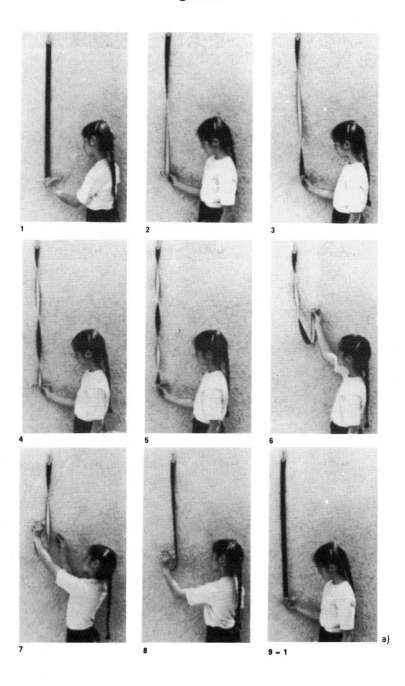

a)

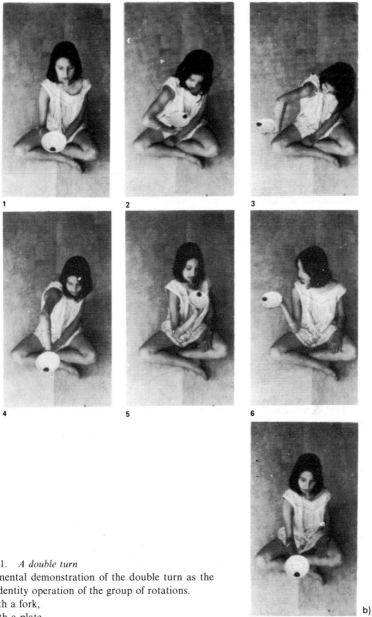

Fig. 2.21. *A double turn*
Experimental demonstration of the double turn as the
'true' identity operation of the group of rotations.
(a) With a fork,
(b) With a plate.
[Photo J.M.L.L.]

to untwist the ribbon by raising the fork and taking it around the ribbon, without changing its orientation at any time, the same is not true in the case of a simple turn. In the latter case, the same manipulation does not untwist the ribbon (but rather, re-twists it), which proves that the two rotations do not produce the same effect on the fork and that it is only the double turn which leaves it unchanged in its relationship to translations in space. There is an analogous experiment which consists of turning, about a vertical axis, a flat object placed in an extended hand, by twisting the arm. One can show that it is possible to return the arm, hand and object to their initial positions, by means of a constant movement in one and the same direction, provided the object is made to go through *two* complete turns about itself (fig. 2.21b).

One has to require, therefore, that the angular period α be a sub-multiple of 4π, the double turn,

$$\alpha = \frac{4\pi}{N}, \qquad N = \text{integer}. \tag{2.3.17}$$

From eq. (2.3.14) it follows that

$$m = \tfrac{1}{2}N. \tag{2.3.18}$$

The components of the quantum angular momentum, in units of $\hbar$, can thus assume either integral values (0, 1, 2, ...) or half-integral values ($\tfrac{1}{2}, \tfrac{3}{2}, ...$). The formal theory of angular momentum confirms this rule.

However, the relative angular momentum of two objects (e.g., that of the electron with respect to the proton in a hydrogen atom), in other words, an orbital angular momentum, cannot assume half-integral values. Indeed, such an angular momentum characterizes a motion which, as quantum as it might be, unfolds itself in geometrical space. Besides, this angular momentum is expressible in terms of the other spatial magnitudes (position, momentum). Under these conditions, a simple turn is certainly equivalent to the identity operation, and a quantization in terms of integers only applies to the orbital angular momentum. The situation is different for an intrinsic angular momentum, or *spin*, such as the intrinsic angular momentum of the electron or the proton in the hydrogen atom. This angular momentum is an independent magnitude, which is no longer reducible to the usual spatial magnitudes and which should not be associated to a 'motion', such as a rotation of a quanton about itself (see exercise 1.19). Not having a classical kinematical or geometrical significance, there is no reason for a spin state to be invariant under a rotation through 2π. Only a rotation by 4π, that is to say, the more general identity in the group of rotations, should leave it invariant – a fact which allows the spin to be half-integral. (Of course, it could also happen that a rotation by 2π leaves the state invariant, which could then have integral spins as well.) Conversely, the existence of quantons with half-integral spins and their behaviour of non-invariance under a rotation by 2π proves the impossibility of interpreting the spin, or the intrinsic angular momentum of quantons in terms of internal motions: the idea that a quanton 'turns about itself' is only a misleading analogy between a composite classical object (an extended ball of matter) and a quantum object. There do actually exist both quantons with integral spins: 0 for pions, α-quantons; 1 for certain mesons (ω, ρ), the deuteron (nucleus of deuterium), etc., and with half-integral spins: $\frac{1}{2}$ for the electron, the proton, $\frac{3}{2}$ for certain baryons (N*, Y*), for the ^{9}Be nucleus, etc.

(c) *Magnitudes*. The quantum theory not only imposes discrete values upon the components of the angular momentum, given by eq. (2.3.16), it also imposes constraints upon the numerical values of the magnitudes of the angular momentum. Classically, the components of a vector A, of a given magnitude $A = |A|$, may assume all numerical values smaller than the absolute value of this number,

$$-A \leqslant A_z \leqslant A. \tag{2.3.19}$$

Conversely, one can consider the magnitude of a vector to be the largest possible value of any one of its components,

$$A = \max A_z \tag{2.3.20}$$

(obtained when the axis of projection coincides with the direction of the vector). If the same property were to hold in the quantum theory, one would expect to have, for the magnitude J of the angular momentum, a relation of the type,

$$J \overset{?}{=} \hbar j, \tag{2.3.21}$$

where the (integral or half-integral) number j is the largest value attained by the number m, which fixes the possible values of a component J_z. But, as we shall see in a moment, this is not what happens.

Let us consider an ensemble of identical systems, having the same magnitude J of the angular momentum, but with all possible 'orientations' of it – in other words, having all possible values for J_z, compatible with the fixed (but yet unknown) value of J. This value depends, in an as yet unknown fashion [and not as in eq. (2.3.21)] on the maximal value of the numbers m. If we denote this value by j, the minimal value of these numbers would be $-j$, since the direction of the axis Oz is arbitrary and since there has to be complete symmetry with respect to a change in the sign of J_z, which would imply the opposite choice of directions. Thus, we have an ensemble of systems such that

$$J_z = \hbar m, \quad -j \leqslant m \leqslant j. \tag{2.3.22}$$

The values of m, like those of j, are either all integers or all half-integers, since the two cases are, as we have seen, of different natures (the invariance being either under an arbitrary or an even number of turns).

We make the crucial hypothesis that for a given value of j, *all* the values of m of the same type (integral or half-integral) and such that $-j \leqslant m \leqslant j$, are possible. For example,

for $j = \frac{1}{2}$, one can have $m = -\frac{1}{2}, +\frac{1}{2}$; ·

for $j = 1$, one can have $m = -1, 0, 1$;

for $j = \frac{3}{2}$, one can have $m = -\frac{3}{2}, -\frac{1}{2}, \frac{1}{2}, \frac{3}{2}$;

for $j = 2$, one can have $m = -2, -1, 0, 1, 2$;

etc.

We easily see that for a given value of j, there exist $2j + 1$ possible values of m. Somewhat loosely, one says that there are $2j + 1$ possible 'orientations' of the vector $\boldsymbol{J}$ (but it should not be forgotten that the values of only *one* of the components are being taken into account here). It is natural now, to extend our hypothesis to state that in the ensemble being considered, all the $2j + 1$ values of m are represented with the same frequency. Stated otherwise, there does not exist any privileged direction: all possible orientations are equivalent [in the classical theory, an isotropic distribution of vectors, all having the same magnitude, indeed leads to a uniform distribution for any one component (see exercise 2.15)]. This is simply a formulation of the fundamental principle of the isotropy of space, or of the invariance under rotations – the invariance which is the very condition for the angular momentum to be conserved. Thus, for our ensemble, the frequency of each value of m is

$1/(2j + 1)$. Let us now calculate the average value of J_z^2 over all orientations,

$$\langle J_z^2 \rangle = \hbar^2 \frac{1}{2j + 1} \sum_{m=-j}^{+j} m^2. \tag{2.3.23}$$

A standard computation for integral j, which easily extends to half-integral j as well, yields

$$\sum_{m=-j}^{+j} m^2 = \tfrac{1}{3}j(j + 1)(2j + 1), \tag{2.3.24}$$

so that

$$\langle J_z^2 \rangle = \tfrac{1}{3}\hbar^2 j(j + 1). \tag{2.3.25}$$

Let us recall that the definition of the magnitude of J,

$$J^2 = J_x^2 + J_y^2 + J_z^2, \tag{2.3.26}$$

itself implies, for our ensemble,

$$J^2 = \langle J_x^2 \rangle + \langle J_y^2 \rangle + \langle J_z^2 \rangle, \tag{2.3.27}$$

(J is fixed, but J_x and J_y, like J_z, are not). Once again, let us invoke the condition of spatial isotropy, which implies the equivalence of the directions of the three axes,

$$\langle J_x^2 \rangle = \langle J_y^2 \rangle = \langle J_z^2 \rangle, \tag{2.3.28}$$

to write

$$J^2 = 3\langle J_z^2 \rangle. \tag{2.3.29}$$

Finally, using eq. (2.3.25), we obtain, for the magnitude of the angular momentum,

$$J^2 = \hbar^2 j(j + 1), \tag{2.3.30}$$

or otherwise (although the previous form is generally preferred),

$$J = \hbar\sqrt{j(j + 1)}, \tag{2.3.31}$$

which ought to be compared to eq. (2.3.21). Writing eq. (2.3.30) in the form

$$J^2 = (\text{max } J_z)^2 + \hbar(\text{max } J_z) \tag{2.3.32}$$

and comparing with eq. (2.3.20), one sees clearly the role played by the constant $\hbar$ in this result. In spite of the difference between the quantum result, eq. (2.3.31) and the erroneous classical expression, eq. (2.3.21), one often says, by an abuse of language, that 'the angular momentum has the value j', or else, one speaks of a system 'having angular momentum j' (in units of $\hbar$, being understood); but j only represents the maximal value of one component.

Let us conclude, therefore, that a quantum angular momentum assumes values which in magnitude are of the form given by eq. (2.3.30), where j is an integer or a half-integer, and that its components can thus assume the $2j + 1$ values of the form given by eq. (2.3.22). Moreover, this quantization furnishes us with a simple means for the experimental determination of j: if it is established that a system, having an angular momentum of fixed magnitude, possesses exactly n different states of 'orientation' of the angular momentum, then we must have $2j + 1 = n$. Thus, from the doubling of the spectral lines of alkaline metals, which possess a single valence electron, one deduces that the electron has two states of spin, and hence, since $2 \times \frac{1}{2} + 1 = 2$, that it possesses a spin $\frac{1}{2}$.

Finally, let us note that if $j \gg 1$, we may approximately write

$$J \simeq \hbar j \quad (j \gg 1), \tag{2.3.33}$$

and recover the classical result, eq. (2.3.21). This is not at all surprising, since the condition $j \gg 1$ can also be written as

$$J \gg \hbar, \tag{2.3.34}$$

and it indeed reveals itself as a condition for the validity of the classical theory with J constituting, under these circumstances, the characteristic action of the problem.

All these results hold for an arbitrary angular momentum. It could, for example, be the angular momentum of one quanton relative to another, or the orbital angular momentum. In general, such an angular momentum is denoted by L and one speaks of a 'momentum l' if $(L_z)_{\text{max}} = \hbar l$. The number l, which gives the value of the orbital momentum, is, necessarily, an integer. One could just as well consider the intrinsic angular momentum, or spin, of a quanton. Such a momentum is denoted by S, in general, and likewise, one speaks of a 'spin s' if $(S_z)_{max} = \hbar s$. The number s, which gives the value of the spin, could be either an integer or a half-integer. The total angular momentum of a system is in general obtained by the addition of the relative and the intrinsic angular momenta of its individual parts, following rules which we shall sketch in ch. 7.

Table 2.3 recapitulates the considerations of the last three sections.

2.3.5. 'Matter' and 'Radiation'

We have insisted heavily on the intrinsic nature of the quantum concepts, synthesized by the Planck–Einstein, de Broglie and their homologous relations. It is not unusual, however, for the quantons characterized by these concepts, to show themselves preferentially as in one or the other of their classical approximations – be it particle or wave. Even if it were only for reasons that are historical on the theoretical level, and experimental on the practical level, one is often led to think of electrons, for example, or of atoms, as particles – i.e., basically as *matter*, and more or less implicitly conceived as being 'solid' – as opposed to the notion of *radiation* which is difficult to avoid when considering photons. Thus, in the process of an interaction between 'matter' and 'radiation', the basic quantum relations (2.3.1), (2.3.3) and (2.3.16) are used with the emphasis being placed on the left-hand member for 'matter' and on the right-hand member for 'radiation'. Provided one remains aware of the phenomenological character of the distinction 'matter'/'radiation' – in fact matter, in a large sense, is made up of quantons (only quantons and nothing but quantons!) – one might accept this

Table 2.3
The classical and quantum concepts, related to the three fundamental invariance properties of the laws of physics.

Invariance under	Classical wave-like concepts		Classical corpuscular concept	(Unified) quantum concept
Translations in time	Period T	Pulsation $\omega = 2\pi/T$	Energy E	$E = \hbar\omega$ (Planck–Einstein)
Translations in space	Wavelength λ	Undulation $k = 2\pi/\lambda$	Momentum $\boldsymbol{p}$	$\boldsymbol{p} = \hbar\boldsymbol{k}$ (de Broglie)
Rotations in space	Rotation α (about a given axis)	$m = 2\pi/\alpha$ (m integral or half-integral)	Component J_z about the Oz-axis of the angular momentum $\boldsymbol{J}$	$J_z = \hbar m$

asymmetry. From the experimental point of view, in particular, it is often easier and more natural, to measure, e.g., the quantum-energy-pulsation as a (classical) energy for an electron, and as a (classical) pulsation for a photon, using measuring instruments that are suitable for the observation of quantons in one or the other of their classical garbs (fig. 2.18).

Thus, let us consider the emission of electromagnetic radiation by a molecule, an atom or a nucleus. In the course of such a process, the atom (say, to fix ideas), possessing an initial energy E_i passes into a state E_f. Suppose that only a single photon is emitted, which by far is the most frequent case, since the relative weakness of the electromagnetic interaction (see ch. 1, sect. 1.4.2) does not favour multiple occurrences of the elementary coupling in the same process (except when, for some particular reason, a single-photon process happens to be suppressed). Let ε be the energy of the photon. No matter how quantum its nature, the energy is conserved, and hence, the initial energy E_i of the atom is divided between its final energy E_f and the energy ε of the photon:

$$E_i = E_f + \varepsilon. \tag{2.3.35}$$

One might just as well consider the equivalent pulsations Ω_i, Ω_f and ω, so that, using the Planck–Einstein relation,

$$E_i = \hbar\Omega_i, \qquad E_f = \hbar\Omega_f, \qquad \varepsilon = \hbar\omega, \tag{2.3.36}$$

one could write the conservation law

$$\Omega_i = \Omega_f + \omega. \tag{2.3.37}$$

But in general, one characterizes the atom by its energy (measured in eV) and the photon by its pulsation, or its frequency (measured in s^{-1} or in Hz). It is thus customary to adopt a mixed notation,

$$E_i = E_f + \hbar\omega. \tag{2.3.38}$$

This is the form in which the Planck–Einstein relation is generally used, to characterize the radiation emitted by an atomic system in the course of a transition. One obtains, e.g., its frequency by means of the formula

$$\nu = \frac{E_i - E_f}{h}. \tag{2.3.39}$$

We might mention here – and as we shall see later – the internal energy of a bound system, i.e., a system whose component parts remain bound to one another, such as an atom, is quantized: only certain discrete values of the energy are accessible to the system. It follows from eq. (2.3.39) that the emitted frequencies of radiation themselves can also assume discrete values only – this is the origin of the 'spectral lines'. Depending on the nature of the state change that the system undergoes, the energy difference, and hence the emitted frequency, could vary, as indicated in table 2.1.

In a completely analogous manner, we would treat the absorption of radiation by an atomic or a molecular system – a phenomenon which is the inverse of that of the emission of radiation, as we have described it. The photoelectric effect, mentioned before, fits into this framework: the ejection of an electron from a metal, under the effect of radiation, was interpreted by Einstein as the absorption of a photon, of energy ε, which then provides the electron with the necessary threshold energy W, to be released from the metal, and with its kinetic energy E. The conservation of energy implies here

$$E = \varepsilon - W, \tag{2.3.40}$$

which, for the same reasons as above, is, in general, written in the mixed form,

$$E = h\nu = W, \tag{2.3.41}$$

where ν is the frequency of the incident radiation.

Let us mention also the Compton effect, which is simply the elastic collision of a photon and an electron. Even though, in the course of the process, it is necessary to consider the electron and the photon as quantons, to which one applies the laws of energy and momentum conservation, the experimental conditions lead one to characterize the photon by means of its wavelength. Measurements of the variation of this wavelength, due to the collision, have historically played an essential role in the consolidation of the very notion of a photon (see exercise 2.17).

Finally, let us underscore the fact that although the processes of emission and absorption of a single photon are the most frequent ones, because of the weakness of the electromagnetic interaction, already mentioned, there do nevertheless exist phenomena involving several photons. These 'multi-photon' phenomena have assumed substantial importance since the advent of lasers – light sources of enormous power. The preceding considerations can easily be generalized. If a system passes from a state of energy E to one of a different energy E', by the absorption of several photons, having frequencies

$v_1, v_2, v_3, \ldots$, respectively, the conservation of energy (in the quantum sense) imposes the condition

$$E' = E + hv_1 + hv_2 + hv_3 + \cdots \qquad (2.3.42)$$

(see exercises 2.18 and 2.19).

It must be understood that, once the quantum concept of energy-pulsation is accepted, the constant $\hbar$ only plays the numerical role of a simple conversion factor. While its existence is essential for the synthesis of the new concept, its numerical value is contingently linked to the empirical convention for the choice of units. In this respect, it compares well to the 'mechanical equivalent of the calorie' – the constant J, in the expression $W = JQ$, of the equivalence of work and heat. It should not be surprising, therefore, if gradually, as the quantum concepts become better assimilated, that the constant $\hbar$ (as previously with the constant J) were to become less and less visible. Very simply, we establish that henceforth we shall treat the energy and the pulsation as two aspects of one and the same magnitude, and that it is neither necessary nor often useful, at the practical level, to retain two different units to express them. On a practical level, the choice of units, and hence the preference for 'energy' or 'pulsation' depends – and we repeat – upon the experimental context. For example, photons in the visible region are, in general, characterized by their frequencies, or wavelengths, to the extent that these quantities are easily measurable directly by means of interference or diffraction experiments. By contrast, γ-ray photons are, in general, described in terms of their energies, since their wavelengths are too small to be easily measurable. But, of course, there is no difference in nature between a photon from the visible domain ($\lambda = 0.6\ \mu\text{m}$) and a γ-ray photon ($E = 2\ \text{MeV}$).

2.3.6. Collective motions and elementary excitations

The classical notion of a field was born, as we have seen, from a description of material media: the propagation of certain perturbations in such a medium led to the notion of a wave. A field, or a wave, initially gave, therefore, a certain picture of the collective motions of the medium and their propagation. Thus, pressure waves in a fluid describe the movements of the collection of its molecules vibrating through their mutual interactions. It is the coherence of all the individual molecular motions, which makes such a global description possible. The fluctuations of the electronic density in a plasma are likewise described by 'plasma waves'. The classical formulism of the theory of fields (see sect. 2.1.2) is the same whether it is applied to 'true' fields, which are thought of as existing by themselves, even in the absence of any material substratum, such as the electromagnetic field, or to 'phenomenological' fields, produced by the collective motions of a propagation medium. The concept of a frequency, for example, plays the same fundamental role in the (Fourier!) analysis of these fields. Now, the quantum synthesis of the concepts of frequency and energy is universal, i.e., they do not depend upon the specific nature of the phenomenon being considered, and characterize quantum theory in all its generality. The Planck–Einstein relation, eq. (2.3.1), as well as its homologues, eqs. (2.3.3) and (2.3.16), hold equally well, therefore, in the quantum description of 'phenomenological' fields. In other words, one must have a discretization of the energy into packets or quanta $E = \hbar\omega$, for an acoustic wave, having pulsation ω, exactly as for an electromagnetic wave.

Thus, the notion of a quanton finds its application, not only in the analysis of the fundamental constituents of matter, but also in that of the collective motions in specific material media. From the classical theory of wave propagation follows a quantum theory in terms of quantons. Just as an electromagnetic wave is replaced by a system of photons in the quantum theory, similarly, an acoustic wave is replaced by a system of quantons which, very naturally, are called 'phonons'. The quantum theory of condensed matter is rich with an entire menagerie of such objects, often with rather exotic names, such as 'polarons', 'magnons', 'plasmons', 'rotons', etc. Naturally, each one of these quantons has specific properties, related to the nature of the particular medium and its collective motion.

Among these properties, one of the most important is the relation between the energy E and the momentum p of the quanton. In the case of a fundamental quanton, such as an electron, propagating freely in vacuum, this relation is determined by the geometry of space–time. We have, according to the Galilean approximation, the fundamental relation $E = p^2/2m$; for a 'phenomenological' quanton, such as a phonon, this relation depends on the medium. In reality, this relation $E(p)$, is just the quantum homologue of the dispersion relation which relates the frequency to the wave number – or the pulsation to the undulation – of a classical wave. The Planck–Einstein relation $E = \hbar\omega$ and the de Broglie relation $p = \hbar k$ allow one to pass immediately from the function $\omega(k)$ to the function $E(p)$.

Nothing illustrates the universal character of the notion of the quanton better than the phenomenon of interaction between 'fundamental' quantons and 'phenomenological' quantons. Thus, (low-energy) neutrons, during an interaction with matter, may give up a part of their energy and momentum to the medium, and thereby excite certain collective motions. One has in fact, in this situation, a creation of phonons together with the scattering of neutrons.

The simplest process is that of the creation of a single phonon of energy ε and momentum q by a neutron, with initial and final energy–momentum (E, p) and (E', p'), respectively (see fig. 2.22). The conservation laws require that

$$E = E' + \varepsilon, \quad \text{and} \quad p = p' + q, \tag{2.3.43}$$

Indeed, the neutron scattering is used to obtain the relation $\varepsilon(q)$: for each

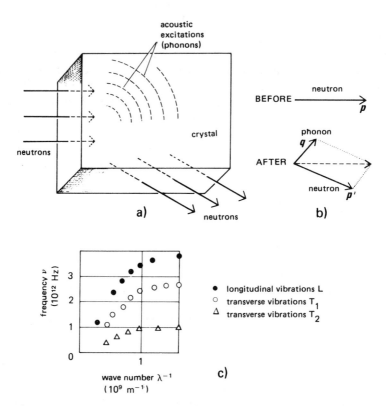

Fig. 2.22. *Neutron scattering and phonon production*
(a) Scheme of experimental principle.
(b) The conservation of momentum, $(p = p' + q)$, and of the energy, $(E = E' + \varepsilon)$, allows one to measure, for each direction of the emerging neutrons, the energy and momentum of the emitted phonons, and hence the 'dispersion relation' $\varepsilon(q)$, or $\omega(k)$.
(c) Results obtained by neutron scattering for the propagation of phonons in a crystal of sodium (for a certain orientation) of the crystal. One verifies that at low frequencies the relation is practically linear: (see exercise 2.20).
[Woods, Brockhouse, March and Bowers, Proc. Phys. Soc. London 79 (1962) 440.]

direction of p', fixed by the detector, a measurement of E' and $|p'|$ gives (E and $|p|$ being known) a point on a curve. This, moreover, is often given in the form of the dispersion relation $\omega(k)$. The fact that the phonon, in this analysis, could be treated exactly as the neutron and that the conservation laws apply [see eq. (2.3.43)] to an arbitrary interaction between two quantons, shows clearly the universality of the notion of a quanton (exercises 2.20, 2.21 and 2.22).

One often speaks of the 'phenomenological' quantons as the 'elementary excitations' of the medium being considered, meaning by it that they constitute the fundamental elements for the description of the collective motions of that medium. The distinction between the two categories of quantons can, however, be completely relative. A quanton which appears as 'fundamental' to us, such as an electron, could in fact be an 'elementary excitation' of some yet unknown underlying medium – a sort of a modern aether. This idea has inspired several theories, currently being discussed, with a view to unifying and deepening the physics of fundamental particles.

Exercises

2.1. Consider the functional equation defining the *harmonic functions*, written as

$$\Phi(t + \tau) = K(\tau)\Phi(t),$$

where, a priori, the factor K depends on the time translation τ. To fix ideas, suppose that the function Φ assumes complex values.

(a) Suppose that Φ is once differentiable, show that one has the differential equation

$$\Phi' - \alpha\Phi = 0,$$

where $\alpha = K'/K$ does not depend either on t or on τ. Hence deduce the function Φ and the factor K.

(b) Devise other techniques to solve the above functional equation.

(c) One could argue that the functional equation, with which we had started, expresses the identity of the form of Φ with its translations in too restrictive a manner; for the graphs 'to have the same form', it should be sufficient that the functions be equal, up to a (multiplicative) factor and an (additive) shift:

$$\Phi(t + \tau) = K(\tau)\Phi(t) + L(\tau).$$

Show that the function Φ is again an exponential, up to an additive constant.

2.2. A regular swell, formed from plane waves, of which two successive crests are separated by a length (wavelength!) λ, is propagating in the ocean. A boat sails, following a rectilinear trajectory, making an angle θ with the direction of propagation of the waves. What is the apparent wavelength in the direction of the boat – i.e., the distance between successive crests along its route? From this deduce that the wavelength λ_n, in the direction of an arbitrary unit vector n is *not* the component, in this direction, of a 'wave vector' λ, $\lambda_n = \lambda \cdot n$ (false). Show that, by contrast, the *undulation* $k_n = 2\pi/\lambda_n$ is indeed the component of the wave vector k, $k_n = k \cdot n$.

2.3. Suppose that we wish to calculate the *diffracted intensity* from an infinitely long slit (see figure).

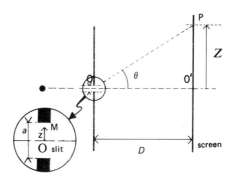

(a) Let M be a point in the slit, at a distance $Z = OM$ from its mid-line. Express the ratio of the phase of the wave emitted by the point M to that of the wave emitted by the central point O, as a function of Z, λ and θ.

(b) Hence deduce the amplitude of the diffracted wave at P.

(c) Obtain the slope of the curve giving the variation of the diffracted intensity as a function of the position $OP' = Z$ of the point of observation.

(d) Express the width d of the central diffraction fringe as a function of λ and a. Examine the two particular cases, $a \simeq \lambda$ and $a \gg \lambda$.

(e) Repeat, using the numbers $\lambda = 0.5$ μm, $D = 1$ m, $a = 1$ mm.

2.4. Calculate the differential *effective cross-sections* for the scattering of a classical point particle in the following cases.

(a) The target is an ellipsoid of revolution, about the direction of incidence, from which the particle rebounds elastically. Let R be the radius of the circular cross-section, perpendicular to the axis of symmetry, and $R' = rR$ the semi-axis, along the axis of symmetry. Establish the formula

$$\chi(\theta) = R^2 \frac{r^2}{[1 + r^2 + (1 - r^2) \cos \theta]^2}.$$

Trace the curves representing $\chi(\theta)$ for some values of r (e.g., $r = 0.1$, 0.2, 0.5, 1, 2). Note that (see fig. 2.12a) $\chi(\theta, 1/r) = \chi(\pi - \theta, r)$.

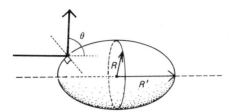

(b) The target is spherical, of radius R, having a constant potential $-V_0$. Show that the particle propagates along a straight line obeying, at the points of entry and exit, the classical laws of refraction in a medium having refractive index $n = (1 + V_0/E)^{1/2}$,

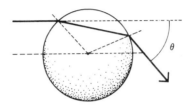

where E is the kinetic energy. Establish the formula

$$\chi(\theta) = R^2 \frac{n^2(n \cos \tfrac{1}{2}\theta - 1)(n - \cos \tfrac{1}{2}\theta)}{4 \cos \tfrac{1}{2}\theta (n^2 - 2n \cos \tfrac{1}{2}\theta + 1)^2}.$$

Trace the curves representing $\chi(\theta)$ for some values of n, and hence of E (see fig. 2.12b).

2.5. Consider again the *Coulomb scattering* of a nucleus of charge $Z_1 q_e$, mass m and velocity v, by a stationary nucleus of charge $Z_2 q_e$ (exercise 1.20).

(a) Compare the classical minimal distance of approach, d, to the quantum wavelength of the incident nucleus, and recover the classical criteria of validity. Obtain the same result by evaluating the relative angular momentum of the two nuclei.

(b) Using dimensional analysis, show that the differential effective cross-section of scattering is necessarily written as

classically:

$$\chi_{\mathrm{cl}}(\theta) = \left(\frac{Z_1 Z_2 e^2}{E}\right)^2 f(\theta);$$

quantically:

$$\chi_{\mathrm{qu}}(\theta) = \left(\frac{Z_1 Z_2 e^2}{E}\right)^2 g[\theta, (Z_1 Z_2 \alpha)^{-2}],$$

where $g(\theta, 0) = f(\theta)$. (A detailed calculation shows that $g(\theta, x)$ does not depend on x: $g(\theta, x) = f(\theta) = (2 \sin \theta/2)^{-4}$. The classical and quantum 'Rutherford' effective cross-sections are thus equal, as we have seen in the text.)

2.6. An experiment for the 'exploration' of the *nucleus* consists in bombarding a thin gold leaf by electrons having an energy of about a few hundred MeV. Show that it is exactly in this range of energies that the electrons can effectively be diffracted by a gold nucleus. For gold, $Z = 79$ and $A = 197$. Recall that the size of a nucleus is $r = A^{1/3} r_0$, with $r_0 = 1.2$ F.

2.7. Calculate the *wavelength* of
(a) the earth (mass 6×10^{24} kg) in its motion around the sun ($V = 3 \times 10^4$ m s^{-1});
(b) a man walking at a normal pace;
(c) a raindrop falling through the atmosphere.
Conclusion?

2.8. In order to have a *wavelength* $\lambda = 1$ Å, what should be the kinetic energy (in electron-volts) of an electron, a photon, a neutron and a lead nucleus? Same question with $\lambda = 1$ F.

2.9. Corresponding to a kinetic energy $E = 1$ MeV, what should be the frequency and the *wavelength* of a photon, an electron, a neutron and a lead nucleus?

2.10. Determine the minimum value of the wavelength of X-rays emitted by a television screen, when the electrons in the tube, accelerated by a voltage of 18 000, are stopped by the screen.

2.11. The *conduction electrons* in copper have a kinetic energy of about 7 eV. Calculate their wavelength. Compare it to the interatomic distance obtained by using the mass density of copper, $\rho = 8.9 \times 10^3$ kg m^{-3} and its mass number $A = 60$.

2.12. The nuclear reactor at the von Laue–Langevin Institute (ILL) in Grenoble is designed to supply a powerful flux of *neutrons*, used in numerous experimental procedures. In it there are three sources of neutrons of different energies, depending upon the temperature of the medium in which the neutrons are brought into thermal equilibrium (by repeated collisions, with the atoms of the medium). Calculate the average energy (in eV) and the average wavelength (in Å) of the neutrons coming from:
(a) the normal source consisting of the moderator (heavy water) at temperature $T_n = 300$ K;
(b) the hot source, made up of graphite at $T_h = 2000$ K;
(c) the cold source, consisting of liquid deuterium at $T_c = 25$ K.
Which source should be used to conduct crystallographic studies of matter by neutron diffraction?

2.13. An artificial satellite having a mass of 5 tons makes a revolution about the earth in $1\frac{1}{2}$ hours, at an altitude of 200 km.
(a) Calculate its orbital *angular momentum* L with respect to the centre of the earth.
(b) What is the corresponding quantum number l ($L^2 = l(l+1)\hbar^2$)? What variation of l would correspond to a change in the altitude by 1 m?

2.14. A nitrogen molecule N_2 is made up of two nitrogen atoms of mass $m = 2.24 \times 10^{-27}$ kg, separated roughly by a distance $d = 1.1$ Å. What is the order of magnitude of the smallest periods of *rotation* of the molecule about itself (i.e., for which its angular momentum assumes the smallest possible values)?

2.15. Consider an ensemble of classical vectors A having fixed magnitude A and random directions, which are isotropically distributed – in other words, the extremities of the vectors A are uniformly distributed on the sphere of radius A. Show that the ensemble of components A_z possesses a uniform distribution in the segment $[-A, +A]$.

2.16. Recently, it has become possible, to produce large fluxes of ultra-cold *neutrons*, having kinetic energies of the order of 10^{-7} eV.

(a) Justify the term 'ultra-cold' by calculating the temperature corresponding to this energy.

(b) At what velocity do these neutrons move? Compare this velocity to the speed of a runner. Explain why one might expect to measure, using these neutrons, the half-life of a neutron (outside the nucleus) which is of the order of 15 min.

(c) What height could these neutrons reach in the gravitational field of the earth? Contrive a simple device for the spectrometry of ultra-cold neutrons, i.e., an apparatus which would be able to separate neutrons spatially from the same beam, according to their energies.

(d) Calculate the order of magnitude of the wavelength of these neutrons. Compare it to the distance between the atoms of a solid, and hence deduce that, to ultra-cold neutrons, a solid appears as a continuous barrier. Explain why this property might facilitate their stockpiling.

2.17. *The Compton effect.* Write down the equations of the conservation of energy and momentum for the elastic collision of a photon and an electron (initially at rest).

(a) Hence deduce the difference between the initial and the final wavelengths of the photon, as a function of its angle of deviation θ:

$$\lambda' - \lambda = \frac{h}{mc}(1 - \cos \theta).$$

(b) Among the first experiments for the verification of this theory, one could cite the results of M. de Broglie and A. Dauvillier, C.R. Acad. Sci. 179 (1924) 11. For the angles of deviation $\frac{1}{4}\pi$, $\frac{1}{2}\pi$ and $\frac{3}{4}\pi$, they observed that 'the shifts are respectively equal to 8.5X, 21X and 31X' (the unit X, which was used, had the value 10^{-3} Å). Taking into account the fact that the X-rays used had a spectrum with a wavelength spread of about 5X, is the agreement between theory and experiment sufficiently good?

2.18. The cesium atom has an ionization energy $E_i = 3.9$ eV. One can observe the ionization of cesium vapour under the effect of a powerful *beam laser* of wavelength $\lambda = 1.06$ μm. Calculate the energy of the photons of such a laser and show that the observed ionization is necessarily a multi-photon phenomenon. How many photons have to be absorbed by an atom in order for it to be ionized?

2.19. Traditional spectrometric measurements of the energy differences between atomic levels have their precisions limited by the *Doppler effect*. Indeed, suppose that an atomic vapour is traversed by a light beam of frequency v, with which we wish to excite the atoms, from an energy level E to a level E'.

(a) If the vapour is at a temperature T, what is the order of magnitude v of the velocity of thermal agitation of an atom (of mass M)? Calculate v numerically for sodium atoms at ordinary temperatures.

(b) If an atom has the velocity v, parallel to the beam, what is the frequency v' of the light in the reference frame of the atom (Doppler effect)? Work only up to the first order in v/c.

(c) Hence deduce that the absorption line of the luminous radiation, which excites the atoms, has a 'Doppler width' given, in order of magnitude, by

$$\Delta v_D = v_0 (kT/Mc^2)^{1/2},$$

where, $hv_0 = E' - E$. Show that, for a line of wavelength $\lambda_0 = c/v_0 = 0.6 \times 10^{-6}$ m, the Doppler broadening is of the order $\Delta v_D \simeq 10^3$ MHz.

(d) A recently developed technique enables one to eliminate the Doppler broadening. It consists of producing the excitation between the energy levels E and E' by the absorption of *two* photons. To do this, one sends the light beam being used back across the vapour, with the help of a mirror. If the frequency of the beam is v and the velocity of an atom v, in the direction of the beam, what are the frequencies v' and v'' of the incident and the reflected beams, respectively, in the reference frame of the atom?

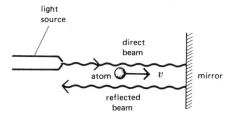

(e) Hence deduce that the beam frequency, exciting the atom through the absorption of a photon from the incident beam and a photon from the reflected beam, is v_1 such that $2hv_1 = E' - E$, *independently* of v, and that the corresponding absorption line does not have a Doppler width. Show that, on the contrary, the absorption of two incident or two reflected photons gives rise to an absorption with the usual Doppler spread.

(f) Finally, show that, with the three cases (incident and reflected photon, two incident photons, two reflected photons) contributing comparably to the absorption, one gets a superposition of two lines having comparable *total* intensity – one broad and weak, and the other narrow and strong. Compare with the figure below (these experiments require strong light intensities and hence are done with the help of lasers). In this figure, the continuous (feeble) background represents the observed spectral line, broadened by the Doppler effect. By suppressing the Doppler effect, one resolves it into the four lines a, b, c and d.

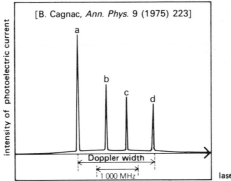

2.20. Use the results of fig. 2.22c to calculate the velocity of the *propagation of sound* inside a sodium crystal. Compare the results with those obtained by the direct measurement of ordinary acoustic frequencies,

$$V_L = 4.8 \times 10^3 \text{ m s}^{-1}, V_{T_1} = 2.5 \times 10^3 \text{ m s}^{-1}, V_{T_2} = 1.2 \times 10^3 \text{ m s}^{-1}.$$

2.21. A neutron of energy E and momentum p interacts with an aluminium crystal, giving rise to a *phonon* of energy ε and momentum q. Let E' and p' be the final energy and momentum of the neutron. Knowing that the pulsation ω (if it is not too large) and the undulation k of the phonon are related by the dispersion relation $\omega = V|k|$, where V is the velocity of sound, express the final energy E' of the neutron, in the case in which it is scattered backwards, and hence the frequency of the emitted phonon. Repeat this with the following numerical data $E = 2$ eV, $V = 6.4 \times 10^3$ m s^{-1}.

2.22. As in the case of neutrons, the inelastic scattering of light inside a solid or liquid can also create phonons. Consider an incident beam of light, of pulsation ω. Upon scattering by an angle θ, there emerges a component having pulsation $\omega' < \omega$. This phenomenon is interpreted as the creation of a phonon of pulsation Ω, by a photon of initial pulsation ω – its final pulsation being ω'.

(a) Knowing that photons obey the dispersion relation $\omega = ck$ (c is the velocity of light, k the undulation of the photon), while the phonons obey $\Omega = VK$ (V is the velocity of sound, K the undulation of the phonon), write down the conservation laws for the energy and the momentum. Hence deduce that $k \simeq k'$ and that $K \simeq 2k \sin \frac{1}{2}\theta$. Show that one can calculate the velocity of sound by means of the expression

$$V = c \frac{1}{2 \sin \frac{1}{2}\theta} \frac{\omega - \omega'}{\omega}.$$

(b) The curve below gives the spectrum of light, scattered through $\theta = \frac{1}{2}\pi$ in water, from a laser beam of wavelength $\lambda = 6328$ Å. The central peak corresponds to light scattered without change in wavelength, by particles suspended in the water. The peak on the left corresponds to light scattered with the creation of phonons. To what does the peak on the right correspond?

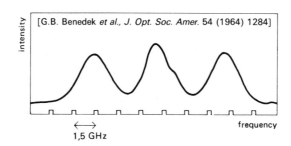

[G.B. Benedek *et al., J. Opt. Soc. Amer.* **54** (1964) 1284]

(c) From these experimental results, deduce the frequency of the phonons excited, and their velocity. Compare it with the velocity of sound in water, $V = 1.48 \times 10^3$ m s^{-1}, for normal sound frequencies.

2.23.   Consider a classical wave, of sound, say, which is 'monochromatic', i.e., having a pulsation ω and undulation k which are well-defined. Its amplitude, in a certain frame of reference R, is written as: $u(x, t) = a \exp i(\omega t - kx)$ (we only consider one spatial dimension in the whole of this problem). Consider this wave in a second frame of reference R', related to the first by a *Galilean transformation* for the velocity V,

$$x' = x - Vt, \qquad t' = t.$$

(a) Write down the amplitude $u'(x', t')$ in the reference frame R'. Establish the transformation formulae, giving the pulsation and undulation (ω', k') in R' in terms of its values (ω, k) in R. Show that the wave is again monochromatic. Hence, deduce that the classical wavelength is a Galilean invariant, and thus recover the classical expression for the frequency shift due to the Doppler effect.

(b) Consider a classical particle of mass m. Given that its momentum, in a reference frame R, is p and its kinetic energy $E = p^2/2m$, establish the transformation formulae giving the energy and momentum (E', p') in a frame of reference R', related to R by a Galilean transformation for the velocity V.

(c) Consider a quanton of mass m. Show that the Planck–Einstein and de Broglie relations are incompatible with the properties of the Galilean transformations for classical waves, established in (a), and for classical particles, established in (b). To make these relations valid, it is necessary, therefore, to adopt one and/or the other of the following solutions:

(i) Modify the Galilean transformation formulae for the energy and the momentum.

(ii) Modify the transformation formulae for the pulsation and the undulation.

Discuss the second alternative. Why should solution (i) be excluded? (Hint: How does the momentum transform now?)

(d) (1) Using (ii), write down the Galilean transformation properties of the *quantum* pulsation and undulation:

$$\omega' = \omega - Vk + \frac{1}{\hbar}\frac{mV^2}{2}, \quad \text{and} \quad k' = k - \frac{1}{\hbar}mV.$$

Show that one recovers the classical results in (a) as $m \to 0$. Comment.

(2) The effective cross-section for the forward transmission of neutrons by an iron crystal shows a sharp discontinuity for a critical value of the wavelength $\lambda_c = 4.046$ Å. If the crystal is in motion, one obtains experimentally a shift in the discontinuity, as indicated in the figure below. Show that this effect qualitatively verifies (particularly the sign) the validity of the formulae established above. Obtain from the experimental curves, for each direction, the theoretical value of the velocity V, imparted to the crystal, and compare it with the experimental value $V = 91.8$ m s^{-1}.

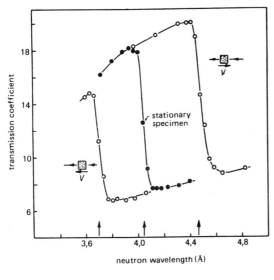

The central curve ($\bullet$) corresponds to a stationary crystal, the two others ($\bigcirc$) to a crystal moving in one or the other direction. [C. G. Shull, K. R. Morashi and J. C. Rodgers, Acta Crystallographica, A24 (1968) 160]

(e) Reconsider the question from the point of view of the *Einsteinian* (no longer Galilean) relativity theory.

(1) Recall the (Lorentz) transformation formulae for the pulsation–undulation of a classical wave and for the energy–momentum of a classical particle.

(2) Show that no inconsistency appears between these formulae and the Planck–Einstein and de Broglie relations.

(3) *The Galilean limit:* Show that, for a classical wave, the Galilean limit (a) of the Lorentz relation is obtained simply by 'making' $c \to \infty$. Show that, one properly recovers the Galilean limit obtained in (b), of the Lorentz transformation for the energy–momentum of a classical particle, *under the condition* that the mass–energy $E_0 = mc^2$ be subtracted *before* passing to the limit $c \to \infty$, so that the transformation works well for the kinetic energy and not for the total energy.

(4) For a quanton, the energy and the pulsation are not independent. Thus, the Galilean limit of the Lorentz transformation for the pulsation–undulation is obtained in the same way as for the energy – by subtracting the 'mass-pulsation' $\omega_0 = mc^2/\hbar$ before passing to the limit $c \to \infty$. Show that one properly recovers, in this way, the results of question (d).

3

Quantum magnitudes and the Heisenberg inequalities

3.1. Spectral extensions and classical wave packets

3.1.1. The temporal spectral inequality

We have seen that a harmonic time dependence, given by $\exp i(\omega t + \varphi)$, corresponds to a phenomenon which is 'always the same', or in other words, remains identical to itself in the course of time. We also say that it represents a *stationary* phenomenon, reproducing itself indefinitely, without modification. The harmonicity, which refers to the physical notion of stationarity expresses in addition a mathematical notion of 'simplicity'. This idea essentially represents the possibility of analyzing any complex temporal evolution as a superposition of harmonic evolutions. This is the same principle as that of the 'Fourier analysis' mentioned already in the previous chapter (sect. 2.1.3). Let us now consider a non-stationary phenomenon, possessing a certain 'time of evolution' or characteristic duration Δt. This means that Δt indicates the order of magnitude of the time interval during which the phenomenon undergoes a change which is appreciable compared to a harmonic (hence stationary) behaviour (fig. 3.1). We shall regard this phenomenon as being due to the superposition of several harmonic phenomena of different pulsations. But then, what is the range of pulsations necessary to reconstruct the phenomenon?

During the time Δt, a particular harmonic component, of pulsation ω, sees its phase ωt (the argument of the harmonic function) change by an amount $\omega \Delta t$. The superposition of all the components undergoes an appreciable modification only if the differences in their phases become sufficiently large during the interval Δt so that the amplitude of the superposition is qualitatively modified. In the simple case, where there are only two harmonic components, of pulsation ω_1 and ω_2 and of the same amplitude, in order for the components to pass from a constructive interference (when they are 'in phase') to a destructive interference (in 'opposite phases'), it is necessary that

function (of time):   spectrum (pulsation):

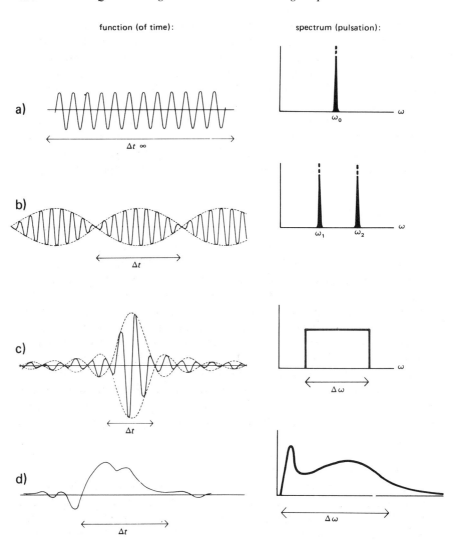

Fig. 3.1.   *Temporal extension and spectral width*

(a) A harmonic function, of infinite extension in time, corresponds to a vanishing spectral width in pulsation.

(b) The superposition of two harmonic functions, of pulsations ω_1 and ω_2, gives rise to beats having a characteristic modulation time: $\Delta t \simeq |\omega_2 - \omega_1|^{-1}$.

(c) A non-periodic phenomenon, evolving with a characteristic time Δt, possesses a continuous pulsation spectrum of width, $\Delta\omega \simeq (\Delta t)^{-1}$, in very simple cases ...

(d) ... and $\Delta\omega > (\Delta t)^{-1}$, in general.

a period of time δt, corresponding to a phase shift of π (so that $\omega_2\,\delta t - \omega_1\,\delta t = \pi$) should elapse. More generally, let ω_M and ω_m be the maximal and minimal pulsations in the phenomenon, and $\Delta\omega = \omega_M - \omega_m$ the width of the pulsation band. During a time interval δt, the phases of the different components acquire increments ranging from $\omega_m\,\delta t$ to $\omega_M\,\delta t$. The difference between the extremal phases thus changes by $(\omega_M - \omega_m)\,\delta t = \Delta\omega\,\delta t$. In order for the shape of the phenomenon to change qualitatively, the character of the interferences must change appreciably, implying that this phase difference must not be negligible compared to unity – more precisely, compared to the unit of phase, the radian. The characteristic time Δt must, therefore, obey the inequality

$$\Delta\omega\,\Delta t \gtrsim 1. \tag{3.1.1}$$

Stated otherwise, the shorter the characteristic duration of the evolution of a phenomenon, the more extended is the range of the harmonic frequencies it spans. For the moment, inequality (3.1.1) is approximate and gives only the minimal order of magnitude of Δt in terms of $\Delta\omega$ (or conversely). We say that the width of the range (or of the band) of pulsations $\Delta\omega$ represents the extension of the pulsation *spectrum* of the phenomenon being analyzed. The form of the spectrum reflects the relative importance of the different harmonic components. This notion is made mathematical with the help of the Fourier transform. Within this formalism, it is possible to give rigorous definitions for the temporal and spectral extensions Δt and $\Delta\omega$, respectively, and hence to prove an inequality of type (3.1.1), justifying thereby our heuristic approach. In the following, we shall often refer to this inequality as the 'spectral inequality'.

To understand the significance of the 'characteristic time' Δt better, it is interesting to consider two particular cases:

(1) If the spectrum reduces to the superposition of two pure pulsations ω_1 and ω_2, we have the well-known phenomenon of beats, and the characteristic time $\Delta t \simeq 1/\Delta\omega$ gives in this case the modulation time (fig. 3.1b), i.e., the time which characterizes the maximal variation, of the total amplitude, compared to that of a purely harmonic phenomenon.

(2) If we have a continuous spectrum, where all frequencies between ω and $\omega + \Delta\omega$ are represented, with equal amplitudes, the resulting phenomenon has a maximum amplitude, the width of which is measured by the characteristic time $\Delta t \simeq 1/\Delta\omega$. Hence, in this case, Δt is a measure of the total duration of the phenomenon (fig. 3.1c). Between these two particular cases, all intermediate situations are possible.

We can also give a simple interpretation to the limiting case, where one of the extensions $\Delta\omega$ or Δt vanishes, while the other one increases indefinitely – since the spectral inequality has to remain valid. Thus, a pure or monochromatic phenomenon, defined by a unique value of the pulsation, evidently has zero spectral extension: $\Delta\omega = 0$. The spectral inequality now demands that $\Delta t = \infty$, so that the modulation period, or the duration of the phenomenon, is unbounded: we recover the stationary character of harmonic phenomena. Conversely, an instantaneous impulsive phenomenon, defined by a vanishing duration $\Delta t = 0$, implies a band width $\Delta\omega = \infty$ – i.e., its spectrum comprises all pulsations, from the lowest to the highest, to infinity. It is for this reason that, during a thunderstorm, a flash of lightning, which is a very brief electromagnetic perturbation – having, therefore, an extremely large spectrum – simultaneously gives rise to a surge of all the radio (short, medium, large and FM) and television wavelengths.

The spectral inequality, eq. (3.1.1), helps us, moreover, to understand in very general terms, the necessity of having available a band of pulsations (or frequencies), around the average operating frequency of a radio transmission station. Indeed, in order to transmit a modulated signal (voice, music), it is necessary to have a certain spectral band, of width $\Delta\omega$, related to the characteristic time of modulation Δt of the signal. Thus, if we wish to obtain a reproduction of average acoustic quality, it is necessary to reproduce a signal which varies appreciably within the time interval $\Delta t = 10^{-4}$ s, and hence, in view of the spectral inequality, to have a pass band of the order $\Delta\omega \simeq 10^4$ Hz. This is the reason why it is not possible to increase indefinitely the number of transmitters within the same spectral zone: in order that their signals would not overlap, the average frequencies must at least be separated by intervals of the order of the band width.

3.1.2. The spatial spectral inequality

Exactly the same considerations apply to spatial variations. The Fourier analysis allows one to similarly consider a phenomenon varying in space – a field – as a superposition of harmonic components in space.

First of all, for a unidimensional phenomenon, depending upon a single spatial coordinate, say x, its analysis into harmonic functions $\exp(ikx)$ introduces a range of undulations, of width Δk, inversely proportional to its spatial extension, or to its characteristic size Δx,

$$\Delta k\, \Delta x \gtrsim 1. \tag{3.1.2}$$

The formation of a 'wave packet' of limited extension, similarly requires the simultaneous presence of harmonic components, with wavelengths extending over a certain region. The smaller the size of the packet, the larger is the region. If we now consider a phenomenon, depending upon the three spatial dimensions, the independence of these dimensions enables us to analyze the phenomenon separately in the three directions of an arbitrary rectangular system of axes. To the characteristic dimensions $(\Delta x, \Delta y, \Delta z)$ of the phenomenon in these three directions, would correspond, respectively, the widths of the ranges for the three components (k_x, k_y, k_z) of the wave vector k, namely, the spectral extensions $(\Delta k_x, \Delta k_y, \Delta k_z)$, in a way such that the following three inequalities would hold:

$$\Delta k_x \, \Delta x \gtrsim 1, \qquad \Delta k_y \, \Delta y \gtrsim 1, \qquad \Delta k_z \, \Delta z \gtrsim 1. \tag{3.1.3}$$

3.1.3. An application: the group velocity

As an example of an application of the classical spectral inequalities given by eqs. (3.1.1) and (3.1.2), let us recapitulate a well-known (or what ought to be so) result on the velocity of propagation of an undulatory phenomenon. A monochromatic wave in space and time, of pulsation ω and undulation k, has (by definition) the phase $\varphi(x, t) = \omega t - kx + \varphi_0$, at a point x at the instant of time t. This phase is the same at all the points of a world line having the equation $x = (\omega/k)t + x_0$, which is characteristic of a uniform displacement with velocity

$$v_{\text{ph}} = \frac{\omega}{k}. \tag{3.1.4}$$

This, as we have already seen (ch. 2, sect. 2.1.3), is called the 'phase velocity'. If we now consider a phenomenon resulting from the superposition of several harmonic components, we see that each one of the latter propagates with a different velocity. This does not preclude, however, their resulting total cooperative effect from maintaining its own particular existence. However, in general, the latter moves with a very specific velocity of propagation. A simple model where the propagation of a collective effect is qualitatively different from that of its individual components, is furnished by the movement of a traffic jam ('pile up') on a highway. The velocity of the jam is not the average velocity of the vehicles which find themselves trapped in the bottleneck for the moment. In the specific case of wave propagation, a classic example of this behaviour is furnished by the pedagogical archetype of wave phenomena: the ripples produced on the surface of a pond due to the impact of a pebble thrown into the water. One observes the appearance of a circular wave packet, moving away from the point of impact. Upon closer examination, it reveals itself as being composed of a multitude of wavelets, which propagate faster than the packet itself – seeming to appear from behind it, then cross it and finally disappear in front of it. It is the velocity of the cooperative phenomenon, termed 'group velocity', arising from a superposition of monochromatic waves, in which we are interested. Let us consider a 'minimal' wave packet, corresponding to equality in the spectral relations, e.g. (3.1.1) and (3.1.2). In other words, the wave packet has a spatial extension Δx and a duration Δt, limited by the effect of the superposition of monochromatic waves, corresponding to the spectra of pulsation and undulation having widths $\Delta \omega \simeq 1/\Delta t$ and $\Delta k \simeq 1/\Delta x$, respectively. This wave train takes a time Δt to move past a given point; this, in turn, means that in this time it covers a spatial distance which is precisely equal to its own extension Δx. This elementary result becomes more transparent if we consider the situation in the reference frame in which the wave packet is stationary: it is now the observation point which moves with a velocity, opposite to that of the wave packet, and it covers therefore a distance Δx in a time Δt. The velocity of the wave packet, $v_{gr} = \Delta x/\Delta t$, may now be expressed as a function of its spectral width,

$$v_{gr} \simeq \frac{\Delta \omega}{\Delta k}. \qquad (3.1.5)$$

The group velocity depends, therefore, on the spectral extensions of the wave packet in pulsation and undulation, i.e., not just on the average pulsation and undulation, but also on the widths $\Delta \omega$ and Δk around these average values. One can arrive at a more specific definition by considering the infinitesimal bands $(\omega, \omega + d\omega)$ and $(k, k + dk)$. Physically, this is the same as considering wave packets of arbitrarily large spatio–temporal extensions, since $\Delta x \simeq 1/\Delta k \to \infty$, as $\Delta k \to 0$, and the same holds for time. Stated otherwise, the group velocity characterizes here a wave packet which approximates a monochromatic wave as closely as we please. The group velocity defined by

$$v_{gr} \triangleq \frac{d\omega}{dk}, \qquad (3.1.6)$$

is then a function of the pulsation ω, or of the undulation k. Indeed, for all wave-like phenomena, in particular, the pulsation and undulation are connected by a relation, often written as $\omega = f(k)$ which is called a 'dispersion relation' since it describes the mutual 'dispersion' of the various harmonic components, by giving their different phase velocities. This relation is characteristic of the physical mechanisms of the propagation phenomenon being studied (see exercise 3.1).

3.1.4. The angular spectral inequality

We can finally establish an 'angular' spectral inequality. We are interested in the dependence of an undulatory physical phenomenon on the azimuthal angle of rotation about a given axis. In general, we may consider this phenomenon as arising from the superposition of several harmonic angular waves, having different angular wave numbers $m, m', m'', \ldots$. These wave numbers are integers or half-integers, as we have seen above, so that we have here a discrete spectrum. We can, nevertheless, characterize it by its dispersion Δm, and relate this latter to the characteristic angular extension $\Delta \varphi$ of the phenomenon. The same reasoning as above, leads to the angular spectral inequality

$$\Delta m \, \Delta \varphi \gtrsim 1. \tag{3.1.7}$$

It is essential to realize, however (and much more so than in the spatial or temporal case), that $\Delta \varphi$ is generally not the angular 'extension' of the phenomenon – most often, this latter is obviously equal to 2π – but rather, it is the order of magnitude of the angular zone in which the phenomenon displays an appreciable modulation, i.e., a qualitative difference compared to ideal harmonicity. The example of an angular wave, given by

$$\Phi(\varphi) = \exp(im_1 \varphi) + \exp(im_2 \varphi)$$
$$= 2 \exp i[\tfrac{1}{2}(m_1 + m_2)\varphi] \cos[\tfrac{1}{2}(m_2 - m_1)\varphi],$$

is in order here. The spectral dispersion is $\Delta m = |m_2 - m_1|$, and the modulation, is appreciable in a sector $\Delta \varphi \simeq \pi/(2|m_2 - m_1|)$, in conformity with inequality (3.1.7).

3.2. The Heisenberg inequalities. Proper states and proper values

We have seen in the preceding chapter that the theoretical description of quantons leads to the identification of pairs of concepts, initially related to classical corpuscles (energy, momentum and angular momentum), on the one hand, and to classical waves (pulsation, undulation and angular wave number), on the other. The Planck–Einstein and de Broglie relations and their angular homologues,

$$E = \hbar\omega, \tag{3.2.1}$$

$$p = \hbar k, \tag{3.2.2}$$

$$J_z = \hbar m, \tag{3.2.3}$$

establish this identification, which then necessarily confers upon the concepts, so synthesized, characteristics which are qualitatively different from those of the original classical concepts. Here, we shall attempt to approach the specific nature of the quantum concepts in a heuristic fashion by utilising, to the extent possible, the classical content of these concepts, precisely in order to make the limits evident. This essentially pedagogical procedure is, in reality, the exact opposite of that which, on the (epistemo-) logical level, leads from the quantum to the classical concepts. The quantum theory has a wider domain of validity than the classical theory, which can be viewed as an approximation. Thus, one has the situation that a single concept in the quantum theory (e.g., $E = \hbar\omega$) breaks up into two classical concepts (E and ω, separately), whose applicability depends upon the specific conditions at hand. If a quantum magnitude carries, in fact, the name of one or the other of its classical homologues, such as 'energy' or 'momentum', it is because of the fact that the epistemological order (quantum → classical) is the reverse of the chronological order. The path that we are retracing here, was followed historically in the direction classical → quantum, by the founders of quantum physics. Yet, there exists a conceptual discontinuity between the quantum and the classical domain. One might speak of the 'singular' nature of the classical limit, in the mathematical sense of the word 'singularity'. One cannot cross this discontinuity without difficulty, and arrive at the quantum theory by a painless passage from the classical theory. Later, it will be necessary for us to leap-frog and parachute ourselves directly into the heart of the quantum territory. But in order to prepare ourselves to take this enormous leap, we shall skirt around the frontier for a while, locating from within the classical territory, the quantum boundaries, and venturing a few exploratory incursions into the frontier region, where the classical language still preserves a relative pertinence, armed with the Planck–Einstein and de Broglie relations and their angular homologues, as a permit for a brief stay.

3.2.1. Energy and time

To this end, let us consider the Planck–Einstein relation, eq. (3.2.1), and focus on the symbol ω. We have seen in detail, in the preceding section, that a classical undulatory phenomenon cannot, in general, be characterized by a unique value of the pulsation, but rather by a spectrum, mostly continuous, the band width of which is related to the characteristic duration of the phenomenon by the spectral inequality given by eq. (3.1.1). We immediately deduce from this that a quantum physical system cannot, in general, be characterized by a unique value of the energy, but actually by an *energy spectrum*. The band width ΔE of this spectrum is clearly related to the band width $\Delta \omega$ of the equivalent pulsation spectrum, as a consequence of relation (3.2.1), by

$$\Delta E = \hbar \, \Delta \omega. \tag{3.2.4}$$

Thus, we have the *temporal Heisenberg inequality*

$$\Delta E \, \Delta t \gtrsim \hbar, \tag{3.2.5}$$

which connects the spectral extension of a quantum system in energy to its characteristic evolution time. The distinctly 'non-classical' content of this inequality resides in the fact that, in general, it prevents us from representing the energy of a phenomenon, even when it is isolated, by a unique number fixed at each instant. In some very special, but important, cases, it is possible however, to attribute to a quantum physical system, a well-defined energy, characterized by a unique numerical value. In such cases, the system possesses this particular energy 'properly'*. More compactly, we say that it is in a *proper state* of the energy and that the value in question is a *proper value* of the energy. This terminology will ultimately be given a formal mathematical meaning. For an energy proper state, the spectral extension vanishes: $\Delta E = 0$. By the Heisenberg inequality, this implies that $\Delta t = \infty$. Such a system has, therefore, a characteristic evolution time of infinite duration, meaning that it does not evolve, and hence that it exists in a *stationary state* (the terminology here parallels that of classical wave physics, where a stationary wave is characterized by a unique value ω of the pulsation and hence by a zero

---

* Let us mention for the already educated readers that the physically defined *proper values* and *proper states* will be identified, but only later on (in the formal developments of the second volume of *Quantics*) with mathematically defined eigenvalues and eigenstates.

dispersion: $\Delta\omega = 0$). One can appreciate the physical importance of stationary states, which play an essential role in the development of the formalism of the quantum theory. From now on we shall identify them with states of well-defined energy, or energy proper states.

3.2.2. Momentum and space

The above considerations can immediately be transcribed to the momentum, quantically identified with the wave vector through the de Broglie relation, eq. (3.2.2). A physical system possesses, in general, a wave vector spectrum. The band width Δk, of one of the components of the wave vector, is related to the characteristic dimension Δx, in the same direction, by the spectral inequality given by eq. (3.1.2). From the point of view of the quantum theory, the phenomenon exhibits a momentum spectrum, as a consequence. Its width Δp, in the direction being considered, is related to that of the undulation spectrum, through the de Broglie relation, by

$$\Delta p = \hbar \, \Delta k. \tag{3.2.6}$$

Thus, we get the *spatial Heisenberg inequality*,

$$\Delta p \, \Delta x \gtrsim \hbar, \tag{3.2.7}$$

connecting the spectral extension of a phenomenon, in momentum, to its characteristic size.

The spatial Heisenberg inequality, eq. (3.2.7), established by considering only one spatial dimension, generalizes immediately to a space of three dimensions, since it is valid in all directions. For a given rectangular system of axes, one can write three spatial Heisenberg inequalities,

$$\Delta p_x \, \Delta x \gtrsim \hbar, \qquad \Delta p_y \, \Delta y \gtrsim \hbar, \qquad \Delta p_z \, \Delta z \gtrsim \hbar. \tag{3.2.8}$$

It ought to be noted that there do not exist any analogous inequalities relating the dispersions in the momentum and the position, corresponding to orthogonal directions, e.g., Δp_x and Δy. On the other hand, such an inequality exists for the dispersions Δp_n and $\Delta r_{n'}$, corresponding to the two non-orthogonal directions n and n' (see exercise 3.2). Considering now the magnitude $p = |\boldsymbol{p}|$ of the momentum, it is natural, in view of the expression $p^2 = p_x^2 + p_y^2 + p_z^2$, to define its dispersion by

$$(\Delta p)^2 = (\Delta p_x)^2 + (\Delta p_y)^2 + (\Delta p_z)^2. \tag{3.2.9}$$

Similarly, the distance $r = |r|$ from the origin has a dispersion Δr, coming from those of its components,

$$(\Delta r)^2 = (\Delta x)^2 + (\Delta y)^2 + (\Delta z)^2. \tag{3.2.10}$$

It is evident, therefore, since, e.g., $\Delta p > \Delta p_x$ and $\Delta r > \Delta x$, that the dispersions of these magnitudes (of p and r) also obey a Heisenberg inequality,

$$\Delta p\, \Delta r \gtrsim \hbar. \tag{3.2.11}$$

One might attempt to improve this inequality by writing

$$\Delta p\, \Delta r > \Delta p_x\, \Delta x + \Delta p_y\, \Delta y + \Delta p_z\, \Delta z > 3\hbar.$$

But in any case, the inequalities given by eqs. (3.2.8) only hold in order of magnitude, up to a numerical coefficient.

For a quantum system to be in a proper state of momentum, i.e., to have a unique and well-defined momentum (in a given direction), so that $\Delta p = 0$, it must have an infinite characteristic dimension, $\Delta x = \infty$, according to eq. (3.2.7). This means, in turn, that it has to be invariant under spatial translations – identical to itself at all points in space. Conversely, if the physical conditions, governing the phenomenon, undergo an appreciable variation from one point in space to another, the system cannot find itself in a proper state of momentum. This is the situation when the system is subject to forces, or more generally – since the term runs the risk of being too precise, in view of its classical connotation (implying a localized action on a material point) – to variable 'interactions' in space, e.g., by being or coming close to other physical systems. We see, therefore, that only a *free* system, without external interactions, can evolve while remaining characterized by a proper value of its momentum. This is a very particular situation. It is important to notice here the difference between momentum and energy. As we have seen the proper states of the latter are the stationary states and hence play a physically essential role. The electron of a hydrogen atom, e.g., in its most stable state, is (almost by definition), in an energy proper state (i.e., characterized by a proper value of the energy which is the smallest one available), but *not* in a proper state of momentum: it possesses a certain spectrum of momentum.

3.2.3. Angular momentum and angle

Finally, let us consider the angular momentum. In the quantum theory, it is related to the angular wave number m, for a given direction, such as Oz. A physical system presents itself, in general, as a superposition of harmonic angular waves and possesses a wave number spectrum, of extension Δm, depending upon its characteristic angular dimension $\Delta \varphi$, through the spectral inequality given by eq. (3.1.7). It follows from this that such a system cannot be characterized by a proper value of the angular momentum about the axis being considered. These proper values, as we have seen above, are given, in units of $\hbar$, by integers or half-integers m. Thus, the width of the angular momentum spectrum is

$$\Delta J_z = \hbar \, \Delta m, \tag{3.2.12}$$

It obeys, therefore, an *angular Heisenberg inequality*,

$$\Delta J_z \, \Delta \varphi_z \gtrsim \hbar. \tag{3.2.13}$$

Of course, the same is true for the other components as well. In order for the system to be in a proper state of one of the components of its angular momentum, it must have a harmonic periodicity about the corresponding axis. It is appropriate to mention here that inequality (3.2.13) has only an approximate meaning, and cannot be formalized exactly as the spatial [eq. (3.2.7)] and temporal [eq. (3.2.5)] inequalities. Indeed, for a proper state of angular momentum, obeying $\Delta J_z = 0$, it leads to an infinite angular extension, $\Delta \varphi_z = \infty$. This is physically absurd, for since $\Delta \varphi_z \leqslant 2\pi$, the spectrum of an angular coordinate cannot extend beyond one complete turn. One can, nevertheless, give a rigorous version of such an angular inequality, which would reduce to the approximate inequality, eq. (3.2.13), when the angular spectrum is not too large ($\Delta \varphi \ll 1$), or the angular momentum spectrum is not too narrow ($\Delta J_z \gg 1$).

Inequalities of the type of eq. (3.2.13) can be written for all components of the angular momentum. For a system to be in an eigenstate of one of these components, it must have a harmonic periodicity about the corresponding axis. Now, it is impossible for the system to display such an invariance under rotations about two different spatial axes, and hence, to be a proper state of two components of the angular momentum at once. Here we have touched upon a fundamental difference between displacements due to translations and those due to rotations, which has important consequences for the momentum and the angular momentum, respectively, in a quantum theory. Translations corresponding to two different spatial directions are, in fact, independent, which is certainly not true for rotations about two different axes: while it is possible to displace a system in a straight line in a certain direction, without involving a displacement along any other direction, it is impossible to turn it about a certain axis, without 'turning' it about another at the same time. In more mathematical terms, this is an expression of the Abelian (commutative) character of the group of translations, as compared to the non-Abelian group of rotations. The few special cases of systems which display rotational symmetry about two or more axes, i.e., which have regular or quasi-regular polyhedral symmetries, may be excluded here. In the language of waves, they correspond in fact to stationary waves, which are superpositions of two harmonic waves, respectively progressing in the two directions of rotation about each axis, and hence are characterized by two opposing angular wave numbers. However, we are only interested in proper states of the components of the angular momentum, and these are linked to single proper values. The only, very special, example of simultaneous invariance under rotations about different axes is that of complete spherical symmetry, in which case there is invariance under rotations about arbitrary axes and through arbitrary angles. We are thus led to conclude that it is not possible for a quantum system to find itself in a state which is a simultaneous proper state of two components of its angular momentum. We shall now make this idea precise, while generalizing it.

Consider a vectorial physical magnitude A, such as the momentum p, the position r, or the angular momentum J itself. The experience we have gathered so far, leads us to assume that as a general rule, for each component of A, a physical system possesses certain spectra of numerical values, which one might characterize using the averages $\langle A_x \rangle$, $\langle A_y \rangle$, $\langle A_z \rangle$ and the dispersions $\Delta A_x, \Delta A_y, \Delta A_z$. During a geometric rotation of the system of axes, through an infinitesimal angle $\delta\varphi_z$ about Oz, the average values of these components undergo the variations (see fig. 3.2):

$$\delta\langle A_x \rangle = \delta\varphi_z \langle A_y \rangle, \qquad \delta\langle A_y \rangle = -\delta\varphi_z \langle A_x \rangle, \qquad \delta\langle A_z \rangle = 0. \qquad (3.2.14)$$

We can characterize the angular dispersion $\Delta\varphi_z$, of the system about the axis Oz, by the condition that a rotation, of the order of this dispersion, does not lead to any variations of the average values $\langle A_x \rangle$ and $\langle A_y \rangle$, larger than the respective dispersions ΔA_x and ΔA_y, of these components. Indeed, if a rotation through the angle $\delta\varphi_z$ does not displace the average values $\langle A_x \rangle$ and $\langle A_y \rangle$ beyond the zones of width $\Delta A_x, \Delta A_y$, then this rotation does not modify the spectral distributions of A_x and A_y (and hence), the characteristics of the system, appreciably. This is the same as saying that the system itself is not defined, in angular terms, any better than up to $\delta\varphi_z$ (similarly, the dispersion Δx of a system, in position, is such that a translation by an amount $\delta\alpha < \Delta x$

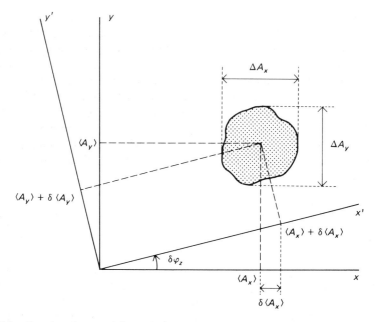

Fig. 3.2. *Rotation of a vectorial magnitude*
Change in the average values of a vectorial magnitude during a rotation.

does not qualitatively modify the average value $\langle x \rangle$). Once again, by eq. (3.2.14),

$$\Delta\varphi_z |\langle A_y \rangle| < \Delta A_x, \qquad \Delta\varphi_z |\langle A_x \rangle| < \Delta A_y. \qquad (3.2.15)$$

These inequalities, together with the angular Heisenberg inequalities given by eq. (3.2.13), imply the following (the last four may be deduced by cyclic permutation):

$$\Delta J_z \, \Delta A_x \gtrsim \hbar |\langle A_y \rangle|, \qquad \Delta J_z \, \Delta A_y \gtrsim \hbar |\langle A_x \rangle|,$$

$$\Delta J_x \, \Delta A_y \gtrsim \hbar |\langle A_z \rangle|, \qquad \Delta J_x \, \Delta A_z \gtrsim \hbar |\langle A_y \rangle|, \qquad (3.2.16)$$

$$\Delta J_y \, \Delta A_z \gtrsim \hbar |\langle A_x \rangle|, \qquad \Delta J_y \, \Delta A_x \gtrsim \hbar |\langle A_z \rangle|.$$

We see that it is not possible to have both $\Delta J_x = 0$ and $\Delta A_y = 0$ at the same time (except in the special case where $\langle A_z \rangle = 0$). In general, therefore, a quantum system cannot be in a state which is at the same time a proper state of one of the components of its angular momentum and a proper state of another component of an arbitrary vectorial magnitude. This should put us on guard against any overly pictorial representation of a quantum vectorial magnitude as an 'arrow', i.e., as a classical vector. We have only utilized the transformation properties of a vector under rotations [see eq. (3.2.14)]. These properties can – and must – be considered as the very definition of what a 'vectorial magnitude' should mean. The inequalities of eqs. (3.2.16) apply, among others, to the angular momentum J itself, when substituted for A, as a particular vectorial magnitude,

$$\Delta J_x \, \Delta J_y \gtrsim \hbar |\langle J_z \rangle|, \qquad \Delta J_y \, \Delta J_z \gtrsim \hbar |\langle J_x \rangle|, \qquad \Delta J_z \, \Delta J_x \gtrsim \hbar |\langle J_y \rangle|.$$

$$(3.2.17)$$

We deduce, in particular, that a system, in a proper state of *one* component of the angular momentum, e.g., J_z, and hence characterized by $\Delta J_z = 0$, will have the other two components assuming zero average values. For a system to be in a proper state of *two* (or three) components, at the same time, we must have, e.g., $\Delta J_x = \Delta J_y = 0$. Then all the average values are zero, hence also the actual values themselves (since their dispersions are zero). In other words, only a system having its angular momentum identically zero, $J = 0$, which also corresponds to complete spherical symmetry, can be a common proper state of several components of the angular momentum.

Finally, let us remark that a rotation does not modify the values of a scalar magnitude S (by definition!). A reasoning, similar to that which had led us to the inequalities of eqs. (3.2.16) is no longer valid, and the dispersion ΔS of this magnitude is not related to those of the components of J. Thus, it is possible for a proper state of an arbitrary component of the angular momentum to be at the same time a proper state of a scalar magnitude. In particular, this applies to the scalar $A = |A|$, which is the modulus of the vectorial magnitude A, for example of the angular momentum, itself. A proper state of a component of J can, therefore, be a proper state of its modulus J. This is what we had implicitly assumed when, in the last chapter, we proved that the possible numerical values – the proper values – of the modulus of an angular momentum are characterized by a number (an integer or a half-integer) j and are given by

$$J^2 = j(j + 1)\hbar^2 \quad (2j \text{ an integer}).$$

$$(3.2.18)$$

Thus, as far as its angular momentum is concerned, a quantum system is, in general, specified by means of the common proper states of one of the components of the angular momentum (conventionally J_z) and its modulus. These states are labelled by two integers or half-integers (m and j) corresponding, respectively, to the proper values $m\hbar$ of J_z and $j(j+1)\hbar^2$ of J^2.

3.2.4. Conclusions

From the above considerations, we shall essentially keep in mind the following ideas:

> The state of a quantum system is, in general, characterized by the numerical values attributed to the various physical magnitudes which characterize it. For each one of these magnitudes, there exist particular states in each of which that magnitude assumes a well-defined value, called the *proper value*: these are the *proper values* of (relative to) this magnitude. In an arbitrary state (not a proper state), the physical magnitude displays a spectrum of numerical values.
>
> In general, the proper states of two physical magnitudes do not coincide. In an arbitrary state, two magnitudes exhibit spectra of numerical values the widths of which may sometimes be correlated through the Heisenberg inequalities.
>
> Two (or more) physical magnitudes are said to be *compatible* if their proper states are the same; in such a case, the corresponding state is characterized by a proper value of each one of these magnitudes, the dispersions of which are thus simultaneously zero.

Thus, p_x and y are compatible (since the product of the dispersions $\Delta p_x \, \Delta y$ is not bounded below) and so also are J_x and J. On the other hand, p_x and x or J_x and J_y are incompatible (see exercise 3.3). We notice the radical nature of the change from classical mechanics where every physical magnitude is describable, mathematically, by a function (of time), having a unique numerical value. The quantum theory, in order to incorporate the notion of extended spectra and incompatible magnitudes, requires a more sophisticated formalism.

The presence of the constant $\hbar$, on the right-hand side of the Heisenberg inequalities, underscores their specifically quantum character. The limiting validity of the classical theory, in its own domain, is justified because the lower (quantum) limit imposed by these inequalities is very small compared to the corresponding product of uncertainties, experimental in origin, of the numerical values assumed by the classical properties. These uncertainties are, in fact, enormously larger than the intrinsic quantum dispersions, and therefore, mask them completely. For example, let us consider a pin-ball weighing 30 g. Its position could possibly be known to $(\delta x)_{exp} = 0.1$ mm, and its velocity to $(\delta v)_{exp} = 1$ mm s^{-1}. From this we obtain an experimental product of uncertainties: $(\delta x)_{exp}(\delta p)_{exp} = 3 \times 10^{-9}$ L $\simeq 3 \times 10^{25}\hbar$, and immediately recognize here an application of the general criterion for the necessity of a quantum theory, given in ch. 2. The product $(\delta x)_{exp}(\delta p)_{exp}$ is a characteristic action of the physical situation. It would, therefore, be necessary to greatly improve the experimental precision and the determination of the classical magnitudes before attaining these limitations, imposed upon them by the intrinsic quantum dispersions. This example should make us understand that quantum dispersions have nothing to do with 'uncertainties', contrary to a still widespread terminology, inherited from the heroic, although confused, beginnings of the quantum theory. Classically, a physical magnitude, such as the position, has a well-determined value x, which might only be imprecisely known, with an uncertainty $(\delta x)_{exp}$. Quantically, this magnitude does *not* have a unique value; Δx represents an intrinsic dispersion of the position and is no reflection of our ignorance.

3.3. Applications

At the heuristic level of this chapter, the Heisenberg inequalities may be looked upon as prosthetic devices, grafted onto the concepts of classical mechanics, in order to enable them to simulate the, yet unknown, concepts of quantics. To use an audacious metaphor (yet another one!), they are comparable to the cylinders of compressed air carried by marine divers: they are not as good as a healthy set of gills, but still enable one to leave the terra firma for the not too deep waters. Let us plunge in, therefore, and explore a few specifically quantum phenomena, of which we can already grasp the essential aspects, without yet having a detailed theory.

3.3.1. Temporal Heisenberg inequalities

3.3.1.1. Life times and energy widths. A system can be in a proper state of well-defined energy, of zero dispersion, $\Delta E = 0$, only in a stationary state, which remains invariant while time passes. Conversely, therefore, a system in evolution, with a characteristic time δt, possesses an energy spectrum ΔE

such that $\Delta E \, \Delta t > h$. In particular, unstable physical systems conform to this description. For example, the nuclei of radioactive atoms, or certain fundamental particles, decay obeying a law, the exact nature of which is unimportant here. One can estimate the duration of the characteristic interval of time, for the evolution of such a system, at the end of which an appreciable fraction (e.g., 50%) of an aggregate sample would have decayed. This is what currently is called the 'radioactive period'. This time is also a measure of the average 'life time' which separates the creation of such an individual system from its disappearance. The intrinsic minimal dispersion of the energy spectrum of the system, or the width of the energy band, which it occupies, is called the 'width of the energy spectrum', and is denoted by Γ. The life time τ, being intrinsically defined, i.e., in the reference frame of the quanton itself, the dispersion in energy, to which it is correlated, concerns therefore its internal (rest) energy – i.e., in view of the Einstein relation, $E_0 = mc^2$, its mass. The two magnitudes, the energy width Γ and the life time τ, are related by

$$\Gamma \tau \simeq h. \tag{3.3.1}$$

In classical physics, not only does this relation not exist, it does not even make sense, since an unstable system (finite τ) has, in principle, a well-defined energy ($\Gamma = 0$).

As a general rule, only one of the magnitudes, either Γ or τ, is experimentally accessible. This is hardly surprising, since an apparatus of macroscopic dimensions can most often be considered as being classical. The experimental set-up defines the time up to $(\delta t)_{\exp}$ and the energy to $(\delta E)_{\exp}$. These quantities must then obey the inequality $(\delta E)_{\exp} (\delta t)_{\exp} \gg h$, for otherwise, as we have seen in the previous section, we would be in the midst of the quantum domain. From eq. (3.3.1) one cannot have, therefore, $(\delta E)_{\exp} \lesssim \Gamma$ and $(\delta t)_{\exp} \lesssim \tau$, simultaneously, which would be indispensable if one wished to measure both quantities. Consequently, one sees that an unstable quantum system is experimentally characterized either by a measure of its life time or of its energy spectrum. Let us consider some specific examples.

The plutonium nucleus, ^{239}Pu. This nucleus disintegrates by radioactive α-decay, has a life time $\tau \simeq 24\,000$ years [this means that no matter what happens, we shall bequeath the better half of our (rapidly increasing) present stock of plutonium, and the waste products from the rest, to our descendants of a thousand generations from now – if, of course, this same plutonium does not actually obliterate the said descendants through explosion or poisoning].

This life time is measured by studying the time rate of decay of a plutonium sample. One deduces that the mass of the plutonium nucleus is only defined up to $\Delta M = \Gamma/c^2 \simeq \hbar/c^2\tau$, where $\Delta M \simeq 4 \times 10^{-63}$ kg. This is to be compared to the mass $M = 4 \times 10^{-25}$ kg of the plutonium nucleus itself. Any direct measurement of this intrinsic mass dispersion is evidently not possible.

The atomic levels. We have already had the occasion to indicate, in the preceding chapter, that a bound quantum system, such as an atom, possesses discrete proper states of energy. The proper values of its energy do not occupy a continuous band. These 'energy levels', except for the lowest or the 'ground level' are unstable, and most often the atom in an 'excited state' is de-excited by the emission of a photon (see sect. 2.3.5). Therefore, these unstable states are, properly speaking, not stationary. They are not true eigenstates of energy, specified by a fixed proper value, but possess a certain dispersion Γ. The typical life time of an excited state being $\tau \simeq 10^{-10}$ s, its energy width $\Gamma \simeq \hbar/\tau$ is of the order of 10^{-5} eV. This width is to be compared to the spacing between the levels, which is typically $\delta E \simeq 1$ eV. We see, therefore, that the excited states, no matter how unstable, have a life time which is sufficiently long, and hence a width which is sufficiently small, for them to be thought of as being approximately stationary. This example shows that problems in atomic physics may, in general, be treated by first neglecting the coupling to the electromagnetic radiation, which renders the atom unstable, and then calculating the positions of its energy levels, as though they were rigorously stationary. It is only at a second stage, then, that the phenomena due to the radiation are taken into consideration which brings about the instability of the energy levels and hence their broadening – and sometimes their displacement; however, these are always very minor corrections (fig. 3.3).

Hadronic resonances. In fundamental 'particle physics' (we use here the terminology which has remained conventional among the practitioners in the field, although evidently, it would be better if we were to speak of fundamental quantons), one studies the energy distribution of final particles, emitted during a collision between a target particle and a projectile particle of fixed energy. If the reaction produces two final particles, the conservation laws of energy and momentum determine, in a unique manner, the energies of these two particles in the centre of mass frame of the whole system. On the other hand, when three or more particles are emitted, the conservation laws do not suffice for a determination of all the energies, since they impose fewer constraints than the number of independent variables. Moreover, different configurations are allowed, and the relative importance of one or the other of

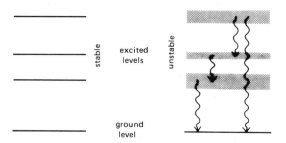

Fig. 3.3. *Level width and life time*
The energy levels of an atom are perfectly well-defined when there is no mechanism present which might enable it, when in an excited state, to lose its energy. In reality, because of the coupling to the electromagnetic field (the electrons are electrically charged!), the atom can radiate. Its excited states are, therefore, unstable and in this more sophisticated description, the energy levels acquire certain natural widths Γ, inversely proportional to their life times τ.

these reflects the characteristics of the interaction, in other words, the dynamical nature of the phenomenon. The distribution of the energy, among the final particles, varies from one collision to the next. Sometimes it is observed that the energy of a pair of particles is concentrated around a particular value. The energy spectrum of this composite system displays a 'peak', centred at this average value and having a certain width. This means that the reaction had first produced an intermediate particle, of a given energy. This metastable particle subsequently disintegrated into the two particles of the couple being examined, and then shared this energy between them. The transitory, hence non-stationary, character of the intermediate particle (quanton!) implies that it possesses a certain dispersion in energy, which explains the width of the peak through which it manifests itself. These metastable states are usually called 'resonances', because of the similarity of their mass spectra to the excitation spectra of resonant phenomena in classical waves, which also correspond to the transitory formation of metastable physical states. Thus, in a reaction involving the scattering of a π meson, $\pi^+ + p \rightarrow \pi^+ + p + \pi^+ + \pi^-$, the energy spectrum of the pair (π^+, π^-) displays a sharp peak, centred at the value $m = 770$ MeV and of width $\Gamma = 160$ MeV (fig. 3.4). From this one infers the formation of a particle, called ρ^0, of *average* mass m. Its life time is now obtained from the width $\tau \simeq \hbar/\Gamma$, giving $\tau \simeq 4 \times 10^{-24}$ s. This time is much too short to be measured directly. As an example, we remark that even at the maximal velocity c, the ρ^0 particle would only be able to cross a distance $d = c\tau = 10^{-15}$ m = 1 F, during its ephemeral existence. It does not even have the time, therefore, to leave the region of the

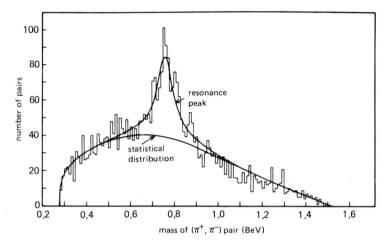

Fig. 3.4. *A hadronic resonance*
The figure gives the energy distribution of the masses of the meson pairs (π^+, π^-) in a reaction $\pi^+ + p \rightarrow \pi^+ + p + \pi^+ + \pi^-$. On the statistical background, which corresponds to a purely random distribution, is superposed a peak, which is interpreted as a manifestation of the existence of an intermediate ρ particle. One has the following two-step reaction mechanism:

$$\pi^+ + p \rightarrow \pi^+ + p + \rho^0$$
$$ \llcorner \rightarrow \pi^+ + \pi^-$$

The energy width of this unstable state, or 'resonance', reflects its short life time. [M. Abolins et al., Phys. Rev. Lett. 11 (1963) 381.]

strong interaction itself, which produced it. This is a phenomenon which is specific to the strong interaction, characterized by times that are comparable to 10^{-24} s (see ch. 1, sect. 1.4.3). This experiment provides us with information regarding the forces between the pions, and we see that, at least within a certain region of energies, this interaction is highly attractive. Such temporary bound states are frequently encountered in particle physics. These resonances may decay into two stable particles, just as well as into three or more. It seems to be necessary, at the present time, to consider them, in spite of their instability, on the same footing as the more stable particles, such as pions and nucleons: whatever might be their relative stability, any hadron (a quanton which interacts strongly) is as much, or perhaps as little 'elementary' as any other.

3.3.1.2. Exchange of virtual particles and ranges of forces. In classical physics, the transmission over a distance of the forces exerted by one particle on another is mediated by a field (wave) – this in itself is the essential role of this concept: the field produced by one particle propagates and then acts on the other. In quantics, only one species of entity is recognized, the quantons. Hence, this scheme is transcribed by means of the idea that the forces which are exerted between two quantons A_1 and A_2, are transmitted by the exchange of a quanton of type B, corresponding, if one so wishes, to the 'quantization' of the classical field (see fig. 3.5):

$A_1 \rightarrow A_1 + B$ creation of the quanton (emission of the field)

                displacement of the quanton (propagation of the field)

$B + A_2 \rightarrow A_2$ annihilation of the quanton (absorption of the field).

The mechanism is evidently reciprocal: A_2 acts on A_1 by a process which is similar, and iterative: A_1 and A_2 can exchange more than one quanton of type B. However, each one of the two elementary phenomena $A_1 \rightarrow A_1 + B$ or $A_2 + B \rightarrow A_2$ ('vertex' of the diagrams) is, in principle, forbidden due to the combined effect of the laws of momentum and energy conservation: a particle cannot disintegrate into itself and another one. Indeed, if we impose the laws of energy and momentum conservation on the reaction $A_1 \rightarrow A_1 + B$, we find out that they are incompatible with the value, fixed a priori, of the mass m of the particle B. To see this, it is enough to write the conservation laws in the form

$$\underline{p} = \underline{p}' + \underline{q}, \tag{3.3.2}$$

where p, p' are, respectively, the energy–momentum four-vectors of A, before and after the reaction and q the corresponding quantity for B. Additionally,

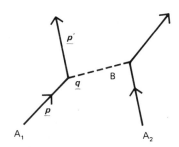

Fig. 3.5. *The quantum 'go-between'*
The action at a distance between two quantons A_1 and A_2 is mediated by a 'virtual' quanton B.

we have

$$p'^2 = p^2 = (Mc^2)^2, \tag{3.3.3}$$

the mass of A, being called M (we say that p and p' are on the 'mass shell' M). From these we see that the four-vector $q = p - p'$ is necessarily space-like and that its Minkowski norm is negative, rather than having a value $(mc^2)^2$. (q is *not* on the mass shell m). Hence, it seems that q cannot be the energy–momentum four-vector for a particle of mass m (see fig. 3.6). However, the particle B only 'lives' for a certain time Δt between its emission by A_1 and absorption by A_2. It is not, therefore, a stationary state, i.e., in an eigenstate of energy: its energy possesses a certain dispersion ΔE, related to Δt by the temporal Heisenberg inequality, $\Delta E \gtrsim \hbar/\Delta t$. It is enough, therefore, that Δt be sufficiently small, so that

$$\Delta t \lesssim \frac{\hbar}{mc^2}, \tag{3.3.4}$$

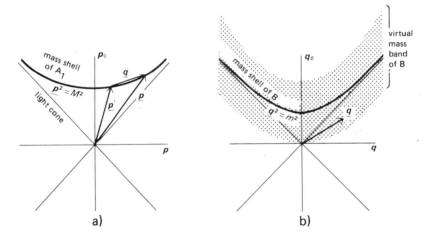

Fig. 3.6. *The energy momentum of a virtual quanton*

In these diagrams, the spatial component of the energy–momentum four-vector has been plotted on the abscissa and the temporal component on the ordinate; the choice of units is such that $c = 1$.

(a) The energy–momentum four-vectors p and p' of A_1, before and after the emission of B, are on the mass shell M; the four-vector of B, fixed by the conservation laws, $q = p - p'$, is space-like.

(b) Thus, q cannot be on the mass shell m of B. But if the state of B is not stationary, its energy possesses a dispersion ΔE. If this quantity is sufficiently large, $\Delta E \gg mc^2$, the energy band becomes large enough to include q.

in order to have $\Delta E \gtrsim mc^2$, and so that the available band of energies, characterized by the dispersion ΔE around the value mc^2 would be large enough to include the four-vector p (fig. 3.6). Consequently, if the propagation of B, between its emission by A_1 and absorption by A_2 does not exceed the duration $\Delta t \simeq \hbar/mc^2$, its dispersion in energy would be sufficiently large, so that its temporary existence would not violate the conservation laws. In the relativistic kinematics which we are using, to the time interval Δt is associated a characteristic spatial interval $\Delta r = c \, \Delta t$, yielding

$$\Delta r \simeq \frac{\hbar}{mc}, \tag{3.3.5}$$

which gives the scale of the distance traversed by B during its propagation, and hence, up to an order of magnitude, the range of the interaction transmitted by the exchange of the particle B.

The nature and the strength of the interaction is unimportant here: one is dealing with a universal mechanism in quantum physics. The case of the electromagnetic interaction is a singular one. The mass of the photon, presumably zero, corresponds to an infinite range, which is a way of describing the very slow decrease of the Coulomb force with distance. On the other hand, relation (3.3.5) may be used to set a limit on the mass of the photon: any verification of the laws of electromagnetism over distances larger than D enables one to put a bound on the mass of the photon: $m_\gamma \simeq \hbar/cD$ (exercise 3.5). In another domain, Yukawa, in 1935, deduced from the observed range of nuclear forces, $a_N \simeq 1$ F, the mass of the hypothetical quantons, which would mediate these forces, to be $m \simeq \hbar/ca_N \simeq 200$ MeV. The discovery of the π mesons confirmed this hypothesis ten years later. We know today that all strongly interacting quantons (hadrons) participate in the exchanges which mediate these nuclear interactions. But, in view of eq. (3.3.5), it is the lightest one among them – the π meson – which determines the maximal range of the forces. Finally, let us remark that the exchange of the intermediate quantons is, by its very nature, unobservable directly: the detection of a quanton B, or more precisely, an effective determination of its presence, would require an energy definition $\Delta'E \ll mc^2$, and hence a time definition $\Delta't \gg \hbar/mc^2$, which is longer than the duration of the phenomenon... For this reason, the exchanged quantons are called 'virtual', as opposed to the 'real' quantons which we detect and measure. However, one should not take this opposing terminology too seriously. From the nature of the process considered, the 'virtual' quantum particles are no less 'real' than the classical fields for which they assume the role of the mediator.

The exchange of 'virtual' quantons gives us a mechanism, which explains the classical notion of a force at a distance. Indeed, momentum is strictly conserved. As a consequence, the momentum carried away by B is taken out of the momentum of A_1 and then given completely to A_2. Thus, between A_1 and A_2, there occurs a momentum transfer. Now, in classical mechanics, one attributes the existence of a 'force' to precisely such changes in the momentum. We see, therefore, that the exchange mechanism which we have just described, furnishes us with a quantum explanation of the forces – one prefers to speak of interactions – between quantum particles.

3.3.1.3. Antiparticles. We have just seen that a virtual quanton – the quantum mediator of an interaction – has a space-like four-vector of energy–momentum. In other words, its four-velocity (the unit four-vector carried by q) lies outside the light cone (see fig. 3.6), and the quanton propagates with a velocity greater than that of light! The spatio–temporal four-vector which connects the event E_1, 'emission of B by A_1' and the event E_2, 'absorption of B by A_2' is, therefore, also space-like. The sign of its time component is, thus, not invariant and the temporal order of the events E_1 and E_2 depends upon the frame of reference used! If in one reference frame, E_2 occurs after E_1, in another equivalent frame it could be E_1 which comes after E_2. In such a frame, the quanton B would, therefore, be absorbed by A_2 *before* being emitted by A_1! It seems as though we had here a violation of the elementary notion of causality. The quantum theory manages to reinterpret the phenomenon, in the second frame of reference, as an emission of E_2 and a subsequent absorption at E_1. However, this reverse exchange must, in general, be considered as being that of a quanton $\bar{B}$, different from B. Let us consider, e.g., the case where A_1 and A_2 are two nucleons which exchange a charged meson. In the first frame of reference, we then have the emission $p \rightarrow n + \pi^+$, at E_1 (the charge state of the nucleon changes, of course, to ensure conservation of the electric charge), and the reabsorption, $n \rightarrow \pi^+ + p$, at E_2 (fig. 3.7a). If now, in the second reference frame, E_2 precedes E_1, the neutron can only be transformed into a proton by the emission of a negatively charged particle, the π^- meson, so that $n \rightarrow p + \pi^-$. This is followed at E_1 by $p + \pi^- \rightarrow n$ (fig. 3.7b). The global nature of the phenomenon remains the same: in each reference frame, an interaction is observed, with exchange of the charges of a neutron and a proton; only its description depends upon the frame of reference (exactly in the same way as one and the same electromagnetic field appears to be more or less electric, or more or less magnetic, depending upon the frame of reference). Thus, relativistic invariance has been re-established. Quite generally, the quantum mechanism of the exchange of

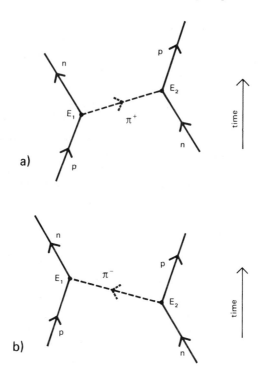

Fig. 3.7. *Antiparticles: the neutron–proton interaction and the π meson*
(a) In the first frame of reference, we have the events:
– first E_1: $p \rightarrow n + \pi^+$
– followed by E_2: $n + \pi^+ \rightarrow p$.
(b) In the second frame of reference, where E_1 and E_2 take place in the reverse order, they are interpreted differently:
– first E_2: $n \rightarrow p + \pi^-$
– followed by E_1: $p + \pi^- \rightarrow n$.

virtual quantons is compatible with Einsteinian relativity, only if for each 'particle' there exists an 'antiparticle' of the same mass but of opposite charge, whether it be an electric charge or any other (baryonic, leptonic) charge, for which an identical argument could be given. Some particles, for which all charges are zero, such as the photon γ or the meson π^0, are identical to their antiparticles. Naturally, there is perfect symmetry between a particle and its antiparticle, and it is only their particular abundance in our part of the universe, which leads us to call the electron e^- and the proton p 'particles', and the positron e^+ and the antiproton $\bar{p}$ '*anti*particles'.

The question, as to whether this symmetry, which is absolute at the microscopic level, is valid also at the macroscopic level, remains open. One tends to believe, today, that there does not exist antimatter on a large scale in the universe, but neither observation, nor theory is as yet convincing on this point.

The necessity for doubling the set of fundamental quantum particles is one of the most important consequences of the marriage between quantum theory and the Einsteinian relativity theory. Sketched in purely qualitative terms here, this result can be demonstrated in great generality and in a rigorous manner, within the framework of quantum field theory. Up to now it has always been confirmed by experiment, which, for each new particle that is discovered, sooner or later brings out its antiparticle as well.

3.3.2. Spatial Heisenberg inequalities

3.3.2.1. The race for high energies.
If we attempt to study the behaviour of matter at smaller and smaller scales, we must explore regions of space of shorter and shorter dimensions. If a quanton is to serve as a probe, in the sense in which we speak of spatial probes, inside a region of dimensions d, it is necessary that it be *localizable* within a volume of smaller size. In other words, its own spatial extension Δx should obey the inequality $\Delta x \ll d$. By the spatial Heisenberg inequality, eq. (3.2.7), this requires that its momentum spectrum have a width $\Delta p > \hbar/d$. This spectrum must, therefore, extend at least up to the momentum $p \simeq \hbar/d$. The larger this is, the smaller d becomes, and the narrower the region to be explored. We see, therefore, why the race towards the infinitely small involves a race towards indefinitely increasing energies (fig. 3.8). It is in a similar way that one justifies physically, if not socially, the construction of particle accelerators of ever increasing performance, size and cost. But then, let us remark that if quantics proves to be expensive, by forcing us to scale the energy ladder, Einsteinian relativity holds the expenses down, since, asymptotically, it allows the energy to grow only linearly in the momentum, $E = (p^2 c^2 + m^2 c^4)^{1/2}$, so that, $E \simeq pc$ as $p \to \infty$ – rather than quadratically ($E = p^2/2m$) as in the Galilean theory. For these high energies, therefore, the relation

$$E \simeq \frac{\hbar c}{d},$$

(3.3.6)

gives the order of magnitude of the minimal energies necessary for the exploration of a region of dimension d. Attention, however! We are concerned

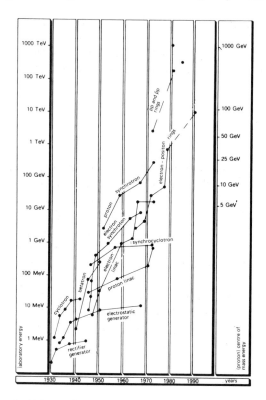

Fig. 3.8.   *The race for high energies*
The maximal energy of the large accelerators has increased regularly by a factor of about 10 every 10 years. [Following M. Crozon, *La Recherche*, February, 1979.]

here with the energy that is physically usable, i.e., which is available in the centre of mass frame of the colliding particles. In the reference frame of the laboratory, where one of the particles is a fixed target, the corresponding energy is much higher.

3.3.2.2. Stability of the hydrogen atom. The stability of atoms is incomprehensible in classical mechanics. It can easily be demonstrated that the energy of a system of charged particles, in pure Coulomb interaction, is not bounded below. Indeed, the potential energy decreases towards infinitely low values as the size of the system tends to zero, while the kinetic energy, which is independent, could be as small as one pleases. Even bound states, for which the kinetic and potential energies are not independent, can have arbitrarily small total energy (see exercise 1.16). In physical terms, the collapse of the system cannot be avoided.

Let us consider the simplest case, that of the hydrogen atom, and suppose, further, that the proton is infinitely heavy, hence immobile. Classically, the energy of the electron is written as

$$E = \frac{p^2}{2m} - \frac{e^2}{r},\tag{3.3.7}$$

where p denotes the absolute value of its momentum and r is its distance from the centre of attraction (the proton). We have used the notation $e^2 = q_e^2/4\pi\varepsilon_0$, already introduced (ch. 1, sect. 1.4.1.) to characterize the strength of the electrostatic attraction. Let us assume that an analogous expression holds quantically. It ought, therefore, to be simulated by the classical expression, eq. (3.3.7), subject to the constraint imposed by the Heisenberg inequality, eq. (3.2.7). Thus, let us denote by r the approximate dimension of the region occupied by the motion of the electron and by p the order of magnitude of the absolute value of its momentum. In fact, r and p characterize, just as well, the widths Δr and Δp of the spectra of position and momentum, respectively (fig. 3.9). We can estimate the order of magnitude of the energy $\tilde{E}$ by writing

$$\tilde{E} \simeq \frac{\tilde{p}^2}{2m} - \frac{e^2}{\tilde{r}},\tag{3.3.8}$$

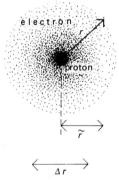

Fig. 3.9. *The hydrogen atom*
The dispersion Δr of the position r of the electron, inside the atom, is of the same order of magnitude as the characteristic radius r of the atom.

subject to the condition

$$\tilde{p}\tilde{r} \gtrsim \hbar. \tag{3.3.9}$$

Thus, we have,

$$\tilde{E} \gtrsim \frac{\tilde{p}^2}{2m} - \frac{e^2\tilde{p}}{\hbar}. \tag{3.3.10}$$

This expression admits a minimum

$$E_0 \simeq -\frac{me^4}{2\hbar^2} \triangleq -E_{\mathrm{H}}, \tag{3.3.11}$$

for the parameters

$$p_0 \simeq \frac{me^2}{\hbar}, \qquad r_0 \simeq \frac{\hbar^2}{me^2} \triangleq a_0. \tag{3.3.12}$$

Stated otherwise, there surely exists a stable state of the hydrogen atom in which its energy attains an absolute minimum. This state is called the 'ground' or 'fundamental' state.* The latter terminology is employed by analogy with classical vibratory systems, the lowest frequencies of which are so named. The relation $E = h\nu$ then allows us to extend the terminology. It seems that we have only recovered the expressions, already established in ch. 1, sect. 1.4.1, by simple dimensional analysis. Although expression (3.3.11), with its numerical factor $\frac{1}{2}$, gives the correct value $E_0 = -13.6\,\mathrm{eV}$, for the energy of the hydrogen atom in its ground state, this coincidence is accidental, given the crudeness of our estimate. However, we have in fact, gone beyond simple dimensional analysis and obtained a better understanding of the stability of the atom. In quantum physics, the kinetic energy of a system increases with its concentration, counteracting in such a situation the drop in the potential energy, which this very concentration would bring about. This mechanism applies generally to all atoms and explains their stability: as in the case of the hydrogen atom, the potential and kinetic energies are anti-correlated. If the size of the atom is made to decrease, leading to a lowering of its potential energy, its kinetic energy necessarily

---

* Actually, English usage tends to prefer the term 'ground state' while in French, the word 'fondamental' is prevalent. (Translator's note.)

increases, because of the Heisenberg inequality, and vice versa. This correlation is specifically quantal. A compromise has to be found between the two opposing tendencies, and the best one characterizes the ground state.

We can further refine this argument and analyze in somewhat more detail the energy structure of the hydrogen atom. Classically, the spherical symmetry of the Coulomb force field implies the conservation of the angular momentum with respect to the centre of attraction. The motion takes place in a plane on which the kinetic energy of the electron, in polar coordinates, is written as

$$E_{kin} = \tfrac{1}{2}m(\dot{r}^2 + r^2\dot{\theta}^2). \tag{3.3.13}$$

In these coordinates, the angular momentum has the form

$$L = mr^2\dot{\theta}^2. \tag{3.3.14}$$

Introducing a radial momentum

$$p_r = \boldsymbol{p} \cdot \boldsymbol{n}_r = m\dot{r}, \tag{3.3.15}$$

(where $\boldsymbol{n}_r = \boldsymbol{r}/r$ denotes the unit vector carried by $\boldsymbol{r}$), we may write

$$E_{kin} = \frac{p_r^2}{2m} + \frac{L^2}{2mr^2}. \tag{3.3.16}$$

Here, the second term just expresses the existence of the centrifugal force. The greater the angular momentum, the larger is this force. One can regard this term as an effective potential, $V_{centr} = L^2/2mr^2$, from which we derive a force

$$F_{centr} = -\frac{\mathrm{d}V_{centr}}{\mathrm{d}r} = \frac{L^2}{mr^3} = mr\dot{\theta}^2,$$

which is just the usual expression. The total energy of the electron is written as

$$E = \frac{p_r^2}{2m} + \frac{L^2}{2mr^2} - \frac{e^2}{r}. \tag{3.3.17}$$

Let us now consider this expression from the quantum point of view. We can separately consider the cases of the different proper states for the modulus L of the angular momentum. Indeed, the scalar magnitudes p_r^2 and r, invariant under rotations, are compatible with L. The eigenvalues of L are given [see eq. (3.2.18)] by

$$L^2 = l(l+1)\hbar^2,\tag{3.3.18}$$

where l is an *integer* (since we are considering an orbital momentum here). For each one of these proper values, we obtain the expression

$$E^{(l)} = \frac{p_r^2}{2m} + \frac{l(l+1)\hbar^2}{2mr^2} - \frac{e^2}{r},\tag{3.3.19}$$

for the energy in the corresponding proper state. It is possible to estimate the minimum value assumed by this energy, in other words, the energy of the most stable state for each value of l, by the same procedure as above. We obtain an evaluation up to an order of magnitude,

$$\tilde{E}^{(l)} \simeq \frac{\tilde{p}_r^2}{2m} + \frac{l(l+1)\hbar^2}{2m\tilde{r}^2} \propto \frac{e^2}{\tilde{r}},\tag{3.3.20}$$

subject to the constraint given by eq. (3.3.9) which expresses the quantum nature of the situation. As before, this inequality is saturated at the minimum that is being sought, so that one can compare the radial kinetic energy $\tilde{T}_r$ and the rotational (centrifugal) energy $\tilde{V}_{centr}$:

$$\tilde{T}_r = \frac{\tilde{p}_r^2}{2m} \simeq \frac{\hbar^2}{2m\tilde{r}^2} < \frac{l(l+1)\hbar^2}{2m\tilde{r}^2} = \tilde{V}_{centr},\tag{3.3.21}$$

and neglect the first term compared to the second, since $l > 0$ (and since we are only interested in orders of magnitude). We would then have, with the approximation $l(l+1) \simeq l^2$,

$$\tilde{E}_0^{(l)} \simeq \frac{l^2\hbar^2}{2m\tilde{r}^2} - \frac{e^2}{\tilde{r}}.\tag{3.3.22}$$

This expression admits a minimum,

$$\tilde{E}_0^{(l)} \simeq -\frac{1}{l^2}\frac{me^4}{2\hbar^2}.\tag{3.3.23}$$

Thus, the stationary states of the hydrogen atom range in series, corresponding to the various integral values of l, each admitting a state of minimum energy given by eq. (3.3.23). The characteristic dimension corresponding to these states is

$$r_0^{(l)} \simeq l^2 \frac{\hbar^2}{me^2}. \tag{3.3.24}$$

This is why one speaks of the concentric electronic 'shells' of these various states.

The same sort of quantitative reasoning can be applied to other systems (exercises 3.10–3.14) as well.

3.3.3. Angular Heisenberg inequalities

Quantization of the angular momentum. Let us go back to the proper values of the modulus J of the angular momentum. For these to correspond to the maximum values of the component J_z, it is necessary, in view of the identity

$$J^2 = J_x^2 + J_y^2 + J_z^2, \tag{3.3.25}$$

that there exists a common proper state of J and J_z, corresponding to these proper states, implying the existence of proper states for J_x and J_y as well, which under the circumstances would have to be zero. However, from the Heisenberg inequalities, (eqs. (3.2.17), for angular momentum, we know that J_x and J_y are incompatible with J_z. Moreover, it is true that if $\Delta J_z = 0$, then by the same inequalities, the average values of J_x and J_y are zero. We are free, therefore, to identify the squares J_x^2 and J_y^2 with the dispersions $(\Delta J_x)^2$ and $(\Delta J_y)^2$, and write

$$J^2 = (\Delta J_x)^2 + (\Delta J_y)^2 + J_z^2. \tag{3.3.26}$$

The Heisenberg inequality

$$\Delta J_x \, \Delta J_y \gtrsim \hbar |J_z|, \tag{3.3.27}$$

(here $\langle J_z \rangle = J_z$, since $\Delta J_z = 0$) implies, in view of the elementary inequality: $a^2 + b^2 = (a - b)^2 + 2ab \geqslant 2ab$,

$$(\Delta J_x)^2 + (\Delta J_y)^2 \gtrsim \eta \hbar |J_z|, \tag{3.3.28}$$

where we have, for once, explicitly included an unknown numerical coefficient η, bearing in mind here that all our inequalities have to be understood in the sense of orders of magnitude. From this we derive

$$J^2 \geqslant J_z^2 + \eta\hbar|J_z|, \tag{3.3.29}$$

where we see that the eigenvalue of J is necessarily strictly greater than the maximal proper value of the component J_z. The presence of the constant $\hbar$ in the additional term underlines its specifically quantum nature. If we denote by j the quantum number associated with the maximal eigenvalue of J_z, so that $(J_z)_{\max} = j\hbar$, we obtain the estimate

$$J^2 = j(j + \eta)\hbar^2. \tag{3.3.30}$$

While this approach does not allow us to fix the numerical constant η, it still confirms the considerations of the preceding chapter, which had led us to the same form [see eq. (2.3.30)] and had fixed the value $\eta = 1$. Furthermore, it sheds additional light on this quantum correction.

Exercises

3.1.   (a) Experiment shows that the velocity V_s of sound in air is independent of the frequency. What can one conclude from this, regarding the *dispersion relation* $\omega = f(k)$, which characterizes the propagation of sound? Compare the group and the phase velocities.

(b) The oscillations which enable waves to propagate on the surface of a liquid, arise due to the gravitational pull (at least for wavelengths larger than a few centimetres: below this, surface tension intervenes). The dispersion relation $\omega = f(k)$ can only involve, therefore, the acceleration due to gravity g and the density ρ of the liquid. Use dimensional analysis to establish such a relation (up to a numerical factor). Explain physically, why ρ does not enter, finally. Hence deduce the simple relation, which holds in this case, between the group velocity and the phase velocity, namely, $v_{\mathrm{ph}} = 2v_{\mathrm{gr}}$. Verify this relation qualitatively by the observation of the wave packet, produced on the surface of a pond, by the impact of a pebble.

(c) For a classical particle, of mass m, the energy and the momentum are connected by the relation $E = p^2/2m$. From this, deduce the dispersion relation $\omega = f(k)$ which connects the pulsation and the undulation of a quanton. Compute the phase and the group velocities as functions of the undulation k. Compare these with the velocity of the classical particle, expressed in terms of the momentum p. Hence draw the conclusion that on the kinematical level, the limiting descriptions of a quanton as a classical particle or a quantum wave are conceptually compatible.

3.2. Show that the dispersion Δp_n of the momentum in a direction specified by the unit vector n, $p_n = p \cdot n$ and the dispersion $\Delta r_{n'}$, of the component of the position in the direction n', $r_{n'} = r \cdot n'$ obey a *Heisenberg inequality*

$$\Delta p_n \Delta r_{n'} \gtrsim \hbar |n \cdot n'|.$$

3.3. Among the following pairs of physical magnitudes for a quanton, which ones correspond to *compatible properties*? x and p_y; x and J_x; p_z and J_y; x and $r = (x^2 + y^2 + z^2)^{1/2}$; J_x and $p = (p_x^2 + p_y^2 + p_z^2)^{1/2}$; r and p; r and J^2; p_z and J^2.

3.4. According to present day cosmology, the universe as a whole is not in a stationary state, but evolves with a characteristic time scale of the order of ten billion years. Deduce from this the existence of an intrinsic dispersion $\Delta_U m$ for the *mass* of *every* physical object. Calculate $\Delta_U m$ in kg as well as in eV.

3.5. Classical electromagnetic theory is only valid if photons have zero mass, implying thereby, not only the invariance of the velocity of light, but also the infinite range of the Coulomb force (varying as $1/r$). Conversely, any verification of the validity of this theory on a scale of distance D sets a lower limit on the range a_γ of the *electromagnetic forces*, transmitted by the photon, and an upper limit on the mass of the photon. Calculate, in kg and in eV, this upper limit, if the electromagnetic theory is verified.
  (a) On the scale of the laboratory (19th century verifications of Coulomb's law);
  (b) on the scale of the earth, of radius $R_E = 6.4 \times 10^3$ km, then of Jupiter, of radius $R_J = 7 \times 10^4$ km (studies of the magnetic fields of these planets, by satellites and space probes, carried out in the past years).

3.6. In quantum physics, the emission of *electromagnetic radiation* by a system – an atom emitting light or a nucleus emitting γ-rays – is described as follows: the system, in an excited state of energy E_1, passes into a state of lower energy – consider, e.g., the ground state – having energy E_0, by emitting a photon which carries away the surplus energy.
  (a) Suppose that the system, of mass M, is initially at rest. Show that the conservation of momentum imparts to it a certain recoil as a consequence of which it carries away a fraction δE of the energy difference $\varepsilon = E_1 - E_0$, leaving the photon with the energy $\varepsilon - \delta E$. Show that, under usual conditions where $\varepsilon \ll Mc^2$, we have $\delta E \simeq \varepsilon^2/2Mc^2$.
  (b) The excited level of energy E_1, possesses a life time τ. Hence deduce that it shows a certain width ΔE_1. What can be said, in this connection, about the ground state? Under what conditions may a photon, emitted as in (a) be re-absorbed by another system of the same type, assumed to be at rest in its ground state?
  (c) Apply these results to the following two examples:
– visible line of light from atomic mercury

$$\varepsilon = 4.86 \text{ eV}, \tau = 10^{-8} \text{ s}, M = 3.4 \times 10^{-25} \text{ kg}$$

– γ-emission from the nucleus of nickel:

$$\varepsilon = 1.33 \text{ MeV}, \tau = 10^{-14} \text{ s}, M = 1.0 \times 10^{-25} \text{ kg}.$$

3.7. In a typical *electron microscope*, the electrons have an average energy $\tilde{E} \simeq 100$ keV, with a dispersion $\Delta E \simeq 1$ eV. Calculate,

(a) the average velocity $\tilde{v}$ of the electrons;

(b) their average waveleigth $\tilde{\lambda}$;

(c) their dispersion in momentum Δp;

(d) their dispersion in position Δx, and compare it to $\tilde{\lambda}$.

3.8. A parallel, monochromatic beam of neutrons falls upon a screen, pierced by a slit of width a. Let $p_x = P$ be the initial momentum of the neutrons, parallel to the axis Ox.

(a) Before crossing the screen, is the neutron in a proper state, of the momentum p_y in the direction Oy of the position coordinate y?

(b) What happens to it after crossing the slit? Evaluate the dispersion Δy in position and, using the spatial Heisenberg inequality, the dispersion Δp_y in momentum.

(c) Hence deduce that the beam now has an *angular dispersion* $\Delta \varphi$ and evaluate it. Express it in terms of the initial wavelength λ of the neutron. Compare the result to the classical wave theory of diffraction by a slit (exercise 2.3).

(d) Recover the dispersion $\Delta \varphi$, using the angular Heisenberg inequalities, after evaluating the dispersion ΔL_z of the angular momentum of a neutron, after it has crossed the slit.

(e) The curve below represents the results of an experimental measurement of the angular distribution of a beam of neutrons, having wavelength $\lambda = 4.43$ Å, after crossing a slit of width $a = 5.6$ μm. Are these results compatible (in order of magnitude) with the foregoing theoretical evaluation?

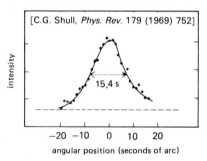

angular position (seconds of arc)

3.9. A beam of light, emitted by a *laser*, is made up of photons having almost identical (in magnitude and direction) wave vectors $\boldsymbol{k}$. This property distinguishes laser radiation from the radiation emitted by an ordinary source, for which the direction of $\boldsymbol{k}$ is random. A laser beam is made to reflect back and forth between two parallel mirrors, forming the two bases of a cylindrical cavity (one of the mirrors is not totally reflecting, which enables the beam to come out of the laser!). The axis Oz of the cylinder is adjusted to lie along the average direction of the vector $\boldsymbol{k}$. Let α be the (small) angle of a wave vector $\boldsymbol{k}$ with the axis Oz, and r the radius of the bases of the cylindrical cavity; r is much smaller than the length of the cylinder.

(a) Show that the components of $\boldsymbol{k}$, perpendicular to Oz are zero, on the average and that so also is the angle α. By applying the Heisenberg inequalities, show that the deviations of α from its average value (zero) are of the order $\delta \alpha \simeq \lambda/r$.

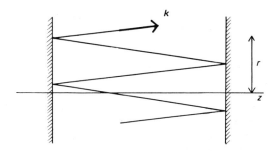

(b) As a result of these deviations, the beam emerging from the cavity has an angular divergence of the order of $\delta\alpha$. Calculate $\delta\alpha$ for $r = 10$ cm and $\lambda = 0.5$ μm.

(c) This laser beam is made to reflect off the moon (of what use could such an experiment be?). What is the width of the luminous spot on the surface of the moon? Comments?

(d) We would like to compare the laser radiation emerging from this cavity with radiation from an ordinary source. In order to obtain a parallel beam of light, with the help of an *ordinary source*, one generally places this source at the focus of a lens having a focal length f. Show that, the smaller the dimension ρ of the source, and the larger the focal length f, the smaller is the angular divergence $\delta\alpha$, by establishing a relation connecting $\delta\alpha$, f and ρ.

(e) What must be the dimensions of an ordinary source, in order to obtain a beam, of angular divergence as small as in (c), with the help of a lens of focal length $f = 10$ cm? Is this technically feasible?

3.10. Use the spatial Heisenberg inequality to evaluate the ground state energy of a *harmonic oscillator* of pulsation ω_0, using the classical expression for its energy

$$E = \frac{p^2}{2m} + \tfrac{1}{2}m\omega_0^2 x^2.$$

3.11. Use the spatial Heisenberg inequality to evaluate the ground state energy of an *atom* having atomic number Z. To do this, first show that the potential energy of the mutual repulsion between the electrons is of the same order of magnitude as the potential energy of the attraction of the electrons by the nucleus. Hence obtain the total potential energy in the form $-Z^2 e^2/\tilde{r}$, where $\tilde{r}$ characterizes the dimensions of the electronic cloud. Now add to this the kinetic energy, suitably evaluated. How do the ground state energy E_0 and the size of the atom r_0 vary with Z? Compare these results with the experimental data, according to which the size of the atoms remains of the order of 1 Å, whatever might be their atomic numbers. We have to wait until ch. 7, to understand this serious discrepancy.

3.12. The *deuteron*, the nucleus of deuterium, is the simplest of the composite nuclei. It is a bound state of a proton and a neutron. It is assumed that the attractive interaction between the proton and the neutron, as a function of their mutual distance r, is sufficiently well represented by the 'Yukawa potential'

$$V_Y(r) = -(g^2/r)\, e^{-r/a_N},$$

where g is the coupling constant of the hadronic interaction (see sect. 1.4.3) and a_N the range of the nuclear forces, which is related to the mass of the pion by $a_N = \hbar/m_\pi c$ (sect. 3.3.1.2).

(a) Calculate a_N numerically, using the value $m_\pi c^2 = 135$ MeV. Compare the Yukawa potential to the attractive Coulomb potential $V_c(r) = -e^2/r$ and comment. We consider the internal energy of the deuteron, which is the sum of the relative kinetic energy of the proton and the neutron and their potential energy,

$$E = \frac{p^2}{2M} + V_Y(r),$$

where p is the relative momentum of the proton and the neutron, and M their reduced mass,

$$M = \frac{m_p m_n}{m_p + m_n} \simeq \tfrac{1}{2} m_n,$$

introducing the average nucleonic mass $m_N \simeq m_p \simeq m_n$.

(b) By means of an argument, similar to that used for the hydrogen atom (sect. 3.3.2.2), minimize E, taking into account the Heisenberg inequalities relating the characteristic values $\tilde{r}$ and $\tilde{p}$. Hence obtain an estimate of the binding energy of the deuteron as a function of its 'size' r_d (optimal value of $\tilde{r}$), as

$$E_d = -\frac{\hbar^2}{2M r_d^2} \frac{a_N - r_d}{a_N + r_d}.$$

From the experimental value, $E_d = -2.2$ MeV, and the known values of M and a_N, deduce the value
– of the radius r_d of the deuteron (compare it to the range a_N),
– of the average attractive potential energy, $V_Y(r_d)$ (compare it to the binding energy E_d),
– of the coupling constant $g^2/\hbar c$ (compare it to $e^2/\hbar c$).

(c) It is implicitly assumed that the deuteron has zero orbital angular momentum. Can it have bound states of orbital angular momentum $l > 0$?

3.13. Consider a quanton of mass m in an attractive potential $V(r) = -K/r^2$. In classical mechanics, the energy of such a potential is not bounded below, and its fall to the centre cannot be avoided. Use the spatial Heisenberg inequalities to evaluate the ground-state energy of the system in quantics. Show that there exists a critical value K_c of the 'coupling constant' K which measures the strength of the potential, such that for $K < K_c$, there do not exist any bound states at all (i.e., of negative energy) of a quanton inside this potential well, while for $K > K_c$, even quantics cannot prevent the fall to the centre. Such a potential, up to an inessential angular dependence, describes the interaction of an electron with a polar molecule (such as CO, NH_3, H_2O, HF, etc.) when the electron is sufficiently far, so that it only feels the dipole component of the field due to the molecule, and not the details of its structure. One has, in this case, $K \simeq q_e D/4\pi\varepsilon_0$, where q_e is the charge of the electron and D the electric dipole moment of the molecule. Estimate the critical value of the dipole moment and compare it with the values of the moments of different molecules.

Was the result anticipated (see exercise 1.10)? We deduce from this, that in the interaction of electrons with polar molecules, the phenomenon differs greatly, depending upon whether the dipole moment of the molecule is smaller or bigger than the critical value.

3.14. We consider a bound system, formed out of different quantons in interaction. This means that the different constituents remain confined inside a common volume. Let R be the order of magnitude of the linear dimensions of this volume. For example, one knows experimentally, that for a light nucleus the value of R is a few fermis.

(a) Show that the size of the system imposes, via the spatial Heisenberg inequality, a lower limit, of the order of magnitude of the momentum p, on each constituent.

(b) Hence obtain a lower limit for the order of magnitude of the kinetic energy of a constituent of mass m. Distinguish between the cases where the kinetic energy can be calculated using the Galilean theory and those where the Einsteinian theory is necessary (see sect. 2.3.3).

(c) Hence deduce – knowing that it is a bound system – a lower limit for the order of magnitude of the absolute value of the potential energy of the constituent, considered as being confined to the interior of the system.

(d) Evaluate this minimal potential energy inside a light nucleus of radius $R = 2$ F, for an electron ($m_e = 0.9 \times 10^{-30}$ kg) and a nucleon ($m_N = 1.6 \times 10^{-27}$ kg). Compare it with the typical nuclear energies of a few MeV. Can electrons be permanent constituents of nuclei?

3.15. In a note, appearing in the Comptes-Rendus de l'Académie des Sciences, two physicists, one of them well-known, were still questioning the *Heisenberg inequalities* in 1971, by proposing 'an experimental scheme enabling one to simultaneously measure the position and momentum of a particle with a precision greater than that allowed by the Heisenberg uncertainty relations'. Read through the essential passage in this note, given below, and discuss the flaw in the reasoning that it contains.

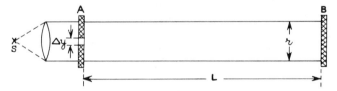

'Let us consider a laser*, having a sufficiently extended length L and finite transverse dimensions r. Suppose that it is sufficiently isolated from external influences so that the only way in which it can be excited is through the arrival of a photon across an aperture of width Δy, cut into one of its extremities. Suppose that the photon is emitted by a source S which we can control.

It is known that the laser is excited only when the photon, which crosses the aperture cut into the mirror A, falls onto the mirror B, i.e., only if it emerges from the aperture Δy at an angle smaller than $\omega = \arctan r/L$. Otherwise, the radiation induced by this photon would

---

* One should not be carried away: the laser, as such, serves no special purpose here (other than to make things look 'modern'). It is enough to assume that one has a diaphragm (of width Δy) and a detector (of diameter r, at a distance L from the diaphragm).

fall upon the inner walls (of the laser)* and be absorbed. Moreover, for a photon crossing the aperture Δy, and impinging at an angle ω, one evidently has, $\Delta p_y \simeq pr/L$, where p is the momentum of the photon, originating in the source S and crossing the aperture Δy.

Suppose that the source S, dispatches through the aperture, a luminous impulse at a given instant t. If, as a result of it, the laser is now excited at an instant $t_0 > t$, this means that at least one photon, having momentum $|p_y| \simeq \Delta p_y$, has passed through the aperture Δy at the instant t. But then, one has,

$$\Delta p_y \, \Delta y \simeq p \frac{r}{L} \, \Delta y.$$

One can now make the length L of the laser, which is independent of the aperture Δy, as large as one wishes, indeed, to even violate the Heisenberg inequality. Hence, from the correlation between the luminous impulse emitted by the source S and the excitation of the laser, it is possible to establish that the photons have crossed, at a fixed instant, an aperture Δy which is as small as one pleases, with an uncertainty Δp_y in the momentum which is also as small as one pleases. In other words, they simultaneously possess a position and a momentum. The fact that the experiment proposed here deals with photons and not with electrons, such as are usually considered in such gedanken experiments, is not of any fundamental importance.' [C.R. Acad. Sci. Paris, 272 (1971) B 1161].

3.16. In classical mechanics, the kinetic energy of a *rotor*, i.e., a rigid body in rotation about itself, is given by the formula $E = L^2/2I$, where I is the moment of inertia of the rotor, and L its angular momentum (this formula is analogous to that which gives the kinetic energy as a function of the momentum, or the 'moment of translation', namely, $E = p^2/2m$). This formula is valid in the quantum theory, on condition that one considers here the orbital angular momentum as a quantum magnitude. One applies this, in particular, to the rotational energies of molecules. We consider here the example of the HCl molecule.

(a) Evaluate the moment of inertia of the HCl molecule, knowing that the distance between the atomic nuclei has the value $d = 1.3$ Å, and that the masses of the atoms are $m_H = 1.6 \times 10^{-27}$ kg and $m_{HCl} = 35 \, m_H$.

(b) Express the proper values of the energy, denoted E_l, in terms of the proper values of the magnitude L^2 and evaluate them numerically for the first few levels ($l = 0, 1, 2, 3$).

(c) The HCl molecule, upon being excited to a higher energy level, de-excites itself by emitting a photon. This latter, having spin 1, can only carry away one unit of angular momentum. Thus, the molecule passes from a level E_l to a level E_{l-1}. Calculate the wavelengths and the frequencies of the lines emitted by this process, and evaluate them numerically for $l = 0, 1, 2, 3$.

(d) Show that a molecule, in a proper state of energy, cannot have a fixed orientation. Evaluate the minimal width of its angular momentum spectrum and of its energy spectrum, if one wishes to define the orientation of its axis with a precision of $\Delta\varphi = 1°$. Show that for 'large quantum numbers' (large values of l), a molecule may approximately be represented as a classical rotor.

(e) In classical mechanics, the electromagnetic radiation emitted by a rotor has a frequency v_{cl} equal to the rotational frequency v_{rot}. Calculate v_{rot} in terms of L and I. Compare v_{cl} to the frequency v_{qu}, calculated quantically in (c) and expressed also in terms of L and I. Recover the conclusion in (d).

---

* Parentheses added by translator.

3.17. A classical electromagnetic wave possesses an energy density

$$\frac{dE}{dV} = \tfrac{1}{2}\varepsilon_0(\mathscr{E} + c^2\mathscr{B}^2)$$

and a momentum (volume) density

$$\frac{dp}{dV} = \varepsilon_0 \mathscr{E} \times \mathscr{B}$$

where $\mathscr{E}$ and $\mathscr{B}$ are the electric and magnetic fields.

Deduce from the Heisenberg inequalities that in the quantum theory, the average values of the electric and magnetic fields inside a volume of linear dimensions δl and over a period of time δt, cannot be defined any better than $\Delta\mathscr{E}$ and $\Delta\mathscr{B}$, where

$$(\Delta\mathscr{E})^2 \gtrsim \frac{1}{\varepsilon_0}\frac{\hbar}{\delta t\,(\delta l)^3},$$

$$(\Delta\mathscr{B})^2 \gtrsim \frac{1}{\varepsilon_0 c^2}\frac{\hbar}{\delta t\,(\delta l)^3},$$

$$\Delta\mathscr{E}\,\Delta\mathscr{B} \gtrsim \frac{1}{\varepsilon_0}\frac{\hbar}{(\delta l)^4}.$$

$\Delta\mathscr{E}$ and $\Delta\mathscr{B}$ are often thought of as giving the *electromagnetic fluctuations* of the vacuum.

3.18. An electron of mass m and electric charge q_e possesses by virtue of its spin (see exercise 1.18) a *magnetic moment* $\mathscr{M}$. We wish to measure $\mathscr{M}$ by measuring the magnetic fields $\mathscr{B}_{mom}$ at a distance r from the electron.

(a) Treating the electron as a classical particle, write down the intensity $\mathscr{B}_{mom}$ at a distance r from the electron. If the electron is displaced, the motion of its charge gives rise to a magnetic field $\mathscr{B}_{ch}$. Estimate its intensity, at the same distance r, as a function of the velocity v of the electron.

(b) Being a quanton, the electron possesses a velocity spectrum of width Δv, from which there results a dispersion $\Delta\mathscr{B}_{ch}$ of the magnetic field, due to its displacement. A measurement of $\mathscr{B}_{mom}$ is only possible if $\mathscr{B}_{mom} \gg \Delta\mathscr{B}_{ch}$. Hence find an upper limit for Δv, as a function of r (for given $\mathscr{M}$ and q_e).

(c) In the same way, the electron also possesses a position spectrum of width Δr. A measurement of $\mathscr{B}_{mom}$ yields a value for $\mathscr{M}$ only if the value of r is sufficiently well-defined, i.e., if $\Delta r \ll r$. Deduce, from this and the condition established in (b), the inequality $\Delta v\,\Delta r \ll \mathscr{M}/q_e$ in order for the measurement of $\mathscr{M}$ to be possible.

(d) Is it possible to satisfy this inequality for an electron, taking into account the order of magnitude $\mathscr{M} = \hbar q_e/m$ of its magnetic moment (established in exercise 1.18)? Explain why the magnetic moment of the atom is used here to obtain that of the electron (see, again, exercise 1.18). Is it possible to measure the magnetic moment of an (isolated) neutron?

(e) In practice, instead of measuring the magnetic field created by the quanton, its motion in an external magnetic field $\mathscr{B}$ is studied. Show that the quanton is then subject to two forces, having orders of magnitude $F_{ch} \simeq qv\mathscr{B}$ and $F_{mom} \simeq \mathscr{M} \, \nabla\mathscr{B}$, respectively where $\nabla\mathscr{B}$ measures the spatial rate of variation of the (inhomogeneous) external magnetic field. Show that F_{ch} is not defined any better than $\delta F_{ch} \simeq q \, \Delta v \, \Delta r \, \nabla\mathscr{B}$, and recover the condition established in (c) for a measurement of $\mathscr{M}$ to be possible.

4

Probabilities and quantum amplitudes

In classical theory, the different physical magnitudes of a system are described mathematically by numbers, or more precisely, by functions of time, which take on numerical values. A *state* of a system is then defined by the data consisting of the set of numerical values assumed by these magnitudes. Thus, for a system consisting of material points, the data of the positions and velocities of the different particles determine the state of the system. Conversely, to each state is associated a unique and well-defined set of numbers: the values assumed by the different physical magnitudes, or more exactly, by the functions which represent them.

On the contrary, the preceding chapter has shown that, as a general rule, it is impossible to characterize a *quantum* system by well-defined values of all its physical magnitudes, since some of them could very well be incompatible.

However, we have also shown that some states of a system are privileged, in being characterized by a unique and well-defined value of one of the physical magnitudes of the system. We have called these states of the system the 'proper states' relative to this magnitude. Since the nature of this special property should be reflected in the formalism of quantum physics, we begin by examining the special role played by these states, amidst the ensemble of states of the system.

The description of a system of quantons, which we now attempt to elaborate, requires a new formalism, appealing to mathematical concepts which are radically different from those used in classical physics, in order to surpass the antinomy between the notions of the wave and the particle. It is for this reason that in the present chapter, the objective of which is to outline – always on a heuristic level – certain fundamental aspects of these new concepts, we shall pay special attention to luminous phenomena, for which we possess both a classical wave theory (Maxwellian electromagnetism) and a rough sketch of a quantum description of the corpuscular type (photons).

4.1. The polarization of light

4.1.1. The classical wave theory

Consider, therefore, a system of photons – a light beam. In the classical theory, it is described by an electromagnetic wave, having an electric component $\mathscr{E}$ and a magnetic component $\mathscr{B}$. For simplicity, let us suppose that the wave is monochromatic and plane, with a well-defined pulsation ω and wave vector k. Let us briefly recall a few classical results, concerning the structure of such a wave (fig. 4.1), which we shall need later in this chapter:

– $\mathscr{E}(r, t)$ and $\mathscr{B}(r, t)$ are sinusoidal functions of the single variable $u = \omega t - k \cdot r$;

– the field is transverse; at each point and at each instant, $\mathscr{E}$ and $\mathscr{B}$ lie on the (wave) plane perpendicular to k:

$$\mathscr{E} \cdot k = \mathscr{B} \cdot k = 0; \tag{4.1.1}$$

– at each point and at each instant of time, $\mathscr{E}$ and $\mathscr{B}$ are related, in an appropriate system of units, by

$$\mathscr{E} = \mathscr{B} \times \hat{k}, \tag{4.1.2}$$

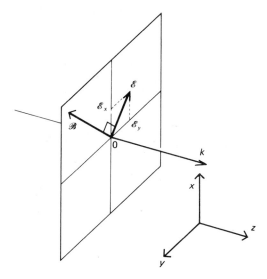

Fig. 4.1. *Structure of a plane wave*
The plane of the wave xOy is perpendicular to the direction of propagation k.

denoting by $\hat{\boldsymbol{k}}$ the unit vector $\hat{\boldsymbol{k}} = \boldsymbol{k}/k$ $(k = |\boldsymbol{k}|)$; $\mathscr{E}$ and $\mathscr{B}$ are orthogonal; – for the description of the system to be complete, it is necessary, in addition, to specify the direction of one of the two vectors in the couple $(\mathscr{E}, \mathscr{B})$ on each wave plane – the direction of the other would then immediately follow from relation (4.1.2). Since it is usual to identify the luminous 'vibration' with the electric vector $\mathscr{E}$, it is the evolution of the direction of $\mathscr{E}$ in the course of time, which completes the characterization of the beam being considered. This physical magnitude – the direction of $\mathscr{E}$ – which is often called, in a rather vague manner, the *polarization* of the wave, is the magnitude that interests us here.

Let us choose a coordinate axis Oz, in the direction of propagation $\boldsymbol{k}$ and let Ox and Oy be two axes forming a rectangular triad together with Oz. In each 'wave plane' ($z = $ constant), the components $\mathscr{E}_x$ and $\mathscr{E}_y$ of $\mathscr{E}$ depend only on the time. These are the harmonic functions (having the same pulsation ω) of the variable t,

$$\mathscr{E}_x(t) = \mathscr{E}_{0x} \sin(\omega t - \varphi_x), \qquad \mathscr{E}_y(t) = \mathscr{E}_{0y} \sin(\omega t - \varphi_y), \qquad \mathscr{E}_z(t) = 0.$$

$$(4.1.3)$$

$\mathscr{E}_{0x}, \mathscr{E}_{0y}, \varphi_x, \varphi_y$ are real numbers, independent of the time. The extremity of the vector $\mathscr{E}$ describes, therefore, an ellipse in the plane xOy (fig. 4.2), the form (namely the eccentricity) and the orientation (e.g., the angle between the

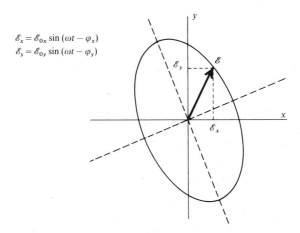

Fig. 4.2. *Elliptic polarization*
In the most general case, the extremity of the vector $\mathscr{E}$ of a plane monochromatic wave describes an ellipse: the wave is said to be in a state of elliptic polarization.

major axis and the axis Ox) of which completely define the polarization of the luminous beam. The characteristics of the elliptical trajectory depend only upon the ratio of the amplitudes, $\mathscr{E}_{0y}/\mathscr{E}_{0x} = r$, and the phase difference of the phases, $\varphi_y - \varphi_x = \delta$ (see exercise 4.1). Two waves, the electric field vectors of which describe ellipses of the same form and orientation (homothetic) differ only in the total intensity (proportional to $\mathscr{E}^2$) and by a constant temporal phase shift. We say that the *polarization state* of the light beam is entirely characterized by the magnitudes r and δ. If δ is a multiple of π, the ellipse reduces to a line segment, for any value of r, and the vector $\mathscr{E}$ maintains a fixed direction: we have a linearly polarized wave. If $\delta = \frac{1}{2}(2p + 1)\pi$ and $r = 1$, the tip of $\mathscr{E}$ describes a circle: we have a circularly polarized wave – right or left depending on whether p is even or odd. These two are particular cases; the most general state is one of elliptic polarization (fig. 4.3).

As shown by relation (4.1.3), the most general state, that of elliptic polarization, can be considered as being a combination – i.e., a linear superposition, since we are dealing with waves – of two states of linear polarization, along the two orthogonal directions Ox and Oy. Indeed, one can write

$$\mathscr{E} = \mathscr{E}' + \mathscr{E}'', \quad \text{with } \mathscr{E}' = \left\{ \begin{array}{c} \mathscr{E}_x \\ 0 \\ 0 \end{array} \right\}, \quad \mathscr{E}'' = \left\{ \begin{array}{c} 0 \\ \mathscr{E}_y \\ 0 \end{array} \right\}. \tag{4.1.4}$$

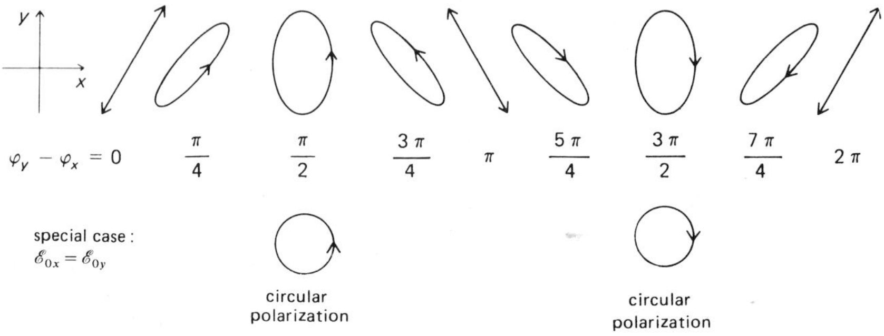

Fig. 4.3. *States of polarization*
Different states of polarization of a plane electromagnetic wave, depending upon the phase difference between the components of $\mathscr{E}$. If $\varphi_y - \varphi_x = 0$ or π, the wave is linearly polarized. If $\varphi_y - \varphi_x = \frac{1}{2}(2p + 1)\pi$ and, in addition $|\mathscr{E}_{0x}| = |\mathscr{E}_{0y}|$, it is circularly polarized – 'left' or 'right' depending upon the parity of p.

These statements, holding for the pair of axes (O*x*, O*y*), also hold for an arbitrary system of axes orthogonal to the plane of the wave. The projections of $\mathcal{E}$ on these axes yield two components which can be considered to be the electric field vectors of the two linearly polarized waves, the superposition of which gives the initial elliptically polarized wave.

Let us note that a 'natural' source of light, such as the sun or an incandescent lamp, does not send out a polarized beam (not even a monochromatic beam), because it consists of a multitude of independent, individual sources – atoms or molecules. These sources emit incoherent wave trains, without any phase relations, in a manner such that the resulting beam displays all possible polarizations. One can 'polarize' this beam by means of a filter, appropriately called a *polarizer*, which only allows a fraction of the beam, having the desired polarization, to pass through. Similarly, one can prepare a linearly polarized beam by means of a polaroid, which is a slab of a special plastic material – also used to make the 'glasses' of certain sun-glasses – and which only transmits light vibrations parallel to a given direction (fig. 4.4).

In the remainder of this chapter, we shall always assume that a polaroid polarizer, or some other similar filter, has been placed in front of the beam of light that is being used. The beam being considered is thus initially in a state

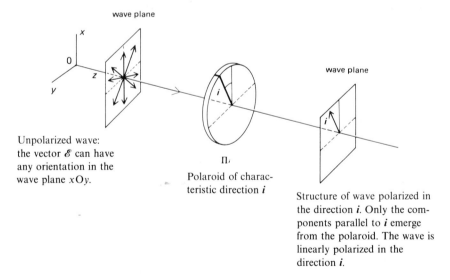

wave plane

Unpolarized wave:
the vector $\mathcal{E}$ can have
any orientation in the
wave plane *x*O*y*.

Π,
Polaroid of charac-
teristic direction *i*

wave plane

Structure of wave polarized in
the direction *i*. Only the components parallel to *i* emerge
from the polaroid. The wave is
linearly polarized in the
direction *i*.

Fig. 4.4. *Polarization of a wave*
One can polarize an arbitrary wave with the help of a polarizing filter (polaroid).

of *definite plane* polarization! The electric field of the electromagnetic wave is parallel to the direction of the unit vector *i* and, hence written as $\mathscr{E} = \mathscr{E}i$.

4.1.2. Analysis of a state of linear polarization

For any pair of orthonormal vectors *a*, *b* in the wave plane $x\mathrm{O}y$, we can, with the help of two polarizers Π_a and Π_b, of characteristic directions *a* and *b*, always extract from a wave, having an electric field vector $\mathscr{E}i$, two waves polarized along the two orthogonal directions *a* and *b*. To do this, it is necessary to proceed in two steps.

(1) In the path of the incident beam is interposed (fig. 4.5a) the polarizer Π_a. From the incident field $\mathscr{E}i$, Π_a only allows the component parallel to *a*, i.e., $\mathscr{E}(i \cdot a)a$ to pass through.

(2) Π_a is replaced by Π_b (fig. 4.5b). Only the component $\mathscr{E}(i \cdot b)b$ of the incident field is transmitted.

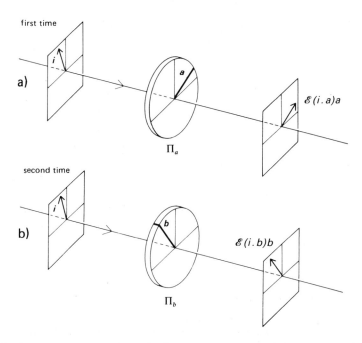

Fig. 4.5. *Decomposition into orthogonal polarizations*
Physically, decomposition (4.1.5) is realized by means of 2 polarizers, having orthogonal characteristic directions *a* and *b*, placed *successively* [a) and b)] in the path of the incident beam.

Since the vector $\mathscr{E}i$ is the sum of the two components $\mathscr{E}(i \cdot a)a$ and $\mathscr{E}(i \cdot b)b$,

$$\mathscr{E}i = \mathscr{E}(i \cdot a)a + \mathscr{E}(i \cdot b)b, \tag{4.1.5}$$

and since the portion of the incident field which is not transmitted by Π_a is precisely the part that is allowed to pass through by Π_b (and vice versa), we obtain in this way a 'complete' decomposition of the initial wave into two component waves, respectively polarized in the directions a and b.

The two polarizers Π_a and Π_b, taken together, do not enable us to analyze, in one step, a linearly polarized wave. To separate the two components it is necessary to proceed in two steps. However, certain devices, which use birefringent slabs, for example (fig. 4.6a), have the property of decomposing a

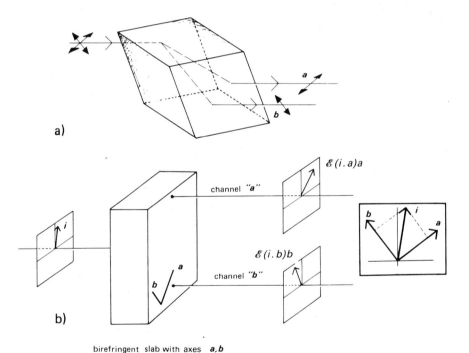

birefringent slab with axes **a, b**

Fig. 4.6. *Separation of the components of polarization*
The birefringent slab realizes the separation of the two orthogonal components of the polarization of a polarized, arbitrary initial state. The index of refraction is not the same for the beam polarized along a as that for the one polarized along b (a and b being the two characteristic directions of the crystal). The two states of polarization find themselves spatially separated upon emerging from the slab.

beam of linearly polarized light into two beams which are spatially separated and are polarized along two orthogonal directions – the principal directions of the slab (fig. 4.6b). Let a and b be these two directions. The birefringent slab, having principal axes (a, b), appears like an analyzer, which realizes in a single set-up, the two projections necessary for the complete analysis of a linearly polarized wave. Since the two resultant components of this analysis are channelled into two distinct 'exits' of the slab, the latter constitutes a 'multichannel' analyzer of linear polarization. In the following, we shall denote the channels by the directions of linear polarization of the beam, as channel 'a', channel 'b', and the corresponding analyzer by the symbol $A_{(a,b)}$. In short, this apparatus establishes a *correlation*, in the emerging beam, between the direction of polarization and the spatial orientation. A beam polarized in the direction a is now recognized as being the beam which only emerges through the exit 'a'. A beam polarized in the direction b emerges only through the exit 'b'. A beam of arbitrary initial polarization i gives rise to two distinct beams which, emerging from the exits a and b, have polarizations a and b, respectively, realizing the decomposition given by eq. (4.1.5).

4.2. Transition probabilities and physical magnitudes

4.2.1. Polarization and photons

We shall now try to describe the above analysis of a state of polarization in terms of *photons*, each carrying an energy $\hbar\omega$ (the pulsation ω is the same for all the photons; we limit ourselves to monochromatic beams).

Let N be the flux of photons in the incident wave (number of photons transported per unit time and by a unit cross-section of the beam). The intensity of the incident beam, i.e., the flux of energy, is

$$I = N\hbar\omega. \tag{4.2.1}$$

In the theory of electromagnetism, it is proved that this energy flux is proportional to the square of the modulus of the electric field $\mathscr{E}$ of the wave. Setting this coefficient of proportionality equal to 1, by a suitable choice of units, we write therefore,

$$I = |\mathscr{E}|^2 = N\hbar\omega. \tag{4.2.2}$$

Upon crossing a birefringent slab, having principal axes a and b, the incident beam finds itself spatially separated. The intensity of the beam which comes out of channel a is

$$I_a = |\mathscr{E}(i \cdot a)|^2 = I \cos^2 \theta, \tag{4.2.3}$$

where θ denotes the angle (i, a), and the intensity of the beam which emerges through channel b is

$$I_b = |\mathscr{E}(i \cdot b)|^2 = I \sin^2 \theta. \tag{4.2.4}$$

The sharing of the energy between the two channels of the analyzer can only be explained as a sharing of the *number* of photons of the incident beam. Indeed, the nature of the photons, in so far as they are 'one and indivisible' quantons, implies the preservation of their integrity, and hence the impossibility of the fragmentation of a single quanton of energy $\hbar\omega$. Two detectors, placed at the exits of the channels a and b, respectively, only register integral numbers of photons! Let N_a and N_b be the fluxes of photons registered, respectively, by these detectors. The energy of each photon being always $\hbar\omega$, we have relations analogous to eq. (4.2.1) between the fluxes and the intensities for each channel,

$$I_a = N_a \hbar\omega, \qquad I_b = N_b \hbar\omega. \tag{4.2.5}$$

Taking relations (4.2.2), (4.2.3) and (4.2.4) into account, this means

$$N_a = N \cos^2 \theta, \qquad N_b = N \sin^2 \theta, \tag{4.2.6}$$

where one verifies that

$$N_a + N_b = N. \tag{4.2.7}$$

Thus, the N incident photons have been sorted out by the apparatus (analyzer plus detectors), N_a having passed through channel a and N_b through channel b. It is difficult to understand just what criterion determines this selection, since all the photons which present themselves at the entrance to the analyzer have the same polarization, the same energy and are subject to the same physical conditions in their interactions with the birefringent crystal! One would expect that they would all be affected in the same way, and would all come out of the analyzer through the same channel, either a or b. A current heterodoxy in quantum physics has attempted to resolve this dilemma by supposing that the photons are not, in fact, identical, but that they are characterized by supplementary physical magnitudes, hidden from our view, the different values of which enable the birefringent slab to make a more natural selection. This hypothesis of 'hidden variables' has proved itself to be sterile and riddled with more difficulties than it has resolved.

The selection effected by the analyzer cannot be explained by a collective effect either, for example, through a mechanism according to which the fate of one photon would condition that of the next one. In fact, if the intensity of the beam is diminished to a point at which only isolated photons pass, one by one, it still is the same proportion, $\cos^2\theta$ (respectively $\sin^2\theta$), that passes through the channel *a* (respectively *b*), *on the average* over a given interval of time (exercise 4.2).

From the classical point of view, the situation described here is completely paradoxical, since the individual fate of each photon cannot be predicted. If, indeed, such a prediction were possible, it would apply to *all* the photons, since they are identical and subject to the same conditions. The initial beam would now generate, at the exit of the analyzer, just one beam, contrary to the facts. This breakdown of classical determinism has caused much ink to flow. For the moment, let us only remark that this paradox surfaces as soon as we try to determine the fate of a single photon, without taking into consideration the changes to which the apparatus itself is subjected during its interaction with the photons of the beam.

We need to introduce here an idea which will enable us to understand the global partition of the intensity of the beam and, at the same time, the fact that for each individual photon, there only exists the dichotomous alternative: be channelled either into the beam *a or* into the beam *b*. The conceptual difficulty arises as a result of the apparent conflict between the two classical representations: one wave-like, which entails a continuous splitting of the intensity and the other corpuscular, which, for each individual photon, implies two discrete possibilities only. It is the notion of *probability* which enables us to get a coherent quantum description for the phenomenon. Indeed, each photon only has a discrete alternative: either to pass through channel *a* or to pass through channel *b*. However, one assigns to each one of these possibilities a probability which can be continuous.

Let us show that one can effectively determine for the photon a probability $\mathscr{P}_a$ of passing through channel *a* and a probability $\mathscr{P}_b$ of passing through channel *b*, which enables one to recover the preceding results. True, these probabilities – attached as they are, to each photon – must describe the evolution of a single photon; but if we consider the passage of many photons, the 'law of large numbers' guarantees that the relative frequencies (in the sense of probability theory) of the photons, which cross the apparatus through channel *a* (respectively *b*) is given by the probability $\mathscr{P}_a$ (respectively $\mathscr{P}_b$), with a precision which gets better the larger the total number,

$$\mathscr{P}_a = \lim_{N \gg 1} N_a/N, \qquad \mathscr{P}_b = \lim_{N \gg 1} N_b/N. \tag{4.2.8}$$

But then, it is precisely as the number N becomes very large ($N \gg 1$), that the characteristic action of a system of N photons becomes very much larger than $\hbar$ and that the classical wave theory becomes valid (ch. 2, sect. 2.2.1). The connection between the quantum and classical descriptions is now made, very naturally, via the Planck–Einstein formula, with the help of eqs. (4.2.2), (4.2.5) and (4.2.6),

$$\lim_{N \gg 1} \frac{N_a}{N} = \lim_{N \gg 1} \frac{N_a}{N} \frac{\hbar\omega}{\hbar\omega} = \frac{I_a}{I} = \cos^2\theta. \tag{4.2.9}$$

Similarly,

$$\lim_{N \gg 1} \frac{N_b}{N} = \frac{I_b}{I} = \sin^2\theta. \tag{4.2.10}$$

Substituting these values into eq. (4.2.8) yields

$$\mathscr{P}_a = \cos^2\theta, \qquad \mathscr{P}_b = \sin^2\theta. \tag{4.2.11}$$

We verify that $\mathscr{P}_a + \mathscr{P}_b = 1$, as is the rule for the probabilities of incompatible events.

We must now make the nature of the probabilities, so introduced, more precise. Probabilities – yes! But probabilities of what? Let us recall that in the experiment being considered, the direction of polarization of each photon, under the effect of the interaction with the analyzer, gets projected onto one or the other of the orthogonal directions *a* and *b*. It is the result of each one of these projections, i.e., the occurrence of each one of the two possibilities, which is determined in a random manner by the probabilities introduced. One often gives to these probabilities the name 'transition probabilities'. By this it is meant that during their interaction with the apparatus, the photon passes from one state to another – making a 'transition' from a certain 'initial' state to a 'final' state. It is, however, essential to note that the word 'transition' has to be understood here in a specific sense: this transition does not a priori have any temporal connotation; it is not at all spontaneous and does not arise from the evolution of an isolated quantum system (the photon). On the contrary, it results from the interaction with another system (the apparatus) and this interaction is of a very particular type, since the first system is brought into a characteristic state of the second and projected, in a certain sense, onto this state. The word 'projection' is in this sense more appropriate than the word transition.

The probabilities introduced here relate to transitions between *two states of the system* and this particularity should appear in the notation adopted to denote them. We shall agree to denote by $(a \leftarrow i)$ the probability of a transition (or projection) from the polarization state i to (or onto) the state of linear polarization a. The sense adopted to denote the initial and final states – from right to left – may not seem too natural. This choice is related to the traditional notation of the mathematical formalism of quantum theory, as we shall see later.

Adopting this convention, eqs. (4.2.11) are written, more explicitly, as

$$\mathscr{P}(a \leftarrow i) = \cos^2(a, i) = \cos^2\theta, \qquad \mathscr{P}(b \leftarrow i) = \cos^2(b, i) = \sin^2\theta. \qquad (4.2.12)$$

4.2.2. The polarization as a physical magnitude

A photon from a beam of initial polarization i passing through an analyzer $A_{(a,b)}$, equipped with detectors, passes, as we have seen, along one or the other of two paths and upon detection, possesses either the polarization direction a or the direction b. These two polarization states of the photon are, therefore, the possible final states in the interaction of the photon with the apparatus, which is designed to study its polarization on the basis of the pair (a, b). We can also say that we *measure* the polarization of the photon in the (a, b) basis by making it pass through the analyzer $A_{(a,b)}$. In other words, the analyzer $A_{(a,b)}$, along with its detectors, is a measuring apparatus. At the end of this measurement, the initial state of the photon is projected either onto the state a or onto the state b. The measurement can give, therefore, two possible results, a or b, with respective probabilities $\mathscr{P}(a \leftarrow i)$ and $\mathscr{P}(b \leftarrow i)$, which depend upon the initial state of polarization, having direction i, as indicated by relations (4.2.12). Only the two cases, in which the *initial* polarization is either a or b, lead to definite unambiguous results. In conformity with the terminology already introduced, we say, therefore, that these states of the photon are its *proper states* of polarization in the (a, b) basis. Any other polarization state $(i \neq a, b)$ possesses a discrete *spectrum* of polarization directions in the (a, b) basis, since its measurement yields two possible directions (those of *a* and *b*) with the indicated probabilities given by eq. (4.1.12). These two orthogonal directions are the proper values of the physical magnitude: 'linear polarization in the (*a*, *b*) basis' – written briefly, the 'polarization/(*a*, *b*)'. One notes the importance of referring the quantum magnitude, called linear polarization, to a pair of orthogonal directions, namely, the basis (*a*, *b*). One cannot speak of the 'linear polarization magnitude' without mentioning with respect to which basis. In fact, two distinct bases (*a*, *b*) and (*a′*, *b′*) define two different physical magnitudes. Not only are their proper values (the directions of the basis vectors) different, but especially their proper states are quite distinct – a proper state of polarization in the direction *a* is not polarized along *a′* or *b′* and, hence, is not a proper state of polarization with respect to the second basis. Physically, this translates into the fact that the beam emerging along the path *a* of the analyzer $A_{(a,b)}$ is separated into two beams by the analyzer $A_{(a′,b′)}$. Thus, polarization with respect to two different bases is an example of two *incompatible* physical magnitudes (sect. 3.2.4).

Let us note, that for two directions of polarization *a* and *b* to define the proper values of a physical magnitude, called 'polarization', it is indispensable that they be orthogonal. Indeed, a state is a proper state of the magnitude, polarization, only if it has a unique and well-defined value of this magnitude, in other words, only if the probability of its transition to another proper state is zero: the analyzer associated to this magnitude should not be able to separate a beam, having the polarization being considered, into two others of different polarizations. Now, a state of polarization *a* has a non-zero probability for the projection onto every polarization state *c* for which $a \cdot c \neq 0$, eq. (4.2.12). Two states whose directions of polarization are not orthogonal cannot, therefore, be the proper states of one single physical magnitude.

In all the above, we have only considered states of linear polarization. The magnitude, 'polarization/(*a*, *b*)' was defined using the states of *linear* polarization along the directions *a* and *b*. Now, there exist devices, called circular polarizers (see exercise 4.3) which are analogous to polaroids, but which, instead of projecting the states of polarization of a photon onto linear polarization states, project them onto one of the two states – left or right – of *circular* polarization (see ch. 4, sect. 4.1.1, and fig. 4.3). Similarly, there exist devices which, like the birefringent slabs, with axes (*a*, *b*), spatially separate the beams, which traverse them, into two beams – one having right circular polarization and the other left circular polarization – and which, by analogy, we denote by $A_{(R,L)}$. Let us recall that the electric field, for these two polarizations, corresponds to the parameters $r = \mathscr{E}_{0y}/\mathscr{E}_{0x} = 1$ and $\delta = \varphi_y - \varphi_x = \pm \frac{1}{2}\pi$.

We can again write

$$R \begin{cases} \mathcal{E}_x = \mathcal{E}_0 \cos(\omega t - \varphi) \\ \mathcal{E}_y = \mathcal{E}_0 \sin(\omega t - \varphi) \end{cases}, \qquad L \begin{cases} \mathcal{E}_x = \mathcal{E}_0 \cos(\omega t - \varphi) \\ \mathcal{E}_y = -\mathcal{E}_0 \sin(\omega t - \varphi) \end{cases}. \qquad (4.2.13)$$

Any wave having a definite polarization given by eq. (4.1.3) can be written as a superposition of waves having right and left circular polarizations. In the case of a wave with linear polarization, along the axis Ox, the decomposition is particularly simple, since one has

$$\begin{cases} \mathcal{E}_x = \mathcal{E}_0 \cos(\omega t - \varphi) = \tfrac{1}{2}\mathcal{E}_0 \cos(\omega t - \varphi) + \tfrac{1}{2}\mathcal{E}_0 \cos(\omega t - \varphi) \\ \mathcal{E}_y = 0 \qquad\qquad\qquad = \underbrace{\tfrac{1}{2}\mathcal{E}_0 \sin(\omega t - \varphi)}_{R} \underbrace{- \tfrac{1}{2}\mathcal{E}_0 \sin(\omega t - \varphi)}_{L} \end{cases}. \qquad (4.2.14)$$

Hence, a linearly polarized wave is formed by the superposition of two waves of opposite circular polarizations, having the *same* amplitude. Finally, let us remark that circular polarization, unlike linear polarization, does not depend upon the geometrical orientation of the axes. Indeed, in a system $x'Oy'$, making an angle δ with the axes xOy, the electric field has its new components written as functions of the old components,

$$\mathcal{E}'_x = \mathcal{E}_x \cos \delta + \mathcal{E}_y \sin \delta, \qquad \mathcal{E}'_y = -\mathcal{E}_x \sin \delta + \mathcal{E}_y \cos \delta. \qquad (4.2.15)$$

Thus, in the new system of axes, the circularly polarized waves given by eq. (4.2.13) are written as

$$R \begin{cases} \mathcal{E}_0 \cos(\omega t - \varphi') \\ \mathcal{E}_0 \sin(\omega t - \varphi') \end{cases}, \qquad L \begin{cases} \mathcal{E}_0 \cos(\omega t - \varphi'') \\ -\mathcal{E}_0 \sin(\omega t - \varphi'') \end{cases}, \qquad (4.2.16)$$

with

$$\varphi' = \varphi + \delta, \qquad \varphi'' = \varphi - \delta. \qquad (4.2.17)$$

Only the arbitrary phases are changed, and not the nature of the polarization. Moreover, this generalizes eq. (4.2.14) to the case of a wave with arbitrary linear polarization [and not just along Ox, as was assumed in eq. (4.2.14)]: it is sufficient to choose a system of axes $x'Oy'$ such that Ox' is in the direction of the linear polarization – which does not affect the definition of the circular polarizations – to be able to write down a decomposition of the type given by eq. (4.2.14).

The interest of these considerations lies in the fact that they provide us with a new physical magnitude, associated to the polarization. Indeed, for *one* photon, the two states of circular polarization, right and left, can be interpreted as being the two proper states of a single magnitude. The transition probability, from one to the other, is indeed zero, since any beam of right circular polarization is completely stopped by a filter of left circular polarization, and these two states enable us to make a complete analysis of any polarization state (see exercise 4.3). If, as is conventional, we associate the numerical value $+1$ to the right circular polarization and -1 to the left circular polarization, we obtain two proper values which, attributed respectively to the two proper states of circular polarization, completely define a quantum physical magnitude, *the* 'circular polarization'. By virtue of the special properties that it possesses, with respect to spatial rotations, there is only one magnitude of 'circular polarization', independent of the choice of axes, unlike the case of the magnitudes of 'linear polarization', which are associated to the different bases. Moreover, it is this invariance of the circular polarization under rotations, which makes for all its interest and which underlies the physical importance of its proper states, as we soon shall see.

4.2.3. *States and quantum magnitudes*

The foregoing considerations on the physical magnitudes of 'polarization', their proper states and proper values are valid in general. To a quantum physical magnitude is associated a set of numerical values – its proper values – and a set of *proper states* – namely, those very special states in which the magnitude has a well-defined numerical value, the corresponding proper value. Conversely, a set of states can be considered to be the set of proper states, relative to a certain quantum magnitude, under two conditions:

(a) It is necessary that the transition probability between two arbitrary states be zero. Indeed, by definition, a proper state of a certain magnitude must have probability one (certainty) of exhibiting the corresponding proper value – i.e., of making a transition to, or of being projected onto, itself – in an analysis (or measurement) of the magnitude. Transitions to other proper states are, therefore, impossible, and their probabilities are all zero. We say that such a state is 'disjoint' from the others. To be considered as being proper states of a single physical magnitude, a set of states must, therefore, have the property of being mutually *disjoint*.

(b) It is necessary that these states should enable one to make an exhaustive analysis of an arbitrary state. Such a state is projected, during a measurement of the considered magnitude, onto one or another of the proper states, with a certain probability, in a manner such that the sum of all these projection probabilities equals unity. A set of states such that the sum of the transition probabilities, of an arbitrary state, onto them is unity, is said to be 'complete'. In order to be considered as the ensemble of proper states of a physical magnitude, a set of states must have the property of *completeness*.

Given a set of disjoint and complete states, a physical magnitude is defined by associating a numerical value to each one of these states, in other words, by the simultaneous data of its proper states and proper values. Two physical magnitudes are compatible if they have common proper states – their proper values could differ arbitrarily.

These ideas, somewhat abstract, are most important for the development of a consistent mathematical formalism for the quantum theory. Let us summarize them in symbolic terms, as shown in table 4.1.

Table 4.1

Physical magnitudes, proper states and proper values

– To a physical magnitude $\mathscr{A}$ corresponds a set of proper states $\{u_n\}$ ($n = 1, 2, ...$), and associated proper values $\{a_n\}$.

– The proper states have, in terms of their transition probabilities, the properties of

$$\text{disjointness:} \quad \mathscr{P}(u_n \leftarrow u_{n'}) = 0, \quad \forall n, n' \, (n \neq n'), \tag{4.2.18}$$

$$\text{completeness:} \quad \sum_n \mathscr{P}(u_n \leftarrow v) = 1, \quad \text{for any state } v. \tag{4.2.19}$$

– A measurement of the magnitude $\mathscr{A}$ on a system in an arbitrary initial state v yields one of the proper values a_n with the transition probability $\mathscr{P}(u_n \leftarrow v)$, and the state of the system is now projected onto the state u_n: the system is in the final state u_n.

4.3. Probability amplitudes

One might believe, at this stage, that the quantum theory has been reduced to a standard probabilistic theory, analogous to classical statistical mechanics. This is not at all so, and it will be necessary for us to develop a more fundamental quantum concept which underlies the probabilities already introduced.

Let us imagine (fig. 4.7a) that after having projected the polarization state i onto two states of linear polarizations a and b, with the help of an analyzer

(a) Experimental set-up

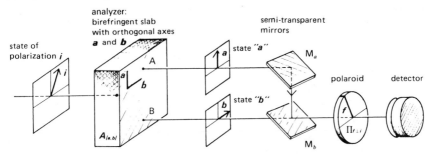

(b) Reasoning based upon the wave concept (electric field of the electromagnetic wave)

$$\mathscr{E}(i.a)(a.f)f + \mathscr{E}(i.b)(b.f)f = \mathscr{E}\cos\theta\sin\theta f - \mathscr{E}\sin\theta\cos\theta f = 0$$

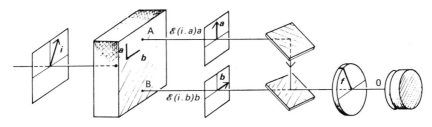

(c) Classical probabilistic reasoning (number of photons in the beam)

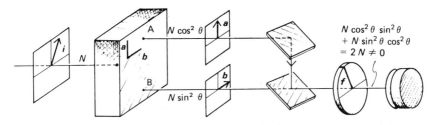

Fig. 4.7. *Reasoning based upon the wave concept and classical probability*
While a reasoning based upon the wave concept enables one to understand why the intensity received by the detector is zero, the classical probabilistic reasoning leads to a manifestly false conclusion.

(birefringent slab) having principal axes a and b, we interpose in the path of the emergent beam two semi-transparent mirrors. These latter are placed, making an angle of 45° to the incident beam and such that the two beams are made to overlap in the same region of space.

To begin with, let us adopt the *undulatory* point of view (fig. 4.7b). In the region where the two beams overlap, the electric field, according to the superposition principle, is

$$\mathscr{E} = \mathscr{E}(i \cdot a)a + \mathscr{E}(i \cdot b)b. \tag{4.3.1}$$

(We assume here that the reflection by the mirrors does not introduce any phase difference between the two waves which come out of the birefringent slab.) Thus, the initial state of polarization along i is reconstituted. If we now interpose, in the path of the reconstituted wave, a polaroid Π_f, having a characteristic direction f which is perpendicular to i, the intensity is zero at the exit to Π_f.

This result, while evident if one adopts the undulatory point of view, nevertheless causes trouble if one tries to adopt the probabilistic corpuscular point of view, developed in the preceding section.

Indeed, let us calculate the probability $\mathscr{P}(f \leftarrow i)$ that a photon will make the transition $f \leftarrow i$ (fig. 4.7c). For this, let us suppose that first the channel b is blocked, with the help of a masking tape, for example. The probability that *one* photon from the incident beam be projected onto the state a is, in accordance with eq. (4.2.12),

$$\mathscr{P}(a \leftarrow i) = \cos^2(a, i) = \cos^2 \theta. \tag{4.3.2}$$

The photons which arrive at the polaroid Π_f are, therefore, all in the state of polarization a and each one has a probability

$$\mathscr{P}(f \leftarrow a) = \cos^2(f, a) = \cos^2(\tfrac{1}{2}\pi - \theta) = \sin^2 \theta, \tag{4.3.3}$$

of being transmitted by Π_f. Thus, a photon from the initial beam has a total probability

$$\mathscr{P}_a(f \leftarrow i) = \mathscr{P}(f \leftarrow a)\,\mathscr{P}(a \leftarrow i) = \sin^2 \theta \cos^2 \theta, \tag{4.3.4}$$

of crossing the experimental set-up along the path a. An analogous reasoning shows that the probability for a photon to cross the apparatus along the channel b (the path a being blocked by a masking tape) is

$$\mathscr{P}_b(f \leftarrow i) = \mathscr{P}(f \leftarrow b)\, \mathscr{P}(b \leftarrow i) = \cos^2(f, b)\, \cos^2(b, i) = \cos^2\theta \sin^2\theta.$$

$$(4.3.5)$$

It seems natural, therefore, to say that if one were to leave both the channels a and b open, then each photon would pass *either* along path a *or* along path b. A proportion, $\mathscr{P}_a(f \leftarrow i) = \sin^2\theta \cos^2\theta$, of the incident photons would cross the apparatus along the path a, while a proportion, $\mathscr{P}_b(f \leftarrow i) = \cos^2\theta \sin^2\theta$ would cross along b. If N is the flux of photons (number of photons per unit time and unit cross-section) in the incident beam, related to the intensity I of the wave through $I = N\hbar\omega$, we expect that the flux of photons at the exit to the apparatus would be

$$N' = N\mathscr{P}_a(f \leftarrow i) + N\mathscr{P}_b(f \leftarrow i) = 2N \cos^2\theta \sin^2\theta.$$

$$(4.3.6)$$

Now, this result is manifestly erroneous since, as shown by the simple argument based upon the wave concept, and as confirmed by experiment, the final intensity is zero and hence *no photon* from the incident beam comes out of Π_f!

Let us insist here on the absurdity of the experimental result vis-à-vis the preceding corpuscular explanation: it is not quite clear which argument, probabilistic or otherwise, would enable one to understand why fewer photons (actually none!) traverse the experimental set-up when both the channels a and b are open, than when only one of them is open.

Let us consider again the explanation of the phenomenon based upon the classical wave concept. Since, at least, it correctly predicts a vanishing intensity, it might tell us how to arrive at a satisfactory quantum explanation. Let us explicitly calculate, as functions of θ, the amplitudes transmitted by the experimental device, when one of the paths, a or b, is closed and when both are open.

When the channel b is blocked, a wave of amplitude $\mathscr{E}(i \cdot a)a$ arrives at Π_f, which transmits the amplitude (see fig. 4.8)

$$\mathscr{E}(i \cdot a)(a \cdot f)f = \mathscr{E} \cos\theta \cos(\tfrac{1}{2}\pi - \theta)f = \mathscr{E} \cos\theta \sin\theta\, f, \qquad (4.3.7)$$

to which corresponds an intensity $I \cos^2\theta \sin^2\theta$, in conformity with what the corpuscular reasoning would anticipate, which, therefore, seems to be valid at least up to this point. Similarly, when the channel a is blocked, the final amplitude, after crossing the polaroid slab Π_f is (fig. 4.8)

$$\mathscr{E}(i \cdot b)(b \cdot f)f = \mathscr{E} \cos\left(\tfrac{1}{2}\pi + \theta\right) \cos\theta\, f = -\mathscr{E} \sin\theta \cos\theta\, f, \qquad (4.3.8)$$

and the corresponding intensity is $I \cos^2\theta \sin^2\theta$, as predicted by the classical corpuscular reasoning.

If now both the channels a and b are open, the intensity at the exit to Π_f is obtained by *first* calculating the final amplitude by the superposition of the amplitudes given by eqs. (4.3.7) and (4.3.8), corresponding to the two paths, and *then* taking the square of its modulus,

$$I = |\mathscr{E} \cos\theta \sin\theta\, f - \mathscr{E} \sin\theta \cos\theta\, f|^2 = 0. \qquad (4.3.9)$$

We find that the intensity at the exit to Π_f is zero, and this is because the corresponding wave *amplitude* is the resultant of the superposition of two opposite amplitudes. Had we believed that we could have obtained the intensity of the final transmitted beam by adding the intensities of each individual beam, which would be $I \cos^2\theta \sin^2\theta + I \sin^2\theta \cos^2\theta$, we would have obtained the same erroneous result as by the probabilistic corpuscular reasoning developed above. But to commit such an error would be tantamount to not having at all understood the wave concept and the superposition principle upon which it is based (see ch. 2, sect. 2.1.2).

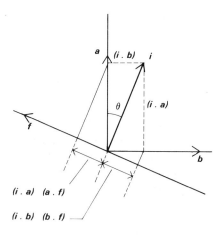

Fig. 4.8. *Projections*
The projection of i onto b and then onto f is equal and opposite to that onto a and then f:
$(i \cdot a)(a \cdot f) + (i \cdot b)(b \cdot f) = \cos\theta \sin\theta + \cos(\tfrac{1}{2}\pi - \theta) \sin(\tfrac{1}{2}\pi - \theta) = 0.$

This tends to make us believe that the probabilistic calculation, which led to eq. (4.3.6), for the number of photons transmitted by the apparatus, also possibly contains the same type of error. We are, therefore, led to extending the analogy, introduced in sect. 4.2, between the intensity of the classical wave and the quantum transition probability, attributed to each photon, and to introduce a *quantum amplitude* analogous to the classical (wave) amplitude: the probability of a quantum transition is thus given by the square of the modulus of the corresponding quantum amplitude. For this reason, quantum amplitudes are often called 'probability amplitudes'. To each quantum transition of *one* photon (and, more generally, of *one* quanton), we associate a quantum amplitude A, such that, the probability $\mathscr{P}$ for this quanton to make a transition is given by the relation

$$\mathscr{P} = |A|^2. \tag{4.3.10}$$

As in the case of its 'model', the wave amplitude, this quantum amplitude, in order to deserve the name amplitude, ought to be governed by a *superposition principle*.

The experiment described in this section is now easily understood. Let us associate to each photon of the incident beam, a quantum amplitude $A_a(f \leftarrow i)$ – the probability amplitude of the quantum transition $f \leftarrow i$ through the channel a – and an amplitude $A_b(f \leftarrow i)$, corresponding to the transition $f \leftarrow i$ through the channel b. The amplitude corresponding to the quantum transition $f \leftarrow i$, when both channels a and b are open, would then be

$$A(f \leftarrow i) = A_a(f \leftarrow i) + A_b(f \leftarrow i), \tag{4.3.11}$$

which we write more simply as

$$A = A_a + A_b, \tag{4.3.12}$$

in the same way as the amplitude along the branch $M_b \Pi_f$ of the apparatus (fig. 4.7), is the sum of the amplitudes along the branches AM_a and BM_b: $\mathscr{E} = \mathscr{E}_a + \mathscr{E}_b$. The probability that each photon of the incident beam would traverse the whole apparatus, when both paths a and b are open, is obtained, as with wave intensities, by taking the square of the modulus of the expression in eq. (4.3.12),

$$\mathscr{P} = |A_a + A_b|^2 = |A_a|^2 + |A_b|^2 + A_a \bar{A}_b + \bar{A}_a A_b. \tag{4.3.13}$$

Here, the overbar denotes complex conjugation, for as we shall soon see, these quantum amplitudes are complex numbers.

Now, the quantity $|A_a|^2$ (respectively $|A_b|^2$) represents [in view of the very definition, given by eq. (4.3.10), of the quantum amplitude] the probability $\mathscr{P}_a(f \leftarrow i)$ (respectively $\mathscr{P}_b(f \leftarrow i)$) that *one* photon would cross the apparatus along the path *a* (respectively *b*). Taking relations (4.3.4) and (4.3.5) into account, it follows that

$$|A_a|^2 = \cos^2\theta \sin^2\theta, \qquad |A_b|^2 = \sin^2\theta \cos^2\theta. \qquad (4.3.14)$$

It is natural, at this stage, to assume in addition that the probability amplitude associated to *one* photon behaves as the amplitude of the classical wave associated to an assembly of a large number of identical and independent photons. We are thus led to attributing opposite values to A_a and A_b: $A_a = \cos\theta\sin\theta$ and $A_b = -\sin\theta\cos\theta$, in conformity with expressions (4.3.7) and (4.3.8) for the electric field of the wave traversing the apparatus along the channels *a* and *b*, respectively. We then verify that the probability given by eq. (4.3.13), for a photon from the incident beam to be transmitted by the entire experimental set-up, is indeed zero, in conformity with experiment. The error that was being made in trying to apply a classical, corpuscular, probabilistic reasoning, as we now see, consisted in not having taken the cross terms (called 'interference' terms) into account, in eq. (4.3.13). Effectively, when $A_a = -A_b$, the interference terms cancel the terms $|A_a|^2$ and $|A_b|^2$. Let us remark, moreover, that the result in eq. (4.3.13) has a general significance, and holds even if the quantities $|A_a|$ and $|A_b|$ are not equal; in particular, this is the case if the direction of transmission *f* of the polaroid is not orthogonal to the direction *i* of the polarization of the incident beam (see exercise 4.4). The total amplitude is now no longer zero, but then, neither is the probability $\mathscr{P}$, of a photon traversing the apparatus when both channels *a* and *b* are open, the sum of the probabilities $\mathscr{P}_a$ and $\mathscr{P}_b$ of its crossing the apparatus when only one of these channels is open at a time.

4.4. Rules for calculating quantum amplitudes

4.4.1. The principle of superposition

There seems to appear, therefore, a radical difference in nature between ordinary probabilities and quantum probabilities. In standard probability theory, the probability of an event, which can take place independently in two (or more) different ways, is given by the sum of the probabilities corresponding to each mode of its realization. This is a fundamental rule in classical theory. Thus, the probability of getting a head and a tail, by tossing a coin twice in succession, is the sum of the probability of obtaining first a head, then a tail, which is $\frac{1}{2} \times \frac{1}{2} = \frac{1}{4}$, and the probability of obtaining first a tail and then a head, which again is $\frac{1}{2} \times \frac{1}{2} = \frac{1}{4}$; finally, the total probability is thus $\frac{1}{2}$.

Now, we have seen that the probability for a photon to pass through when both paths *a* and *b* are open, is not the sum of the probabilities corresponding to each one of the paths. Is it, as is sometimes said, that ordinary probability theory does not apply to quantum theory? No! It is not the conclusion, in this reasoning, which is wrong, rather it is the premise: it is impossible to affirm that during its passage through the apparatus, the photon passes *either* along path *a or* along path *b*. Making such a statement amounts to adopting a classical, corpuscular point of view which, as we have seen, is not adequate. In other words, it is impossible, in the present case, to specify two *different* modalities by which the same event (that of the photon traversing the apparatus) could be realized. It is hardly surprising, therefore, that the rule of the addition of probabilities does not apply ..., since we do not even have the conditions under which they apply.

Let us go back to the definition of quantum amplitudes. To each quantum transition, in other words, to each pair of states – an initial state v and a final state w – is associated an amplitude, which we denote by $A(w \leftarrow v)$, in a way which makes the relationship between the states of the pair (v, w) obvious. As with probabilities, by convention, the initial state is written on the right and the final state on the left, whence the sense of the arrow. The transition probability is given by

$$\mathscr{P}(w \leftarrow v) = |A(w \leftarrow v)|^2. \tag{4.4.1}$$

The quantum amplitude $A(w \leftarrow v)$ is only defined to the extent that the states v and w are completely specified, i.e., they exhaustively describe the entire system. Let us consider a phenomenon where the same final situation could be achieved by going through several possible intermediate states. It could be either that one part of the system retains one or another feature of this intermediate transition, so that we are, in fact, dealing with different final states (even if the other part of the system, and the one in which we are interested in general, does not depend upon the intermediate transition); or else, it could be that the final state is uniquely and completely specified (for all parts of the system) and, hence, it is not possible to know what the intermediate states were: we are dealing with one single quantum transition.

In the first case (in which at least a part of the whole system retains some trace of the intermediate state), one is dealing with as many final states as there are distinct intermediate states, and hence with as many different quantum transitions. Each one possesses a certain probability and their sum yields – in conformity with the usual rule, recapitulated above – the total probability. In the second case, the intermediate states are not physically realized and remain *virtual* – to use a word, already employed in a similar context (ch. 3, sect. 3.3.1.2). The intermediate states do not correspond to different processes now, since there is only one single final state. There is only one transition and only one amplitude.

It remains to express the amplitude in terms of the partial amplitudes, each corresponding to the specification of one intermediate state. If we have a complete set of disjoint intermediate states $\{u_n\}$, the fundamental rule of quantum superposition is stated as

$$A(w \leftarrow v) = \sum_n A_n(w \leftarrow v), \tag{4.4.2}$$

where the amplitudes A_n correspond to the different intermediate states (see exercise 4.7).

That quantum amplitudes obey a superposition principle, analogous to that for classical wave amplitudes (to the extent that in the case of the photons they behave in an identical manner, as we have assumed a moment ago), should not surprise us: the logical relation between the two concepts is the opposite of the pedagogical relation utilized here. In fact, as we shall see in ch. 7, the wave amplitudes correspond to the classical manifestations of the quantum superposition principle, under conditions where the wave-like approximation is valid.

We emphasize once more the fact that the rule for the superposition of quantum amplitudes is valid only for transitions between an initial state and a final state which are specified completely, to the extent of being unique. This necessary condition is also sufficient. We shall now see that whenever it is possible to distinguish between two modalities for the realization of the same phenomenon – whenever a secondary phenomenon is produced, which is different for the two cases, leading therefore to different final states for the *entire* systems being considered – the usual rule for the addition of probabilities is recovered.

4.4.2. *The superposition rule and the Heisenberg inequalities*

Let us consider again the experiment in fig. 4.7, and let us introduce into one of the channels, for example *a*, a device which enables us to distinguish between the two beams. To do this, it is enough to allow the mirror M_a to move – for instance to recoil as a result of the impact of a photon which might pass through this channel. For each photon picked up beyond Π_f, the observation of a recoil (or its absence) of the mirror would enable us to say whether the photon had passed through channel a or through channel b.

However, for this recoil to be detected, it is necessary that the momentum absorbed by the mirror, from the impact of a photon (a quantity which is of the order of the momentum p_γ of the photon), be greater than the intrinsic width Δp_M of the momentum spectrum of the mirror,

$$p_\gamma > \Delta p_M. \tag{4.4.3}$$

We shall show that it is impossible to satisfy this condition without modifying at the same time, the experimental conditions to such an extent that the resulting experiment is no longer that of fig. 4.7. Indeed, if eq. (4.4.3) is satisfied, the mirror, having absorbed a momentum larger than the width Δp_M of its spectrum, would have the average value p_M of its momentum modified: under these circumstances, p_M, which would be zero, under the experimental conditions of fig. 4.7, now has a non-zero value, of the order of p_γ. Instead of permitting the reflection of a photon from a stationary mirror M_a, the experimental set-up now corresponds to a qualitatively different experiment, in which the photon is made to reflect off a moving mirror.

Stated otherwise, while it is true that one can observe the passage of a photon along the channel a (and hence to distinguish between the two possible paths a and b), one is, nevertheless, dealing now with an experiment which is no longer the same as before. The proof that this is, in fact, quite a different experiment, lies in the fact that the intensity picked up by the detector is no longer zero. Indeed, applying the Heisenberg inequality to the mirror (which, after all, is itself also a quantum system), it follows that

$$\Delta p_M \, \Delta l_M \gtrsim \hbar, \tag{4.4.4}$$

where Δl_M denotes the width of the position spectrum of the mirror. In view of inequality (4.4.3) and the de Broglie relation for the photon, the undulation k_γ is

$$\hbar k_\gamma \geqslant \hbar/\Delta l_M, \quad \text{or} \quad \Delta l_M \geqslant 1/k_\gamma. \tag{4.4.5}$$

This amounts to saying that the passage of a photon along the channel a can only be detected if the width of the position spectrum of the mirror is greater than the reciprocal of the undulation of the photon. But now, the difference in the optical paths between the two possible routes (channel a and channel b) is only defined up to Δl_M, and the phase difference between the two corresponding partial amplitudes, which become superimposed beyond the mirror M_b displays a certain dispersion

$$\Delta\varphi = k_y \, \Delta l_M, \tag{4.4.6}$$

which, in view of eq. (4.4.5) becomes

$$\Delta\varphi \gtrsim 1. \tag{4.4.7}$$

Stated otherwise, the relative phase of the two waves, which are superimposed along the branch $M_b \Pi_f$, is not only zero, it is not even defined! The situation classically is, therefore, that of two waves, whose phase difference displays random fluctuations and whose interference is not detectable. One is dealing in this case with the 'incoherent' superposition of two light waves, the intensities, and not the amplitudes, of which add up to give a total intensity $2I \cos^2\theta \sin^2\theta$, which is different from zero. From the quantum point of view, this corresponds to the addition of probabilities – the probability that a photon will cross the entire apparatus is just the sum of the probabilities corresponding to the two paths, i.e., $|A_a|^2 + |A_b|^2 = 2I \cos^2\theta \sin^2\theta$. In other words, in this case, the passage of the photon along one *or* the other of the paths a and b, properly corresponds to distinct phenomena, detectable by the recording (or the absence) of an impact on the mirror M_a. We note that the final states are quite different, depending upon whether one considers only the photon, beyond the polarizer Π_f, or the entire system: it is the state of the mirror M_a which is not the same in the two cases.

We find out once again that the rule of the addition of probabilities becomes applicable, now that the experimental conditions have been modified [relation (4.4.3) indicates to what extent]. We remark, moreover, that since condition (4.4.5) is satisfied, the characteristic action of the interaction between the photon and the mirror, $A = p_y \Delta l_M$, is greater than $\hbar$. It is not surprising, therefore, to find that the photon exhibits a corpuscular behaviour, following a well-defined trajectory, whether a or b. The photon–mirror interaction is, therefore, sufficiently strong to oblige the photon to be projected onto one of the two beams (see also exercises 4.9 and 4.10).

Let us keep in mind, from this analysis, the consistency of the quantum notions introduced up to now. The Heisenberg inequalities were applied (and this is important) to the mirror, hence to the experimental apparatus, considered as a quantum system on the same footing as the photons, which are the very objects of the experiment. These inequalities, for which we have shown how they characterize the original nature of quantum objects – *neither* waves *nor* corpuscles – have enabled us to avoid any contradiction between the case where the simple addition of probabilities holds (in short, an already classical case) and the specifically quantum case, where the probability amplitudes are superposed.

4.4.3. The factorization rule

It is useful to introduce a currently used symbol to denote quantum amplitudes. Let us denote the amplitude for the quantum transition probability $w \leftarrow v$ as

$$A(w \leftarrow v) = \langle w|v \rangle. \tag{4.4.8}$$

In this symbolism too, the initial state is, by convention, placed on the right and the final state on the left. The corresponding probability is now written as

$$\mathscr{P}(w \leftarrow v) = |\langle w|v \rangle|^2. \tag{4.4.9}$$

We can have an idea of the reason for introducing such a notation – which is analogous to that of the scalar product – by applying it to the case of the projection of a state of linear polarization i onto a state of linear polarization along f, just as what takes place inside a polaroid Π_f (see fig. 4.5a). The probability of this transition $\mathscr{P}(f \leftarrow i)$ is given, as in eq. (4.2.12), by

$$|\langle f|i \rangle|^2 = \cos^2\theta = (f \cdot i)^2, \tag{4.4.10}$$

where θ is the angle between the two directions f and i. The quantum amplitude is thus equal to cos θ, *up to a phase factor*. One generally agrees to strictly identify the projection amplitude of a linear state of polarization onto another, with the cosine of the angle between the two directions of polarization. In other words, we set

$$\langle f|i \rangle = (f \cdot i). \tag{4.4.11}$$

The similarity of the notation, obvious from eq. (4.4.11), between quantum amplitudes and scalar products is not fortuitous, as the subsequent development of the mathematical formalism of the quantum theory will show.

For the moment, this notation will be useful to us in extracting from the experiment described above, and illustrated in fig. 4.7, the laws for combining quantum amplitudes.

Let us denote by $\langle f|i\rangle_a$ and $\langle f|i\rangle_b$ the two partial amplitudes A_a and A_b, corresponding to the two paths *a* and *b*. The fundamental quantum rule of superposition, stated above, is written, using these notations as

$$\langle f|i\rangle = \langle f|i\rangle_a + \langle f|i\rangle_b. \qquad (4.4.12)$$

However, the transition $f \leftarrow i$ along path *a* is completed in two successive steps: crossing the birefringent slab, with the projection of the state *i* onto the state of polarization *a*, followed by the projection across Π_f, of this state onto the state of polarization *f*. Along the way, the amplitude of the classical wave (and hence, also the transition amplitude) is multiplied first by $\cos(i, a)$ and then by $\cos(a, f)$. This, in view of eq. (4.4.11), may symbolically be denoted by

$$\langle f|i\rangle_a = \langle f|a\rangle \langle a|i\rangle. \qquad (4.4.13)$$

In an analogous fashion,

$$\langle f|i\rangle_b = \langle f|b\rangle \langle b|i\rangle. \qquad (4.4.14)$$

From the above we obtain the second fundamental quantum rule, the so-called *factorization rule*: if a transition is effected in a way such that, between the initial and the final states, the system passes through a certain well-defined, intermediate state, the probability amplitude factorizes into the product of the probability amplitudes corresponding to the two stages of the transition. Upon taking the square of the modulus of the preceding expression, we obtain the equality between probabilities:

$$\mathcal{P}_a(f \leftarrow i) = \mathcal{P}(f \leftarrow a) \, \mathcal{P}(a \leftarrow i), \qquad (4.4.15)$$

where $\mathcal{P}_a(f \leftarrow i)$ denotes the transition probability of the state of polarization *i* to the state *f*, *via* the intermediate state of polarization *a*. This equality is in complete accordance with the classical law for the computation of joint probabilities. We observe, however, that the quantum rule for the factorization of amplitudes is stronger than the classical rule for the factorization of probabilities. In fact, the latter only requires the multiplicativity of the moduli of these amplitudes (in general, complex; see sect. 4.4.4), without demanding anything from their phases.

We have treated in detail the quantum description of polarization, because the relations that we have extracted between the transition amplitudes have a general validity. We summarize them in table 4.2, where the symbols v and w denote arbitrary states of an arbitrary quantum system and $\{u_n\}$ a complete set of disjoint states.

Table 4.2

Rules for computing quantum amplitudes

Principle of superposition
    When a quantum transition involves a *complete set of disjoint intermediate states*, the probability amplitude of the transition is the sum of the amplitudes corresponding to each intermediate state,

$$\langle w|v \rangle = \sum_n \langle w|v \rangle_n. \tag{4.4.16}$$

Principle of sequential factorization
    When a quantum transition is effected by the passage through a *well-defined intermediate state*, the probability amplitude of the transition is the product of the amplitudes corresponding to the two stages of the transition (from the initial state to the intermediate state, from the intermediate state to the final state).

$$\langle w|v \rangle_n = \langle w|u_n \rangle \langle u_n|v \rangle \tag{4.4.17}$$

4.4.4. Circular polarization and complex quantum amplitudes

The preceding rules find an immediate application in the computation of the probability amplitude of the projection of a state of linear polarization, which we denote by r (r is the unit vector in the direction of polarization), onto a state of circular (for example, right) polarization, denoted by R. This transition, the amplitude of which is denoted by $\langle R|r \rangle$, is effected by traversing a circular polarizer (see sect. 4.2.2), denoted by Π_R. The motion of the extremity of the light vector r of the incoming wave, can be analyzed as a superposition of two circular motions, one to the right, and the other to the left, of the same amplitude, so that the intensity transmitted by the polarizer Π_R represents half the intensity of the incident beam, the other half being absorbed by Π_R. Here r denotes a state of arbitrary, linear polarization. It follows from this that the transition probability of a state of linear polarization to a state of circular polarization, has the value $\frac{1}{2}$, *whatever r might be*,

$$|\langle R|r \rangle|^2 = \tfrac{1}{2} \qquad \forall r, \tag{4.4.18}$$

which fixes the value of the modulus of $\langle R|r \rangle$ to be

$$|\langle R|r \rangle| = 1/\sqrt{2}. \qquad (4.4.19)$$

Now let us imagine that instead of projecting directly, with the aid of Π_R, the state of linear polarization r onto the state of circular polarization R, we place, in front of Π_R, an analyzer $A_{(a,b)}$. Also suppose that a system of semi-transparent mirrors, of the type used in the experiment in fig. 4.7, enables the two waves coming out of the slab, along the two paths a and b (fig. 4.9) to superpose (without introducing any phase shift). As in sect. 4.3, the quantum transition $R \leftarrow r$ can be realized *via* one or the other of the intermediate states a and b. The corresponding partial quantum amplitudes, $\langle R|r \rangle_a$ and $\langle R|r \rangle_b$, are themselves factorizable, whence, the expression for $\langle R|r \rangle$:

$$\langle R|r \rangle = \langle R|a \rangle \langle a|r \rangle + \langle R|b \rangle \langle b|r \rangle. \qquad (4.4.20)$$

The two amplitudes $\langle a|r \rangle$ and $\langle b|r \rangle$ of the projection of a state of linear

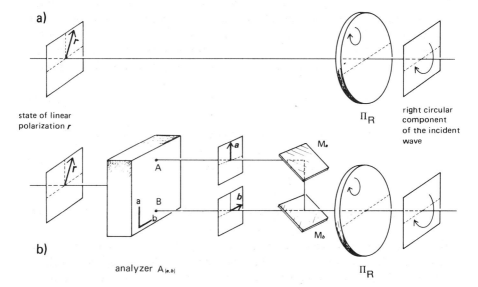

Fig. 4.9. *Circular polarization*

(a) Π_R projects the state r onto the state R of circular polarization. The corresponding quantum amplitude is $\langle R|r \rangle$.

(b) Due to the presence of the analyzer $A_{(a,b)}$, this same projection of the state r onto the state R, proceeds through the intermediate states a and b.

polarization onto another state of linear polarization have, according to the convention established in eq. (4.4.11), the values

$$\langle a|r\rangle = \cos\theta, \qquad \langle b|r\rangle = \sin\theta, \tag{4.4.21}$$

where θ denotes the angle between the directions of r and a.

As regards the three other quantum amplitudes appearing in eq. (4.4.20), which are amplitudes for the projection of a state of linear polarization onto a state of circular polarization, they all have the same modulus $1/\sqrt{2}$, following eq. (4.4.19),

$$|\langle R|r\rangle| = |\langle R|a\rangle| = |\langle R|b\rangle| = 1/\sqrt{2}. \tag{4.4.22}$$

We see now that it is impossible to satisfy relation (4.4.20) using real numbers: a real number, of absolute value $1/\sqrt{2}$, cannot be obtained as a sum or a difference of the quantities $(\cos\theta)/\sqrt{2}$ and $(\sin\theta)/\sqrt{2}$. We are thus led to considering the possibility of giving complex values to quantum amplitudes.

Is it possible to assign to the amplitudes $\langle R|r\rangle$, $\langle R|a\rangle$ and $\langle R|b\rangle$, with the known modulus, eq. (4.4.22), phases such that relation (4.4.20) would be satisfied? We note that this relation cannot fix the phase of each one of the three amplitudes $\langle R|r\rangle$, $\langle R|a\rangle$ and $\langle R|b\rangle$, but only their relative phases. Let us, therefore, arbitrarily fix the phase of one of them, and set for example

$$\langle R|a\rangle = 1/\sqrt{2}. \tag{4.4.23}$$

The only possible solutions satisfying eq. (4.4.20) are, therefore, either

$$\langle R|b\rangle = i/\sqrt{2} = e^{i\pi/2}/\sqrt{2}, \qquad \langle R|r\rangle = e^{i\theta}/\sqrt{2}, \tag{4.4.24}$$

or

$$\langle R|b\rangle = -i/\sqrt{2} = e^{-i\pi/2}/\sqrt{2}, \qquad \langle R|r\rangle = e^{-i\theta}/\sqrt{2}. \tag{4.4.25}$$

We verify the consistency of these results by checking that for $\theta = 0$ and $\theta = \pi/2$, $\langle R|r\rangle$ actually does become identical to $\langle R|a\rangle$ and $\langle R|b\rangle$, respectively. As to the choice between the two solutions, it is arbitrary. Generally, the convention given by eq. (4.4.25) is adopted, which means taking $\langle R|r\rangle = e^{-i\theta}/\sqrt{2}$. It could be shown, following an argument analogous to that which led to the solutions given by eqs. (4.4.24) and (4.4.25), that the amplitude of the projection of a state of linear polarization r onto a state of left circular polarization, denoted L, is also of the form $\langle L|r \rangle = e^{\pm i\theta}/\sqrt{2}$. Since we have chosen $\langle R|r \rangle = e^{-i\theta}/\sqrt{2}$, it is appropriate to take $\langle L|r \rangle = e^{i\theta}/\sqrt{2}$, since to the projections of the arbitrary state r onto two different states R and L, there could not correspond the same quantum amplitude for all values of θ. Finally,

$$\langle R|r \rangle = e^{-i\theta}/\sqrt{2}, \qquad \langle L|r \rangle = e^{i\theta}/\sqrt{2}. \tag{4.4.26}$$

The same reasoning could have been applied in the opposite direction. By considering an initial state of right circular polarization R, projected onto a final state of linear polarization, and by interposing (or not interposing) an analyzer $A_{(a,b)}$, we would end up with the relation

$$\langle r|R \rangle = \langle r|a \rangle \langle a|R \rangle + \langle r|b \rangle \langle b|R \rangle. \tag{4.4.27}$$

From this it follows, as previously with $\langle R|r \rangle$, that the amplitudes of the type $\langle r|R \rangle$ are complex, that their moduli have the value $1/\sqrt{2}$ and that their angular dependence is of the form $e^{\pm i\theta}$, if θ is the angle made by the direction of r with the direction of a. Evidently, the same is true for $\langle r|L \rangle$. But once the conventional choice, eq. (4.4.25), for $\langle R|r \rangle$ is adopted, the amplitudes $\langle r|R \rangle$ and $\langle r|L \rangle$ are fixed unambiguously.

Let us consider the effect of the projection of a state of linear polarization s, the direction s of which makes an angle θ' with a, onto a state of linear polarization r. The amplitude $\langle r|s \rangle$ can be written assuming the interposition of a circular analyzer $A_{(L,R)}$, as

$$\langle r|s \rangle = \langle r|R \rangle \langle R|s \rangle + \langle r|L \rangle \langle L|s \rangle. \tag{4.4.28}$$

Again, since by eq. (4.4.11), $\langle r|s \rangle = (r \cdot s) = \cos(\theta' - \theta)$, and upon utilizing known definitions (4.4.26), this becomes

$$\cos(\theta - \theta') = \langle r|R \rangle\, e^{-i\theta'}/\sqrt{2} + \langle r|L \rangle\, e^{i\theta'}/\sqrt{2}. \tag{4.4.29}$$

The amplitudes $\langle r|R \rangle$ and $\langle r|L \rangle$ are thus determined by inspection,

$$\langle r|R \rangle = e^{+i\theta}/\sqrt{2}, \qquad \langle r|L \rangle = e^{-i\theta}/\sqrt{2}. \tag{4.4.30}$$

Let us remark that the projection amplitudes, eqs. (4.4.30), of a state of circular polarization onto a state of linear polarization have a *harmonic* angular dependence. We have seen in chs. 2 and 3 that such a harmonicity characterizes the proper states of a component of the angular momentum. We see, therefore, that one should identify the proper states of circular polarization of the photon with the proper states of its angular momentum. To a proper value $m\hbar$ of the component of the angular momentum under consideration, there corresponds a harmonic function with argument $m\theta$. Equations (4.4.30) lead us to believe that the states R and L, proper states of circular polarization, for which $m = +1$ and $m = -1$, respectively, are also proper states of the z-component of the angular momentum of the photon, having proper values $+\hbar$ and $-\hbar$, respectively.

Moreover, we can verify that the states R and L remain unchanged by a rotation about the direction of propagation of the photon, as was the case with the proper states of the angular momentum in this direction. Indeed, we already know that the transition probabilities $r \leftarrow R$ and $r \leftarrow L$ have the values [see, for example, eq. (4.4.18)],

$$\mathscr{P}(r \leftarrow R) = \tfrac{1}{2}, \qquad \mathscr{P}(r \leftarrow L) = \tfrac{1}{2}, \tag{4.4.31}$$

and do not depend, therefore, on the angle θ, i.e., on the particular direction of the final state r of linear polarization. Thus, the proper states of circular polarization do not favour any particular spatial direction, and their physical character is, indeed, invariant under rotations. We have already emphasized, in sect. 4.2.2, the invariance under rotations of the circular polarization, from which follows the invariance of its proper states. This situation is in contrast with that for the linear polarization. The expression

$$\mathscr{P}(r \leftarrow s) = \cos^2(\theta - \theta'), \tag{4.4.32}$$

for the transition $r \leftarrow s$, readily shows that there exists, in the state s, a privileged direction – evidently that of s, characterized by the angle θ'. From the point of view of transition amplitudes, and not just from that of the probabilities, if the system of axes is made to undergo a rotation through an angle δ, thereby transforming the final linear polarization r into r', characterized by the angle $(\theta - \delta)$, we can write

$$\langle r'|R\rangle = e^{i(\theta - \delta)}/\sqrt{2}, \qquad \langle r'|L\rangle = e^{-i(\theta - \delta)}/\sqrt{2}, \tag{4.4.33}$$

so that,

$$\langle r'|R\rangle = e^{-i\delta}\langle r|R\rangle, \qquad \langle r'|L\rangle = e^{i\delta}\langle r|L\rangle. \tag{4.4.34}$$

The transition amplitudes of the states of circular polarization are thus only modified by a *phase factor*, which would evidently disappear if one were to calculate the transition probabilities. By contrast, for the transition amplitudes, from two basis states of linear polarization (a, b) to a state r, or its transform r' by rotation, we have

$$\langle r'|a\rangle = \cos(\theta - \delta), \qquad \langle r'|b\rangle = \sin(\theta - \delta), \tag{4.4.35}$$

so that

$$\langle r'|a\rangle = \cos \delta \, \langle r|a\rangle + \sin \delta \, \langle r|b\rangle,$$
$$\langle r'|b\rangle = -\sin \delta \, \langle r|a\rangle + \cos \delta \, \langle r|b\rangle. \tag{4.4.36}$$

The transformation formulae, eqs. (4.4.34) and (4.4.35), for the quantum probability amplitudes are evidently in keeping with the transformation formulae, eqs. (4.2.13), (4.2.16), (4.2.17) and (4.2.15), respectively, for the classical wave amplitudes.

4.4.5. *Conjugation of amplitudes, symmetry of probabilities*

Comparing expressions (4.4.26) and (4.4.30) for the two opposite transitions $R \leftarrow r$ and $r \leftarrow R$, or just as well, $L \leftarrow r$ and $r \leftarrow L$, we see that the corresponding amplitudes are complex conjugates of one another,

$$\langle r|R\rangle = \overline{\langle R|r\rangle}, \qquad \langle r|L\rangle = \overline{\langle L|r\rangle}. \tag{4.4.37}$$

These conjugate relations are only a very particular case of a general rule, which is valid for any quantum amplitude whatsoever. First of all, let us agree to set the amplitude of the transition, between an initial state v and an identical final state, equal to unity,

$$\langle v|v\rangle = 1. \tag{4.4.38}$$

This is a natural choice for the phase of this amplitude, the modulus of which is obviously fixed by the implied value of the corresponding probability

$$\mathscr{P}(v \leftarrow v) = |\langle v|v\rangle|^2 = 1. \tag{4.4.39}$$

With this in mind, let us consider a complete set of disjoint states $\{u_n\}$, and let us carry out an analysis of the state v with respect to this set. We may write, on the one hand,

$$1 = \langle v|v \rangle = \sum_n \langle v|u_n \rangle \langle u_n|v \rangle, \tag{4.4.40}$$

according to the general rules of computation of amplitudes (table 4.2). On the other hand, we know that

$$1 = \sum_n \mathscr{P}(u_n \leftarrow v) = \sum_n |\langle u_n|v \rangle|^2. \tag{4.4.41}$$

A comparison of relations (4.4.40) and (4.4.41) which, as we recall, hold for the states of any system whatsoever, renders at least plausible the relation

$$\langle v|u_n \rangle = \overline{\langle u_n|v \rangle}. \tag{4.4.42}$$

Finally, the relation of conjugation (4.4.42), which is valid for an arbitrary state v and one of the states $\{u_n\}$, in fact generalizes to two arbitrary states v and w if we exploit the various rules for the computation of amplitudes:

$$\begin{aligned}
\langle w|v \rangle &= \sum_n \langle w|u_n \rangle \langle u_n|v \rangle, &&\text{by eqs. (4.4.16) and (4.4.17),} \\
&= \sum_n \overline{\langle u_n|w \rangle} \, \overline{\langle v|u_n \rangle}, &&\text{by eq. (4.4.42),} \\
&= \sum_n \overline{\langle v|u_n \rangle \langle u_n|w \rangle}, && \\
&= \overline{\langle v|w \rangle}, &&\text{by eqs. (4.4.16) and (4.4.17).} \tag{4.4.43}
\end{aligned}$$

We are thus led to accepting the general validity of the rule according to which two quantum amplitudes, characterizing transitions in which the initial and final states are interchanged, are obtained from one another by complex conjugation. It follows immediately, that their moduli are equal, and consequently, that the probabilities of the two inverse transformations are equal. This result, as natural and satisfactory as it may appear, is nevertheless not at all evident. One can consider it as being witness to the consistency of the notions developed here. Let us, therefore, summarize the results of this section in table 4.3.

Table 4.3

Inverse transitions

Conjugate relations for quantum amplitudes
　The quantum amplitudes of two inverse transitions are complex conjugates of one another,

$$\langle w|v \rangle = \overline{\langle v|w \rangle}, \tag{4.4.44}$$

or, written differently,

$$A(w \leftarrow v) = \overline{A(v \leftarrow w)}. \tag{4.4.45}$$

Symmetry of probabilities
　The probabilities of two inverse transitions are equal,

$$\mathscr{P}(w \leftarrow v) = \mathscr{P}(v \leftarrow w). \tag{4.4.46}$$

4.4.6. Classical trajectories and quantum amplitudes

　When a quantum transition is characterized by several, disjoint intermediate states, its amplitude is expressible as a sum of the amplitudes corresponding to these various states,

$$A(w \leftarrow v) = \sum_n A_n(w \leftarrow v). \tag{4.4.47}$$

We have just seen that, in general, amplitudes assume complex values. The total amplitude A, resulting from a superposition such as in eq. (4.4.47), depends, therefore, in a crucial manner, upon the relative phases of the different amplitudes A_n. A knowledge of the individual transition probabilities when one of the intermediate states is selected by a physical device, i.e., of the numbers

$$\mathscr{P}_n(w \leftarrow v) = |A_n(w \leftarrow v)|^2, \tag{4.4.48}$$

only fixes the moduli $|A_n|$. This is altogether insufficient for an evaluation of the total amplitude A.

　The rigorous computation of quantum amplitudes is the very object of the formalism of quantum theory. However, there exist numerous situations where an approximate evaluation of these amplitudes, in particular of their phases, is possible, using a clever compromise between classical and quantum ideas. The procedure is analogous to the use of the notion of the optical path in the theory of light, which combines geometrical optics with wave optics. Let us consider, for example, the classical interference experiment – Young's double slit experiment. A source S illuminates two slits T_1 and T_2, from where the light falls on the screen E, on which we observe an interference pattern (see fig. 4.10). A completely rigorous theoretical description of this experiment requires a solution of the Maxwell equations, in all space with complicated boundary conditions fixed by the plate carrying the slits T_1 and T_2. However, if the width of the slit is small compared to the wavelength of the light, which itself is small compared to all the geometrical dimensions of the system, we can, following the approximation due to Huygens, replace T_1 and T_2 by two identical point sources (we neglect absorption or reflection by the plate). Each point on the screen E receives, then, a luminous wave, the amplitude of which is the sum of the amplitudes of the two waves originating at T_1 and T_2, respectively. If the distance from the plate to the screen is large compared to the separation between the slits, the amplitudes of these waves are practically equal and the total amplitude depends only on their phase difference. This latter is now easily calculated, at a point M, using the difference in the optical paths along the 'rays' $T_1 M$ and $T_2 M$ of geometrical optics.

Exactly the same type of calculation is valid in quantum theory, if one takes into account the fact that light waves are, in reality, a system of photons. But now it is quantum amplitudes and not electric fields (or classical wave amplitudes) that are evaluated. The probability amplitude for *one photon* to arrive at the screen at M, when the two slits T_1 and T_2 are open, is the sum of the partial amplitudes corresponding to the intermediate passage through each one of the slits. With the previous notation, adapted in an obvious way,

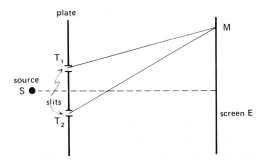

Fig. 4.10. *Young's double slit experiment*

this becomes

$$A(M \leftarrow S) = A_{T_1}(M \leftarrow S) + A_{T_2}(M \leftarrow S). \tag{4.4.49}$$

Indeed, it is clear that the final state (photon at M) is well-defined and that one should not add the probabilities corresponding to the two intermediate states here. To say that the photon has a wavelength λ (or an undulation $k = 2\pi/\lambda$) and a pulsation ω, is the same as saying that its propagation is a harmonic phenomenon in space and time. It is natural to assume that this harmonicity characterizes the probability amplitude which describes the propagation. The spatio–temporal dependence of quantum probability amplitudes is, therefore, quite analogous to that of classical wave amplitudes. Between two points, a short distance Δr apart, and two instants, separated by Δt, the phase of such an amplitude varies, in conformity with eq. (2.1.6), by

$$\Delta\varphi = \omega\,\Delta t - \boldsymbol{k} \cdot \Delta r. \tag{4.4.50}$$

Let us consider a quantum transition between two states of localization, that is, the propagation of a quanton between two points A and B, such that several well separated paths, $\mathscr{C}_1, \mathscr{C}_2$, etc., are available (see fig. 4.11). Then,

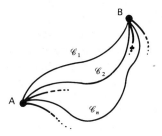

Fig. 4.11.   *Classical trajectories and quantum amplitudes*
If a quanton is able to make a quantum transition between two states of localization (at A and at B), following several separate paths ($\mathscr{C}_1, \mathscr{C}_2, ..., \mathscr{C}_n, ...$), the transition amplitude B ← A depends on the relative phases:

$$\varphi_n = -\int_{\mathscr{C}_n} \frac{2\pi}{\lambda}\,\mathrm{d}r.$$

(λ = wavelength of the quanton), calculated along the various classical 'trajectories'.

following the general rules, the transition amplitude is written as

$$\langle A|B\rangle = \sum_n \langle A|B\rangle_{\mathscr{C}_n},$$ (4.4.51)

and each partial amplitude $\langle A|B\rangle_{\mathscr{C}_n}$ has a phase (up to a temporal phase factor, common to all)

$$\varphi_n = -\int_{\mathscr{C}_n} k\,dr.$$ (4.4.52)

If the wavelength (or the momentum of the quanton) remains constant on $\mathscr{C}_n$ we have, more simply,

$$\varphi_n = -2\pi\frac{L_n}{\lambda},$$ (4.4.53)

where L_n is the length of the path $\mathscr{C}_n$. From the relative phases calculated in this way, for the different partial amplitudes, we can deduce the form of their sum – i.e., the total amplitude – as a function of the position of the arrival point B, for example. Thus, the interference pattern calculated in the quantum case would be identical to that in the classical case, even though it would have a completely different significance. It is the *probability* of the arrival of a photon at the screen, displaying maxima and minima, which we calculate and not the luminous intensity. Naturally, for a very large number of photons, the probability gives the frequency of arrival at a point, and hence the relative intensity of the classical theory. Such situations are quite frequent, in which one can distinguish, for a given quantum transition, different possible classical 'trajectories' which define as many intermediate states, or different realizations of the final state. The fact that the corresponding amplitudes can be added means, of course, that one cannot attribute to the quanton *one* well-defined trajectory. These trajectories maintain their utility, however, by allowing us to define a 'quantum path', analogous to the optical path, in terms of which we can calculate the total amplitude, which depends on the relative phases of the amplitude corresponding to each modality. As in the classical theory, this implies an approximation. In order to be valid, this approximation demands that the wavelength of the quantons be small compared to the various dimensions of the experimental set-up, in such a way that the different quantum paths are properly separated. This condition signifies, moreover, that the characteristic action of the system, which is the product of the momentum of the quanton, $p = \hbar/\lambda$, with a typical dimension of the apparatus $l \gg \lambda$, is much larger than the quantum constant ($A = pl \gg \hbar$).

We note in particular that, in view of these considerations, an experiment on the interference of two sources conducted with electrons must, for a given geometry, lead to the same interference pattern as with the experiment using photons (see exercise 4.13). This is all the more interesting, for there does not exist any classical wave theory for electrons (for reasons that we shall see in ch. 7), and hence no classical computation analogous to that for the interference of light. These ideas are completely confirmed by experiment. Moreover, we had anticipated this result in ch. 2 where, while discussing the phenomenon of the interference of light and photons, we showed a figure (fig. 2.8), arising from an interference experiment ... for electrons! These optical analogies enable us to give a simple explanation of numerous experiments (see exercise 4.14).

The preceding ideas find a concrete elementary illustration, in a technique using a high-resolution electron microscope, which enables one to 'see' individual atoms, as shown by fig. 4.12. A beam of high energy (a few hundred keV) monoenergetic electrons is passed through the atoms to be detected. The final detection probability amplitude of an electron results from the superposition of the amplitudes corresponding to the different possible paths of the electron, across the material being explored. For a path passing through an atom, the interaction between the electron and the atom modifies the kinetic energy of the electron, hence its momentum or its undulation, and, consequently, the phase of the corresponding amplitude. Thus, the amplitude A', corresponding to the crossing of an atom, is phase shifted with respect to the value A which would have corresponded to the same trajectory in the absence of the atom. We can write

$$A' = e^{i\delta\varphi} A, \tag{4.4.54}$$

where the phase shift $\delta\varphi$ is directly related to the strength of the interaction between the electron and the atom. One can show that in the physical situations encountered, the phase shift is small, i.e., $\delta\varphi \ll 1$ (exercise 4.17). If electrons are detected on a screen perpendicular to the beam, the intensity of the flux of electrons at a point is proportional to the probability of the individual detection of an electron at this point. This probability is given by the square of the modulus of the detection probability amplitude, which is A' for the points corresponding to the crossing of an atom and A for the others. However, the respective probabilities $|A'|^2$ and $|A|^2$ are equal, by eq. (4.4.54), and it does not seem to be possible to detect the atoms. However, a technique

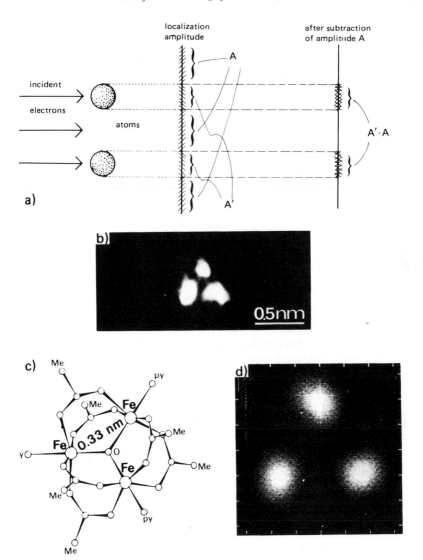

Fig. 4.12. *Images of atoms from electron microscopy*
(a) The principle of the method of 'dark field imaging'.
(b) Image of an individual molecule.
(c) Observed stereochemical composition scheme, showing three atoms of iron.
(d) Average image obtained by numerical treatment of several individual images.
[D. Dorignac, Ecole d'Eté de Microscopie Electronique en Science des Matériaux, CNRS,
(Private communication 1981).]

analogous to optical microscopy using phase contrast, permits such a detection. It involves subtracting from the amplitude corresponding to A', or A following the points of the image, the amplitude A which corresponds to the continuous background in the absence of the atoms. One thus gets a zero amplitude $(A - A = 0)$, for points of the image which do not correspond to atoms (this is why this technique is called 'dark field imaging'), and an amplitude

$$\delta A = A' - A = (e^{i\delta\varphi} - 1)A, \tag{4.4.55}$$

for the region corresponding to the presence of atoms. Hence, one gets a probability

$$|\delta A|^2 = |e^{i\delta\varphi} - 1|^2 \, |A|^2 \simeq |\delta\varphi|^2 \, |A|^2, \tag{4.4.56}$$

since $\delta\varphi \ll 1$ (exercise 4.17). The intensity of the electron flux at each point, therefore, directly gives the phase shift $\delta\varphi$, which is a measure of the electron–atom interaction, i.e., yet another representation of the atomic structure. Naturally, this distribution of the intensity is exactly on the atomic scale, and, hence, can only be observed after magnification by an electron microscope, in which the device is placed. One obtains in this way a veritable visual image of individual atoms (fig. 4.12).

4.5. *The diffraction of neutrons*

Let us deepen the ideas introduced in this chapter by employing them in the analysis of a new physical phenomenon. It deals with experiments on the 'diffraction' of neutrons by a crystal. We recall that the crystalline solid state is characterized by the regularity of the spatial arrangement of the atoms (which we shall suppose to be all of the same species). The positions of the atoms, or the atomic sites, form a three-dimensional lattice, the determination of the geometric characteristics of which, and in particular of the interatomic distances, is the objective of the diffraction technique.

To do this, a parallel beam of monoenergetic neutrons is sent through the crystal and the deflected neutrons are collected by a detector, the position of which is marked by its direction $\Omega = (\theta, \varphi)$ with respect to the beam direction. In order for the neutrons to be 'diffracted' by the atoms of the crystal, it is necessary that their energy correspond to a de Broglie wavelength λ, of the same order of magnitude as the interatomic distance, which is generally of the order of a few Å. Thus, one would choose neutrons having an energy around $E \simeq 0.01$ eV. The phenomenon which we shall now study closely, is in all respects analogous to the diffraction of light by a crystal (figs. 4.13 and 4.14).

4.5.1. The neutron–atom interaction

At such energies, the neutrons are incapable of displacing the atoms out of their sites (for this, an energy of the order of a few eV, which is the order of magnitude of the atomic and molecular binding energies, would be necessary). The impact of the neutrons on the atoms of the crystal does not, therefore, give rise to a kinetic energy transfer – it is most often an elastic collision. The interaction is not of an electrostatic type, since the neutrons, as their name indicates, are electrically neutral. In particular, the neutrons do not 'feel' the electrons of the atoms, and essentially, interact only with the nuclei. The neutron–nucleus interaction is essentially of the 'strong nuclear' type (see ch. 1, sect. 1.4.3). The description of this interaction, generally difficult, is greatly simplified in the present case by the fact that the energy of the neutrons is so low. Indeed:

(1) The quantum wavelength of neutrons ($\simeq 10^{-10}$ m) is much larger than the range of the nuclear forces ($\simeq 10^{-15}$ m). To the neutrons, the nuclei appear, therefore, as point scatterers.

(2) The collision between a neutron and a nucleus is elastic. The neutron has the same energy before and after the scattering by a nucleus; only its

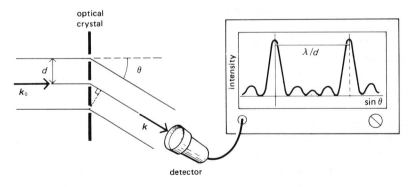

Fig. 4.13. *Diffraction of light by a crystal*
The intensity of light diffracted by a periodic crystal shows, as a function of the angular direction, sharp diffraction peaks, corresponding to the constructive interference of the partial wave amplitudes diffracted by each opening. In other words, the intensity is maximal when the difference of two successive optical paths is a multiple of the wavelength: $d \sin \theta = p\lambda$.

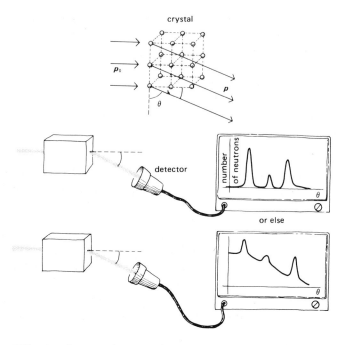

Fig. 4.14. *Diffraction of neutrons by a crystal*
The phenomenon is analogous to that of the diffraction of light by a lattice, except that here one
is dealing with the interference of the quantum amplitudes of neutrons. The angular distributions
(or spectra) of the diffracted neutrons display sharp peaks, but sometimes also a continuous
background.

direction changes. After the scattering, the momentum p of the neutron has
the same modulus as before

$$|p| = |p_0|, \tag{4.5.1}$$

but p has a direction $(\theta, \varphi) = \Omega$ with respect to p_0 (fig. 4.14).

(3) The energy of the incident neutrons is sufficiently weak so that the
spherical symmetry of a situation, where the neutron and the nucleus are
stationary, is not modified. The angular momentum of the neutron with
respect to the nucleus is zero – none of the directions Ω is preferred – and the
scattering is isotropic (exercise 4.19).

(4) The effective scattering cross-section of a neutron and an isolated atom
is, at the energies considered, of the order of magnitude of the geometric

cross-section of the nuclei, which is $\sigma \simeq 10^{-28}$ m^2. It can now be proved that the probability for each neutron to be scattered by a nucleus, in a crystalline sample of reasonable size, is very small, and hence that one can neglect the probability that the neutron would undergo two or more collisions. In other words, those neutrons which interact within the crystal (a very small minority), do so essentially only once.

4.5.2. Peaks of the diffracted intensity

Under these conditions, the number $N(\Omega)$ of neutrons collected over a certain interval of time, in a given direction Ω, is proportional to the probability of a single neutron being scattered by the crystal.

Let us calculate this probability. The scattering of a neutron corresponds to a quantum transition – a transition from an initial state, in which the neutron finds itself in the incident beam, with momentum p_0, to a final state in which the neutron enters the detector, placed in the direction Ω, with momentum p. These momenta, initially p_0 and finally p, define the initial and final states of the neutron, and hence the transition which it undergoes. Adopting the symbolism introduced in sect. 4.4.3, we write this as

$$\mathscr{P}(\Omega) = |\langle p|p_0 \rangle|^2. \qquad (4.5.2)$$

Now, the scattering of the neutron by one of the atoms of the crystal is a transition which can be realized through as many intermediate states as there are atoms in the crystal, which we set equal to $\mathscr{N}$. As we have explicitly seen in sect. 4.5.1, the energy of the neutrons is too weak for the atom, against which it scatters, to retain any effects of the scattering. Thus, there is a final state which is completely specified, and the rules of the superposition of amplitudes apply here:

$$\mathscr{P}(\Omega) = \left| \sum_{k=1}^{\mathscr{N}} \langle p|p_0 \rangle_k \right|^2, \qquad (4.5.3)$$

where k denotes the ordinal number of the nucleus against which the scattering takes place.

Now, the transition of a neutron from the incident beam into the direction Ω, by scattering off the kth nucleus, breaks up into three stages, or successive transitions, each adequately described by an amplitude:

– the neutron from the incident beam reaches the site k with a transition amplitude from the proper state of momentum p_0 to a proper state of position r_k, given by $\langle r_k | p_0 \rangle$;

– another transition amplitude corresponds to its (isotropic) scattering at the site r_k. Since the nuclei are identical, this amplitude does not depend on the number k of the scattering nucleus, and since the scattering is isotropic, it does not depend on the angle Ω of deviation. It is a constant which we denote by a;

– the neutron is detected following a transition to a proper state of momentum p, given by $\langle p | r_k \rangle$.

We have, therefore, for the complete amplitude, following the principle of sequential factorization,

$$\langle p | p_0 \rangle_k = \langle p | r_k \rangle a \langle r_k | p_0 \rangle, \tag{4.5.4}$$

hence, as a consequence,

$$\mathscr{P}(\Omega) = |a|^2 \left| \sum_{k=1}^{\mathscr{N}} \langle p | r_k \rangle \langle r_k | p_0 \rangle \right|^2. \tag{4.5.5}$$

Each one of the terms $\langle p | r_k \rangle \langle r_k | p_0 \rangle$ corresponds to a well-defined transition – a certain 'quantum path'. Each one is distinguishable from the others by the intermediate site against which the scattering of the neutron is effected. This 'quantum path' does not correspond to a physical trajectory in the case of neutrons, any more than in the case of photons. It only has a meaning in connection with the computation of the relative phases of the quantum amplitude which are superposed in eq. (4.5.5).

Indeed, the computation of the resultant quantum amplitude which appears in eq. (4.5.5) is formally analogous to that of the amplitude of a plane, monochromatic, light wave, scattered by an optical 'lattice', i.e., exactly as in that case, by a regular distribution of the scattering centres, separated by distances of the order of the wavelength of the light.

As in the case of the optical lattice, the amplitudes have phase differences which are well-defined and correspond to the regularity of the geometrical distribution of the scattering centres. Under these circumstances, the total intensity displays alternating maxima and minima (fig. 4.14). It can be verified, in the case of a one-dimensional lattice (exercise 4.18), that in certain directions the amplitudes add up completely (constructive interference), while in other directions their sum is zero (destructive interference) and that the separation between the diffraction 'peaks' depends on the lattice 'spacing' (distance between two neighbouring scatterers).

We see, by extending to the three-dimensional case, that the curve of $N(\Omega)$ shows the same kind of variation, as a function of Ω, and that it is possible to deduce the interatomic distances in the crystalline lattice from the separation of the peaks of the curve $N(\Omega)$.

4.5.3. *The continuous background and incoherent superposition*

It is found, however, that for certain crystals, the curves $N(\Omega)$ display a 'continuous background', i.e., the peaks of constructive interference do not stand out against a background of zero intensity, but appear as being superimposed on a nearly uniform intensity distribution (fig. 4.14). This phenomenon can also be explained by the simple combination of quantum amplitudes, mentioned in sect. 4.4. We now proceed to prove it.

Indeed, in the course of the reasoning above, we have assumed that the state of a scattered neutron is entirely defined by its momentum. However, the neutron possesses internal degrees of freedom, due to its intrinsic angular momentum or spin. The specification of this physical magnitude is, therefore, necessary if we want to give a complete description of the state of a neutron. Hence, it is necessary for us to know if the spin state of a neutron could change during its scattering by a nucleus, since under these conditions, it would not be enough to know the initial and final momenta of the neutron in order to define the transition: *several* quantum transitions might be possible and hence several amplitudes could be necessary to describe the scattering. Such a modification of the spin state of the neutron is plausible, since the nuclear forces depend on the spin state of the nucleons and act on this state. However, in the process of the scattering the *total* angular momentum of the (isolated) nucleus + neutron system is conserved. This angular momentum J is the sum of the following three quantities: the intrinsic angular momentum S_n of the neutron, the intrinsic angular momentum S_N of the nucleus and the orbital angular momentum $L_{N/n}$ due to the relative motion of the neutron and the nucleus:

$$J = S_n + S_N + L_{N/n}. \tag{4.5.6}$$

The orbital angular momentum $L_{N/n}$ is given by the (vector) product of the distance of the neutron from the nucleus and the momentum of the neutron. Now, for the low-energy scattering process which is of interest to us here, the momentum of the neutron is small enough, so that, considering the short range of the nuclear forces, which limits the distance of the neutron from the nucleus, the orbital angular momentum $L_{N/n}$ can be taken to be zero (exercise

4.19). The law of the conservation of the total angular momentum applies, therefore, only to the sum of the spins of the neutron and the nucleus.

Certain nuclei, such as ^{12}C or ^{56}Fe, have zero spin. Under these conditions, the total angular momentum, as given by eq. (4.5.6) reduces just to the spin of the neutron, and its conservation implies that the spin state of the neutron cannot change in the course of the scattering by the nucleus. Hence, for each initial state of the neutron, there exists only one possible final state and the change of its momentum suffices to describe the transition completely. Nothing changes, therefore, either in the preceding analysis or (fortunately) in the experimental results.

There are other nuclei that have a non-zero spin (we shall see in ch. 7 that this is necessarily the case, at least for all nuclei which consist of an odd number of nucleons). In particular, let us consider the spin-$\frac{1}{2}$ nuclei such as ^{1}H, ^{23}N, ^{51}V or ^{59}Co, and let us analyze more particularly the spin states of the neutron + nucleus system. We know that it is only possible to consider proper states of one component of the angular momentum at a time, in the direction Oz, let us say. In the case of spin-$\frac{1}{2}$, this component can only assume the values $\pm\frac{1}{2}\hbar$, corresponding to the two proper states, which we denote as $\uparrow$ and $\downarrow$, respectively. For simplicity, let us suppose at first that the neutrons come out of the source filtered in a way such that they always appear in the *same* spin state, e.g., the state $\uparrow$. Likewise, let us suppose that at the start of the experiment the target crystal is 'polarized', i.e., that its nuclei are also all in the same state of spin. Two possibilities present themselves, according to whether this is the $\uparrow$ or the $\downarrow$ state.

(1) *All the nuclei are initially in the state of spin* $\uparrow$

The component J_z of the total angular momentum, before the neutron–nucleus interaction, has in this case the value

$$J_z = S_{nz} + S_{Nz} = \tfrac{1}{2}\hbar + \tfrac{1}{2}\hbar = \hbar. \tag{4.5.7}$$

Its conservation implies that J_z has the same value after the interaction. The components S_{nz} and S_{Nz} must, therefore, also preserve their initial values, since the total value (4.5.7) can only be obtained in one way, using the proper values $S_{nz} = \tfrac{1}{2}\hbar$ and $S_{Nz} = \tfrac{1}{2}\hbar$. Once again, the specification of the spin states does not involve any change in the argument developed above.

(2) *All the nuclei are initially in the state of spin* $\downarrow$

The component J_z of the total angular momentum now has the value

$$J_z = S_{nz} + S_{Nz} = \tfrac{1}{2}\hbar - \tfrac{1}{2}\hbar = 0, \tag{4.5.8}$$

both before and after the scattering. But this time, such a value could correspond to *two* different situations:

(a) the neutron and the nucleus each remains in its spin proper state, i.e., $\uparrow$ and $\downarrow$, respectively;

(b) the neutron and the nucleus each has its spin changed to become $\downarrow$ and $\uparrow$, respectively. There is a 'spin-flip', i.e., a reversal of the individual spins, but it is clear that

$$J_z = S_{nz} + S_{Nz} = -\tfrac{1}{2}\hbar + \tfrac{1}{2}\hbar = 0, \tag{4.5.9}$$

still retains the same value.

The scattering of a neutron by the crystal can, therefore, lead now to different final states and accordingly, it can make as many different transitions, of which it is necessary to add the respective probabilities to obtain the total scattering probability:

(a) The scattering proceeds without a reversal of the spins. In the final state, all the nuclei have the same spin $\downarrow$ and the probability is calculated as above, by adding the amplitudes corresponding to the scattering by each nucleus,

$$\mathscr{P}_a(\Omega) = |a|^2 \left| \sum_{k=1}^{\mathscr{N}} \langle p|r_k \rangle \langle r_k|p_0 \rangle \right|^2. \tag{4.5.10}$$

(b) The scattering is effected with a spin-flip. In the final state, *one* nucleus of the crystal has its spin reversed. It is now in the state $\uparrow$. Thus, there are as many different final states possible as there are nuclear scatterers, namely, $\mathscr{N}$. For each one of them, we easily calculate the scattering probability. Denoting by b the scattering amplitude, at the site of a neutron, for a nucleus undergoing spin-flip, the quantum amplitude for the scattering by the kth nucleus is written as

$$\langle p|p_0 \rangle_k^{\text{flip}} = \langle p|r_k \rangle \, b \langle r_k|p_0 \rangle, \tag{4.5.11}$$

and one has the corresponding probability

$$\mathscr{P}_{b,k}(\Omega) = |b|^2 \, |\langle p|r_k \rangle \langle r_k|p_0 \rangle|^2. \tag{4.5.12}$$

The total probability of scattering is the sum of the probabilities corresponding to the $\mathscr{N} + 1$ possible final states: one with and $\mathscr{N}$ without spin-flip.

From this there follows the expression

$$\mathscr{P} = \mathscr{P}_a + \sum_{k=1}^{\mathscr{N}} \mathscr{P}_{b,k},$$
(4.5.13)

which can be written as

$$\mathscr{P} = \mathscr{P}_a + \mathscr{P}_b,$$
(4.5.14)

by defining a 'scattering probability with spin-flip off an arbitrary nucleus'

$$\mathscr{P}_b = \sum_{k=1}^{\mathscr{N}} \mathscr{P}_{b,k}(\Omega) = |b|^2 \sum_{k=1}^{\mathscr{N}} \langle p|r_k \rangle \langle r_k|p_0 \rangle|^2.$$
(4.5.15)

There is an essential difference between the two terms in eq. (4.5.14). The first term, $\mathscr{P}_a$, given by eq. (4.5.10), results from the addition of amplitudes and displays the phenomenon of interference already described. By contrast, $\mathscr{P}_b$ is calculated by forming the sum, *after* having taken the square of the moduli of the individual amplitudes ... Hence, the phases of each one of these amplitudes lose all their relevance, and there does not appear any interference phenomenon. The probability $\mathscr{P}_b$ reduces to a sum of $\mathscr{N}$ equal terms. In the total probability $\mathscr{P}$, therefore, one adds to an oscillatory term $\mathscr{P}_a$, an almost constant term $\mathscr{P}_b$. It is this term which corresponds to the observed continuous background (fig. 4.14). One often says that the scattering of neutrons consists of a 'coherent' scattering, described by $\mathscr{P}_a$, and an 'incoherent' scattering, described by $\mathscr{P}_b$. In the first case, the principle of the superposition of quantum amplitudes applies and governs the phenomenon, while in the second case, the condition for its applicability (a *single* final state) is not fulfilled. Finally, let us add, that the crystal is said to be unpolarized, if half of its nuclei are in the ↑ state and half in the ↓ state. The scattering probability is easily calculated by treating each case separately (see exercise 4.20).

The phenomenon which we have just analyzed, clearly shows the essential role, played by the principle of the superposition of amplitudes, in the quantum theory, and sheds light on the necessity of being vigilant regarding the conditions of its validity (exercises 4.5, 4.10 and 4.20). The terminology introduced here is used in a general way, and one speaks about 'coherence' in the case of the addition of amplitudes and of 'incoherence' in the case of the addition of probabilities. Many different causes could introduce an incoherent background into a coherent phenomenon.

In the case of diffraction spectra, *any* effect producing spatial disorder with respect to the crystalline order (thermal agitation, contamination by impurities, etc.), translates into a loss of coherence. Indeed, the interference term in the sum given by eq. (4.5.10) becomes zero, on the average, as the fluctuations of r_k become sufficiently large ($\gtrsim \hbar/p$).

Figure 4.15 illustrates a situation of this type. One observes here a modification of the spectrum of neutrons, for a crystal of manganese oxide (MnO), as the temperature passes from 80 to 293 K: the peaks M_1, M_2, ..., show up against a continuous background between the peaks (N_1, N_2, ...) which are common to both the spectra (and also common to the spectra of X-ray diffraction).

The order which is revealed in this way at low temperatures is a magnetic order. Indeed, the Mn^{2+} ions are the carriers of a magnetic moment, to which the neutron – itself the carrier of a magnetic moment – is sensitive. The scattering amplitude of a neutron by a Mn^{2+} ion includes, besides the nuclear term which alone has been considered up to now, a term of magnetic origin as well.

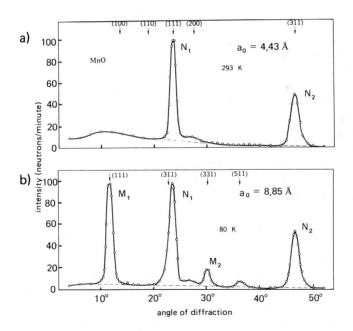

Fig. 4.15 (a, b)

c)

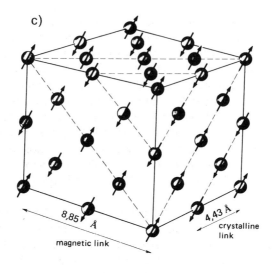

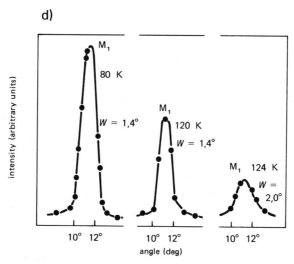

Fig. 4.15. *Neutron diffraction and magnetic order*
(a) and (b): Diffraction spectra of neutrons, for an MnO crystal, below (b) and above (a) the transition temperature, respectively.
(c) Magnetic link and crystalline link.
(d) Broadening of the 'magnetic' peak, in the neighbourhood of the Curie point [C. G. Shull, W. A. Strauser, E. O. Wollan, Phys. Rev. 83 (1951) 333].

Now, it is found that MnO displays the property of antiferromagnetism. This means, that below a certain temperature – the so-called Curie point, which, in the present case, has the value of 120 K – the magnetic moments of the Mn^{2+} ions suddenly arrange themselves in a fashion such that two neighbouring atoms carry anti-parallel moments (fig. 4.15c). The 'link' of the 'magnetic' lattice of the MnO crystal is thus double the length of that of the 'nuclear' lattice. Thus, a diffraction spectrum at 80 K exhibits peaks which are twice as close as in a nuclear spectrum. At 293 K, on the other hand, there is no magnetic order any more, and one sees a superposition, on the coherent nuclear spectrum (the peaks N_1, N_2, ...), of the spectrum of an incoherent magnetic scattering of the continuous background type (represented by the dashed lines in fig. 4.15a).

The passage, from a situation of coherence to a situation of incoherence, is particularly appreciable when one follows the evolution of a magnetic peak (e.g., M_1) as a function of temperature. One observes (fig. 4.15d) a sudden broadening of the peak in the neighbourhood of 120 K. This effect is exploited in the determination of the Curie temperature of a given material.

Exercises

4.1. Consider a monochromatic and plane *electromagnetic wave*, the electric field of which has the components

$$\mathscr{E}_x = \mathscr{E}_{0x} \sin(\omega t - kz - \varphi_x), \qquad \mathscr{E}_y = \mathscr{E}_{0y} \sin(\omega t - kz - \varphi_y), \qquad \mathscr{E}_z = 0.$$

On each wave plane ($z = $ const.), the extremity of the vector $\mathscr{E}$ describes an ellipse.

(a) Calculate, as a function of $(\mathscr{E}_{0x}, \mathscr{E}_{0y})$ and (φ_x, φ_y), the orientation of the principal axes of the ellipse (that is, for example, the angle χ of its major axis with respect to Ox) and its flatness (that is, for example, the ratio of the lengths of its two principal axes $\mathscr{E}_1$ and $\mathscr{E}_2$). To do this, one could express the components $(\mathscr{E}_{x'}, \mathscr{E}_{y'})$ of $\mathscr{E}$ in a system of axes (Ox', Oy'), obtained from (Ox, Oy) by a rotation through an angle α, and to set this angle $\alpha = \chi$, so as to bring the equation of the ellipse directly into canonical form,

$$\mathscr{E}_{x'} = \mathscr{E}_1 \cos(\omega t - kz - \varphi), \qquad \mathscr{E}_{y'} = \mathscr{E}_2 \sin(\omega t - kz - \varphi).$$

(b) Set $\mathscr{E}_1/\mathscr{E}_2 = \tan \psi$, and on a sphere of unit radius (the 'Poincaré sphere'), consider the point having azimuthal angular coordinate $\varphi = 2\chi$ and polar coordinate $\theta = \frac{1}{2}\pi - 2\psi$. Show that each state of polarization corresponds to exactly one point on the sphere, and vice versa. To what sort of points do the poles of the Poincaré sphere correspond? What about the equatorial points? For an arbitrary polarization, how does a rotation of the coordinate system, through an angle α, affect the position of the representative point on the Poincaré sphere?

4.2. To conduct an experiment on the interference of light, a source of power $P = 1$ W, emitting a monochromatic radiation of wavelength $\lambda = 0.6$ μm is used.

(a) How many *photons* reach, during the photographic exposure time $\tau = 1$ s, a surface $S = 1$ cm^2 of the photographic plate, situated at a distance $L = 2$ m from the source? What is the average time interval separating the arrival of two successive photons and the distance which separates them within the beam?

(b) By means of absorbing filters, the beam intensity is reduced by a factor of 10^7. What is the exposure time necessary to capture the same luminous energy on the plate? Experiment shows [G. I. Taylor, Proc. Cambridge Phil. Soc. 15 (1909) 114] that the interference pattern is not changed. What are the values then, of the time intervals separating the successive arrivals of two photons at the surface S of the plate, and what is the average distance between two successive photons in the beam? Conclusions?

4.3. Consider a slab, having parallel faces, and cut out of a birefringent material, which has two different indices, depending on whether the incident light vibration has a linear polarization parallel to one or the other of its principal directions Ox and Oy. This difference in the indices leads to a difference between the optical paths and hence between the phases, of the two light vibrations polarized in these two directions. The slab is a 'quarter-wave' plate, i.e., it introduces a phase difference of $\frac{1}{2}\pi$ between the 'slow' vibration (of polarization parallel to x) and the 'rapid' vibration (parallel to y). We shall study the effect of such a slab on a beam of light and find direct evidence for the *angular momentum* transported by a photon.

(a) Consider a plane monochromatic wave, of pulsation ω having a direction which is perpendicular to the plate and a linear polarization which makes an angle θ with the 'slow' direction Ox. In the system of axes (Ox, Oy), write down the components of the electric field of the outcoming wave. Show that this wave is elliptically polarized, and can be considered to be the superposition of two circularly polarized waves – left and right. From this, calculate the amplitudes, $\mathscr{E}_R$ and $\mathscr{E}_L$, of the electric fields. Show that, if $\theta = \pm\frac{1}{4}\pi$, the quarter-wave plate converts a linearly polarized wave completely into a circularly polarized wave.

(b) What are the quantum probability amplitudes, A_R and A_L, and the probabilities, $\mathscr{P}_R$ and $\mathscr{P}_L$, for a photon from the incident beam to emerge in a state of right and left circular polarization, respectively? It should turn out that

$$\mathscr{P}_R = \tfrac{1}{2}(1 + \sin 2\theta), \qquad \mathscr{P}_L = \tfrac{1}{2}(1 - \sin 2\theta).$$

(c) If the luminous power received by the slab is W, what is the number n of photons received by it per second? If one measures the circular polarization of the emerging photons, what are the average numbers n_R and n_L, of 'right' and 'left' photons, respectively, received per second?

(d) A photon of right (respectively left) circular polarization has a component of angular momentum, parallel to its direction, of value $\hbar$ (respectively $-\hbar$). What is the average angular momentum transported per second by the incident beam? By the outcoming beam? Hence deduce that the quarter-wave plate is subject to a torque,

$$C = (\sin 2\theta)\,W/\omega.$$

Calculate the maximal numerical value of this torque for a radiation of power $W = 7$ W, and of wavelength $\lambda = 1.2$ μm – values which approximately correspond to the

characteristics of an experiment performed by R. A. Beth [Phys. Rev. 50 (1936) 27].
Compare the theoretical expression for the couple, as a function of the angle θ, with the
measured experimental values of this experiment. [This measurement is carried out by
suspending the slab from the end of a very fine quartz torsion fibre and by measuring its
rotation. The experiment has to be performed in a vacuum and is riddled with difficulties.
One appreciates its great delicateness on reading the original article.]

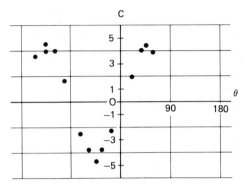

4.4. Consider again the experiment of sect. 4.3 in which a polarized beam is decomposed, by
means of an analyzer, into two other perpendicularly polarized beams, before being
reconstituted and then passed through a polaroid. However, now the final polaroid is no
longer orthogonal to the initial polarization, and makes an angle α with it.

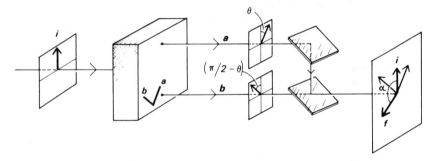

What are the intensities I_a and I_b, emerging from the polaroid, in the cases where only one
of the two beams *a* and *b* is allowed to pass through the apparatus? What is the total
intensity I' when both the beams pass through? Give both a classical wave-like and a
quantum explanation of the phenomenon.

4.5. Consider the same experiment as in exercise 4.4, modified in the following way: before the
recombination of the beams by the action of the mirrors, an *absorber* is placed in the path
of beam *a*, which only allows $\frac{1}{4}$ of the incident intensity to pass through. Discuss the
polarization of the reconstituted final beam. Give a quantum explanation. Could one not

argue that it is possible to distinguish between the two beams, from the presence of the
absorber in one of them and that one should, therefore, apply the rule of the addition of
probabilities here?

4.6.   Consider an ensemble of linearly *polarized photons*, along the direction i, which makes an
angle of 30° with the axis Ox of a system of reference xOy on the wave plane.

(a) Give the expression for the x- and y-components of the electric field vector $\mathscr{E}i$ of the
corresponding wave. Write this wave as the superposition of two circularly polarized
waves – one right and the other left. What is the probability for *one* photon from the
ensemble to be in a state of circular polarization?

This ensemble of photons is made to pass through a polaroid Π_y, the direction of
transmission of which is along Oy, and it is intended to determine the probability $\mathscr{P}_R$ of
having *one* photon pass through Π_y, with a left circular polarization. In order to do this, a
circular polarizer Π_R is placed between the source of photons, polarized along i, and Π_y,
which only allows photons of right circular polarization to pass through. $\mathscr{P}_R$ is then
determined by counting the number of photons which come out of Π_y. In the same way
$\mathscr{P}_L$ is determined by replacing Π_R by a left circular polarizer Π_L.

(b) Show by means of an explicit computation that the sum of the two probabilities so
determined *is not* equal to the probability of a photon from the ensemble to be
transmitted across Π_y in the absence of the circular polarizer.

4.7.   *The Stern–Gerlach experiment.* A monochromatic beam of (electrically neutral) silver
atoms, with velocity v, crosses the gap, of length d, between the poles of a magnet in which
there exists a strong magnetic induction gradient $\partial\mathscr{B}/\partial z$. A diaphragm (of radius r) selects
those atoms, the velocities of which are exactly along Oy. The atoms are paramagnetic –
carriers of an intrinsic spin magnetic moment.

(a) Show that each atom is subject to a force

$$F = \mu_z \frac{\partial\mathscr{B}}{\partial z} \hat{z},$$

where, μ_z is the projection of the spin magnetic moment of the atom onto the unit vector $\hat{z}$.

(b) One observes the impact of the atoms, coming out of the apparatus, on a glass plate
placed perpendicular to Oy at a distance D from the end of the gap. Evaluate the
characteristic action of the beam of atoms. Hence, deduce that in this experiment, the
atoms may be assigned a trajectory in the classical sense of the term, and that the position
z, of the impact of an atom on the glass plate, is a measure of the component μ_z of the
magnetic moment of this atom. Estimate the resolution power of this measuring
apparatus. It is assumed, that two impacts, one millimetre apart, can be separated.
The numerical values are $v = 500$ m s^{-1}, $\partial\mathscr{B}/\partial z = 10^3$ T m^{-1}, $d = 5$ cm, $r = 0.1$ mm and
$D = 10$ cm.

(c) It is found that the impacts of the atoms (although identical) of the beam, separate
into two nearly point-like spots, $T_\uparrow$ and $T_\downarrow$, of equal intensity at the same distance (fig. 1).
How can one understand this result (see sect. 2.3.4)? Show that the apparatus (called a
Stern–Gerlach apparatus) functions as a 'polarizer'. Let $\uparrow$ and $\downarrow$ denote the states of the
atoms impinging at $T_\uparrow$ and $T_\downarrow$, respectively.

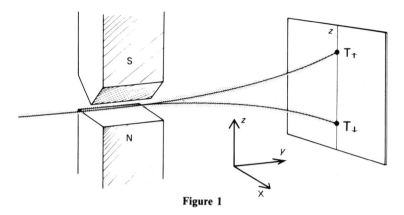

Figure 1

(d) By intercepting, at the exit to one Stern–Gerlach apparatus, the atoms corresponding to the spot $T_\downarrow$, a beam of atoms of polarization $\uparrow$ is selected. This beam is then made to pass through a second Stern–Gerlach apparatus, identical to the first, except that it is rotated through an angle θ, around the direction (fig. 2) of the beam emerging from the first apparatus.

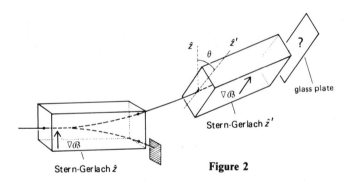

Figure 2

How many spots are now observed on a plate placed perpendicular to the axis of the second apparatus? Are the observed spots of the same intensity?

(e) The second apparatus is slightly modified, and it is arranged (by means of a judicious combination of magnets) so that the beams, separated by this apparatus, are superimposed along the axis, after having followed paths of the type shown in fig. 3.

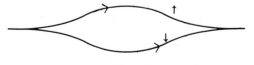

Figure 3

Now a third Stern–Gerlach of the type z (fig. 4) is placed at the exit to this apparatus. Describe what is observed on the plate.

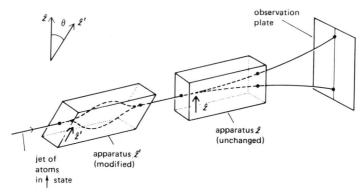

Figure 4

(f) Instead of leaving both the paths in the second apparatus open, one of them (for example, ↓) is blocked (fig. 5). It is observed that more atoms come out of the third apparatus, along the path ↓, now than in the preceding case. Explain why.

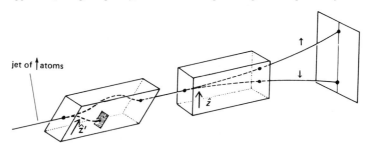

Figure 5

4.8. In a *Young's double hole* experiment, a third hole is made
– just between the other two;
– very close to one of the two holes;
– in a way such that the three form an equilateral triangle.
Describe the changes observed in the diffraction fringe pattern in the three cases.

4.9. A *Young's double hole* experiment is performed with a parallel and monochromatic light (of wavelength λ). Let $\boldsymbol{p}_0 = \hbar k_0$ be the momentum of the incident photons, with $|\boldsymbol{p}_0| = h/\lambda$.
(a) Express the interfringe distance i as a function of the quantities λ, d and D.
(b) The experimental conditions are changed by allowing a certain mobility to the plate Π, in the $x'x$-direction and by introducing a mechanism for the measurement of the component p_x, along the direction x, of the momentum transferred to the plate Π by each photon during its passage through one of the slits. Let $p_x^{(1)}$ (respectively $p_x^{(2)}$) denote

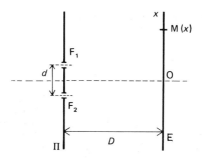

the value of p_x if the photon has passed through F_1 (respectively F_2). Express $|p_x^{(1)} - p_x^{(2)}|$ as a function of λ, and D.

(c) Show that for a measurement of p_x which should enable one to determine the hole through which the photon has passed, it is necessary that the position spectrum of the plate Π have a minimal width. Calculate this width and compare it to i. Conclusion?

4.10. A *Young's double hole* experiment is performed using electrons. A is a source of monoenergetic electrons. The plate Π is pierced by two identical holes 1 and 2 at a distance d from each other, and D is an electron detector which can be moved in a plane parallel to Π; x denotes the position of D.

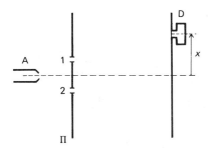

(a) Using the symbolism introduced in sect. 4.4.3, write down the probability $|\langle x|A\rangle|^2$ for an electron to make a transition, from the source A to the detector D at x. Display explicitly, the 'quantum interference' terms.

(b) An attempt is made to 'see' through which hole the electron passes. To do this, a light source S is placed on the axis of the experimental device, S emits isotropically. A photomultiplier P picks up the photons which get deflected during the passage of the electrons through the space between P and Π. Let $|a|^2 = |\langle P|S\rangle_1|^2$ denote the probability that a photon, emitted by the source S, is scattered into the detector P, by the passage of an electron coming out of hole 1. Similarly, let $|b|^2 = |\langle P|S\rangle_2|^2$ denote the probability of having a photon scattered into P by an electron coming out of hole 2. Using the same symbolism, write down the probability $\mathscr{P}(x; P \leftarrow A; S)$ for recording, simultaneously, an electron at D and a photon at P. Under what conditions on a and b would the resolving power of the experimental device enable one to 'see' through which hole the registered electron had passed? What conditions should be put on the wavelength λ of the photons for this? Show that, in this case the characteristic action of the system is $\gg h$. Discuss the consistency of this result.

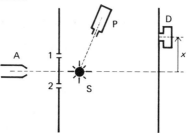

(c) A second photomultiplier is now introduced, which is placed at P′ symmetrically with respect to P, about the axis of the device.

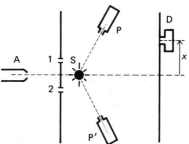

Write down the probability $\mathscr{P}(x; \text{P or } \text{P}' \leftarrow \text{A}; \text{S})$ for detecting, at the same time, an electron at D and a photon at one *or* the other of the two photomultipliers. Does this experiment enable one to decide better through which hole the detected electron had passed, than the previous one?

(d) The holes, in this experiment, behave like two sources of electrons. One often says that two sources are incoherent when their 'intensities' add and coherent when it is their 'amplitudes' that add. Discuss the notion of coherence for this example and show that there are no intrinsically coherent (or incoherent) sources, and that this property depends on the kind of experiment in which the sources enter.

(e) The plate Π is now pierced by three holes, 1, 2 and 3 and we wish to see through which hole an electron passes, with the help of a source, placed between 1 and 2; and a photomultiplier placed at P.

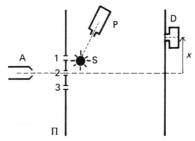

Let $|c|^2 = |\langle P|S\rangle_3|^2$ denote the probability that a photon would be scattered into P by an electron coming out of hole 3.

Suppose that $|a| \simeq |b| \gg |c|$. Calculate the probability $\mathcal{P}(x; P \leftarrow A; S)$ in this experiment. Show that it is the result of a partially coherent superposition.

4.11. A *Young's double hole* experiment is performed with a monochromatic light of wavelength λ.

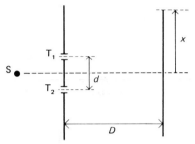

(a) Adopting the notation of the above figure, express the intensity $I(x)$, of the light wave at a point x on the observation screen. Calculate the interfringe distance.

(b) An absorber is placed over T_1, which reduces the luminous intensity by a half, without modifying the wave amplitude. What happens to the interference pattern and to $I(x)$?

(c) The absorber is replaced by a transparent plate with parallel faces, of refractive index $n \neq 1$ and thickness e. Same question as in (b).

(d) The individual behaviour of each photon is to be analyzed. Does the presence of the absorber or the slab enable one to know through which hole the photon has passed? With the help of the symbolism introduced in sect. 4.4.3, write down the probability $\mathcal{P}(x \leftarrow S)$ for a photon to effect a transition from the source to the point x, in each of the cases corresponding to questions (a), (b) and (c). Let τ_1 and τ_2 denote the amplitudes transmitted through the holes T_1 and T_2. Suppose that in question (a), $\tau_1 = \tau_2 = 1$. Show that the presence of the absorber, or the slab, multiplies τ_1 by a certain factor. Evaluate it explicitly in both cases.

4.12. (a) The interference pattern produced by a *Young's double hole* arrangement, illuminated by two independent point sources S and S', is observed. With the help of the symbolism introduced in this chapter, express the probability $\mathcal{P}(x)$ for the arrival of a photon at x.

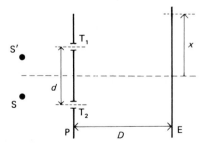

(b) Instead of illuminating the holes with two independent sources, they are illuminated using the arrangement shown in the figure below. S and S' are two pin holes (separated by a distance d), pierced into an opaque screen and illuminated by a single point source S_0. How would you express the transition probability of a photon, from S_0, at x?

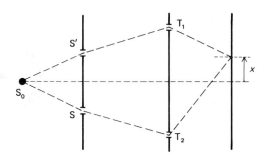

(c) In front of the pin hole S', is placed a slab [of the same type as in question (c) of the preceding exercise]. Let θ denote the phase difference introduced by this plate. Express the transition probability, from S_0 to x as a function of θ.

(d) Show that, by taking the average over θ $(0 \leqslant \theta \leqslant 2\pi)$, of the result so obtained, one retrieves the probability $\mathscr{P}(x)$, found in (a). Hence deduce that two independent sources produce the same effect as two sources, of the type in the figure above, emitting with a random relative phase difference.

4.13. It is possible to obtain an *electron interference* by means of an ingenious device, invented and constructed for the first time by Möllenstedt and Düker (1954). One uses a beam from an electron microscope, consisting of electrons emitted by a source S and accelerated in the vertical direction by a potential $U = 8 \times 10^4$ V. At a distance $a = 10$ cm from S, a positively charged cylindrical wire F deflects the electrons (fig. 1a), just as a double prism ('Fresnel biprism') would deflect the rays from a beam of light (fig. 1b).

(a) Show that the electric field in the vicinity of the wire is $\mathscr{E} = Kr/r^2$, at a distance $r = |\mathbf{r}|$ from the axis of the wire, where $K = \gamma/2\pi\varepsilon_0$, γ being the linear, electric charge density carried by the wire. What is the order of magnitude of the potential difference W established between the wire and an external armature, used to obtain a charge density $\gamma = 10^{-10}$ C m^{-1}? The exact value of W evidently depends upon the detailed geometry of these armatures, but not sensitively. This is proved, e.g., by considering a cylindrical armature (fig. 2) of radius R, which is concentric with the wire, of radius r (something that is evidently impossible for the experiment being envisaged, since it is, of course, necessary to let the electrons pass through!) Show that, for a given geometry, the charge density γ is proportional to the potential W.

(b) We want to calculate the angle of deviation α of the trajectory of an electron (fig. 3) considered here as a classical point particle. Evidently, this angle can only depend upon the charge q_e of the electron, its initial kinetic energy E, the impact parameter h (the distance of the asymptotically free trajectory from the axis of the wire) and the intensity of the electric field, i.e., the constant K. Establish the dimensional equations of these magnitudes and

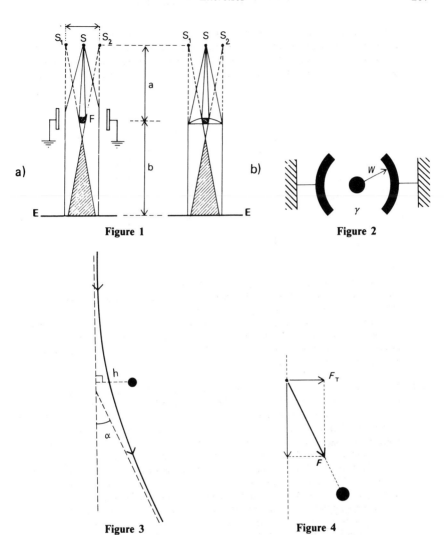

Figure 1

Figure 2

Figure 3

Figure 4

deduce that the angle of deviation does *not* depend on the impact parameter h, and hence that it is the same for all the electronic trajectories. Show that, if the deviating potential W is not too large, we must have

$$\alpha = C q_e K / E,$$

where C is a purely numerical constant, of the order of unity, which will be taken to be equal to $\frac{1}{2}\pi$ in what follows [see optional question (g)]. Calculate numerically, the angle of deviation α using the given values of the parameters. Show that the independence of α from h implies that the electrons deviated by the wire F – depending upon whether they had passed along one side or the other – seem to come out of one or the other of two virtual sources, S_1 and S_2, situated on either side of S. Calculate the distance d between these two sources.

(c) Calculate the wavelength of the electrons. Show that one can treat the motion of the electrons in the approximation of geometrical optics.

(d) We want to calculate the probability for an electron to arrive at a point M, of abscissa x, on the screen E, situated at a distance $b = 6$ cm from the wire. Is it possible to know whether an electron, which reaches M, has passed by one side of the wire or the other? Hence deduce that the probability amplitude for the arrival of an electron at M is the sum of two probabilities corresponding to these two possibilities. Show that the relative phase of these two amplitudes depends on the position of M and hence that an interference phenomenon is observed on the screen. Express this relative phase and the interfringe distance ξ as a function of λ and the geometrical parameters a, b and d. What is the value of ξ? How could one observe these fringes visually?

(e) Figure 5 indicates (with the scale) the fringes observed for different deviating potentials W. Evaluate the variation of the interfringe distance with W for cases (c) to (f) and compare with the theory. How would one explain fig. 5a and, partly, 5b? Hint: Compute the radius r of the wire F by proving (using geometrical considerations) the following relationship between the width Δ of the interference region and the other dimensions of the problem:

$$bd = \Delta a + 2r(a + b).$$

From a measurement of Δ (from fig. 5e, for example), deduce the value of the radius r.

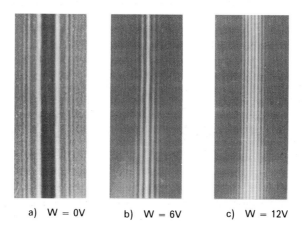

a)　W = 0V　　　　　b)　W = 6V　　　　　c)　W = 12V

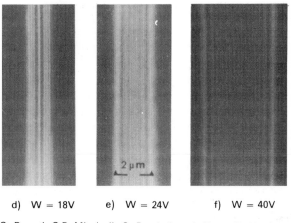

d) W = 18V   e) W = 24V   f) W = 40V

[O. Donati, G.F. Missiroli, G. Pozzi, Am. J. Phys. *41*, 639 (1973)].

Figure 5

(f) The electronic current has a value $I = 6 \times 10^{-14}$ A. From this deduce the number of electrons arriving at the screen E_1 per unit time, the interval of time separating, on the average, the emission of two successive electrons and their average distance. Hence deduce that, on the average, there is *at most* only one electron in the set-up and that the phenomenon of interference is not of a collective nature: 'every electron interferes with itself' (Dirac).

(g) *Optional*. In classical mechanics, calculate the transverse momentum acquired by an electron in the electric field of the charged wire, having a trajectory comparable to that of free motion, hence rectilinear, and moving with a constant velocity, so that

$$\delta p_T = \int_{-\infty}^{\infty} F_T \, dt, \quad \text{with } dx = v \, dt.$$

Hence deduce the deviation $\alpha = \delta p_T / p$, where $p = mv$ is the initial momentum, and show that one also has $C = \frac{1}{2}\pi$.

(h) A device inserted into *one* of the two branches of the interferometer, gives rise to an increase in the energy E of the electrons. Write down the spatio–temporal dependence at a point M on the screen. Hence deduce that the electronic current, registered at such a point, of abscissa x, and at time t is of the form

$$I = I_0 \sin^2(\pi x/\xi + \pi v t),$$

where ξ is the interfringe distance obtained earlier and v a beat frequency which has to be calculated.

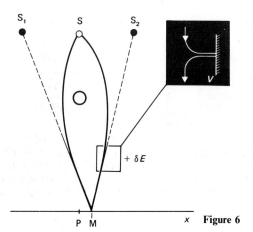

P M   x   **Figure 6**

(i) The device in question, in a simplified form, consists of reflecting the electron off a mobile mirror, which is moving with a velocity V. Assuming an elastic reflection off the mirror, calculate the increase in energy ε and show that the frequency of the beats can be expressed in the form: $v = V/\lambda$, where λ is the wavelength of the electrons.

The curve (fig. 7) shows the current, registered in the course of time, across a slit situated at a certain point of the screen. The mirror moves first in one direction, then in the other, with a velocity $V = 0.04$ Å s^{-1} (this extremely low velocity is obtained by controlling the displacement of the mirror by piezo-electric effect). Compare the theoretical and experimental beat frequencies. Calculate, in eV, the energy difference ε thus brought out.

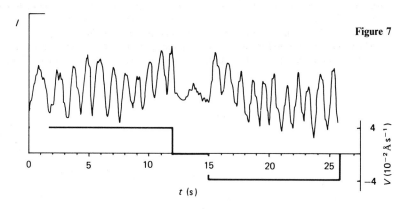

Figure 7

t (s)

4.14. Consider a simple model of a *nuclear reaction*, depending upon a quantum approximation of a geometro-optical nature, applied to elastic collisions from the surfaces of nuclei. The incident projectile – for example, an α-particle – possesses an initial momentum $\boldsymbol{p}_i$. The nucleus is assumed to be infinitely heavy. After the reaction, the particle is deviated through an angle θ. Let $\boldsymbol{p}_f$ be its final momentum; $|\boldsymbol{p}_f| = |\boldsymbol{p}_i|$, since the collision is elastic.

Suppose that the interaction is point like, at a point P on the surface of the nucleus. Let $OP = r$ and $p_f - p_i = q$.

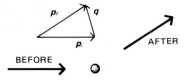

(a) Calculate the angular momentum given up by the projectile to the nucleus. The modulus of this angular momentum is quantized and has the value $l\hbar$ (l integer). Hence deduce that the point P lies on the surface of a cylinder, the axis of which is the diameter of the nucleus in the direction q, and having radius $a = l\hbar/|q|$. Compute numerically, the value

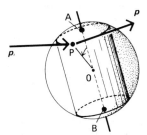

of $|p_i|$ and the order of magnitude of a for α-particles of kinetic energy $E = 64$ MeV (it is assumed that $l \simeq 1$ and $\theta \lesssim \frac{1}{2}\pi$). Compare this order of magnitude to nuclear radii. Hence deduce that such an interaction can only be produced in a neighbourhood of the two poles, A and B, of the nucleus, defined by the diameter in the direction q.

(b) Consider the simplest case, in which the angular momentum exchanged is zero ($l = 0$) and where the interaction only takes place at A and B. For an emerging α-particle,

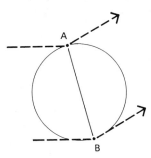

are the processes of scattering at A and at B, distinct? Hence deduce that the scattering amplitude is the sum of two amplitudes, corresponding to the interactions at A and B, respectively. Calculate the phase difference between these two amplitudes, in the approximation of geometrical optics, by considering the optical paths corresponding to

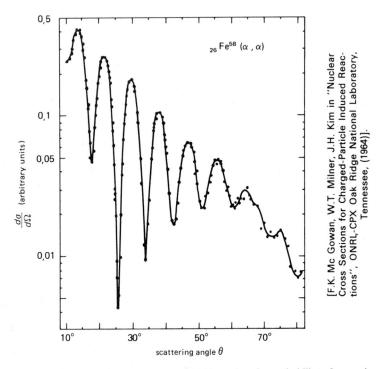

[F.K. Mc Gowan, W.T. Milner, J.H. Kim in "Nuclear Cross Sections for Charged-Particle Induced Reactions", ONRL-CPX Oak Ridge National Laboratory, Tennessee, (1964)].

the trajectories passing through A and B. Show that the probability of scattering, and hence the effective cross-section, displays, as a function of the scattering angle θ, alternating minima and maxima. The figure above shows this result in a measurement of the effective cross-section for the scattering of α-particles, of kinetic energy $E = 64$ MeV, on $^{58}_{26}$Fe nuclei. Compare this result with the theoretical predictions. Hence, deduce a value for the radius of the nucleus being considered.

4.15. The scheme shown below is that of a *neutron interferometer*, with the help of which one can find evidence for the action of gravity upon neutrons. A beam, falling upon a suitably adjusted crystal (at 'Bragg incidence') is separated into two beams – one transmitted directly and the other reflected. Consider the set-up shown in the figure, where a first crystal separates the incident beam at A and the third crystal recombines at C, the beams

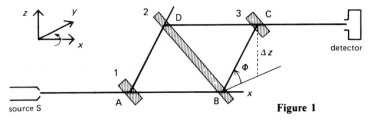

Figure 1

reflected by the second at B and D. The plane ABCD makes an angle Φ with the horizontal plane xOy. The neutrons issuing from the source S are monochromatic. The detector picks up the neutrons which cross the apparatus.

(a) The vertical axis, Oz, of the figure marks the direction of the gravitational field, in which the interferometer is placed. Is the energy of neutrons along the branch AB the same as along the branch CD? Do they have the same de Broglie wavelength?

(b) Using the symbolism introduced in this chapter (sect. 4.4.3), express the transition probability $\mathcal{P}(\text{detector} \leftarrow \text{source})$, for *one* neutron, from the source to the detector. Denote, for example, by $\langle C|D \rangle$, the amplitude of propagation from D to C, and assume that in each crystal, the amplitudes of transmission in the direct and the reflected beams are equal. In what way does the difference Δz in altitude between the two branches of the apparatus affect the probability?

(c) At first the apparatus is arranged so that the plane ABCD is vertical ($\Phi = \frac{1}{2}\pi$). Assuming that the relative phase of the amplitudes, which are superimposed at the entrance to the detector, may be calculated as in optics (i.e., by means of the difference of the 'optical paths'), show that the phase differences along the routes AD and BC are the same, while on the other hand, the phase difference along the route AB is greater than that along DC, by an amount

$$\Delta\varphi = 2\pi M^2 gS\lambda/h^2,$$

where λ is the wavelength of the neutrons being used, $\lambda = 1.42$ Å; M is the mass of the neutron, g the acceleration due to gravity and $S = 10.1$ cm^2, the area of the parallelogram ABCD.

(d) The apparatus is now rotated about the axis Ox. Interpret the curve in fig. 2, which reproduces the variation in the number of neutrons picked up by the detector, as a function of the angle Φ made by the plane ABCD with the horizontal. Calculate the phase difference $\Delta\varphi$ as a function of the angle of rotation Φ. Verify quantitatively, the accord between theory and experiment.

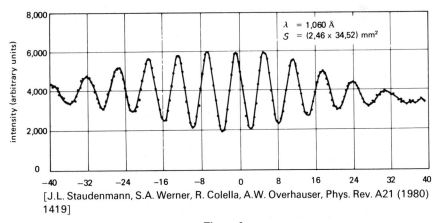

[J.L. Staudenmann, S.A. Werner, R. Colella, A.W. Overhauser, Phys. Rev. A21 (1980) 1419]

Figure 2

4.16.   It is often said that 'a quanton interferes only with itself'. How should this expression be understood? It is known that charged particles (e.g., protons) traverse a crystal along certain preferred directions. This gives rise to 'peaks' of the transmitted intensity, comparable to diffraction peaks. One might attempt, by analogy, to look for a quantum explanation, of the 'undulatory' type, for this phenomenon. The protons are emitted by a thermal source ($T = 300$ K) and the divergence of the beam cannot be reduced below $\alpha = 10^{-4}$ rad. Calculate the transverse momentum of the proton. From this obtain the transverse dimension of the localization region Δh, of the proton. Compare it to the distance between atoms in the interior of the crystal and hence deduce that there is *no* need for a quantum explanation of this phenomenon, called '*channelling*'. Give a possible classical explanation.

4.17.   In a high-resolution *electron microscope*, electrons, accelerated by a potential $U = 100$ kV, are used to visualize individual atoms.

   (a) Compute the wavelength of the electrons and compare it to the atomic dimensions. Hence, deduce that the semiclassical approximation, which enables one to evaluate the quantum amplitudes, with the help of trajectories, is justified.

   (b) We schematize below the interaction of the electron with the atom by an average constant potential of value $-V$, depending, evidently, upon the distance d from the centre of the atom. Let l be the length of the path in the interior of the atom. Show that the phase difference undergone by the localization amplitude of the electron, upon crossing the atom, is

$$\delta\varphi = q_e Vl/hv,$$

where v is the velocity of the electron.

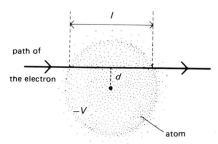

   (c) Evaluate $\delta\varphi$ numerically by taking for V and l, reasonable values corresponding to an iron atom.

4.18.   An *optical grating* consists of a series of N parallel lines (perpendicular to the plane of the figure) which are equidistant, and separated by a distance d. This lattice is illuminated by a monochromatic, parallel light of wavelength λ.

   (a) Show, without calculating the intensity, that the waves diffracted by each line interfere constructively in the directions θ_n such that $d \sin \theta_n = n\lambda$.

   (b) Show, again without calculating the intensity, that for each direction $\theta \neq \theta_n$, the interference of the diffracted waves is zero as $N \to \infty$.

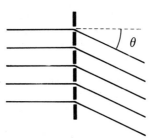

(c) Calculate the intensity $I(\theta)$ diffracted in the direction θ. Plot the curve $I(\theta)$. Retrieve the results of the qualitative discussions in points (a) and (b).

4.19. Consider the *diffraction of neutrons*, of kinetic energy $E = 0.025$ eV, by a crystal. For a neutron to interact effectively with a nucleus in the crystal, the maximum distance from the centre of the nucleus, at which it could pass, must be comparable to the nuclear radius, which could, e.g., be $R = 5$ F. Hence obtain a classical evaluation of the angular momentum of the neutron with respect to the centre of the nucleus. Compare it to the possible proper values of the angular momentum $L = [l(l+1)]^{1/2}\hbar$ (see sect. 3.3.3), and hence deduce that the relative angular momentum of the neutron and the nucleus may be taken to be identically zero.

4.20. The *diffraction of neutrons*, of momentum p_0, by the nuclei of a crystal is being studied. To simplify the problem, consider a unidimensional crystal made up of a chain of N identical atoms, separated by a distance a (fig. 1).

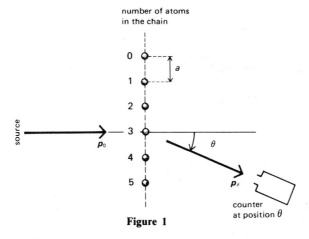

Figure 1

This 'crystal' is perpendicular to p_0. The nuclei are considered to be infinitely heavy, and remain at rest during the scattering, which is elastic. The neutrons are picked up by a detector pointing in a direction which makes an angle θ with the incident beam.

(a) With the help of the notation introduced in sect. 4.4.3, write down the probability

$\mathscr{P}(\theta)$ for a neutron to be scattered by the whole crystal in the direction θ. Denote by α the scattering amplitude of a neutron by a nucleus and assume that it is independent of θ.

(b) Calculate the phase difference between the quantum amplitudes corresponding to the scattering by two consecutive nuclei in the chain as a function of N and θ and give the shape of the curve $\mathscr{P}(\theta)$. Recall why it displays the variations in the intensity of the beam of neutrons scattered in the direction θ.

(c) Consider now a crystal, the nuclei of which have spin-$\frac{1}{2}$, like the neutrons. Let $\uparrow$ and $\downarrow$ symbolically denote the two spin states of one single particle (nucleus or neutron). Suppose that all the nuclei are in the $\uparrow$ state, and so also are the neutrons. Denote by α the scattering amplitude. Once again, using the symbolism of sect. 4.4.3, express the probability $\mathscr{P}(\theta)$ of the scattering of a neutron by the crystal.

(d) Suppose now, that the beam of neutrons is polarized in the $\uparrow$ state, but that the nuclei are in the $\downarrow$ state. Denote by β the scattering amplitude with spin reversal. What does $\mathscr{P}(\theta)$ become? Plot the curve of the scattered intensity.

(e) What does $\mathscr{P}(\theta)$ become when half the target consists of $\uparrow$ nuclei and the other half of $\downarrow$ nuclei, distributed randomly, the neutron beam being still polarized in the $\uparrow$ state? What is the shape of the curve of the scattered intensity? What happens to $\mathscr{P}(\theta)$ if the beam is no longer polarized?

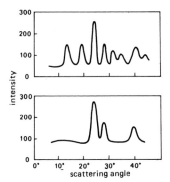

[From M.J. Marcinkowski and
N. Brown, J. Appl. Phys. 32 (1961) 375.]

Figure 2

(f) Consider now a crystal in which the atoms, of zero spin, belong to two different species A_1 and A_2 of equal numbers $N_0 = \frac{1}{2}N$. Denote by α_1 and α_2 the scattering amplitudes for a neutron from an A_1 and an A_2 atom, respectively. Suppose that α_1 and α_2 are real. Calculate $\mathscr{P}(\theta)$ in the following three cases:
– the atoms A_1 and A_2 alternate: $A_1 A_2 A_1 A_2 A_1 \ldots$,
– the atoms alternate in pairs: $A_1 A_1 A_2 A_2 A_1 A_1 \ldots$,
– the atoms are distributed randomly.
Give the shape of the variation of the scattered intensity as a function of θ. Specify the position and the height of the scattering peaks in each case.

(g) Interpret, using the results of the previous question, the two curves of fig. 2, which give the spectra of neutron diffraction from an alloy of $Ni_3 Mn$, in two phases – one ordered and the other disordered.

4.21. *The Davisson–Germer experiment:* 'The investigation reported in this paper was begun as the result of an accident which occurred in this laboratory in April 1925. At that time we were continuing an investigation, first reported on in 1921, of the distribution-in-angle of electrons scattered by a target of ordinary (polycrystalline) nickel. During the course of this work a liquid-air bottle exploded at a time when the target was at a high temperature; the experimental tube was broken, and the target heavily oxidized by the inrushing air. The oxide was eventually reduced and a layer of the target removed by vaporization, but only after prolonged heating at various high temperatures in hydrogen and in vacuum.

When the experiments were continued it was found that the distribution-in-angle of the scattered electrons had been completely changed. Specimen curves exhibiting this alteration are shown in fig. 1. These curves are all for a bombarding potential of 75 V. The electron beam is incident on the target from the right, and the intensities of scattering in different directions are proportional to the vectors from the point of bombardment to the curves. The curve on top was obtained before the accident, the one below after it.'

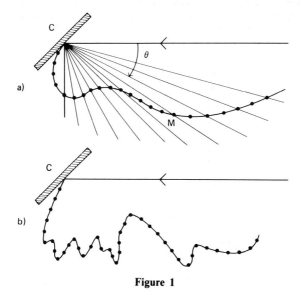

Figure 1

[The horizontal arrow, from right to left, indicates the direction of the incident beam. The shaded bar C represents the crystal. Along each direction of scattering, marked by the angle θ which it makes with the incident direction, Davisson and Germer drew a length, proportional to the electrical intensity indicated by the galvanometer, which they used to collect the electrons scattered during a given interval of time in this direction. They obtained in this way a polar representation of the angular distribution of the scattered electrons (this is also called a scattering indicator).]

'We must admit that the results obtained in these experiments have proved to be quite at variance with our expectations. To us, it seemed likely that strong beams would be found issuing from the crystal along what may be termed its transparent directions – the directions in which the atoms in the lattice are arranged along the smallest number of lines per unit area. Strong beams are indeed found issuing from the crystal, but only when the speed of bombardment lies near one or another of a series of critical values, and then in directions quite unrelated to crystal transparency.' [C. Davisson, L. H. Germer, Phys. Rev. 30 (1927) 705.]

This is how the article of Davisson and Germer begins, in which they describe their famous experiment on the diffraction of electrons – an experiment which appeared at that time to be the experimental proof of the validity of the de Broglie relation.

(a) By examining the changes brought about in the crystal during the 'accident', show that the observed phenomenon concerns essentially the atoms on the surface of the crystal.

(b) Why do low-energy (< 100 eV) electrons not penetrate the crystal beyond the first layer (or the first few layers)?

(c) Show that the shape of the curve (1b) cannot be explained if the electrons are assumed to be classical particles. What would be the angular distribution of the electrons, scattered by the surface of the crystal, under this hypothesis? It is understandable now, why the indicator (1a) would not have had to come as a surprise to Davisson and Germer, while the consequences of the 'accident', on the other hand, could have surprised them.

(d) In what way, on the other hand, does the shape of the curve (1b) manifest immediately, to the eye of the present day physicist, the non-classical (i.e., quantum) nature of the phenomenon?

When in 1926, during a conference held in Oxford, Davisson gave an account of the curves in fig. 1, some of his colleagues, who had read the memoir of de Broglie, where he had established the relation $\lambda = h/mv$, suggested the hypothesis that the maxima, observed on the curve 1b, could be the diffraction peaks of the 'wave associated to the electrons'. It was to a verification of this hypothesis that the article cited had been devoted.

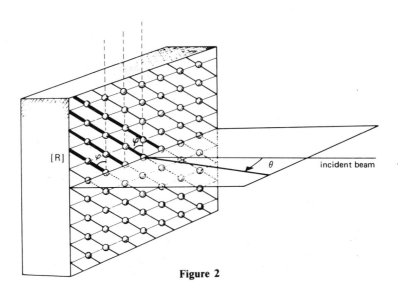

Figure 2

'From the experimental results, and having once posed the hypothesis on the wave nature of the phenomenon, it would have been possible for us to proceed as follows: first show that it is possible, in a simple and natural way, to associate to the beam of incident electrons a wavelength, then calculate this wavelength and finally show that it correctly verifies the relation $\lambda = h/mv$. It seemed preferable to us to start with the idea that a beam of electrons, of velocity v, is in a way equivalent to a radiation of wavelength h/mv and to show to what extent our observations could be accounted for on the basis of this hypothesis!'

On the plane of the surface of the crystal, the atoms, which are regularly spaced, can be regrouped into equidistant rows which are parallel to the straight line joining two of them. Each one of these families of rows (let us denote them by [R]) is characterized by a angle φ – which it makes with a reference direction on the plane of the surface. Once having accepted the undulatory character of the phenomenon, it was natural for Davisson and Germer to assume that each family [R] behaves like a grating vis-à-vis the incident electrons, similarly to an optical grating (obtained by etching on a glass plate a series of parallel and equidistant lines).

(e) Show that the diffraction peaks, corresponding to a given family [R], occur on the plane perpendicular to [R] and containing the direction of the incident beam, and that, in this plane, their position φ with respect to the incident direction is such that $d_R \sin \varphi = n\lambda$ (n integral); d_R denotes the distance separating two consecutive rows of the family [R] and λ the wavelength of the electrons. Suppose that the incident beam is normal to the surface of the crystal, which was the case in the experiment of Davisson and Germer.

(f) Figure 4 has been taken from the article cited. Figure 3 indicates the atomic arrangement on the surface of crystal of nickel used by Davisson and Germer, which displays a periodicity of $120°$ in φ.

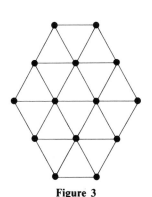

Figure 3

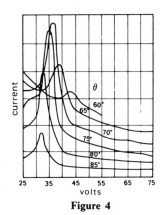

Figure 4

The curves in fig. 4 are obtained by moving the galvanometer (collector for the current) in a plane perpendicular to the surface of the crystal, and by noting, for each position φ of the galvanometer on this plane, the variation of the diffracted intensity as a function of the accelerating voltage V. From these curves, Davisson and Germer obtained the following table:

φ (deg.)	V (volts)	$V^{1/2} \sin \varphi$
85	32.0	5.64
80	33.0	5.66
75	35.0	5.72
70	36.0	5.64
65	38.5	5.63
60	42.5	5.65

In what way does this table constitute a numerical verification of the relation $\lambda = h/mv$? What is the precision of this verification? Can one deduce from here the distance which separates two rows of atoms in the family being considered?

(g) Imagine that the surface of the crystal is contaminated by foreign atoms, distributed randomly on the surface. An electron from the incident beam can be scattered just as well by a nickel atom as by a foreign atom. How should one calculate the probability of the scattering of an electron by the surface of the crystal? Now explain why fig. 1 does not show any peaks.

4.22. In the experiment schematized in fig. 1, a mono-energetic *beam of atoms* of argon, of energy E_0, is sent onto the surface of a crystal of silicon. The silicon crystal is shaped in a way such that the atoms on its surface are distributed at the nodes of a square lattice of side $a = 5.43$ Å. The plane of incidence is parallel to one of the directions of the lattice. The variation of the flux $\mathscr{F}(\theta_r)$ of argon atoms, reflected by the crystal, in the direction θ_r, is measured. In this way, a curve is obtained, which is analogous to that in fig. 2, where the experimental conditions are $\theta_i = 67°$, $E = 0.075$ eV.

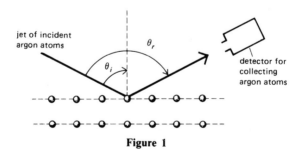

jet of incident argon atoms

θ_r

θ_i

detector for collecting argon atoms

Figure 1

(a) How would you interpret the existence and the regularity of the peaks observed on the curve of fig. 2?

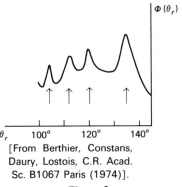

[From Berthier, Constans,
Daury, Lostois, C.R. Acad.
Sc. B1067 Paris (1974)].

Figure 2

(b) One might be tempted to say, a priori, that the atoms of argon only interact with the atoms on the surface of the crystal, and do not 'see' the ones that are inside. In order to verify this point, one studies the interaction potential of *one* argon atom with *one* atom of silicon, as a function of the distance separating their centres. This potential, which can be calculated, is represented in fig. 3. Justify *qualitatively* the shape of its variation. For an atom of argon to penetrate the silicon crystal, it is necessary that it be able to pass between two atoms. Evaluate from the curve in fig. 2, the minimal energy that this argon atom must have. Hence, deduce that under the experimental conditions of fig. 2, the argon atoms do not actually interact with the atoms on the surface of the crystal.

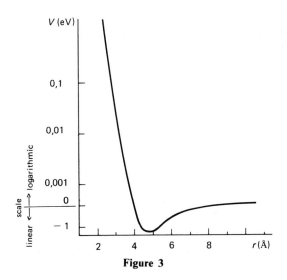

Figure 3

(c) Calculate the theoretical position of the diffraction peaks, so as to yield the relation $2d \sin \theta = n\lambda$, established in exercise 4.21, during the analysis of the Davisson–Germer experiment. Compare it to the experimental result of fig. 2. Hence conclude that the atoms on the surface of the crystal interact collectively with each atom of argon.

(d) The diffraction of atoms, just as the diffraction of slow electrons (of energy < 300 eV), allows one to perform 'surface crystallography'. Which of these two methods seems preferable to you for studying the surface and just the surface?

4.23. Neutrons coming out of a nuclear reactor display a rather large energy spectrum (Maxwellian distribution in thermal equilibrium). For many experiments, it is necessary to make this spectrum much narrower, so as to obtain, at least approximately, a monochromatic beam.

(a) Show, using the relation $2d \sin \theta = n\lambda$, that a polycrystalline block of matter (fig. 1) behaves as a *filter for neutrons*, allowing to pass through only neutrons of wavelength larger than $\lambda_0 = 2d_0$, where d_0 is the maximal distance between two crystalline planes.

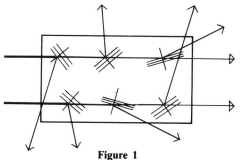

Figure 1

(b) Thus, interpret fig. 2 which gives the spectrum of neutrons that have crossed a polycrystalline block of beryllium.

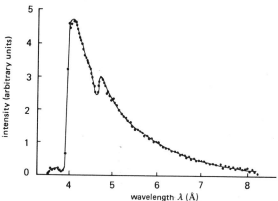

[From M.G. Zemlyanov and N.A. Chernoplekov, *Pribory Tekh, Ekp.* n° 5 (1962)].

Figure 2

(c) The crystal parameter d_0 is given for the following materials. Deduce from it the critical wavelength λ_0 and the corresponding energy E_0 (minimal or maximal?) of the filtered neutrons:

	Material				
	Beryllium	Beryllium oxide	Lead	Graphite	Bismuth
d_0 (Å)	2.0	2.2	2.9	3.4	4.0

5

Quantons in space and time

We have shown in ch. 4 that in quantum theory, the state of a system cannot be defined by the numerical values of the physical magnitudes of the system, as in classical theory. It is the set of projection amplitudes, denoted by $\langle v|u \rangle$, of a state u onto the set of all the other states v of the system, which characterizes the system in the state u.

As with any physical theory, quantum theory undertakes to study the time evolution of the system, taking into account the various interactions to which it could be subjected. In classical theory, this evolution is described by a functional dependence on time, of the different physical magnitudes of the system. In quantum theory, it is in the fundamental concept of the amplitude that the temporal evolution of the system must appear. It is to a clarification – again on a heuristic level – of the essential characteristics of this time dependence of quantum amplitudes, that the first section of this chapter is devoted.

Furthermore, while it is true that a state u of the system is completely determined by the set of its projections $\langle v|u \rangle$ onto the other states v, some of these projections are – because of their physical interpretation – of greater interest than the rest. This is the case with the projections (or transitions) of the state u, being considered, onto the states of localization in space. These 'localization amplitudes'– generally called 'wavefunctions'– will be the subject of the second section of this chapter.

In the third section, we shall show how the notions introduced enable us to give a quantum formulation of collision phenomena – something which is of considerable practical importance. In particular, we shall sketch the quantum interpretation of the notion of an effective cross-section.

The fourth section will collect the results of the first two sections, to carry out a simultaneous spatial and temporal description of the behaviour of quantons.

5.1. *The time dependence of quantum amplitudes*

Generally, a quantum system changes its state in the course of time. Let us suppose, that we read off at each instant t, the state u in which the system finds itself. The set of all these states u constitutes a very particular family of states, among all the possible states of the system. By following them from one instant to the next, we obtain some sort of a 'trajectory' of the system – a 'trajectory' in the set of its states – which we symbolically denote by $\{u(t)\}$.

Let v be an arbitrary state of the system, i.e. – pursuing the metaphor – a 'point' in the set of states (fig. 5.1). The quantum projection amplitude $\langle v|u(t)\rangle$, from the state occupied, at the instant t, by the system describing the trajectory $\{u(t)\}$, onto the reference state v, is a (complex) number which varies in the course of time. The variation of $\langle v|u(t)\rangle$, as a function of time, reflects the temporal evolution of the system. The complete knowledge of this evolution is equivalent, therefore, to the knowledge of the set of functions, $\langle v|u(t)\rangle$, of the time, for all possible reference states v.

5.1.1. *Stationary states*

There is, however, a situation where the time dependence of the amplitude $\langle v|u(t)\rangle$ does not depend essentially on the reference state v and is, by the

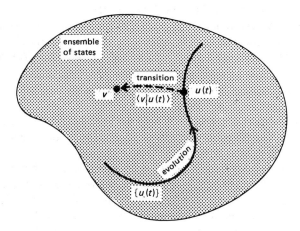

Fig. 5.1. *The quantum 'trajectory'*
In the set of possible states of a quantum system, the states successively occupied by a system, in the course of its evolution, define a trajectory $\{u(t)\}$ parametrized by the time. The time dependence of the transition amplitudes $\langle v|u(t)\rangle$, onto all the possible states v, describes the evolution of the system.

same token, particularly simple to characterize. This is the case where the state u is a proper state of the energy of the system.

Consider such a proper state, characterized by the energy proper value E. As we have seen in sect. 3.2.1, this state is stationary, and remains an energy proper state, with the same proper value E, in the course of the temporal evolution of the system. If, therefore, the system is 'prepared' at the instant t_0 in one of its energy proper states, it is still, at a later instant $t_1 > t_0$, in the same proper state, characterized by the same energy proper value. In other words, for a given quantum system, the trajectory of a stationary state is, from start to finish, characterized by one and the same energy proper value, which can therefore be used to label it. Thus, we denote this trajectory by $\{u_E(t)\}$. Each one of its points (i.e., each one of the states, successively occupied by the system during its 'stationary' evolution in time), is marked by a value of the coordinate t, which plays, therefore, the role of a curvilinear abscissa here.

Along the same stationary trajectory $\{u_E(t)\}$, the projection probabilities, from a point of this trajectory onto an arbitrary given state v, preserve their values in the course of time, since by the very definition of a stationary state, its physical properties are independent of the time. Stated otherwise, the condition

$$|\langle v|u_E(t)\rangle|^2 \quad \text{independent of } t, \text{ and } v, \text{ arbitrary,} \tag{5.1.1}$$

conveys the stationarity of the trajectory $\{u_E(t)\}$.

This condition imposes on the quantities $\langle v|u_E(t)\rangle$ a certain type of dependence upon t. Indeed, $\langle v|u_E(t)\rangle$ – which must maintain a constant modulus – can only depend on the time through its phase. We, thus, have

$$\langle v|u_E(t)\rangle = K\, e^{i\varphi(t)}, \tag{5.1.2}$$

where K is a real, positive constant which depends both on v and u_E. As for $\varphi(t)$, it is real-valued function, the analytic form of which depends, a priori, on both v and E.

However, $\langle v|u_E(t)\rangle$ refers to a stationary phenomenon, characterized by a unique value of the energy. Now, in the quantum theory, the concepts of energy and pulsation are one and the same thing. The amplitude $\langle v|u_E(t)\rangle$, characterized by the unique and well-defined value $\omega = E/\hbar$ of its pulsation, is, therefore, a harmonic function of time. In other words, the phase $\varphi(t)$ is a linear function of the form

$$\varphi(t) = \pm \frac{E}{\hbar} t + \varphi_0, \tag{5.1.3}$$

where φ_0 is a real constant. The choice of the sign $\pm$, before the argument ωt is a matter of convention. In conformity with the most current usage, we choose the $-$ sign.

Furthermore, we shall choose from now on, a system of units appropriate to the quantum theory, in which the quantum constant $\hbar$ is taken to be unity, and hence becomes invisible. However, to reassure the reader, who might be disturbed by this disappearance, we shall signal, by means of an 'ħ' in the left-hand margin, those expressions in which the quantum constant is implied in this manner (and which can be made to reappear, using a dimensional argument).

Finally, absorbing the fixed (and arbitrary) phase φ_0 in eq. (5.1.3) into a new complex constant $C = K\,e^{i\Phi_0}$, we write eq. (5.1.2) in the form

ħ $\qquad \langle v|u_E(t)\rangle = C\,e^{-iEt}.$ $\hfill (5.1.4)$

From this we may deduce

ħ $\qquad \langle v|u_E(t_2)\rangle = e^{-iE(t_2-t_1)}\,\langle v|u_E(t_1)\rangle,$ $\hfill (5.1.5)$

In going from one 'point' to another on the 'trajectory' $\{u_E(t)\}$, the projection amplitudes, $\langle v|u_E(t)\rangle$, are all multiplied by the same phase factor $e^{-iE(t_2-t_1)}$. This phase factor depends only upon the instants of time t_1 and t_2, at which the state u_E is being considered, and the energy E. Furthermore, it only depends on the difference $t_2 - t_1$ of the two times. This means that the evolution of the state u_E depends only on the interval of time which has elapsed, and not, independently, on the initial or final instants. We recognize here a form of invariance under time translations.

Relation (5.1.5) can also be interpreted by saying that the result of a change $\tau = t_2 - t_1$ of the origin of time, is to multiply all the transition amplitudes, of a stationary state onto an arbitrary state, by the same phase factor $e^{-iE\tau}$ (the same for all states v). If, therefore, we know the state of a system at an instant t, i.e., if we know all the amplitudes at this instant, we can immediately deduce from it the amplitudes for all later time t_2. In particular, setting $t_1 = 0$ and $t_2 = t$, we obtain

ħ $\qquad \langle v|u_E(t)\rangle = e^{-iEt}\,\langle v|u_E(0)\rangle.$ $\hfill (5.1.6)$

Consider now, an inverse quantum transition – of an arbitrary state v onto a stationary state of energy E. Applying to eq. (5.1.4) or (5.1.6), the relation of conjugation,

$$\langle v|u\rangle = \overline{\langle u|v\rangle},$$

we get

$\hbar$ $\qquad \langle u_E(t)|v\rangle = e^{iEt}\langle u_E(0)|v\rangle = \bar{C}\,e^{+iEt}.$ (5.1.7)

This relation fixes the temporal dependence of the projection amplitudes of an arbitrary state onto a stationary state. In conformity with the stationary nature of the final state of such a transition, the transition *probabilities* are themselves independent of the time.

Everything that has been said so far, confirms the necessity of having complex values for quantum amplitudes. If, indeed, we had limited ourselves to real values only, it would have been necessary to take a sinusoidal function, for the harmonic function $\langle v|u_E(t)\rangle$, but then the probability $|\langle v|u_E(t)\rangle|^2$ would have oscillated in the course of time, contrary to the condition given by eq. (5.1.1). Similarly, we see from this example, the insufficiency of the concept of probability for distinguishing between different stationary states: it is in the probability *amplitude* that we mark the difference between two stationary states corresponding to different energies (and hence to different pulsations).

5.1.2. Arbitrary states

Let us proceed to study now the evolution, in time, of a system prepared in a non-stationary state w – in other words, the time dependence of the projection amplitudes $\langle v|w(t)\rangle$, of the state $w(t)$ onto all the other states v of the system.

Unlike the stationary states u_E (in the following we shall reserve the letter u for denoting the proper states of a physical magnitude, which we shall index by the corresponding proper value), the non-stationary states w are not characterized by a single value – a single proper value – of the energy. They each have an energy *spectrum*, consisting of many different values.

To each one of these values is assigned a certain weight – namely, the probability of the system being projected from its state w onto the proper state of energy, u_E, corresponding to the proper value in question. This probability, which has the value

$\qquad \mathscr{P}_E = |\langle u_E|w\rangle|^2,$ (5.1.8)

in view of the symmetry rules, eqs (4.4.46), for probabilities and preceding results [see eq. (5.1.7)], is independent of the time [which justifies why the latter does not figure in the notation of eq. (5.1.8)].

The stationary states $\{u_E(t)\}$ which, to the extent that proper states form a disjoint and complete set, can be considered as being intermediaries for the transition $v \leftarrow w(t)$, the dependence of which on t we shall try to characterize. In accordance with the factorization rules of sect. 4.4, we write,

$$\langle v|w(t)\rangle = \sum_E \langle v|u_E(t)\rangle \langle u_E(t)|w(t)\rangle. \tag{5.1.9}$$

This way of writing has the merit of bringing in the projection amplitudes, from and to a stationary state, the time dependence of which has been made explicit in the preceding section. Now, the amplitude $\langle u_E(t)|w(t)\rangle$ does not depend on t. Indeed, the invariance under time translations demands that the origin, which serves to mark the points on the time axis of space–time, be irrelevant (in that whether the particular instant being considered be called t or t' is of no consequence). We have, therefore,

$$\langle u_E(t)|w(t)\rangle = \langle u_E(0)|w(0)\rangle. \tag{5.1.10}$$

Hence, the entire time dependence of eq. (5.1.9) is concentrated in the amplitudes $\langle v|u_E(t)\rangle$, the form of which we have established in eq. (5.1.6). It follows, therefore, that

$\hbar$ $\quad\quad \langle v|w(t)\rangle = \sum_E e^{-iEt} \langle v|u_E(0)\rangle \langle u_E(0)|w(0)\rangle. \tag{5.1.11}$

The time dependence of the quantum amplitude $\langle v|w(t)\rangle$ is thus expressed by means of a sum of harmonic terms. For $t = 0$, eq. (5.1.11) reduces immediately to

$\hbar$ $\quad\quad \langle v|w(0)\rangle = \sum_E \langle v|u_E(0)\rangle \langle u_E(0)|w(0)\rangle. \tag{5.1.12}$

Let us remark that eq. (5.1.11) is just a 'Fourier' expansion of the amplitude $\langle v|w(t)\rangle$ as a function of time.

Finally, it is interesting to note that, in expansion eq. (5.1.11), we could just as well insert the intermediate energy proper states $\{u_E(t')\}$ taken at any instant t' – by virtue of the stationarity of these states (exercise 5.1). Thus, we see the essential role played by the intermediate stationary states, enabling us to write the time-dependent amplitudes, given by eq. (5.1.11), as a sum of constant amplitudes multiplied by a harmonic factor, having a pulsation which is one of the proper pulsations of the system.

As soon as the sum contains at least two proper values, i.e., whenever w itself is not stationary (in which case all the amplitudes $\langle u_E | w \rangle$ would be zero, except for one), the time dependence of $\langle v | w(t) \rangle$ becomes non-trivial. There is a very illuminating particular case of eq. (5.1.11) – the case $v = w(0)$. Stated differently, the state, initially $w(0)$, evolves until it becomes $w(t)$ at the instant t, when we ask ourselves what the probability of a transition to the initial state $w(0)$ would be. Even more simply, what is the probability of recovering the same state at the time t? By eq. (5.1.11), we have the probability amplitude

$$\hbar \qquad \langle w(0) | w(t) \rangle = \sum_E e^{-iEt} \langle w(0) | u_E(0) \rangle \langle u_E(0) | w(0) \rangle, \qquad (5.1.13)$$

which, by virtue of the conjugation relation, can simply be written as

$$\hbar \qquad \langle w(0) | w(t) \rangle = \sum_E e^{-iEt} |\langle u_E(0) | w(0) \rangle|^2. \qquad (5.1.14)$$

The coefficients $|\langle u_E(0) | w(0) \rangle|^2$ are just the probabilities $\mathscr{P}_E$ for the transitions $u_E \leftarrow w(0)$, of the initial state to the stationary states [see eq. (5.1.8)]. One verifies that

$$\langle w(0) | w(t) \rangle = 1 \quad \text{for } t = 0,$$
$$|\langle w(0) | w(t) \rangle| \leqslant 1 \quad \text{for } t \neq 0. \qquad (5.1.15)$$

Thus, the transition probability appears as a sum of constant and of oscillating terms:

$$\hbar \qquad \mathscr{P}(w(0) \leftarrow w(t)) = |\langle w(0) | w(t) \rangle|^2$$

$$= \sum_E \mathscr{P}_E^2 + 2 \sum_{E \neq E'} \sum \mathscr{P}_E \mathscr{P}_{E'} \cos(E - E')t$$

$$= 1 - 4 \sum_{E \neq E'} \sum \mathscr{P}_E \mathscr{P}_{E'} \sin^2(E - E')t/2. \qquad (5.1.16)$$

Let us make an important remark in this connection, which applies equally well to the most general form, eq. (5.1.11), of a quantum amplitude: the temporal dependence of the amplitude $\langle v | w(t) \rangle$, as of the corresponding probability, can be very different, depending upon the nature of the energy spectrum of the state w, i.e., depending upon the nature of the set of proper values of the energy, which appear in the sum given by eq. (5.1.11). The simplest case is that where only two values of the energy, E_1 and E_2, have to be taken into account. The existence of only two harmonic terms in the amplitude leads in this case to temporal oscillations – a typical beat phenomenon, of which we shall study a few examples. On the other hand, if the sum of eq. (5.1.11) involves a large number of values of the energy, and, in particular, if the energy spectrum is continuous, we can have a completely different phenomenon, with a monotonic time dependence. This is the case of unstable states (e.g., radioactive disintegration), which exhibit a continuous decrease with time (see exercise 5.2).

Finally, let us mention that, in order not to make the notation too clumsy, we have denoted by Σ_E a summation which in fact runs over all the stationary *states* – and not just over the energy proper values alone, any one of which could appear several times in the sum in the case of the degeneracy of a level. In any event, it only represents a formal notation – the sum could be discrete or continuous (an integral).

5.1.3. Quantum beats

Let us imagine that the state w only has non-zero transition probabilities to two stationary energy proper states E_1 and E_2. The general transition amplitude is written, therefore, following eq. (5.1.11) as

$$\hbar \qquad \langle v|w(t)\rangle = \alpha_1 \, \mathrm{e}^{-iE_1 t} + \alpha_2 \, \mathrm{e}^{-iE_2 t}, \qquad (5.1.17)$$

where we have set,

$$\alpha_i = \langle v|u_{E_i}\rangle \langle u_{E_i}|w(0)\rangle \quad (i = 1, 2). \qquad (5.1.18)$$

The corresponding transition probability has the form

$$\hbar \qquad \mathcal{P}(v \leftarrow w(t)) = |\alpha_1|^2 + |\alpha_2|^2 + 2 \, \mathrm{Re} \, \alpha_1 \bar{\alpha}_2 \, \mathrm{e}^{i(E_2 - E_1)t}$$
$$= a + b \cos[(E_2 - E_1)t + \varphi], \qquad (5.1.19)$$

where we have introduced the notations

$$a = |\alpha_1|^2 + |\alpha_2|^2, \qquad b = 2|\alpha_1||\alpha_2|, \qquad \varphi = \mathrm{Arg}\, \alpha_1 - \mathrm{Arg}\, \alpha_2. \qquad (5.1.20)$$

The transition probability exhibits, therefore, oscillations in time – due to the beats of the two harmonic amplitudes – with frequency

ℏ $\qquad v_{\text{beat}} = (E_2 - E_1)/2\pi.$ $\qquad\qquad$ (5.1.21)

Let us look at two examples (one of them we have already seen, by the way, in exercise 4.13).

(a) *Beat spectroscopy*. When an atomic or nuclear system possesses two (or more) very closely spaced energy levels, it is necessary, if we wish to measure the corresponding proper energies directly and to distinguish between them experimentally, to use a detector having an energy resolution which is better than the separation between the levels. It is, therefore, quite difficult to measure in this way very small energy differences. However, instead of measuring the energies, ε_1 and ε_2, of the levels concerned, separately, one can directly measure the difference $\varepsilon_2 - \varepsilon_1$, for the corresponding frequencies, by means of beats. For this, it is necessary to excite the system in a way such that its state is *not* one of the stationary states for the energy ε_1 or ε_2, but rather a non-stationary state, in which both probabilities of occupying these two states are different from zero. For an atomic system, this can be done in different ways: by the excitation of an atomic target through the impact of pulses from an optical laser, or of a beam of electrons; by the scattering of an atomic beam against a thin target, etc. The radiation emitted by the system, as it de-excites itself into the ground state, of energy ε_0, does not correspond to a well-defined energy difference, $E_1 = \varepsilon_1 - \varepsilon_0$ or $E_2 = \varepsilon_2 - \varepsilon_0$, and hence is not a collection of photons of well-defined energy (fig. 5.2a). The photons are not in a stationary state, and their state w is such that the transition amplitudes $\langle u_1 | w \rangle$ and $\langle u_2 | w \rangle$, to the states of the proper energies E_1 and E_2, are both different from zero. The amplitude and the probability of the detection of such a photon is, therefore, written in precisely the forms given in eqs. (5.1.17), (5.1.19). The measured intensity, which is proportional to the probability of detection, manifests beats at the frequency in eq. (5.1.21), given by

ℏ $\qquad v_{\text{beat}} = (\varepsilon_2 + \varepsilon_1)/2\pi,$ $\qquad\qquad$ (5.1.22)

since $E_2 - E_1 = (\varepsilon_2 - \varepsilon_0) - (\varepsilon_1 - \varepsilon_0) = \varepsilon_2 - \varepsilon_1$. The *temporal* measurement of the beat frequency directly gives in this way the *energy* difference between the levels. Modern electronic techniques allow one to measure temporal frequencies in the GHz(10^9 Hz) region, corresponding to a scale of energy ($\varepsilon = hv$) of the order of a μeV(10^{-6} eV). This is exactly the order of magnitude

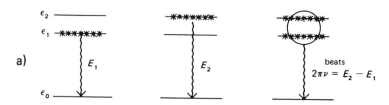

a)

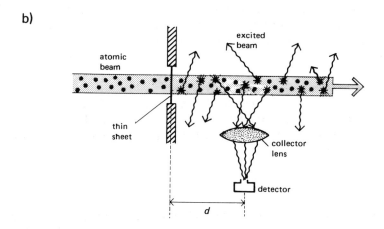

b)

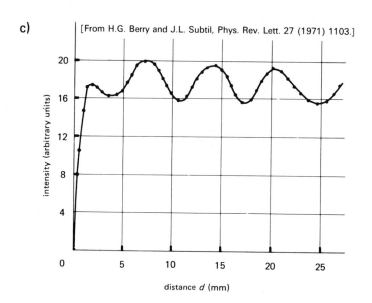

c)   [From H.G. Berry and J.L. Subtil, Phys. Rev. Lett. 27 (1971) 1103.]

of the energy difference between certain atomic levels, due to the interaction of the magnetic moments of the electrons with the Coulomb electric field of the nucleus (fine structure of the atomic nucleus), or with the nuclear magnetic moment (hyperfine structure). It is in this domain that beat spectroscopy is currently in use (fig. 5.2, exercise 5.3).

(b) *Evolution of a spin state.* A particularly simple example of beats is furnished by a system of spin-$\frac{1}{2}$, possessing a magnetic moment and placed in a field of uniform magnetic induction $\mathscr{B}$. It can reasonably be assumed that the magnetic moment $\mathscr{M}$ is collinear with the spin S,

$$\mathscr{M} = \mu S, \tag{5.1.23}$$

as in the classical theory. However, the components of S, and hence of $\mathscr{M}$, are non-compatible quantum magnitudes (see sect. 3.3.3). Let us choose the direction of the field $\mathscr{B}$ as the Oz axis. The magnetic potential energy of the system is now written as

$$E = -\mathscr{M} \cdot \mathscr{B} = -\mu \mathscr{B} S_z. \tag{5.1.24}$$

The energy proper states coincide with those of the component S_z. For spin-$\frac{1}{2}$, there exist two such proper states, which we denote symbolically as ↑ and ↓, with the respective proper values $+\frac{1}{2}$ and $-\frac{1}{2}$. The stationary states are, therefore,
– the ↑ state with proper value $E_\uparrow = -\frac{1}{2}\mu\mathscr{B}$.
– the ↓ state with proper value $E_\downarrow = +\frac{1}{2}\mu\mathscr{B}$.
Let us now suppose that, at the instant $t = 0$, the system is in a state of spin

---

Fig. 5.2. *Beat spectroscopy*
(a) An atomic system, excited to the energy level ε_1 (or ε_2) emits a photon of proper energy $E_1 = \varepsilon_1 - \varepsilon_0$ (or $E_2 = \varepsilon_2 - \varepsilon_0$). If the excited state is not stationary, the emitted radiation is not monochromatic and displays beats with a frequency

$$v_{\text{beat}} = (E_2 - E_1)/2\pi = (\varepsilon_2 - \varepsilon_1)/2\pi.$$

(b) In a typical experiment, a beam of helium atoms interacts with a thin sheet of carbon which projects the atoms onto a large spectrum of excited states. The luminous emission of the beam is observed, at the wavelength corresponding to the average of the levels of interest, ε_1 and ε_2, as a function of the distance from the sheet, i.e., as a function of the time that has elapsed after the excitation.
(c) The measurement of the intensity displays the characteristic beats, from which the separation between the levels can be deduced (exercise 5.3).

which is a proper state of a component of S orthogonal to $\mathscr{B}$, for example, S_x. We denote this state by $\rightarrow$. This is evidently not a stationary state. We thus ask ourselves about its evolution, i.e., about the state $w(t)$ such that $w(0) = \rightarrow$. In particular, we would like to know the probability – a priori different from unity – with which, at the instant t, the system would be once again exactly in the state $\rightarrow$, or, more precisely, with which the system effects a transition from the state $w(t)$ to the state $\rightarrow$. Thus, we are trying to calculate the amplitude $\langle \rightarrow | w(t) \rangle$. Let us apply the general expression, eq. (5.1.11), where the sum is taken over the proper values $E_\uparrow$ and $E_\downarrow$, and the corresponding proper states, and where $v = w(0) = \rightarrow$:

$\hbar$

$$\langle \rightarrow | w(t) \rangle = \langle \rightarrow | \uparrow \rangle \langle \uparrow | \rightarrow \rangle \, e^{-iE_\uparrow t} + \langle \rightarrow | \downarrow \rangle \langle \downarrow | \rightarrow \rangle \, e^{-iE_\downarrow t}$$

$$= |\langle \rightarrow | \uparrow \rangle|^2 \, e^{-iE_\uparrow t} + |\langle \rightarrow | \downarrow \rangle|^2 \, e^{-iE_\downarrow t}. \tag{5.1.25}$$

[We have recovered a particular case of eq. (5.1.14).] For reasons of symmetry, the proper state $\rightarrow$ of S_x must have the same projection probabilities onto $\uparrow$ and $\downarrow$ (the orientation of Oz is arbitrary). Thus, we have

$$|\langle \rightarrow | \uparrow \rangle|^2 = |\langle \rightarrow | \downarrow \rangle|^2. \tag{5.1.26}$$

However, on the other hand, the two states $\uparrow$ and $\downarrow$ form a disjoint and complete set. Hence, by sect. 4.2, we have

$$|\langle \rightarrow | \uparrow \rangle|^2 + |\langle \rightarrow | \downarrow \rangle|^2 = 1. \tag{5.1.27}$$

Consequently,

$$|\langle \rightarrow | \uparrow \rangle|^2 = |\langle \rightarrow | \downarrow \rangle|^2 = \tfrac{1}{2}, \tag{5.1.28}$$

and the amplitude, eq. (5.1.25), becomes

$\hbar$

$$\langle \rightarrow | w(t) \rangle = \tfrac{1}{2} e^{i\mu\mathscr{B}t/2} + \tfrac{1}{2} e^{-i\mu\mathscr{B}t/2}$$

$$= \cos(\mu\mathscr{B}t/2). \tag{5.1.29}$$

The probability with which, at the instant t, the system comes back to its initial state of spin $\rightarrow$ is thus written as

$\hbar$

$$\mathscr{P}_{\rightarrow}(t) = |\langle \rightarrow | w(t) \rangle|^2 = \cos^2(\mu\mathscr{B}t/2). \tag{5.1.30}$$

The probability $\mathscr{P}_{\leftarrow}(t)$ of finding the system, at the instant t, in the other proper state $\leftarrow$ of S_x, is evidently such that $\mathscr{P}_{\leftarrow}(t) + \mathscr{P}_{\rightarrow}(t) = 1$. Its value is, therefore,

$$\hbar \qquad \mathscr{P}_{\leftarrow}(t) = |\langle \leftarrow | w(t) \rangle|^2 = \sin^2(\mu\mathscr{B}t/2). \qquad (5.1.31)$$

These probability oscillations are often interpreted by referring to the evolution of the average values of the spin components. If we calculate, e.g., the average value, at the instant t, of the component S_x, it has, by definition the value

$$\hbar \qquad \langle S_x(t) \rangle = \tfrac{1}{2}\mathscr{P}_{\rightarrow}(t) + (-\tfrac{1}{2})\mathscr{P}_{\leftarrow}(t); \qquad (5.1.32)$$

by eqs. (5.1.30) and (5.1.31), this yields

$$\hbar \qquad \langle S_x(t) \rangle = \tfrac{1}{2}\cos(\mu\mathscr{B}t). \qquad (5.1.33)$$

Similarly, it can be shown (exercise 5.4) that

$$\hbar \qquad \langle S_y(t) \rangle = \tfrac{1}{2}\sin(\mu\mathscr{B}t). \qquad (5.1.34)$$

The vector, having the average values of the spin for its components, or the 'average spin', goes through, therefore, a uniform rotational motion, called the 'Larmor precession', around the magnetic field. Its angular velocity is

$$\hbar \qquad \omega_L = \mu\mathscr{B}. \qquad (5.1.35)$$

It can be seen, without much difficulty, that the behaviour displayed here for the initial state $w(0) = \rightarrow$, is the same for any initial spin state: in a uniform magnetic field, the average value of the spin precesses with the velocity given by eq. (5.1.35) (... if the quanton possesses a magnetic moment which couples its spin to the field!). This property is at the basis of 'magnetic resonance' experiments. These techniques exploit the evolution of the magnetic moments of electronic spins (EPR: 'electron paramagnetic resonance') or nuclear spins (NMR: 'nuclear magnetic resonance'), in variable magnetic fields. The Larmor precession is at their very foundation. The Larmor precession can be detected by a direct measurement of the average value of the spin component of a beam of neutrons which has passed through a uniform magnetic field (fig. 5.3 and exercises 5.5 and 5.6).

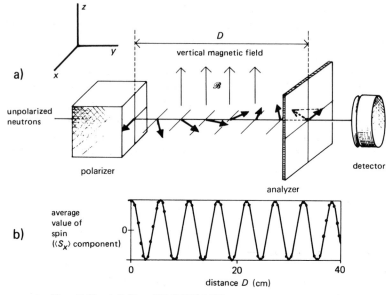

[From F. Mesei, Z. Phys. 225 (1972) 146.]

Fig. 5.3. *Larmor precession of neutrons*
(a) A beam of neutrons is polarized in a manner such that the average spin of the neutrons is parallel to the direction Ox, at the entrance to a region where there exists a uniform magnetic field, parallel to Oz. The spin of the neutrons precesses in the field, where the neutrons cover a distance D, before being picked up by a detector–analyzer, which measures the final average component of their spin in the direction Ox.
(b) The experimental results display a sinusoidal variation of the spin as a function of the distance D (i.e., of the time of flight). From the precession frequency, so measured, the magnetic moment of the neutron can be deduced (exercise 5.5).

5.2. *Localization amplitudes and wavefunctions*

5.2.1. *The wavefunction and the probability density*

(a) *The notion of the wavefunction.* In the present section, we shall no longer be interested in the temporal dependence, and shall consider a state w at an instant fixed once and for all. Among all the amplitudes $\langle v|w \rangle$ which characterize this state, we shall particularize those which are the projections of the state w onto the localized states of the quanton. These amplitudes give the localization probabilities of the quanton.

They correspond to the customary experimental situation, where the position of a particle is observed with the help of an appropriate detector. For example, the screen of an oscilloscope is built to interact with an electron in a way such as to 'force' it, so to say, to be localized to a sharp point in space, realizing in this way such a projection onto a localized state.

In order to simplify both the understanding and the description, we begin by considering the particular case of a one-dimensional quanton, constrained to be on one axis. We denote by u_x, the proper state of localization to the point x on the abscissa, i.e., the proper state of the position magnitude, having the proper value x. It is the amplitude $\langle u_x | w \rangle$ which is of interest to us here. The mapping, which to each point x on the abscissa associates the amplitude $\langle u_x | w \rangle$, defines a function of x which characterizes the state w. It is called the 'wavefunction' of the quanton and is denoted by ψ_w,

$$\psi_w(x) = \langle u_x | w \rangle. \tag{5.2.1}$$

This name, the 'wavefunction', is explained by the fact that the function $\psi_w(x)$, in so far as it is a spatial amplitude, resembles classical wave amplitudes. However, the physical interpretation of a 'wavefunction', a probability *amplitude*, differs radically from the interpretation of a classical field amplitude. In order not to be misled by a semantic analogy, it is essential to always bear in mind that the 'wavefunction' $\psi_w(x)$ denotes an amplitude of spatial localization.

(b) *The probability density and the normalization condition.* At the present state, we assume that the square of the modulus of the wavefunction, $|\psi_w(x)|^2$, represents the probability of localization, at the point x, of the system in the state w. We expect, therefore, that the sum of these probabilities, over all the proper values of position, considered as having a complete set of proper states, is unity, in conformity with eq. (4.2.19), so that

$$1 = \sum_x |\langle u_x | w \rangle|^2 = \sum_x |\psi_w(x)|^2. \tag{5.2.2}$$

However, now we have stumbled into the difficulty of having to give a meaning to a sum, such as in eq. (5.2.2), over a continuous magnitude! Naturally, only the notion of the integral could enable us to formalize such an idea and hence we ought to write it as an integral over the entire (continuous!) extension of the spectrum of the position magnitude:

$$1 = \int_{-\infty}^{\infty} |\psi_w(x)|^2 \, dx. \tag{5.2.3}$$

From now on, we are led to modifying somewhat the physical interpretation of the wavefunction. For reasons of homogeneity alone, the magnitude $|\psi_w(x)|^2$, which has the dimensions of an inverse length, cannot be a probability. Moreover, the continuous nature of the spectrum of position prevents us from defining a probability of localization at a sharply defined point. We can only define an infinitesimal probability of localization between x and $x + dx$, which is necessarily written as $d\mathscr{P}(x, x + dx) = \rho(x)\,dx$, where $\rho(x)$ is a probability *density*. Since the probability of being localized somewhere on the whole line must evidently be equal to one, we should have

$$1 = \int_{-\infty}^{\infty} \rho(x)\,dx. \tag{5.2.4}$$

This justifies, upon comparison with eq. (5.2.3), the identification of the probability density of localization with the square of the modulus of the wavefunction,

$$\rho(x) = |\psi_w(x)|^2. \tag{5.2.5}$$

The infinitesimal probability of localization between x and $x + dx$, in the state w, is, therefore,

$$d\mathscr{P}_w(x, x + dx) = |\psi_w(x)|^2\,dx, \tag{5.2.6}$$

and, more generally, the probability of localization in the interval $[a, b]$ for the state w is

$$\mathscr{P}_w([a, b] \leftarrow w) = \int_a^b |\psi_w(x)|^2\,dx. \tag{5.2.7}$$

If the interval $[a, b]$ extends to the entire line $(-\infty, +\infty)$, the probability tends to unity and we recover eq. (5.2.3), and hence the physical meaning is now clear. We may consider this so-called *normalization* condition, as a requirement which a function $\psi(x)$ must satisfy, in order to be the wavefunction of a quantum state (exercises 5.7 and 5.8). The normalization condition, which plays an essential role in quantum theory, clearly marks the limit of the analogy suggested by the term *wave*-function, since there is nothing in the classical theory of fields which corresponds to it, and for good reason: in classical theory, the amplitude of a wave has nothing to do with the characteristically corpuscular notion of pointwise localization.

Let us conclude this section by showing, in fig. 5.4, two particular wavefunctions and their associated probability densities. It is a good idea to familiarize oneself with the rough construction, by looking and by drawing, of the graph of a density from that of the corresponding amplitude (exercise 5.8). [Beware, in particular, of an elementary difficulty: where the graph of a function cuts the abscissa, the graph of the square of its modulus at the same point is a curve which is tangent to the axis and does not have a sharp corner: if the function behaves as $k(x - a)$ for $x \simeq a$, the square of its modulus goes as $k^2(x - a)^2$, which is parabolic.]

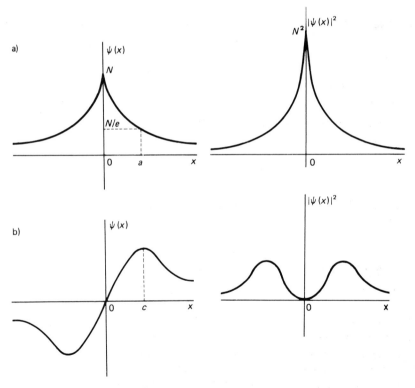

Fig. 5.4. *Wavefunctions and probability densities*
(a) The wavefunction $\psi(x) = N\,e^{-|x|/a}$ and the corresponding density $|\psi(x)|^2$.
(b) The wavefunction $\psi(x) = Nx\,e^{-x^2/2c^2}$ and the corresponding density $|\psi(x)|^2$.

(c) *Average values.* The position X of a quanton having wavefunction $\psi_w(x)$ can thus be considered to be a random variable, governed by the probability law defined by the density $\rho_w(x) = |\psi_w(x)|^2$. It is, therefore, only natural to consider averages, in the sense of probabilities. Hence, the average of a magnitude $f(X)$ in the state w of the quanton is the number

$$\langle f(X) \rangle_w = \int_{-\infty}^{\infty} f(x)\, \rho_w(x)\, \mathrm{d}x = \int_{-\infty}^{\infty} f(x)\, |\psi(x)|^2\, \mathrm{d}x \tag{5.2.8}$$

(assuming that the integral converges!).

For example, the 'average position' of the quanton is very naturally defined as

$$\langle X \rangle = \int x|\psi(x)|^2\, \mathrm{d}x \tag{5.2.9}$$

(the state label w is often omitted in such expressions, and we shall not indicate the limits of integration any longer). The probabilistic notion of the mean-square variation, enables us to give a strict definition of the dispersion ΔX in the position of the quanton,

$$(\Delta X)^2 = \langle (X - \langle X \rangle)^2 \rangle = \langle X^2 \rangle - \langle X \rangle^2, \tag{5.2.10}$$

which, in turn, is

$$(\Delta X)^2 = \int x^2|\psi(x)|^2\, \mathrm{d}x - \left(\int x|\psi(x)|^2\, \mathrm{d}x \right)^2. \tag{5.2.11}$$

Let us particularly note that we can define in this manner the average values of functions of position only, to the exclusion (for the time being) of the other physical magnitudes (momentum, angular momentum, etc.).

5.2.2. *Generalization: magnitudes having continuous spectra*

The foregoing considerations are not particular to the physical magnitude of position, but extend to all magnitudes, the proper values of which form a set which is continuous rather than discrete (finite or denumerable), as we had implicitly assumed the case to have been, in ch. 4. Such magnitudes are quite commonly used: not only the position, but also the momentum and, in certain cases, the energy are examples in point. The considerations of ch. 4 can now be extended as follows. Let $\mathscr{A}$ be a physical magnitude, having a continuous spectrum of proper values $\{\alpha\}$ and the corresponding proper states $\{u_\alpha\}$. The properties of the proper states are then written as (compare tables 4.1 and 4.2)

$$\langle u_\alpha | u_\beta \rangle = 0 \quad \text{if } \alpha \neq \beta \qquad \text{disjointness,} \qquad (5.2.12)$$

$$\int_{\{\alpha\}} \langle v | u_\alpha \rangle \langle u_\alpha | v' \rangle \, d\alpha = \langle v | v' \rangle \qquad \text{completeness.} \qquad (5.2.13)$$

This last relation can also be written as,

$$\int |\langle u_\alpha | v \rangle|^2 \, d\alpha = 1. \qquad (5.2.14)$$

We recognize here a normalization condition, analogous to eq. (5.2.3), which justifies the interpretation of the function

$$\rho_v(\alpha) = |\langle u_\alpha | v \rangle|^2, \qquad (5.2.15)$$

as a probability density for the physical magnitude $\mathscr{A}$. More precisely,

$$\rho_v(\alpha) \, d\alpha = |\langle u_\alpha | v \rangle|^2 \, d\alpha,$$

is the probability of transition from the state v to one of the proper states of $\mathscr{A}$, of proper values lying between $\alpha + d\alpha$. We can also say that $\rho_v(\alpha) \, d\alpha$ is the probability for a magnitude $\mathscr{A}$ of a quanton, in the state v, to assume a value lying between α and $\alpha + d\alpha$. The magnitude $\mathscr{A}$ is, therefore, considered as being a random variable with probability density $\rho_v(\alpha)$, which allows one to define average values of the type

$$\langle f(\mathscr{A}) \rangle_v = \int f(\alpha) \, \rho_v(\alpha) \, d\alpha = \int f(\alpha) |\langle u_\alpha | v \rangle|^2 \, d\alpha. \qquad (5.2.16)$$

Let us finally mention that a physical magnitude may display a mixed spectrum, partly discrete and partly continuous, requiring in that case, both methods of writing presented so far.

5.2.3. Wavefunctions of momentum proper states

(a) *Harmonic wavefunctions.* We shall not establish the form of the wave-functions of some rather special states – the proper states of momentum. Let u_p be such a state, of proper value p. We shall try to find an expression for the amplitude, which we simply denote by

$$\psi_p(x) \triangleq \langle u_x | u_p \rangle. \tag{5.2.17}$$

From the spatial Heisenberg inequalities (sect. 3.2.2), a proper state of momentum ($\Delta p = 0$) has an infinite spatial extension ($\Delta x = \infty$) – it is completely unlocalized. Its probability density $|\psi_p(x)|^2$ is, therefore, a constant, independent of x. Hence the wavefunction $\psi_p(x)$ has a modulus independent of x. Furthermore, since the concepts of momentum and undulation are one and the same thing in the quantum theory, saying that the system is in a proper state of p, is the same as saying that its spatial localization ought to display a harmonic dependence on x. The harmonicity in question is characterized precisely by the undulation $k = p/\hbar$, or rather $k = p$, choosing, as in eq. (5.1.4), a system of units suitable for quantum theory, in which $\hbar = 1$; $\psi_p(x)$ is thus of the form

$$\hbar \qquad \psi_p(x) = A \, e^{\pm ipx}, \tag{5.2.18}$$

where A is a constant, in general complex, independent of x, and for the moment arbitrary. However, since $\psi_p(x)$, being a transition amplitude, is only defined up to a phase, we shall often adopt in the following the convention – to the extent that there is nothing to the contrary – of taking A to be a real number. As to the choice of sign, $+$ or $-$, in the exponential, it is a matter of convention. We choose the usual option

$$\hbar \qquad \psi_p(x) = \langle u_x | u_p \rangle = A \, e^{ipx}. \tag{5.2.19}$$

(b) *The problem of normalization.* It would be appropriate now, in conformity with the conclusions reached in sect. 5.2.1, to impose on the wavefunction of eq. (5.2.19) the normalization condition, eq. (5.2.3), so that

$$1 \overset{?}{=} \int_{-\infty}^{\infty} |\psi_p(x)|^2 \, dx \overset{?}{=} \int_{-\infty}^{\infty} |A|^2 \, dx. \tag{5.2.20}$$

Evidently, it is impossible to satisfy this condition, since the integral diverges if $A \neq 0$. This obstacle arises from the fact that we are attempting to attribute to a quanton a probability density of localization which is uniform throughout the whole of space (since the quanton is completely unlocalized), although this space itself is infinite. But, of course, such a state is only a limiting case: all the electrons that we produce and study on earth, however poorly localized they may be, still have a much smaller probability of appearing on Sirius than in our laboratory. The proper states of momentum are not, therefore, 'true' physical states of quantons. They provide us with a very useful abstraction to them, and for this reason alone they have the right to be included in the formalism.

It ought to be particularly noted that the idealization which leads us to consider the proper states u_p of momentum, are of exactly the same nature as that in which the proper states u_x, of position, are used. Quantons prepared in the laboratory always have non-zero dispersions, Δx in position and Δp in momentum. The condition $\Delta x = 0$ (the u_x states) or $\Delta p = 0$ (the u_p states) can only be limiting situations. We have already seen (sect. 5.2.1) the extension of the formalism which the use of the amplitudes $\langle u_x | w \rangle$ requires. But now we stumble upon the difficulty of extending the theory to amplitudes of the type $\langle u_x | u_p \rangle$, where *both* the states are proper states of continuous magnitudes. Moreover, there is another difficulty, of the same order, with amplitudes such as $\langle u_x | u_{x'} \rangle$ or $\langle u_p | u_{p'} \rangle$. By contrast, amplitudes such as $\langle u_p | w \rangle$ are treated entirely following the generalized rules of sect. 5.2.2 (on condition that w be *not* a proper state of a magnitude with a continuous spectrum!). It is, therefore, just our desire for a *double* idealization which is bringing about this problem with $\langle u_x | u_p \rangle$: once was all right, but twice ....

There are different ways for getting around this obstacle with the normalization. First, only relative probabilities can be considered and different amplitudes compared, without normalizing them. One can equally envisage the physical space as the limit of a finite domain of size L. Upon integrating the probability density of localization, and taking into account eq. (5.2.19), one obtains in this way, in the one-dimensional case, a relation between A and L,

$$1 = \int_L |A|^2 \, \mathrm{d}x = |A|^2 L, \qquad (5.2.21)$$

which allows for the normalization of the amplitude

$$A = L^{-1/2}. \qquad (5.2.22)$$

The trick lies, therefore, in keeping all through the calculations a finite size L, however large that might be, and then letting it tend to infinity in the final result. Indeed, a credible physical result should not depend on the size of the domain used for the computation if this domain is very large: the value of the proton–proton effective scattering cross-section at CERN in Geneva, must be practically identical whether it is calculated inside a sphere having for its radius the earth-to-moon distance or having for its diameter that of the galaxy .... Finally, there exist elaborate mathematical schemes which allow one to integrate (!) this difficulty into a coherent formalism (in the same way as the notion of the integral, replacing that of the summation, allowed one to absorb the first difficulty with the idealization arising from the continuous spectrum).

(c) *The 'wavefunction in p' and the Fourier transform.* The proper states of momentum, although idealizations, play no less a fundamental role in the development of the theory, in that they enable us to give a simple description of the wavefunction of the system when the latter is in a completely arbitrary state w. Indeed, for a transition from w to a state of localization u_x, the set of all proper states u_p of momentum define as many possible modes for this transition, and provide us with a complete and disjoint set. The rule of sequential factorization, together with that for the addition of quantum amplitudes allows us now to write the amplitude $\psi_w(x) = \langle u_x | w \rangle$ in the form

$$\langle u_x | w \rangle = \sum_p \langle u_x | u_p \rangle \langle u_p | w \rangle, \tag{5.2.23}$$

or rather, since in general, the proper values of momentum form a continuous $\hat{\psi}_w(p) = \langle u_p | w \rangle$

$$\langle u_x | w \rangle = \int \langle u_x | u_p \rangle \langle u_p | w \rangle \, \mathrm{d}p. \tag{5.2.24}$$

By analogy with the definition of the wavefunction, one often adopts the notation

$$\hat{\psi}_w(p) = \langle u_p | w \rangle, \tag{5.2.25}$$

and calls $\hat{\psi}_w(p)$ the 'momentum wavefunction' of the state w (by a double abuse of language, since it is not a wave and since a wave amplitude is a spatial function). We can thus write relation (5.2.24), taking eq. (5.2.19) into account, in the form

$\hbar$ $$\psi_w(x) = A \int e^{ipx} \, \hat{\psi}_w(p) \, dp. \qquad (5.2.26)$$

For the moment we leave the numerical coefficient A undetermined and work 'up to a constant' (see, however, exercises 5.10 and 5.11).

The wavefunction of the arbitrary state w, which is a linear combination of the functions $\langle u_x | u_p \rangle = A \, e^{ipx}$, appears in this way as a superposition of 'monochromatic waves'. The coefficients $\langle u_p | w \rangle = \hat{\psi}_w(p)$ of this combination are projection amplitudes of w onto the proper states of momentum.

Relation (5.2.26) can also be inverted. The same reasoning, based on the rules for the computation of quantum amplitudes which led us to eq. (5.2.24), allows us to write

$$\langle u_p | w \rangle = \int \langle u_p | u_x \rangle \, \langle u_x | w \rangle \, dx, \qquad (5.2.27)$$

by using, this time, the localization proper states as the complete and disjoint set. Furthermore, following the rule of conjugation, we write

$\hbar$ $$\langle u_p | u_x \rangle = \overline{\langle u_x | u_p \rangle} = A \, e^{-ipx} \qquad (5.2.28)$$

(without loss of generality, A may be taken to be real), whence,

$\hbar$ $$\hat{\psi}_w(p) = A \int e^{-ipx} \, \psi_w(x) \, dx. \qquad (5.2.29)$$

Thus, the position and momentum wavefunctions are related by the Fourier transform. It is remarkable that from a blind manipulation of our formalism, there suddenly appeared the result according to which the Fourier transform in eq. (5.2.26) could be inverted into eq. (5.2.29). This is a very encouraging indication of the coherence of the formalism. The full mathematization of the theory will bring to light the profound duality, thus revealed, between the physical magnitudes of position and momentum. For the moment, we start by familiarizing ourselves with this idea through several examples (exercises 5.10, 5.11, 5.12).

If the function $\hat{\psi}_w(p)$ is concentrated around a value p_0, i.e., if the state w is close to a proper state u_{p_0}, the wavefunction $\psi_w(x)$ approaches the corresponding wavefunction $A\,e^{ip_0x}$, and this is all the more so the narrower the width Δp of the momentum spectrum. Thus, the more concentrated the function $\hat{\psi}_w(p)$ is around p_0, the more poorly localized is the state described by $\psi_w(x)$ (fig. 5.5). Conversely, the larger Δp is, the more sharply localized is the phenomenon described by $\psi_w(x)$. The widths of the localization domain of the quanton and of its momentum spectrum, Δx and Δp, vary oppositely (exercise 5.13). This is in conformity with the Heisenberg inequalities, for which the mathematical theory of the Fourier transform allows us to give a rigorous proof (exercises 5.14, 5.15).

For the moment, it is sufficient for us to establish that a 'wave packet', namely a superposition of plane 'waves', eq. (5.2.26), concentrated in a band Δp around a value p_0 – and hence of width $\Delta x \gtrsim 1/\Delta p$ – constitutes the quantum representation of something which in the classical limit would be defined as a particle occupying a region of width Δx and endowed with a momentum p_0 (fig. 5.5).

The momentum being a continuous magnitude, the square of the modulus of the amplitude $\langle u_p | w \rangle$ yields the probability density, in momentum, for the state w,

$$\sigma_w(p) = |\langle u_p | w \rangle|^2 = |\hat{\psi}_w(p)|^2. \tag{5.2.30}$$

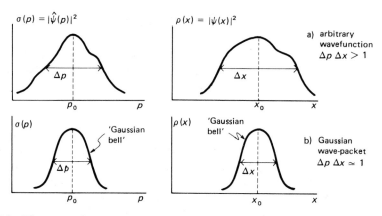

Fig. 5.5.   *The wave packet*
(a) To obtain a relatively localized quanton, the spatial extension of which is to be of order Δx, it is necessary to construct its wavefunction as a superposition of plane waves – a wave packet – with a momentum spectrum of width $\Delta p > 1/\Delta x$.
(b) Wave packets which are Gaussians in both x and p have dispersions with minimal product: $\Delta p \Delta x \simeq 1$, and hence allow for the best possible representation of a state with a practically well-defined momentum and position: p_0 (to within Δp) and x_0 (to within Δx).

The total probability in momentum must evidently be unity. This means that the function $|\hat{\psi}_w(p)|^2$ must have its integral equal to unity and that $\hat{\psi}_w$, like ψ_w, must, in principle, be normalized,

$$1 = \int \sigma_w(p) \, dp = \int |\hat{\psi}_w(p)|^2 \, dp. \tag{5.2.31}$$

The problem of normalization is identical to that which we have discussed for the position. Finally, we can calculate, with the help of $\hat{\psi}_w(p)$, the average value, in the state w, of any function of the momentum P,

$$\langle g(P) \rangle_w = \int g(p) \, \sigma_w(p) \, dp = \int g(p) \, |\hat{\psi}_w(p)|^2 \, dp. \tag{5.2.32}$$

In particular, we define the average momentum

$$\langle P \rangle = \int p\sigma(p) \, dp, \tag{5.2.33}$$

and its quadratic dispersion

$$(\Delta P)^2 = \langle P^2 \rangle - \langle P \rangle^2 = \int p^2 \sigma(p) \, dp - \left(\int p\sigma(p) \, dp \right)^2. \tag{5.2.34}$$

This last expression, together with that of the quadratic dispersion in position, eq. (5.2.11), enables us to give a rigorous proof of the Heisenberg inequality (exercise 5.15).

5.2.4. In three dimensions

(a) *Wavefunctions.* The generalization of the foregoing considerations to three-dimensional space is easy. Let u_r be the proper states of the position, where the coordinates (x, y, z) of r are the components of the position magnitude. An arbitrary state w possesses a wavefunction

$$\psi_w(r) = \langle u_r | w \rangle, \tag{5.2.35}$$

which obeys the normalization condition

$$1 = \int |\psi_w(r)|^2 \, d^3r, \tag{5.2.36}$$

and the square of the modulus of which gives the probability density of localization

$$\rho(r) = |\psi_w(r)|^2. \tag{5.2.37}$$

A particularly important case is that of a spherically symmetric state, the wavefunction of which depends only on the modulus r of the vector r. The normalization condition is now written, having first carried out the integration over the solid angle, as:

$$1 = \int_0^\infty 4\pi r^2 |\psi_w(r)|^2 \, dr. \tag{5.2.38}$$

Upon defining a new function

$$\varphi_w(r) = (4\pi)^{1/2} r \psi_w(r), \tag{5.2.39}$$

it takes the form

$$1 = \int_0^\infty |\varphi_w(r)|^2 \, dr. \tag{5.2.40}$$

Thus, the function φ_w is normalized as though it were the wavefunction of a one-dimensional quanton, constrained to stay on the half-line $r > 0$ (or, just as well, $\varphi_w(r) = 0$ for $r \leqslant 0$). Indeed, it is possible, when the physical conditions (e.g., the external potential) display a spherical symmetry, to confine ourselves to a one-dimensional problem on the half-line, by considering φ_w instead of ψ_w.

(b) *Plane waves.* The proper states u_p of the momentum vector have the components (p_x, p_y, p_z) of p, as the proper values of the three components of momentum – compatible physical magnitudes, as we recall (see exercise 5.17). The wavefunctions of these states u_p have the form

$$\hbar \qquad \langle u_r | u_p \rangle = A \, e^{ip \cdot r}. \tag{5.2.41}$$

These are sometimes termed 'plane waves', using a classical analogy. The same difficulty of normalization as in the one-dimensional case appears here, and has the same solution. These amplitudes enable us to describe the most general wavefunction in the form of a Fourier integral

$\hbar$ $\qquad \langle u_r|w\rangle = A \int e^{i p \cdot r} \langle u_p|w\rangle \, d^3p,$ $\qquad\qquad$ (5.2.42)

or,

$\hbar$ $\qquad \psi_w(r) = A \int e^{i p \cdot r} \, \hat{\psi}_w(p) \, d^3p,$ $\qquad\qquad$ (5.2.43)

if we define the 'momentum wavefunction'

$\qquad \hat{\psi}_w(p) = \langle u_p|w\rangle.$ $\qquad\qquad$ (5.2.44)

(c) *Centered spherical waves.* It often happens that we have to consider the proper states of momentum in a radial direction, emanating from a certain point O. Among such states are, e.g., those of quantons emitted or scattered by a source placed at O and the modulus, but not the direction of the momentum of which is fixed. They are no longer the proper states of the three independent components of the momentum, but just of its modulus p, and possibly, of other physical magnitudes – related to the angular momentum, for example. Let us denote these states by $u_{p,\eta}$, the index η specifying the proper values of additional magnitudes necessary to determine the state completely. [We realize that the proper states of momentum being determined by the *three* values (p_x, p_y, p_z), the specification of *one* value p is not sufficient to determine a proper state unambiguously.] It is natural, for expressing the wavefunction

$\qquad \psi_{p,\eta}(r) = \langle u_r|u_{p,\eta}\rangle,$ $\qquad\qquad$ (5.2.45)

of these states, to use a spherical system of coordinates centered at O, in which the vector r is marked by its modulus r and its direction $\Omega = (\theta, \varphi)$, θ and φ being the usual angular coordinates.

For $u_{p,\eta}$ to be the proper state of the radial component of momentum, means that this state ought to be completely unlocalized along every 'radius' emanating from the centre O. In the three-dimensional case, the square of the modulus of the wavefunction $|\psi(r)|^2$, should be identified with a *volume* density of localization, providing us with an infinitesimal probability

$\qquad d^3\mathscr{P} = |\psi(r)|^2 \, d^3r,$ $\qquad\qquad$ (5.2.46)

of localization inside the volume element d^3r around r. Now, in spherical coordinates, the differential element is given by

$$d^3r = r^2 \, dr \, d^2\Omega, \tag{5.2.47}$$

$d^2\Omega = d(\cos\theta) \, d\varphi$ denoting the differential solid angle. From this, in the case at hand, there follows the expression for the infinitesimal probability $d^3\mathscr{P}$ in polar coordinates,

$$d^3\mathscr{P}(r, \Omega) = |\psi_{p,\eta}(r, \Omega)|^2 \, r^2 \, dr \, d^2\Omega. \tag{5.2.48}$$

For this probability to be independent of r for a 'radially unlocalized' state, it is necessary that the quantity

$$\mu(r, \Omega) = |\psi_{p,\eta}(r, \Omega)|^2 \, r^2, \tag{5.2.49}$$

be also constant (in r), and hence that the modulus of the wavefunction be of the form $1/r$.

The form of the *phase* of the wavefunction $\psi_{p,\eta}(r, \Omega)$ is now determined by the fact that the state $u_{p,\eta}$ is a proper state of the modulus of the momentum, having the proper value p. Hence, $\psi_{p,\eta}(r, \Omega)$ is harmonic in r, with undulation p. As in the preceding case, only a harmonicity expressed as an imaginary exponential can make the quantity $\mu(r, \Omega)$ be independent of r. From this, we finally obtain the form of $\psi_{p,\eta}(r, \Omega)$ and which we call a 'centered wave':

$$\hbar \qquad \psi_{p,\eta}(r, \Omega) = \frac{B_\eta(\Omega)}{r} \, e^{ipr}. \tag{5.2.50}$$

We verify that

$$\mu(r, \Omega) = |B_\eta(\Omega)|^2, \tag{5.2.51}$$

which gives the radial probability density in the direction Ω, is independent of r. Thus, the probability for the localization of the quanton inside a cone of opening $d^2\Omega$, between the distances r and $r + dr$ from the origin, depends only on dr and not on the distance r itself (fig. 5.6). An important special case is when there exists a complete spherical symmetry around O. The state, which we denote by $u_{p,\text{sph}}$, now possesses a wavefunction which is independent of the direction of r,

$$\hbar \qquad \psi_{p,\text{sph}}(r, \Omega) = \frac{B}{r} \, e^{ipr}, \tag{5.2.52}$$

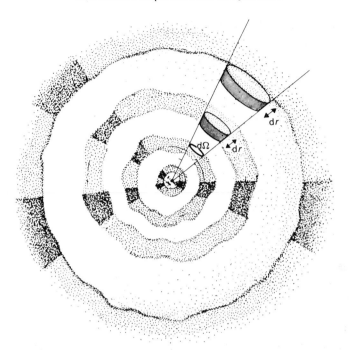

Fig. 5.6. *The centered wave*
The spatial amplitude of a proper state of the radial momentum is written as

$$\psi_{p,\eta}(r, \Omega) = \frac{B_\eta(\Omega)}{r} \, e^{ipr},$$

where p characterizes the radial harmonicity, $B_\eta(\Omega)$ the dependence on the direction Ω (which is indicated in the diagram by the unequal shadings of the regions, depending upon the solid angle). The factor $1/r$ ensures the uniformity of the localization probability density, as a function of the distance for a given direction: in the cone, having opening $d^2\Omega$, the probability of localization is the same for two volume elements which have the same thickness dr.

where B is just a normalization constant. One speaks now of a '*spherical wave*'. Let us remark that the function

$$\varphi_{p,\mathrm{sph}} = (4\pi)^{1/2} r \psi_{p,\mathrm{sph}}, \tag{5.2.53}$$

defined in conformity with eq. (5.2.39), can be written as

$$\hbar \qquad \varphi_{p,\mathrm{sph}}(r) \propto e^{ipr}, \tag{5.2.54}$$

and, just as expected, has the form of the wavefunction of a quanton in *one dimension* of proper momentum p (see exercises 5.18 and 5.19 for further discussions).

5.3. Amplitudes and effective scattering cross-sections

5.3.1. Wavefunction, flux and effective cross-section

We now show how the notion of the wavefunction enables us to express the idea of the effective cross-section in quantum theory. In ch. 2, it was the non-classical interference effect, manifested by the effective cross-sections of certain scattering processes, which had led us to the idea of the quanton (*neither* wave, *nor* corpuscle). Now, in the classical theory, the effective cross-section is calculated using the trajectories of the colliding particles (ch. 2, sect. 2.2.2). The quantons, as far as they are concerned, do not have trajectories and are only described by transition amplitudes. It is, therefore, necessary for us to arrive at an expression for the effective cross-section with the help of these amplitudes.

Let us consider a parallel and monoenergetic beam of quantons, having momentum p. If these incident quantons are not subject to any interaction, their wavefunction is of the form of a plane wave

$$\psi_{\text{inc}}(r) = \psi_p(r) = A\,e^{ip \cdot r}. \tag{5.3.1}$$

The objective of a scattering experiment is to study the interactions between the projectile quantons of the beam and the target quantons. For the sake of simplicity, we consider the latter as being fixed. Let us assume that the interaction has a short range, i.e., it only shows up if the two interacting quantons are at a distance smaller than a certain finite range, from each other, and, furthermore, that this interaction is elastic, i.e., it does not modify either the kinetic energy or, hence, the modulus of its momentum. A quanton, scattered by a target quanton at O, is in a state described by a centered wavefunction, the proper value of its radial momentum being equal to the modulus of its initial momentum, $p = |p|$. This wavefunction is written as

$$\hbar \qquad \psi_{\text{scat}}(r) = \psi_{p,\eta}(r) = \frac{B(\Omega)}{r}\,e^{ipr}, \tag{5.3.2}$$

as soon as the quanton has effectively regained its initial kinetic energy, i.e., outside the interaction region, where the potential energy of the interaction is zero. But there are two possibilities which could lead to the same final state of the localization, in r, of the projectile quanton inside a detector, far away from the interaction region: the quanton may have interacted or it may not have done so. Its quantum amplitude of localization, i.e., its complete wave-function, is therefore the sum of two amplitudes, ψ_{inc} and ψ_{scat},

$\hbar$ $\qquad \psi(r) = \psi_{inc}(r) + \psi_{scat}(r)$

$$= A\, e^{ip \cdot r} + \frac{B(\Omega)}{r}\, e^{ipr} \quad (r \gg \text{range of interaction}), \qquad (5.3.3)$$

which we often write as

$\hbar$ $\qquad \psi(r) = A\left(e^{ip \cdot r} + \frac{f(\Omega)}{r}\, e^{ipr} \right), \qquad (5.3.4)$

by defining

$$f(\Omega) = B(\Omega)/A. \qquad (5.3.5)$$

Let us stress the fact that this expression only represents the function at large distances – but it is just there at the detector, that our interest in it lies. Let us now forget, for the moment, the absolute normalization condition, eq. (5.2.36), which is impossible to satisfy and let us try to interpret the amplitudes, given by eqs. (5.3.1) and (5.3.2), 'as though' the square of its amplitude were a proper probability density. The incident flux is calculated as the number of particles crossing a unit surface in unit time. These particles are evidently the particles which are contained inside a prismatic volume standing on the unit surface $\Delta S = 1$, and having for its height the distance traversed over $\Delta t = 1$, namely v (fig. 5.7); $|A|^2$ is the probability density for such a quanton. The average number of these inside the volume ΔV, which is just the flux, is evidently

$$\mathscr{F} = v|A|^2. \qquad (5.3.6)$$

In fact, relation (5.3.6) furnishes us with an interpretation of the quantity $|A|^2$, since

$$|A|^2 = \frac{\mathscr{F}}{v} = \frac{\mathscr{F}\,\Delta t\,\Delta S}{v\Delta t\,\Delta S}, \qquad (5.3.7)$$

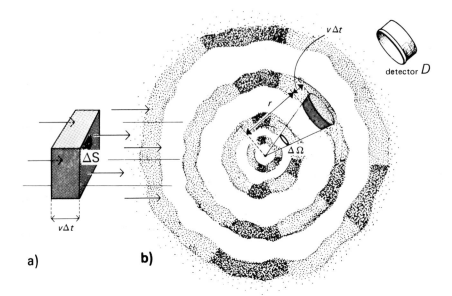

Fig. 5.7. *Flux and effective cross-section*

(a) Consider a parallel incident beam having a density $\rho_{\text{inc}} = |A|^2$ of quantons of velocity v. The number of quantons crossing the surface ΔS in the time Δt is given by $\mathscr{F} \Delta S \, \Delta t$, where $\mathscr{F}$ is the flux. This is also the number of quantons inside the parallelepiped of height $v \, \Delta t$, standing on the surface ΔS, which is $\rho_{\text{inc}} \Delta S \, \Delta t$. Hence, $\mathscr{F} = \rho_{\text{inc}} v = |A|^2 v$.

(b) In the scattered beam, diverging from the scattering centre, the density is $\rho_{\text{scat}} = |B|^2 / r^2$. The quantons crossing a detector of solid angle $\Delta\Omega$, in time Δt are contained in the cylinder of height $v \, \Delta t$, standing on the surface $r^2 \, \Delta\Omega$. Their number is $\Delta N = \rho_{\text{scat}} \, r^2 \, \Delta\Omega \, v \, \Delta t = |B|^2 v \, \Delta\Omega \, \Delta t$.

The differential effective cross-section, defined by the ratio of the number of quantons, received per unit of time and of solid angle, $\Delta N / \Delta\Omega \, \Delta t = |B|^2 v$, to the incident flux, $\mathscr{F} = |A|^2 v$, is, therefore, $\chi(\Omega) = |B|^2 / |A|^2$.

is just the average number of particles per unit volume in the incident beam. For an ensemble of a large number of identical quantons, all described by the wavefunction of eq. (5.3.4), which is impossible to normalize, the quantity $|A|^2$ can, therefore, be identified with the density (in the usual sense) of particles. We have just seen here another means for getting around the difficulty with the normalization of 'doubly idealized' amplitudes, $\psi_p(\mathbf{r}) = \langle u_r | u_p \rangle$. This procedure, which, in short, amounts to replacing the description of a single object by that of an ensemble of a large number of identical objects, is not uncommon in physics.

On the other hand, the particles reaching the detector D, of opening $\Delta\Omega$, at a distance r from the target, during the time Δt, are contained in the truncated conic volume of base $\Delta S = r^2 \Delta\Omega$ and of height $v\Delta t$ (fig. 5.7). Since the probability density of a scattered quanton is

$$\rho_{\text{scat}} = \frac{|B(\Omega)|^2}{r^2},\qquad(5.3.8)$$

the average number ΔN of quantons reaching the detector during the time Δt is,

$$\Delta N = \rho_{\text{scat}} r^2 \, \Delta\Omega \, v \, \Delta t = |B(\Omega)|^2 \, v \, \Delta\Omega \, \Delta t$$
$$= |A|^2 \, |f(\Omega)|^2 \, v \, \Delta\Omega \, \Delta t.\qquad(5.3.9)$$

The differential effective cross-section, experimentally defined (see ch. 2, sect. 2.2.2) by

$$\chi(\Omega) = \frac{\Delta N}{\mathscr{F} \, \Delta\Omega \, \Delta t},\qquad(5.3.10)$$

is therefore, very simply given, following eqs. (5.3.6) and (5.3.7), by

$$\chi(\Omega) = |f(\Omega)|^2.\qquad(5.3.11)$$

The quantity $f(\Omega)$, which has the dimensions of a length, is often called the 'scattering amplitude'. The square of its modulus directly yields the differential effective cross-section. We remark that in reality, the 'scattering amplitude' is a *ratio* of amplitudes [see eq. (5.3.5)], which clearly proves that its physical interpretation does not depend on the particular solution of the delicate problem of the normalization of plane waves. The scattering amplitude for a collision process is a function of the scattering direction Ω. Naturally, it depends upon the initial kinetic energy, or what amounts to the same thing, on the modulus p of the momentum. More rigorously, we ought to write, therefore, $f_p(\Omega)$, something which we shall sometimes do.

When the forces responsible for the phenomenon of scattering are spherically symmetric (about the scattering centre O), the scattering amplitude does not depend on the azimuthal angle φ, but in general it depends on the zenithal angle θ: there exists in the physical situation, a privileged spatial direction – that of the incident beam – which distinguishes between the different values of θ, but there is no preferred plane which would enable us to distinguish between the various values of φ. Thus, the scattering amplitude only depends on the zenithal angle and is written as $f(\theta)$.

The quantum theory of a scattering process can now be built. We establish – using methods to be discussed later – the form of the wavefunction $\psi(r)$ of a scattered quanton. It evidently depends on the nature – assumed to be known – of the interactions. Then we consider the asymptotic form of this wavefunction, which has to be of the type given by eq. (5.3.4). Identifying the scattering amplitude $f(\Omega)$, we obtain finally the differential effective cross-section, using eq. (5.3.11).

The formalism developed here can be extended to certain long-range interactions, such as the extremely important Coulomb interaction. It also extends to inelastic processes (exercise 5.19).

5.3.2. Coulomb scattering

In order to give an example of the scattering amplitude, as well as to underline its intrinsic importance, we indicate here the form of the amplitude for the pure Coulomb scattering of the charges $Z_1 q_e$ and $Z_2 q_e$. An elaborate computation, relying on theoretical methods, which we do not have at our disposal yet, yields the expression – which only depends on θ, by virtue of the spherical symmetry of the Coulomb force –

$$f_{\text{Coul}}(\theta) = \frac{Z_1 Z_2 e^2}{4E} \frac{1}{\sin^2 \frac{1}{2}\theta} e^{i\zeta}. \tag{5.3.12}$$

The phase ζ, which is the only (although not a minor) complexity in this expression is written as

$$\zeta = \gamma \ln(\sin^2 \tfrac{1}{2}\theta) + \pi + 2 \, \text{Arg} \, \Gamma(1 + i\gamma), \tag{5.3.13}$$

where

$$\gamma = \frac{Z_1 Z_2 e^2}{\hbar v} = \frac{Z_1 Z_2 \alpha c}{v}, \tag{5.3.14}$$

is a dimensionless characteristic parameter, which depends on the velocity; Γ is the generalized factorial function. It can be shown that the differential effective cross-section, calculated in the quantum theory,

$$\chi_{\text{Coul}}(\theta) = |f_{\text{Coul}}(\theta)|^2 = \left(\frac{Z_1 Z_2 e^2}{4E}\right)^2 \frac{1}{\sin^4 \frac{1}{2}\theta}, \tag{5.3.15}$$

is completely identical to the classical expression, the so-called Rutherford cross-section, as we have already pointed out (sect. 2.2.2.4; see also, exercise 2.5). The 'quantumness' of the Coulomb scattering is completely contained in the phase, eq. (5.3.13), of the amplitude, which alone displays the presence of the quantum constant $\hbar$ [shown explicitly, for once, in eq. (5.3.14)]. We shall see in ch. 7, under what particular circumstances one might try to get this phase to manifest itself, leading to strictly quantum effects in the Coulomb scattering. We observe that for Coulomb scattering, being of long range, the total effective cross-section, which is the integral of $\chi_{\text{Coul}}(\theta)$ over the solid angle $\Omega = (\theta, \varphi)$ [see eq. (5.3.15)], is infinite. This fact is related to the slow decrease (in $1/r$) of the Coulomb force with distance: being a long-range force, it affects more or less any incident projectile, no matter at what distance it passes.

5.3.3. The optical theorem

The attentive reader has probably detected a gap – and not just a mere mistake – in the reasoning which had led us to the form of the effective cross-section. We have actually calculated the number of particles detected, by using only the density $\rho_{\text{scat}} = |\psi_{\text{scat}}|^2$. But the scattering amplitude ψ_{scat} in eq. (5.3.3) cannot be separated from the incident amplitude ψ_{inc}. There is only one amplitude – their sum – and the true probability density is given by

$$
\begin{aligned}
\rho = |\psi|^2 &= |A\,e^{ip\cdot r} + B(\Omega)\,e^{ipr}/r|^2 \\
&= |A|^2 + (2/r)\,\text{Re}[\bar{A}B(\Omega)\,e^{i(pr-p\cdot r)}] \\
&\quad + |B(\Omega)|^2/r^2,
\end{aligned}
\tag{5.3.16}
$$

where we note, in particular, the presence of the interference term. We have used an idealized amplitude here, described by a plane wave and a centred wave, the singular nature of which we have already underscored. However, the concrete physical situation corresponds to an *almost* plane wave, the transverse extension of which is certainly not infinite, since the beam of quantons is collimated by the very way in which it is produced (fig. 5.8). From this it follows that if we were to adopt a more realistic description of the amplitudes, in terms of three-dimensional wave packets, the term ψ_{inc}, in the total amplitude, would only assume appreciable values inside a tubular region corresponding to the collimation of the beam. It is only the term ψ_{scat} which can, therefore, show up in the directions which are not the initial direction, and it alone enters only the computation of the probability density at the level of the detector, provided this latter is placed sufficiently far from

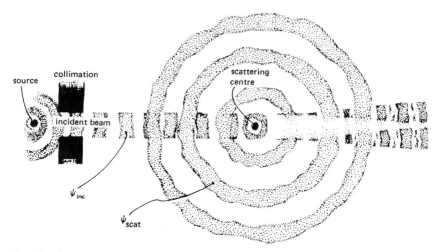

Fig. 5.8. *Quantum scattering and the optical theorem*
The wavefunction of a quanton in a scattering process is the sum of two terms. The one, ψ_{inc}, describes the state of the quanton in the incident beam (after emission and collimation) and the other, ψ_{scat}, is the localization amplitude produced by the scattering. The latter is a wave centered at the scattering centre. The two terms in the wavefunction interfere destructively within a narrow forward zone. Indeed, the localization probability of the quanton in the initial direction must be diminished in a manner so as to compensate, exactly, its scattering probability (*optical theorem*).

the scattering centre and in a direction sufficiently oblique to the axis of the beam (fig. 5.8). This justifies, therefore, why we could calculate the effective cross-section uniquely using ψ_{scat}. For the rest, it is only along the forward direction that the two terms ψ_{scat} and ψ_{inc} contribute together to the probability density and there the two cannot be separated! Thus, the interference term in eq. (5.3.16), namely,

$$\rho_{interf} = (2/r)\, \mathrm{Re}\left[\bar{A}B(\theta, \varphi)\, e^{ipr(1-\cos\theta)}\right] \tag{5.3.17}$$

(where θ is the angle between the direction of scattering – that of $\boldsymbol{r}$ – and the initial direction – that of $\boldsymbol{p}$), has now to be taken into account. The spatial integral of the interference term ρ_{interf}, receives appreciable contributions only from the regions where the exponential does not oscillate too much, i.e., where its phase $pr(1 - \cos\theta)$ is stationary. At large distances $(r \gg 1/p)$, therefore, these contributions come only from a zone of angular opening

$$\theta \lesssim (pr)^{-1/2} \lesssim 1 \tag{5.3.18}$$

(using $1 - \cos\theta \simeq \frac{1}{2}\theta^2$) and hence are overwhelmingly concentrated in the forward direction. We can now estimate the contribution of ρ_{interf} to the total probability, by first integrating over the angular variables,

$\hbar$
$$\int d^2\Omega \, \rho_{\text{interf}} = (2/r) \, \text{Re}\left[\bar{A} \int d^2\Omega \, B(\Omega) \, e^{ipr(1 - \cos\theta)} \right]$$

$$= (2/r) \, \text{Re}\left[\bar{A} \int\int d\cos\theta \, d\phi \, B(\theta, \varphi) \, e^{ipr(1 - \cos\theta)} \right]$$

$$= (4\pi/r) \, \text{Re}\left[\bar{A}B(0) \int_{-1}^{1} d\cos\theta \, e^{ipr(1 - \cos\theta)} \right], \qquad (5.3.19)$$

where we have substituted the 'forward' value $B(0)$ for $B(\Omega)$, since in the limit $r \to \infty$, in which we are interested, the only contribution comes from the region $\theta \simeq 0$ (our computation here is not rigorous, but it can easily, although at some length, be justified). Thus we have, for the 'purely radial' (i.e., integrated over the angles) probability density

$\hbar$
$$\int d^2\Omega \, \rho_{\text{interf}} = (4\pi/r) \, \text{Re}\left[\bar{A}B(0) \frac{1 - e^{2ipr}}{-ipr} \right]. \qquad (5.3.20)$$

For a detector, placed at a distance large compared to the wavelength, i.e., $r \gg \lambda$, or, just as well, for $pr \gg 1$, the rapidly oscillating exponential e^{-2ipr} does not contribute to the radial integration which remains to be done, so that in fact we may write

$\hbar$
$$\int d^2\Omega \, \rho_{\text{interf}} = -\frac{4\pi}{r^2} \frac{1}{p} \, \text{Im} \left[\bar{A}B(0) \right]. \qquad (5.3.21)$$

The interference term, eq. (5.3.17), contributes, therefore, to the total probability.

However, all this makes perfect sense! The *total probability* of finding the quanton in a scattering state having wavefunction $\psi(r) = \psi_{\text{inc}}(r) + \psi_{\text{scat}}(r)$ [see eq. (5.3.3)] must equal unity ... exactly as for the total probability corresponding to the single incident wave $\psi_{\text{inc}}(r)$, since this latter is assumed to be correctly normalized. This means that the total probability, calculated using the density given by eq. (5.3.16), which we may write as

$$\rho = \rho_{\text{inc}} + \rho_{\text{interf}} + \rho_{\text{scat}}, \qquad (5.3.22)$$

must be the same as the total probability calculated using ρ_{inc} alone.
This condition of the 'conservation of probability' is written as

$$\int \rho_{\text{interf}} \, r^2 \, dr \, d^2\Omega + \int \rho_{\text{scat}} \, r^2 \, dr \, d^2\Omega = 0, \tag{5.3.23}$$

which, taking expressions (5.3.21) and (5.3.8) into account, is

$\hbar$
$$\int r^2 \, dr \left[-\frac{4\pi}{r^2} \frac{1}{p} \operatorname{Im} \bar{A}B(0) \right] + \int r^2 \, dr \, \frac{1}{r^2} \int d^2\Omega |B(\Omega)|^2 = 0. \tag{5.3.24}$$

We verify that this compensation is possible, since both terms have the same
dependence (in $1/r^2$). Using the fact – already noted – that ρ_{interf} contributes
appreciably only in the forward direction, relation (5.3.24) like eq. (5.3.23), of
which it is a consequence, can easily be interpreted: the number of particles
scattered out of the incident beam, in all directions, is equal to the diminution
of the number of particles in the initial direction, after the scattering. Stated
otherwise: on the one hand, the scattering out of the initial direction takes
place at the expense of the incident beam, which is a triviality, but on the
other hand, this depletion of the incident beam results from the destructive
interference between the incident and the scattered amplitudes in the forward
direction – something which is much less trivial.

This compensating effect ought to lead to the existence of a relation
between the total effective scattering cross-section σ_{tot} and the forward ($\theta = 0$)
scattering amplitude, f_{for}. This is exactly what is expressed in eq. (5.3.24);
indeed, we obtain

$\hbar$
$$\frac{4\pi}{p} \operatorname{Im} \bar{A}B(0) = \int d^2\Omega \, |B(\Omega)|^2. \tag{5.3.25}$$

Again, introducing the scattering amplitude $f = B/A$ and the total effective
cross-section, this becomes

$\hbar$
$$\int d^2\Omega \, |f(\Omega)|^2 = \frac{4\pi}{p} \operatorname{Im} f(0), \tag{5.3.26}$$

or finally,

$\hbar$
$$\sigma_{\text{tot}} = \frac{4\pi}{p} \operatorname{Im} f_{\text{for}}. \tag{5.3.27}$$

Thus, the effective cross-section (for any energy) is related to the imaginary part of the forward scattering amplitude, something which imposes very severe constraints on the scattering amplitude (exercise 5.18). This result – extremely important, as we shall see – goes by the name of the 'optical theorem'. It has in fact a classical analogue in the wave theory of light. After all, what we have here is a shadow problem .... The optical theorem has numerous applications, and we shall study one such elementary application. Finally, let us note that the optical theorem rests on the hypothesis that the number of quantons is conserved, i.e., there is no absorption during the scattering. However, it is possible to generalize it to the case of processes in which there exists a certain probability of absorption or of inelastic scattering (exercise 5.19).

5.3.4. Scattering length and 'quantum index of refraction'

(a) *Scattering length.* There is a particular case of a scattering amplitude that occurs frequently. Let us consider a short-range interaction, i.e., one which manifests itself only over a distance between the projectile quanton and the target quanton, which is smaller than a certain length d, the range of the interaction. In this case, the relative angular momentum L of the two quantons, during their interaction, would be of the order of

$$L \lesssim pd, \tag{5.3.28}$$

where p is the momentum of the projectile. Taking the quantization of the angular momentum into account, we see that if $pd \ll 1$, we can only have an interaction in the state of angular momentum $l = 0$ (the values $l = 1, 2, ...$ require a relative distance greater than the range of the interaction). Such a state corresponds to a complete spherical symmetry. Hence, it is not marked by any privileged spatial direction and the scattering takes place isotropically. Figuratively, we may say that the projectile moves so slowly that, in the course of the scattering, its initial direction is of no consequence and so all final directions are equivalent – as though it had been immobile from the beginning. From this it follows, that at low energies, that is, more precisely, when

$$\hbar \qquad E \ll \frac{1}{md^2}, \tag{5.3.29}$$

the scattering amplitude is independent of the direction Ω. Its limit at zero energy is, therefore, also constant (in Ω). In general, one defines the quantity (independent of Ω)

$\hbar$ $$\lim_{p \to 0} f_p(\Omega) = -a. \tag{5.3.30}$$

Here is the anticipated application of the optical theorem. When written in the form

$\hbar$ $$\operatorname{Im} f_p(0) = p \frac{\sigma_{\text{tot}}}{4\pi}, \tag{5.3.31}$$

the theorem shows that at zero energy $(p = 0)$, the forward scattering amplitude, $f_p(0)|_{p=0}$, is real. In the present situation, the scattering amplitude at zero energy, $-a = f_p|_{p=0}$, does not depend on the direction, in view of eq. (5.3.30). Hence, a is a *real* quantity, having the dimensions of a length. It is called the 'scattering length'. Caution! It could be positive as well as negative (exercise 5.21). The differential effective cross-section, at zero energy, is constant (in direction)

$$\chi_p(\Omega)|_{p=0} = a^2, \tag{5.3.32}$$

which is another way of expressing the isotropy of the scattering, and the total effective cross-section has the value

$$\sigma|_{p=0} = 4\pi a^2, \tag{5.3.33}$$

'as though' the target were a classical sphere of radius $2a$. The scattering length characterizes, therefore, the apparent size of the target – for the projectile being considered. It ought not to be forgotten, however, that the 'scattering length' is not a geometrical parameter, but rather a quantum amplitude, and by virtue of it, obeys the principle of superposition. From this follow the characteristic interference effects in the scattering by composite objects, the effective cross-sections of which could very well be smaller than the sum of the effective cross-sections of their constituents.

We have seen this phenomenon at the macroscopic level in the preceding chapter, when we were discussing the scattering of neutrons by crystals: the amplitude a, which we had used (see sect. 4.5) was nothing other than the scattering length.

The restrictions so imposed on the scattering amplitude, by the optical theorem, are general and they manifest themselves just as well at non-zero energies (exercises 5.18 and 5.19).

Let us note, however, that the preceding considerations are only valid so long as the scattering potential has a finite range. In the case of Coulomb scattering, for example, the total effective cross-section is infinite and the scattering length loses its significance.

(b) *The quantum index of refraction.* The notion of scattering length is particularly useful in discussing the behaviour of low-energy quantons inside matter. Consider the concrete case of neutrons. Their principal interaction with matter is of the nuclear type, having a short range. When a mono-energetic neutron, described by a plane-wave type localization amplitude, $e^{ip \cdot r}$, interacts with a thin layer of some material of thickness δ (fig. 5.9), the transmitted amplitude at a certain point P, at a distance l from the material, can be calculated following the quantum principle of superposition,

$$\varphi_{tr}(l) = \varphi_{inc}(l) + \varphi_{scat}(l), \tag{5.3.34}$$

where, $\varphi_{inc}(l) = \exp(ipl)$ is the incident amplitude and where, φ_{scat} is the total

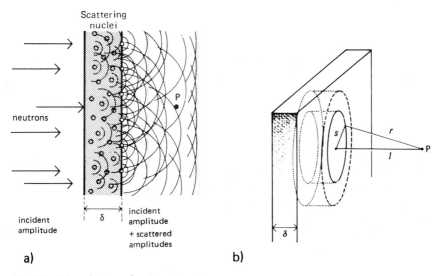

a)   b)

Fig. 5.9. *The refraction of neutrons in matter*
(a) The localization amplitude of neutrons, after having crossed a slab of matter, is the result of the superposition of the incident amplitude (of the 'plane wave' type) and the amplitudes due to the scattering by the nuclei (of the 'spherical wave' type).
(b) The geometry of the integration allows us to compute the index of refraction.

scattering amplitude, resulting from the superposition of the amplitudes scattered individually by each nucleus, i.e.,

$\hbar$

$$\varphi_{\text{scat}} = \sum_{\text{nuclei}} \frac{-a}{r} e^{ipr}, \tag{5.3.35}$$

r being the distance of the nucleus from the point P, and $-a$ the individual scattering amplitude of each nucleus. Calling the density of nuclei (number of scattering centres per unit volume) $\mathcal{N}$, the total scattering amplitude is written as

$\hbar$

$$\varphi_{\text{scat}} = \int \frac{-a}{r} e^{ipr} \mathcal{N} \, d^3 r, \tag{5.3.36}$$

where the integral extends over the volume of the layer of matter traversed. Furthermore, assuming that $l \gg \delta$,

$\hbar$

$$\varphi_{\text{scat}} = -\mathcal{N} a\delta \int_0^\infty \frac{e^{ipr}}{r} 2\pi s \, ds, \quad \text{where } r = \sqrt{s^2 + l^2}, \tag{5.3.37}$$

with the notation of fig. 5.9. Again, since $s \, ds = r \, dr$, this becomes

$\hbar$

$$\varphi_{\text{scat}} = -2\pi \mathcal{N} a\delta \int_l^\infty e^{ipr} \, dr = -2\pi \mathcal{N} a\delta \frac{e^{ipr}}{ip} \Big|_l^\infty. \tag{5.3.38}$$

The divergence of the integral at infinity does not make physical sense: there always exists a certain amount of absorption of the scattered wave – or otherwise, the average over the oscillations at large distances is zero due to cancellations. We simply have, therefore,

$\hbar$

$$\varphi_{\text{scat}} = -i \, 2\pi \mathcal{N} a\delta \frac{e^{ipl}}{p}, \tag{5.3.39}$$

and the total amplitude at P can be written as

$\hbar$

$$\varphi_{\text{tr}}(l) = \left(1 - i \, 2\pi \frac{\mathcal{N} a}{p} \delta\right) e^{ipl}. \tag{5.3.40}$$

This expression is to be compared to that which would have been obtained by attributing a 'quantum index of refraction' n, to the layer of material, such that inside the material the wavelength would have become

$$\lambda' = \frac{1}{n}\lambda, \quad \text{hence} \quad p' = np. \tag{5.3.41}$$

Upon emerging from the layer of matter, the amplitude would have acquired in this way an additional phase shift

$\hbar$
$$\Delta\alpha = (p' - p)\delta = (n - 1)p\delta, \tag{5.3.42}$$

and we would have written

$\hbar$
$$\varphi_{tr}(l) = \exp(ipl + i\Delta\alpha) = \exp[ipl + i(n - 1)p\delta]$$
$$\simeq [1 + i(n - 1)p\delta]\,e^{ipl}, \tag{5.3.43}$$

since we have assumed that the layer was thin and hence $\Delta\alpha \ll 1$. It is enough to identify eqs. (5.3.43) and (5.3.40) to see that we can effectively attribute to the material a '(quantum) index of refraction for neutrons' n, such that

$\hbar$
$$n = 1 - 2\pi\mathcal{N}\frac{a}{p^2}, \tag{5.3.44}$$

which is often written, as a function of the wavelength $\lambda = 2\pi/p$,

$$n = 1 - \frac{1}{2\pi}\mathcal{N}a\lambda^2. \tag{5.3.45}$$

Thus, for slow neutrons, the material behaves like an optical medium of refractive index n. All sorts of experiments and apparatus utilize this property, in fact, to an extent that one often speaks about 'neutron optics'. This is a domain in which the quantum theory (we should not forget that the quantum constant is hidden, in eq. (5.3.44), under $p = \hbar k = h/\lambda$) manifests itself macroscopically (through the density $\mathcal{N}$).

The neutronic index of refraction, calculated in this way, differs only slightly from unity. Typical numerical values, for thermal neutrons and an ordinary material, are

$$\lambda \simeq 1\,\text{Å}, \quad \mathcal{N} \simeq 10^{29}\,\text{m}^{-3}, \quad a \simeq 10\,\text{F}, \tag{5.3.46}$$

from which we get

$$n - 1 \simeq -2 \times 10^{-6}. \qquad (5.3.47)$$

Nevertheless, the effects of neutronic refraction can easily be observed and utilized. Thus, let us consider a material which is less refractive than the vacuum ($n < 1$, $a > 0$). Neutrons incident on its surface would be refracted towards the interior, only if their angle of incidence is smaller than a certain limiting angle, failing which there will be a total reflection. Since the refractive index is close to unity, this phenomenon will take place at grazing incidence, and for this

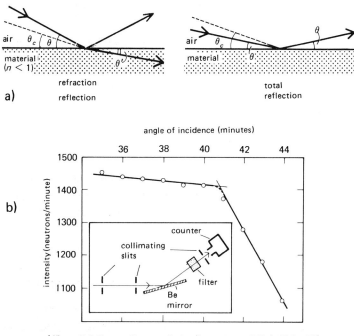

[(From D.J. Hughes, Neutron Optics (Interscience, N.Y. (1954)p. 58]

Fig. 5.10. *The total reflection of neutrons*
(a) A beam of neutrons, arriving at the surface of a material, of refractive index $n < 1$, is either refracted, or not, depending upon whether the incidence is greater, or smaller, than the critical incidence.
(b) A measurement of the intensity of the beam reflected from a beryllium surface, for neutrons of wavelength $\lambda = 8.9$ Å, shows the existence of critical incidence (just as in optics, reflection occurs here, but it is only partial for $\theta > \theta_c$). This experiment enables us to measure the scattering length of neutrons for beryllium nuclei (exercise 5.20).

reason, it is preferable in neutron optics to measure the angles of incidence and refraction with respect to the surface of the material, rather than with respect to the normal (fig. 5.10). Hence, the Snell–Descartes law is now written as

$$\cos \theta = n \cos \theta', \tag{5.3.48}$$

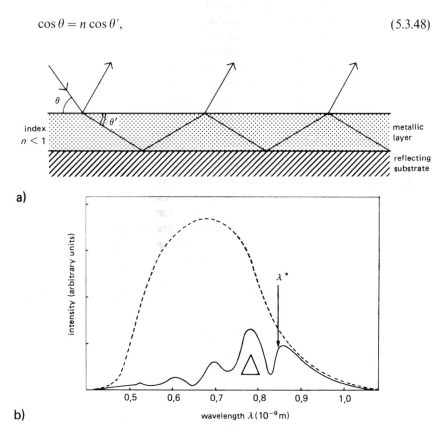

a)

b)

wavelength $\lambda\,(10^{-9}\,\text{m})$

[From J.B. Hayter, J Penfold, W.G. Williams, Nature, 262 (1976) 569.]

Fig. 5.11. *Interference of neutrons by reflection from a thin layer*
(a) A beam of (*non*-monochromatic) neutrons impinges, at an angle of incidence $\theta = 1.5 \times 10^{-2}$ rad, on a thin metallic layer. There is interference between the waves, emerging from the thin layer after zero, one or more reflections.
(b) The interference is not constructive except for certain values of the wavelength. The spectrum of the emergent neutrons exhibits, therefore, a series of maxima in wavelength (exercise 5.22). The dashed curve is the spectrum of incident neutrons, λ^* denotes the critical wavelength, above which there is total reflection (at the incidence θ being considered), and Δ indicates the experimental resolution.

where θ and θ' are the angles so measured. Thus, there will be total internal reflection below the critical incidence θ_c, such that

$$\cos \theta_c = n. \tag{5.3.49}$$

Since $n \simeq 1$ and hence $\theta_c \ll 1$, this becomes

$$1 - \tfrac{1}{2}\theta_c^2 = 1 - \frac{1}{2\pi} \mathscr{N} a\lambda^2, \tag{5.3.50}$$

and finally,

$$\theta_c = \lambda(\mathscr{N} a/\pi)^{1/2}. \tag{5.3.51}$$

In order of magnitude, this is

$$\theta_c \simeq 2 \times 10^{-3} \simeq 6'. \tag{5.3.52}$$

The existence of total reflection thus enables us to make a measurement of the scattering length (at least if $a > 0$). Thus we obtain a microscopic magnitude (the nuclear scattering length – of the order of a fermi), by means of a genuinely macroscopic optical experiment (the critical angle of incidence is of the order of a tenth of a degree or so) (figs. 5.10, 5.11 and exercises 5.20. 5.21, 5.22).

5.4. Quantons 'in motion'

5.4.1. Spatio–temporal amplitudes

We can now collect the results of the two preceding sections and study the *temporal* dependence of *spatial* amplitudes. We shall be able to describe, in this way, the evolution of the localization amplitudes in the course of time, i.e., the changes in the probability of localization at any point in space. In the absence of the notion of a trajectory, it is only in these terms that we may speak of the 'motion' of a quanton. We consider, therefore, the set $w(t)$ of the states of the quanton in the course of time, and focus our interest on their transitions to the proper states of localization, characterized by the amplitudes $\langle u_x | w(t) \rangle$ (we restrict ourselves to the one-dimensional case). Considered as functions of x, these are wavefunctions parametrized by the time, which may be denoted as

$$\psi_{w(t)}(x) = \langle u_x | w(t) \rangle. \tag{5.4.1}$$

One often adopts the notation

$$\psi_w(x; t) = \langle u_x | w(t) \rangle, \tag{5.4.2}$$

and speaks of 'time-dependent' wavefunctions. But the apparent similarity of the notation, for the spatial argument x and the temporal argument t, of the wavefunction $\psi_w(x; t)$, should not hide, or make us forget, the profound difference between their physical status (we try to remind ourselves of this difference, at least provisionally, by separating the x and the t by means of a semicolon rather than by a simple comma). In fact, t parametrizes the evolution of the family of states $\{w(t)\}$ of the system, while x only labels the particular family of proper states of localization $\{u_x\}$, onto which we have chosen to study the projections of the state $w(t)$. While the evolution of the state $w(t)$, in the course of time is unavoidable, its localization in x does not occur by itself. It requires the intermediary of an apparatus (a detector, a photographic plate, an oscilloscope screen, etc.), which forces the system to be localized.

Given the evident physical importance of the position magnitude, and its proper states, the choice of the $\{u_x\}$ and the corresponding amplitudes $\langle u_x | w(t) \rangle$ to concretize the description of the temporal evolution of the system, following the trajectory $\{w(t)\}$, (we sometimes speak of the 'x-representation' to indicate this choice) seems very natural. We are by no means forced to do this, however, for we might just as well have been interested – and indeed, we have at times had the occasion to be so – in the projections $w(t)$ onto another family of proper states, e.g., those of one component of the angular momentum or those of the momentum, $\{u_p\}$. In this last case we would have been led to defining a 'time-dependent momentum wavefunction' (or a 'p-representation'):

$$\hat{\psi}_w(p; t) = \langle u_p | w(t) \rangle. \tag{5.4.3}$$

It is clear that the dependence on time exists for both eq. (5.4.2) and eq. (5.4.3), as well as for any other family of amplitudes, while the argument x or p depends on the family considered. Another way to express this idea, is to say that the argument x in eq. (5.4.2) is a proper value of a certain quantum physical magnitude of the system under consideration, namely, its position [just as p in eq. (5.4.3) is a proper value of its momentum]. By contrast, the time t is *not* the proper value of a physical magnitude: the time is not a property of every system, but a common and universal parameter for the states of all systems. There does not exist for it, therefore, an analogue of the normalization condition, eq. (5.2.3). Let us remark that we cannot measure 'the position of a system on the time axis'. For time, there does not exist an apparatus, such as the photographic plate for space, which would force the system to be 'localized on the time axis', by projecting it onto a fixed point on this axis. There only exist apparatus (clocks) which serve to graduate the coordinate t.

From now onwards we shall be interested, therefore, in the wavefunctions $\psi(x;t)$. Let us recall their physical significance. The quantity

$$\rho(x;t) = |\psi(x;t)|^2, \tag{5.4.4}$$

is the probability density of localization of the quanton at the point x and at the time t.

There is one particular type of spatio–temporal wavefunction which deserves special attention, namely, that of a stationary wave. We have already indicated in sect. 5.1 that the use of stationary states enables us to write [see eq. (5.1.11)] any time-dependent projection amplitude as a sum of amplitudes which are independent of t, multiplied by a harmonic factor, having for its pulsation one of the proper values of the energy of the system.

Applying this idea to the case of the localization amplitudes, we get

$$\psi_w(x;t) = \langle u_x|w(t)\rangle = \sum_E \langle u_x|u_E(t)\rangle \langle u_E(t)|w(t)\rangle, \tag{5.4.5}$$

in which there appear the functions of the type,

$$\psi_E(x;t) = \langle u_x|u_E(t)\rangle. \tag{5.4.6}$$

Being the wavefunction of a stationary state, with a proper value E for the energy, $\psi_E(x;t)$ effectively factorizes, upon the application of eq. (5.1.6), into a harmonic dependence on t and a purely spatial function:

$$\hbar \qquad \psi_E(x;t) = e^{-iEt} \langle u_x|u_E(0)\rangle = e^{-iEt}\psi_E(x;0). \tag{5.4.7}$$

The function $\psi_E(x;0)$, which we denote as $\varphi_E(x)$, is called the 'time-independent wavefunction' of the stationary state with the energy proper value

E. The probability density of the localization of such a state,

$$\rho_E(x; t) = |\psi_E(x; t)|^2 = |\varphi_E(x)|^2, \tag{5.4.8}$$

is independent of the time, in conformity with the stationary nature of the state. We note also, that the function $\varphi_E(x)$ is, in principle, subject to the normalization condition

$$\int_{-\infty}^{\infty} |\varphi_E(x)|^2 \, dx = 1. \tag{5.4.9}$$

Introducing eq. (5.4.7) into eq. (5.4.5), we finally get for the spatio–temporal wavefunction of the state $w(t)$,

ħ
$$\psi_w(x; t) = \sum_E c_E \, e^{-iEt} \, \varphi_E(x), \tag{5.4.10}$$

where the coefficients

$$c_E = \langle u_E(t)|w(t)\rangle = \langle u_E(0)|w(0)\rangle, \tag{5.4.11}$$

are constants determined by the state w.

We see, in this way, that once the stationary states of the system and their time-independent wavefunctions $\varphi_E(x)$ have been determined, we are in a position to describe any 'motion' of the quanton. It is sufficient, for this purpose, to know the coefficients $c_E = \langle u_E|w\rangle$, which are the projections of the state of the quanton at a given instant ($t = 0$, for example, but any other instant will do just as well [see eq. (5.4.11)]), onto the various stationary states.

It is this general procedure which we shall apply to some simple physical situations in the following sections.

It is easy to understand why the stationary states are able to play a privileged role in the determination of the 'motion' of a quanton. Indeed, we often deal with a situation where a quanton is subject to interactions which do not evolve in time, and they have no reason to display a particular behaviour under spatial invariance. It is hardly surprising that stationary states happen to figure in the above expressions.

5.4.2. Forces and quantum potentials

We consider now the case of a quanton of mass m, subject to a 'field of external force'. This expression, borrowed from the language of the classical theory, implies that although the physical system is composed of mutually interacting particles, we have nevertheless agreed to represent the situation by the effect on one particle of all the rest, by means of a force which only depends on the position of this particle. In other words, we neglect the reciprocal effect of the particle on all the others, and we reduce the actual problem of N particles to one involving just a single particle. However, the notion of a force in the Newtonian sense of the term is correlative to that of a particle. A force acts pointwise on a (localized) particle, by modifying its momentum in accordance with the fundamental equation, $F = \mathrm{d}p/\mathrm{d}t$, of classical dynamics. As soon as one considers objects (such as quantons), for which neither the position nor the momentum is well-defined, the notion of the force has definitely to be modified.

To tell the truth, the notion of a force even loses its pertinence in the quantum theory. By contrast, the notion of a potential, from which the force is classically derived, survives and is well adapted to the quantum theory, where it is sufficient, moreover, to enable one to describe the interactions. This is not particularly surprising if one considers the fact that the notion of the potential energy is related to the general idea of the laws of conservation (of energy, in the present case), and hence of invariance, the efficacy of which, in the quantum theory, has already been seen.

In fact, we assume that the energy of a quanton, placed in a potential $V(x)$ (we restrict ourselves to the one-dimensional case), is once again – as in classical theory – the sum of its kinetic energy $p^2/2m$ and its potential energy $V(x)$,

$$`E = \frac{p^2}{2m} + V(x)`. \tag{5.4.12}$$

Obviously, this expression ought to be correctly, i.e., quantically, reinterpreted (whence the quotation marks!). Since for a quantum state, neither the energy, nor the momentum, nor the position have in general well-defined numerical values, the equality in eq. (5.4.12) cannot be an expression of a relationship between such values. Additionally, it could not even relate the proper values of these three magnitudes to each other. A proper state of momentum can surely not be a proper state of position. If, therefore, the magnitude $p^2/2m$ corresponds to a unique value, the same cannot be true of x

or of $V(x)$. All the same, however, if for the reasons laid out in the preceding section, we are interested in the stationary states of the quanton, it is clear that these are neither the proper states of momentum nor the proper states of position. A more complete formalism is necessary to understand the meaning to be attributed to the equality in eq. (5.4.12).

There is, however, a situation where things simplify. This occurs when the potential $V(x)$ is a constant and does not depend on x. With the potential no longer intervening, the relation

$$E = \frac{p^2}{2m} + V_0, \tag{5.4.13}$$

may be looked upon as being a numerical relationship between the proper values of the energy and the momentum – magnitudes which in the present situation are compatible.

5.4.3. Stationary states in a constant potential. Density and current

Let us determine, therefore, the stationary states of a quanton, placed in a constant potential $V(x) = V_0$. Relation (5.4.13) connects the spectrum of the energy to that of the momentum. Thus, to a proper value E of the energy, there correspond two proper values of the momentum

$$p = \pm p_E, \quad \text{where} \quad p_E \triangleq [2m(E - V_0)]^{1/2}. \tag{5.4.14}$$

In other words, while a stationary state of the system is not a proper state of momentum, its spectrum in p reduces to only the two values p_E and $-p_E$, corresponding evidently to the two possible orientations of the motion along the axis, either from right to left or from left to right. If we apply now the rules of superposition and of sequential factorization of amplitudes to the computation of the wavefunction $\varphi_E(x)$, we get

$$\varphi_E(x) = \langle u_x | u_E \rangle$$

$$= \langle u_x | u_{p_E} \rangle \langle u_{p_E} | u_E \rangle + \langle u_x | u_{-p_E} \rangle \langle u_{-p_E} | u_E \rangle. \tag{5.4.15}$$

Replacing $\langle u_x | u_{p_E} \rangle = \psi_{p_E}(x)$ and $\langle u_x | u_{-p_E} \rangle = \psi_{-p_E}(x)$ by eq. (5.2.19), this becomes

$\hbar$ $\quad\quad \varphi_E(x) = a_+ \exp(ip_E x) + a_- \exp(-ip_E x), \tag{5.4.16}$

where we have set

$$a_\pm = A\langle u_{\pm p_E} | u_E \rangle. \tag{5.4.17}$$

The wavefunction of eq. (5.4.16) is a superposition of two 'plane waves'. The state which it describes is delocalized, like all proper states u_p. This should not surprise us, since this wavefunction is not normalizable [i.e., it cannot be subjected to the normalization condition given by eq. (5.2.3)]. It can be verified that the probability density

ℏ
$$\rho_E(x) = |\varphi_E(x)|^2$$
$$= |a_+|^2 + |a_-|^2 + 2|a_+ \bar{a}_-| \cos(2p_E x + \text{Arg } a_+/a_-), \tag{5.4.18}$$

is not an integrable function. It is interesting to note that this probability density is not a constant, but displays a sinusoidal modulation (fig. 5.12c) with a characteristic length

ℏ
$$\Delta x = (2p_E)^{-1}. \tag{5.4.19}$$

Furthermore, the width of the momentum spectrum, consisting of the two values $(+p_E, -p_E)$ is

$$\Delta p = 2p_E. \tag{5.4.20}$$

Clearly, $\Delta p \, \Delta x$ satisfy the Heisenberg inequality

ℏ
$$\Delta x \, \Delta p \gtrsim 1. \tag{5.4.21}$$

Up to a coefficient of normalization, A, the quantities

$$\rho_\pm = |a_\pm|^2 = |A|^2 |\langle u_{\pm p_E} | u_E \rangle|^2, \tag{5.4.22}$$

represent the transition probability densities from the stationary state u_E to the proper states of momentum $u_{\pm p_E}$. These measure, therefore, the relative probabilities for a quanton of energy E to move to the left $(-p_E)$ or to the right $(+p_E)$. But beware! Equation (5.4.18) clearly demonstrates that one cannot consider the probability of localization of the quanton in a certain interval to be the sum $(\rho_+ + \rho_-)$ of the probabilities of finding it there, moving to the left and to the right! It is the amplitudes which are added in eq.

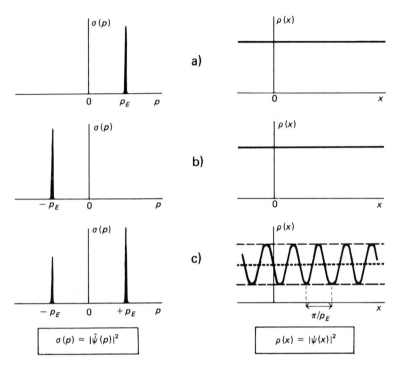

Fig. 5.12. *Stationary states of a free quanton*
To the left, the momentum spectrum; to the right, the probability density of localization
(arbitrarily normalized) for:
(a) A stationary state of energy E which is also a proper state of momentum, with proper value
$p_E = \sqrt{2mE}$. The state is completely unlocalized and its localization density is a constant.
(b) The same, for $p = -p_E$.
(c) The most general stationary state of the energy E has a wavefunction which is a
superposition of the two preceding ones. Its localization density is oscillatory (the more
symmetric the spectrum in p, the larger is the 'contrast' $\rho_{max} - \rho_{min}$).
The states considered here being stationary, the probability density $\rho(x)$ does not depend on the
time.

(5.4.15), according to the principle of superposition, whence the interference
term in the local density, eq. (5.4.18).

By contrast, and as is usual, if the average value $\tilde{\rho}_E$ of $\rho_E(x)$ is taken, over
an interval which is sufficiently large compared to the characteristic length,
eq. (5.4.19), the interference effects disappear

$$\tilde{\rho}_E = |a_+|^2 + |a_-|^2 = \rho_+ + \rho_-, \tag{5.4.23}$$

and one has a simple addition of probabilities, just as in classical theory. This is a situation which is encountered when dealing with a detector which has a spatial resolution larger than Δx. In any event, the wavefunctions in eq. (5.4.16) describe idealized states; in a more realistic description, which we shall develop later (sect. 5.4.4), these interference effects are no longer present. Rather than considering individual quantons, we could consider a (statistical) ensemble of quantons, as in sect. 5.3.1. In this case, the quantities ρ_E [eq. (5.4.18)] and $\rho_\pm$ [eq. (5.4.22)] should be interpreted as being relative densities (in view of the problem of normalization) in the usual sense of the term: the (relative) number of quantons per unit of length. Thus, the system is considered to be a double beam of quantons, of energy E, one moving from left to right, with the density ρ_+, and the other in the opposite direction, with the density ρ_-. This interpretation allows us to introduce an interesting quantity, the *current* of quantons, or in other words, the number of quantons passing per unit time at each point of the axis. For a velocity v, the current of a beam of density ρ is ρv (this is the number of quantons in a segment of length $v \Delta t$ with $\Delta t = 1$). Here, since the quantons have velocities $v = p_E/m$, with the density ρ_+, or $-v = -p_E/m$, with the density ρ_-, we arrive at the following expression for the net current (difference of the currents in the two directions):

$$j = v(\rho_+ - \rho_-);\qquad\qquad (5.4.24)$$

which, furthermore, is

$$j = \frac{p_E}{m}(|a_+|^2 - |a_-|^2).\qquad\qquad (5.4.25)$$

Returning to the case of a single quanton, where the quantities $|a_\pm|^2$ are interpreted as being *probability* densities, eq. (5.4.25) can be thought of as giving the definition of a *probability current*. Like any current, it expresses the notion of a transport. In this case, the transport is not one of charge or of matter, but rather of probability. We shall see later the usefulness of this notion, here reduced to its simplest form, to the extent that the state being stationary and the physical situation invariant under spatial translations, the current is automatically a constant and is identical at all points. Finally, we observe that the quantity of physical interest is

$$\mathscr{R} = \frac{|a_+|^2}{|a_-|^2},\qquad\qquad (5.4.26)$$

which gives the ratio of the currents corresponding to the two directions, and has the advantage of being independent of the absolute normalization of the wavefunction and of having the same value, whatever might be the solution adopted for the problem of normalization.

5.4.4. The free quanton and its wave packet

A quanton is said to be 'free' when it is not subject to any forces or, since we prefer to use the notion of a potential, when it moves in a constant potential. We proceed now to establish the form, eq. (5.4.16), of the wavefunction $\varphi_E(x)$ of the stationary states of such a quanton. Following eq. (5.4.10), we are in a position to write down the spatio–temporal wavefunction of an arbitrary state of this quanton in the form

$$\hbar \qquad \psi(x;t) = \int_{V_0}^{+\infty} [C_+(E)\, e^{ip_E x} + C_-(E)\, e^{-ip_E x}] e^{-iEt}\, dE, \qquad (5.4.27)$$

where we have set $C_\pm(E) = c_E a_\pm(E)$ and written the sum over the energies more precisely as an integration [starting at V_0, in view of eq. (5.4.13)]. It is often preferable to write this expansion as an integral over the momentum. The change of variable,

$$p = \pm p_E = \pm [2m(E - V_0)]^{1/2}, \qquad (5.4.28)$$

in the first and the second terms, respectively, in the integral of eq. (5.4.27) allows us to write

$$\hbar \qquad \psi(x;t) = \int_{-\infty}^{+\infty} F(p) \exp i(px - E_p t)\, dp, \quad \text{with } E_p = \frac{p^2}{2m} + V_0, \qquad (5.4.29)$$

where the function F is determined by $C_\pm(E_p)$ for $p \gtrless 0$. Moreover, we point out that in this form we recover, for $t = 0$, eq. (5.2.26) of the spatial wavefunction $\psi(x)$ in terms of the wavefunction $\hat{\psi}(p)$ with which, therefore, we identify $F(p)$. This result is hardly surprising. It is simply due to the fact that the proper states of momentum u_p being, for a free quanton, also the stationary states u_E, expansion (5.2.23) in momentum amplitudes coincides with the development, in eq. (5.4.5), in amplitudes of the energy at $t = 0$, and it generalizes immediately to an arbitrary time t by the simple insertion of the evolution factor e^{iEt}. We shall write, therefore,

$\hbar$
$$\psi(x; t) = A \int_{-\infty}^{\infty} \hat{\psi}(p)\, e^{i(px - E_p t)}\, dp,$$ (5.4.30)

which manifestly displays the probability amplitude $\hat{\psi}(p)$ in momentum.

Let us now return to the question of the 'motion' of a 'free' quanton. The wavefunction of a stationary state of energy E of such a quanton is given by eq. (5.4.16). However, strictly speaking, it is impossible to talk about the 'motion' of such a state of the quanton in the sense of a displacement of its localization. Indeed, we have seen that the probability density of localization $|\varphi_E(x)|^2$ is a regularly oscillating function, all along the axis, and is independent of the time. There is nothing in it which resembles a 'motion' comparable to that of a particle.

Thus, it is meaningful to speak of the motion of a quanton only if it is 'reasonably' localized, i.e., has a spatial extension Δx which is not too large. This, in view of the Heisenberg inequality, $\Delta x \, \Delta p \geq 1$, forces the quanton to have a momentum spectrum which gets larger the more localized it becomes: $\Delta p \geq 1/\Delta x$. In other words, the motion of a quanton cannot be represented by means of a single stationary state. As we have already seen, in sect. 5.2.3, it is only by using a superposition of many plane waves, characterized by a continuous sequence of values of p_E, contained within an interval Δp, that one might expect to describe a motion in the usual sense. However, in order to be able to speak of the velocity of this motion, it is necessary that the momentum of the quanton be sufficiently well-defined, and that the width Δp of its spectrum be not too large either. Thus, we are led to trying to approach the limit imposed by the Heisenberg inequality as closely as possible and to consider states for which $\Delta p \, \Delta x \simeq 1$, to the extent that this can be done! (exercise 5.23).

The wavefunction of our quanton, obtained by taking the 'packet' of plane waves so defined, has, therefore, the general form of eq. (5.4.30), in which the amplitude $\hat{\psi}(p)$ assumes appreciable values only in a region of width Δp (fig. 5.13). Let us assume that it displays a sharp maximum for a certain value p_0, which is the average value of the momentum. The contributions to the integral in eq. (5.4.30) come, therefore, only from values close to $p = p_0$. However, the integral being complex, it is also necessary that its phase be not too rapidly varying in this neighbourhood. Otherwise, the different contributions, in spite of their large moduli, would compensate each other and their sum would be greatly reduced – as any sum of complex numbers of nearly

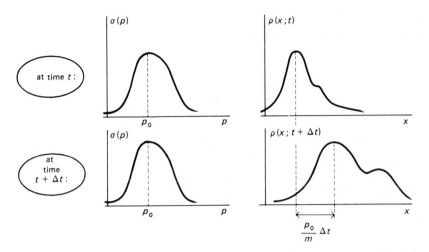

Fig. 5.13. *The evolution of a wave packet*
Suppose, that at the instant of time t, a quanton is in a state represented by a sufficiently well-localized wave packet, given by a momentum spectrum centred at the value p_0. At a later time $t + \Delta t$, the wave packet is displaced by an amount $(p_0/m)\,\Delta t$. In general, this displacement is accompanied by a deformation of the wave packet, and in particular, by a broadening (exercise 5.23).
The state of the quanton, not being a stationary one, the probability density in position $\rho(x, t)$ depends explicitly on the time – even though the density in position $\sigma(p)$ is constant (for a free quanton).

equal moduli but widely dispersed phases. In other words, the interference between the proper states of momentum, which generates the superposition in eq. (5.4.30), has to be constructive, and for this the different contributions to the sum must have phases which are as close to each other as possible. Displaying this phase explicitly in the form,

$$\psi(x; t) = A \int_{-\infty}^{\infty} |\hat{\psi}(p)| \exp i\,[px - E_p t + \delta(p)]\,\mathrm{d}p, \qquad (5.4.31)$$

where δ is the phase of $\hat{\psi}$, we arrive at the conclusion that the phase $[px - E_p t + \delta(p)]$ must vary as little as possible in the domain of integration surrounding the value p_0. We are thus led to requiring that this phase be stationary, as a function of p, for the average value $p = p_0$,

$$\frac{\mathrm{d}}{\mathrm{d}p}[px - E_p t + \delta(p)]\bigg|_{p = p_0} = 0. \qquad (5.4.32)$$

Thus,

$$\hbar \qquad x - \frac{dE_p}{dp}\bigg|_{p_0} t + \frac{d\delta}{dp}\bigg|_{p_0} = 0. \qquad (5.4.33)$$

This is a relationship which has to be fulfilled in order for the integral to assume an appreciable value. It expresses, therefore, the motion of the quanton by showing that the point at which its probability amplitude is maximal, i.e., the vicinity in which it is most likely to be found, moves uniformly according to

$$x - vt = \text{const.} \qquad (5.4.34)$$

The velocity of this motion,

$$\hbar \qquad v_{p_0} = \frac{dE_p}{dp}\bigg|_{p_0} = \frac{p_0}{m}, \qquad (5.4.35)$$

is both the group velocity, in the sense of waves ($v_g = d\omega/dk$, since $E = \omega$ and $p = k$ – see sect. 3.1.3), as well as the velocity of a classical particle, the momentum of which is the average momentum p_0 of the quanton – a fact which expresses rather nicely the quantum synthesis and the passage back and forth between the notions of the wave and the particle. Finally, we note that the equation of motion can be written more precisely as

$$x - v(t - \tau) = 0, \qquad (5.4.36)$$

where,

$$\hbar \qquad \tau = \frac{d\delta}{dE}\bigg|_{p_0}, \qquad (5.4.37)$$

is the (average) time at which the quanton passes the origin. Let us bear in mind this temporal interpretation of the derivative, with respect to energy, of the phase of the amplitude in momentum.

The preceding considerations are applicable to a wave packet having a spectrum centred at a well-defined value p_0 of the momentum, or in other words, to the wavefunction of an 'almost monochromatic' state, which is more realistic (since it is not completely unlocalized) than the pure plane wave. Now, the most general stationary state, of energy E, has a wavefunction which is the sum of two progressive plane waves, moving in opposite directions, eq. (5.4.27). Its spectrum in momentum consists, therefore, of the two values $p_0 = p_E$ and $-p_0$. A more realistic, 'almost stationary' state would have a spectrum consisting of two sufficiently narrow, but not pointlike distributions, centred at $\tilde{p}_0$ and $-\tilde{p}_0$ (fig. 5.14). The wavefunction would, therefore, be a sum of two terms, each corresponding to one of these two distributions. Thus, it would be the sum of two wavefunctions, of the type of two 'almost monochromatic' wave packets $\psi_{\tilde{p}_0}(x; t)$,

$$\psi(x; t) = \psi_{\tilde{p}_0}(x; t) + \psi_{-\tilde{p}_0}(x; t). \tag{5.4.38}$$

This means that it has two regions of maximal amplitude, which move with

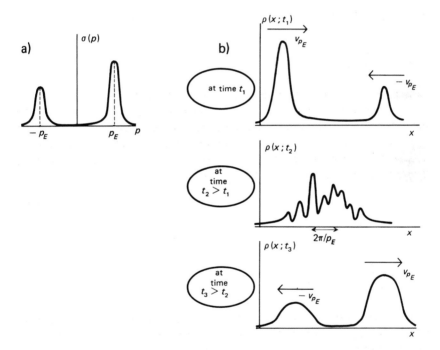

Fig. 5.14. *An 'almost' stationary state*

(a) The distribution in momentum consists of two peaks, around $p_E = \sqrt{2mE}$ and $-p_E$, where E is the average energy of the state.

(b) The distribution in position, which depends on the time, displays two 'humps' moving with velocities v_{p_E} and $-v_{p_E}$, respectively, and interfering only in the region of their mutual crossing. Compare this diagram with that of fig. 5.12c.

the (group) velocities v_{p_0} and $-v_{p_0}$. The probability density in position is, therefore,

$$|\psi(x; t)|^2 = |\psi_{\bar{p}_0}(x; t) + \psi_{\bar{p}_0}(x; t)|^2$$

$$= |\psi_{\bar{p}_0}(x; t)|^2 + |\psi_{-\bar{p}_0}(x; t)|^2$$

$$+ \text{Re } \overline{\psi_{\bar{p}_0}(x; t)} \, \psi_{-\bar{p}_0}(x; t). \tag{5.4.39}$$

However, most of the time, the two wave packets $\psi_{\bar{p}_0}$ and $\psi_{-\bar{p}_0}$ do not overlap (at any point x, at least one of the two vanishes), so that *the interference term is zero*: the probability density only displays two 'humps' moving in opposite directions. It is only during the (short) period of their 'crossing', that the two wave packets have a certain overlap, and that interference effects are manifested (fig. 5.14).

Exercises

5.1. Show that in the expansion of a transition amplitude with the help of a (complete and disjoint) set of intermediate states, consisting of the *stationary states* at a given instant, namely in the expression

$$\langle v|w \rangle = \sum_E \langle v|u_E(t) \rangle \langle u_E(t)|w \rangle,$$

each term in the expansion is independent of the intermediate time t.

5.2. We shall see later (sect. 6.4.2) the great importance, for certain systems, of states possessing a *continuous* energy spectrum, consisting of all values $E \geqslant 0$ and having a probability density in energy of the special form,

$$\mu(E) = C \frac{1}{(E - E_0)^2 + (\Gamma/2)^2},$$

where $E_0 > 0$ [$\mu(E) \, dE$ is the probability that the system would have an energy lying between E and $E + dE$].

(a) Show that the probability amplitude for the system, in the state $w(t)$, to make a transition to its initial state $w(0)$, which we shall write as $h(t) = \langle w(0)|w(t) \rangle$, can be written, by generalizing eq. (5.1.14), as

$$h(t) = \int\limits_0^{+\infty} \mu(E) \, e^{-iEt} \, dE.$$

(b) Sketch graphically, the shape of the function $\mu(E)$. What is the physical meaning of

E_0 and of Γ? Show that if $E_0 \gg \Gamma$, the spectrum may be extended to $-\infty$, and $h(t)$ replaced by its approximation,

$$g(t) = \int_{-\infty}^{\infty} \mu(E)\, e^{-iEt}\, dE.$$

(c) By changing variables: $u = 2(E - E_0)/\Gamma$, reduce the computation of $g(t)$ to that of the integral

$$G(s) = 2\,\mathrm{Re} \int_0^{+\infty} \frac{e^{-isu}}{u^2 + 1}\, du.$$

(d) Assuming, in order to ensure the convergence of the integrals, that s has a small imaginary part, show that G obeys the differential equation $G'' = G$, and from its behaviour at infinity and at the origin, deduce that $G(s) = \pi\, e^{-|s|}$.

(e) Hence deduce that the probability of finding the system in its initial state, at the time t, is (up to a constant)

$$\mathscr{P}(t) \propto e^{-\Gamma t} \quad (t > 0).$$

The energy spectrum $\mu(E)$ – called the 'Breit–Wigner spectrum' – is, therefore, that of an *unstable system*, undergoing an exponential decay.

(f) Express the dispersion in the energy ΔE and the characteristic time τ in terms of Γ and show that there is a Heisenberg type of correlation, $\tau \Delta E \simeq 1$.

5.3. We want to analyze the *quantum beats* experiment, described by fig. 5.2. It deals with the separation between a pair of levels of an ionized helium atom, for which the average radiated wavelength is $\lambda = 3188$ Å.

(a) What is the average value of the energies of the levels in question (in eV)?

(b) The helium atoms have a kinetic energy of 300 keV. What is their velocity?

(c) Decide, from the experimental results (fig. 5.2c), the beat frequency and then the energy separation of the levels (in eV).

5.4. Consider a *spin-$\frac{1}{2}$* quanton (see sect. 5.1.3., point (b)). Let $\uparrow$ and $\downarrow$ be, as usual, the proper states of S_z. Consider an arbitrary 'state of spin', u, characterized by the transition amplitudes

$$\langle \uparrow | u \rangle = \alpha, \qquad \langle \downarrow | u \rangle = \beta.$$

(a) What are the transition probabilities $\uparrow \leftarrow u$ and $\downarrow \leftarrow u$? What relationship should connect these two quantities?

(b) Show that a state u can be parametrized by two numbers (θ, φ) such that,

$$\alpha = \cos \theta, \qquad \beta = \sin \theta\, e^{i\varphi}.$$

(c) Compute the transition amplitude between two states $u(\theta, \varphi)$ and $u(\theta', \varphi')$, namely, $\langle u(\theta', \varphi') | u(\theta, \varphi) \rangle$. Do there exist states for which the transitions to the proper states of

$S_z(\uparrow, \downarrow)$ and the proper states of $S_x(\rightarrow, \leftarrow)$ are equiprobable? Denote these states by $\odot$ and $\oplus$. What could be their physical interpretation?

(d) Assume next that the quanton possesses a magnetic moment $\mathcal{M}$ and that it is placed in a magnetic field $\mathcal{B}$ having the direction Oz (see sect. 5.1.3, point (b)). Let $w(t)$ be a time-dependent state such that $w(0) = u$, characterized above. Calculate the amplitude $\langle u | w(t) \rangle$ as a function of θ and φ, and the probability for finding the system, at the time t, in the same state as at the time $t = 0$. Calculate the average values $\langle S_x(t) \rangle$, $\langle S_y(t) \rangle$, $\langle S_z(t) \rangle$.

5.5. The experiment shown in fig. 5.3 displays the *Larmor precession* of the spin of the neutrons in a magnetic field. The neutrons used in the experiment, have a wavelength $\lambda = 1.55$ Å.

(a) Calculate the velocity of the neutrons.

(b) Deduce from fig. 5.3b the value of the angular velocity of precession for the spin.

(c) The magnetic field has the value $\mathcal{B} = 15.5 \times 10^{-4}$ T. Hence deduce the value of the magnetic moment of the neutrons.

5.6. We wish to study the quantum properties of the *rotation of the spin* of a neutron, under the effect of a magnetic field. This field, having the value $\mathcal{B}$, acts in a region of length δl, traversed by the neutron with an average velocity v and hence in a period of time $\delta t = \delta l / v$. Recall that the neutron possesses a magnetic moment $\mathcal{M} = \mu S$, where S is the spin.

(a) Suppose that a neutron is in the proper state $\uparrow$ of S_z, where the Oz axis is parallel to $\mathcal{B}$. Show that the crossing of the magnetic field produces a phase shift

$$\delta\varphi = \tfrac{1}{2}\mu\mathcal{B}\,\delta t$$

of its wavefunction.

(b) This phase shift is demonstrated by an interferometric experiment (fig. 1), in which the magnetic field can only be traversed along one of two paths inside the interferometer. Show that the intensity of the neutron beam at the exit to the interferometer is of the form

$$I = \tfrac{1}{2}I_0(1 + \cos\delta\varphi).$$

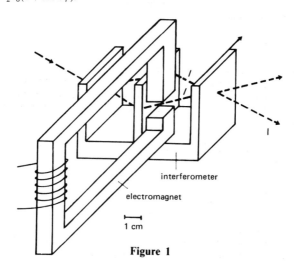

interferometer

electromagnet

1 cm

Figure 1

(c) What would be the corresponding results had the neutron been in the proper state $\downarrow$ of S_z? Finally, what would these results look like for a neutron in an arbitrary state?

(d) We have seen (sect. 5.1.3) that in the later case, we may describe the *average* motion of the spin vector as a (Larmor) rotation around $\mathcal{B}$. Calculate the angle, $\delta\alpha$, of rotation and show that $\delta\alpha = 2\,\delta\varphi$. How large is the phase shift for a Larmor rotation of 2π?

(e) Fig. 2 gives the results of the measurement of the intensity of the neutron beam at the exit to the interferometer (arbitrary units) as a function of the quantity $\int \mathcal{B}\,dl \simeq \mathcal{B}\,\delta l$, evaluated in 10^{-6} T m. The neutrons had a wavelength $\lambda = 1.82$ Å. The magnetic moment of the neutron is $\mathcal{M} = -0.97 \times 10^{-26}$ MKSA units. Interpret these results in the light of the preceding questions. In what sense can it be said that this experiment demonstrates the period of rotation 4π (the double turn), specific to quantons of half-integral spins? (see sect. 2.3.4).

[From H. Rauch, A. Zeilinger, G. Badurek, A. Wilfing, W. Bauspiess, U. Bonse, Physics Letters A *54* (1975) 425.]

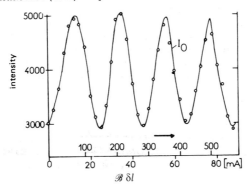

Figure 2

5.7. Consider the wavefunction

$$\psi(x) = N\,e^{-\kappa|x|}.$$

(a) Calculate the *normalization constant N*.

(b) Calculate the probability, $\mathcal{P}(c - \tfrac{1}{2}a, c + \tfrac{1}{2}a)$, of localization inside an interval of width a, centred at the point c on the abscissa. We want this probability to equal $\tfrac{1}{2}$. How should c be chosen so that the interval a be as small as possible? What is the value of a in this case?

5.8. Are the following wavefunctions normalizable? If so, compute the value of the *normalization constant N*,

(a) $N(x^2 + a^2)^{-1/2}$ (consider also the case $a = 0$);

(b) $N \exp[-\alpha(x - b)]$;

(c) $N \exp[-x^2/2c^2]$;

(d) $N(1 + \beta x) \exp[-x^2/2c^2]$;

(e) $Nx^{-1} \sin \gamma x$.

In the cases in which the function is normalizable, sketch its graph and also, side by side, that of the corresponding probability density $\rho(x) = |\psi(x)|^2$.

5.9. For the (normalized) wavefunctions of exercise 5.8, compute the *average values* $\langle X \rangle$, $\langle X^2 \rangle$ and the *dispersion* ΔX.

5.10. Consider the wavefunction

$$\psi(x) = N\, e^{-\kappa|x|},$$

appropriately normalized (see exercise 5.7).

(a) Compute the *wavefunction in momentum* $\hat{\psi}(p)$.

(b) By requiring that $\hat{\psi}(p)$ be normalized, calculate the value of the numerical coefficient A, in the reciprocal transformations between ψ and $\hat{\psi}$, and valid for *all* wavefunctions [see eqs. (5.2.26) and (5.2.29)].

5.11. Consider the wavefunction in momentum

$$\hat{\psi}(p) = N\, e^{-p^2 a^2/2},$$

appropriately normalized (see exercise 5.8c, suitably adapted).

(a) What is the corresponding spatial wavefunction $\psi(x)$?

(b) Having normalized $\psi(x)$, calculate the value of the constant A in the reciprocal *Fourier transforms* [eqs. (5.2.26) and (5.2.29)] between $\psi(x)$ and $\hat{\psi}(p)$ (valid for *all* wavefunctions).

5.12. Let $\hat{\psi}(p)$ be the momentum wavefunction, corresponding to the spatial wavefunction $\psi(x)$. Express, in terms of $\hat{\psi}$, the momentum wavefunctions corresponding to the spatial wavefunctions, modified according to the following *transformations*:

(a) $\psi'(x) = \psi(x - a)$ (translation)
(b) $\psi^m(x) = e^{iqx}\, \psi(x)$ (modulation)
(c) $\psi^d(x) = \lambda^{1/2}\psi(\lambda x),\ \ \lambda > 0$ (dilation)

(It should be verified at first that the modified functions are still normalizable). How do you judge the results?

5.13. For each of the following wavefunctions,

(a) $N\, e^{-\kappa|x|}$,
(b) $N\, e^{-x^2/2c^2}$,
(c) $N(1 + \beta x)\, e^{-x^2/2c^2}$,

appropriately normalized (see exercises 5.8c, d), first compute the wavefunction in momentum $\hat{\psi}(p)$, and the densities $\rho(x)$ and $\sigma(p)$, and then calculate the *dispersions* ΔX and ΔP, and their product.

5.14. For an arbitrary positive function f, integrable on the line, we define the 'x-integral' width L_α by the following condition

$$L_\alpha(f) = \min\left\{ \Delta \subset R \Big| \int_\Delta f(x)\, dx = \alpha \int_{-\infty}^{+\infty} f(x)\, dx \right\}.$$

In other words, L_α is the measure of the smallest domain on which the integral of f is a fraction α of its integral over the entire line.

(a) Calculate L_α for $f(x) = e^{-\kappa|x|}$ and $f(x) = (x^2 + a^2)^{-1}$. The 'α-integral dispersions' in position and momentum, of a quanton in the state W, are defined by

$$(\delta_\alpha X)_W \triangleq L_\alpha(|\psi_W|), \qquad (\delta_\alpha P)_W \triangleq L_\alpha(|\check\psi_W|)$$

(b) Using the inequalities (to be established)

$$\int_\Delta |\psi(x)|\, dx \leqslant \Delta \sup |\psi|, \qquad \int_{-\infty}^{\infty} |\psi(x)|\, dx \geqslant \sqrt{2\pi} \sup |\check\psi|,$$

show that

$$\delta_\alpha X \geqslant \sqrt{2\pi}\, \alpha \frac{\sup |\check\psi|}{\sup |\psi|}.$$

(c) Hence deduce the *'Heisenberg-like'* inequality

$$\delta_\alpha X\, \delta_\alpha P \geqslant 2\pi\alpha^2.$$

[Following D. N. Williams, Am. J. Phys. 47 (1979) 606.]

5.15. Consider a quanton, in the state w, having wavefunction $\psi_w(x)$ and the corresponding momentum wavefunction $\check\psi_w(p)$. Let $\langle X \rangle_w$ and $\langle P \rangle_w$ be the average values of its position and momentum.

(a) Show that, in the state V, the wavefunction of which is *defined* by

$$\psi_V(x) = e^{-i\langle P \rangle_w x}\psi_w(x + \langle X \rangle_w),$$

one has

$$\langle X \rangle_V = 0.$$

More generally, show that

$$\langle f(X) \rangle_V = \langle f(X - \langle X \rangle_w) \rangle_w,$$

and hence that

$$(\Delta X)_W^2 = \langle X^2 \rangle_V.$$

(b) Show that the momentum wavefunction of the state V is

$$\check\psi_V(p) = e^{i(\langle P \rangle_w \langle X \rangle_w)} e^{i\langle X \rangle_w p} x \check\psi_w(p + \langle P \rangle_w),$$

whence, by symmetry,

$$\langle P \rangle_V = 0.$$

More generally, show that

$$\langle g(P) \rangle_V = \langle g(P - \langle P \rangle_W) \rangle_W,$$

whence,

$$(\Delta P)_W^2 = \langle P^2 \rangle_V.$$

(c) For an arbitrary wavefunction ψ, show that

$$\int p^2 |\hat\psi(p)|^2 \, dp = \int \left| \frac{d\psi}{dx} \right|^2 dx.$$

(d) Show that,

$$\int \left| x\psi + \lambda \frac{d\psi}{dx} \right|^2 dx = \int x^2 |\psi|^2 \, dx + \lambda + \lambda^2 \int \left| \frac{d\psi}{dx} \right|^2 dx.$$

for any real number λ. Hence deduce that for any quantum state

$$\langle X^2 \rangle + \lambda + \lambda^2 \langle P^2 \rangle \geqslant 0, \text{ for all } \lambda,$$

and thus, that

$$\langle X^2 \rangle \langle P^2 \rangle \geqslant \tfrac{1}{4}.$$

(e) With the help of the results in (a) and (b) establish the *Heisenberg inequalities* for standard deviations

$$\Delta X \, \Delta P \geqslant \tfrac{1}{2}.$$

5.16. (a) Show that for a quanton, the radial momentum p_r, the square of the modulus of the orbital *angular momentum*, L^2, and the component L_z are three compatible magnitudes. (b) Let $u_{p,l,m}$ be a common proper state of these three magnitudes [with the respective proper values p, $l(l+1)$ and m]. Consider its spatial wavefunction $\psi_{p,l,m}(r) = \langle u_r | u_{p,l,m} \rangle$. Show that it can be written, in spherical coordinates, in the form

$$\psi_{p,l,m}(r, \theta, \varphi) = e^{im\varphi} P_{l,m}(\theta) \, e^{ipr}/r,$$

where $P_{l,m}(\theta)$ is an unknown function, which can only be determined using a more elaborate formalism.

5.17. Why is it not possible to define a *wavefunction 'in angular momentum'* $\tilde\psi(l_x, l_y, l_z)$, analogously to the wavefunction in momentum $\hat\psi(p_x, p_y, p_z)$, the square of the modulus of which might then yield the probability density for the components L_x, L_y, L_z of the orbital angular momentum L to assume the numerical values l_x, l_y, l_z, respectively?

5.18. We are interested in scattering phenomena for which the scattering amplitude is independent of the direction (*low-energy scattering* by a finite range potential; see sect. 5.3.4, point (a)).

(a) Show that the optical theorem, eq. (5.3.26), enables us to express the scattering amplitude f – a complex number – as a function of a single real parameter δ (called the 'phase shift') through the formula

$$f = \frac{1}{2ip}(e^{2i\delta} - 1) = \frac{1}{p}e^{i\delta}\sin\delta.$$

Find the curve, on the complex plane, on which the extremity of the vector, representing f (for a given p), lies.

(b) How can the effective cross-section σ be expressed as a function of δ? Show that $\sigma \leqslant 4\pi/p^2$.

5.19. The elastic collision of two quantons, leading to their scattering, can also be accompanied by '*inelastic*' *processes*: various rearrangements (or reactions), absorption, etc. In order to make this idea concrete, consider the collision of a neutron, of initial momentum p, with a nucleus, assumed fixed, which can scatter *and* absorb the neutron. We proceed to demonstrate that the two processes are correctly described if the complex scattering amplitude, which we write as

$$f_p = \operatorname{Re} f_p + i \operatorname{Im} f_p,$$

satisfies certain conditions.

(I) Suppose that the energy is sufficiently low so that following the arguments already developed, f_p depends only on the scattering angles (θ, φ).

(a) Show that the effective cross-section for *elastic* scattering is given by

$$\sigma_{\text{el}} = 4\pi|f_p|^2.$$

(b) Show that in order for the probability to be conserved, the argument behind the optical theorem leads one to the notion that there ought to exist, in addition, reaction phenomena, described by the effective cross-section,

$$\sigma_{\text{re}} = 4\pi\left(\frac{1}{p}\operatorname{Im} f_p - |f_p|^2\right).$$

(c) Hence deduce that the amplitude f_p is represented, on the complex plane, by a vector, the extremity of which lies in the interior of a circle. Compare with the results of exercise 5.18 and show that f_p may be parametrized as

$$f_p = \frac{1}{2ip}(e^{2i\delta} - 1),$$

where the phase shift δ is complex, with a positive imaginary part.

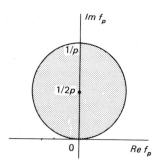

(d) For what value(s) of the amplitude f_p [which point(s) of the domain is (are) admissible?] is the scattering purely elastic ($\sigma_{re} = 0$)? Show that one always has $\sigma_{el} \leqslant 4\pi/p^2$.

(e) Show that a reaction cannot take place without there being an elastic scattering as well (it is not possible that $\sigma_{re} \neq 0$ and $\sigma_{el} = 0$ simultaneously) and that one always has

$$\sigma_{el} \geqslant \frac{1}{4\pi} p^2 (\sigma_{re})^2.$$

For what value(s) of f_p is the reaction cross-section maximal? Show that $\sigma_{re} \leqslant \pi/p^2$ and that, when the effective cross-section for reaction is maximal,

$$\sigma_{el} = (\sigma_{re})_{max} = \pi/p^2.$$

(II) Consider now the situation at very low energies, i.e., in a region where $p|f_p| \ll 1$. Set $f_p \simeq f_0 = a$ and define the 'complex scattering length' $a = a_r + ia_i$.

(a) Show that this condition is effectively fulfilled in the nuclear interactions of thermal neutrons, for which

$$\lambda_p = h/p \simeq 1 \text{ Å}, \qquad f_p \simeq 1\text{--}10 \text{ F}.$$

(b) Show that, under these conditions, the preceding inequalities set very mild restrictions and that $\sigma_{re} \simeq 4\pi a_i/p \gg \sigma_{el}$, since a_i is not negligible compared to a_r. Hence, deduce that the total effective cross-section for the interaction of the neutrons with the nuclei, at very low energies, varies in inverse proportion to their velocity (the so-called '1/v' law).

(c) Cadmium has a scattering length of

$$a = (3.8 + 1.2i) \text{ F}$$

for neutrons. Calculate the elastic scattering cross-section σ_{el} and the reaction cross-section σ_{re} of cadmium for neutrons of wavelength $\lambda = 1$ Å. Hence deduce that cadmium is an excellent absorber of neutrons (hence its use in the control rods of nuclear reactors).

5.20. The intensity of a beam of thermal neutrons, of wavelength $\lambda = 8.9$ Å, is measured after reflection from a polished beryllium surface. Deduce from the experimental results (fig. 5.10) the critical angle for total reflection and then the neutronic *scattering length* of beryllium, using the mass density $\mu = 1.85$ g cm^{-3} and the atomic mass $M = 8$ amu of beryllium.

5.21. Consider a composite material, consisting of several species of atoms and hence of nuclei. Let a_k be the scattering length of neutrons by the kth nuclear species, and $\mathcal{N}_k$ the density of neutrons.

(a) Show that the *quantum index of refraction* for neutrons of wavelength λ can be written as

$$n = 1 - \frac{1}{2\pi}\left(\sum_k \mathcal{N}_k a_k\right)\lambda^2.$$

(b) A measurement of the angle of total reflection for triethyl-benzene $(C_{12}H_{18})$ gives a value of $\theta_c = 6'$, for neutrons of wavelength $\lambda = 8$ Å. Knowing that the mass density of triethyl-benzene is $\mu = 6.7$ g cm^{-3}, show that this experiment yields the weighted average, $a = \frac{2}{5}a_C + \frac{3}{5}a_H$, of the scattering lengths of carbon (a_C) and of hydrogen (a_H).

(c) Previous measurements have furnished the value $a_C = 6.63$ F. Calculate a_H. Why is it not possible to measure a_H directly, by the method of total reflection (e.g., from a liquid hydrogen mirror)? [Following D. J. Hughes, M. T. Burgy and G. R. Ringo, Phys. Rev. 77, (1950) 291.]

5.22. A beam of neutrons, having a wide spectrum of wavelengths, impinges at a grazing incidence θ, on a thin metallic film of thickness d, deposited upon a reflecting material. The quantum amplitudes corresponding to the reflection from the upper surface of the film and to the successive reflections from its lower surface interfere (fig. 5.11a). The index of refraction for the metallic film is written as

$$n = 1 - \tfrac{1}{2}\theta_c^2, \quad \text{where} \quad \theta_c = (\mathcal{N}a/\pi)^{1/2}\lambda$$

is the angle of *total reflection* for the wavelength λ. Set

$$L = (\mathcal{N}a/\pi)^{-1/2}$$

(a) Using the fact that $\theta \ll 1$ and $\theta_c \ll 1$, show that the angle of refraction θ' inside the thin film is related to θ by

$$\theta'^2 = \theta^2 - \theta_c^2.$$

(b) Evaluate the phase difference between the successively reflected amplitudes for a given value of λ, and show that the condition for having constructive interference is $2\,d\theta' = s\lambda$, where s is an integer.

(c) Taking into account the dependence of θ' on λ (via θ_c), show that, for fixed θ, there is resonance for discrete values λ_s $(s = 1, 2, ...)$, which can be expressed in the form

$$\lambda = \lambda^*[1 + s^2(L/2d)^2]^{-1/2},$$

where λ^* is the critical wavelength above which there is total reflection (at the incidence θ under consideration).

(d) Show that the spectrum of neutrons, obtained after reflection from the thin film, can be understood in terms of the interferences considered, and compare the theoretical values of λ_s/λ^* with the experimental results of fig. 5.11b.

5.23. A quanton of mass m possesses, at the instant $t = 0$, a Gaussian 'wavefunction in momentum':

$$\hat{\psi}(p) = N \exp[-(p - p_0)^2/2(\Delta p)^2].$$

(a) At $t = 0$, calculate its wavefunction $\psi(x; 0)$ and its dispersion in position $\Delta x|_{t=0}$.

(b) Calculate its wavefunction $\psi(x; t)$ and its dispersion in position $\Delta x|_t$, at an arbitrary time t. Show that there is a '*spreading of the wave packet*'.

(c) For a proton of average energy $E_0 = 1$ MeV, such that $\Delta x|_{t=0} = 1$ Å, calculate p_0 and Δp, and then $\Delta x|_t$ at the time t, such that the proton would have travelled (starting at $t = 0$), an average distance $\langle x \rangle = 1$ m.

6

Steps, wells, barriers and battlements (flat potentials)

6.1. Flat potentials

6.1.1. Model building

The description arrived at, in the last section of the preceding chapter, concerned a *free* quanton, i.e., a quanton which moves in a constant potential. Besides the intrinsic interest of this case, which after all is essential (one has, of course, to begin with it!), the theory developed there has the advantage of being easily generalizable to any situation where there is a *piecewise* (constant) flat segmented potential (fig. 6.1a). In classical terms, this corresponds to 'impulsive' forces, which only act at certain specific points, namely, at the discontinuities of the potential (fig. 6.1b). Between these points a classical particle behaves as though it were free. Nevertheless, the non-local character of quantum phenomena has the effect that even away from these discontinuities, a quanton 'feels' their existence. More precisely stated, the wavefunction has to obey certain global conditions (of continuity, under the present circumstances) which distinguish it from the wavefunction of a free quanton. In spite of the very specific nature of segmented potentials, the conclusions of this chapter have a very general sweep, in so far as it is true that any potential can in principle be approximated by a piecewise constant function (fig. 6.1c).

We shall successively be interested in different physical situations, characterized by potentials which typically have the shapes of a 'step', a 'well', a 'barrier', a 'double well' and a 'battlement'. We shall study these through their corresponding models, which we shall build using piecewise flat potentials (fig. 6.2). Naturally, we shall put the emphasis on those physical conclusions, of general interest, which do not specifically depend upon the flatness of the potentials... .

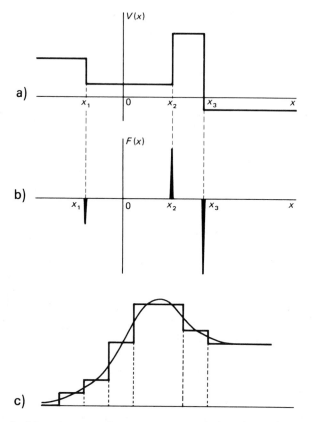

Fig. 6.1. *Flat (or 'piecewise constant') potentials*
(a) On each segment ($[x_1, x_2], [x_2, x_3],...$) the quanton is 'free'. The absolute value of $V(x)$ is fixed only up to an additive constant. It does not matter at what level the origin $V = 0$ is set. The only important things are the positions and the heights of the discontinuities.
(b) The corresponding force is 'impulsive': to each discontinuity of the potential corresponds a force given by a Dirac 'function'.
(c) Any function $V(x)$, which is 'not too singular' can be approximated by a succession of steps.

6.1.2. *Actual realizations*

Let us not lose sight of the fact, however, that flat potentials can describe real physical situations (and hence should not be relegated to being just schematic models). As an example, let us consider the thermal neutrons, the behavior of which, inside a homogeneous material, has been analyzed (sect.

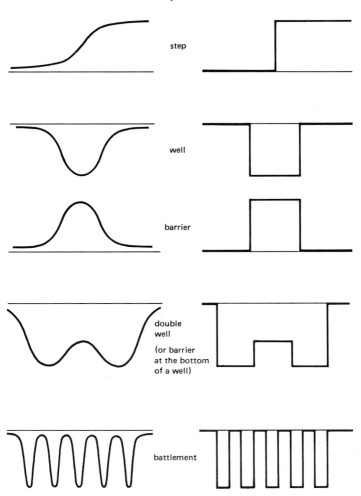

step

well

barrier

double
well

(or barrier
at the bottom
of a well)

battlement

Fig. 6.2. Typical potentials and their 'segmented' models.

5.3.4) in terms of an optical analogy, using a quantum index of refraction

ħ

$$n = 1 - 2\pi \mathcal{N} \frac{a}{p^2}. \tag{6.1.1}$$

This phenomenon can be differently described if we focus on the corpuscular, rather than the wave-like, aspect of the neutron. The index n connects the wavelength λ of the neutron outside the material (free neutron) to λ', the wavelength inside the material, and hence also the corresponding momenta $p = h/\lambda$ and $p' = h/\lambda'$,

$$\lambda' = \lambda/n, \qquad p' = np. \tag{6.1.2}$$

In view of eq. (6.1.1) for the refractive index, this becomes

$$\hbar \qquad p' = p - 2\pi\mathcal{N}\frac{a}{p}. \tag{6.1.3}$$

From this we obtain, finally, a relationship between the energies $E = p^2/2m$ and $E' = p'^2/2m$,

$$\hbar \qquad E' = E - 2\pi\mathcal{N}\frac{a}{m}, \tag{6.1.4}$$

where we have taken into account the fact that $|n - 1| = 2\pi\mathcal{N}a/p^2 \ll 1$. We see that the energy E' of the quanton inside the material differs from its (purely kinetic) energy outside by a constant term, which represents a uniform potential in the interior of the material,

$$\hbar \qquad V_0 = 2\pi\mathcal{N}\frac{a}{m}. \tag{6.1.5}$$

By juxtaposing films of different materials, having different nuclear scattering lengths, we can effectively create situations in which the neutrons are subjected to piecewise constant potentials.

Similarly, one constructs today, layers of thin films (a few angstroms) of material, in which the electrons have different propagation properties. These sandwiches, with alternating layers of, e.g., AsGa and AlAsGa, have remarkable electronic properties. They are being used more and more in the electronic components industry.

We shall see, at the end of this chapter and the next, interesting applications of the study of piecewise constant potentials, to such actual physical situations (exercises 6.4, 6.8, 6.17, 6.21).

6.2. The infinite flat well

6.2.1. In one dimension

The simplest case of a physical situation, described by such a 'flat' potential, is that of a quanton trapped in a potential 'well', with walls so steep and high and bottom so flat that it might be approximated by a constant value over a certain interval (the 'width' a of the well), and an infinitely large value outside this interval. Let us fix the constant value of the potential to be zero, which just amounts to taking this value as the origin of the energy. We denote by $x = 0$ and $x = a$, the extremities of the confinement region (fig. 6.3). A classical particle, under these conditions, would bounce back and forth between the two extremities with a constant (in magnitude) velocity.

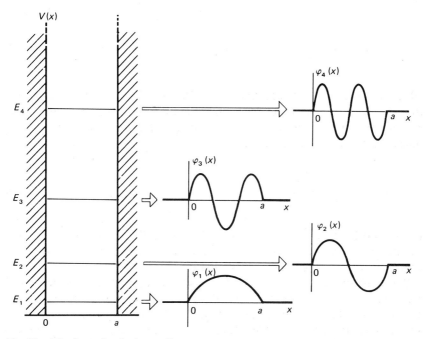

Fig. 6.3. *The flat, infinitely deep well*
An infinitely deep potential well, with a flat bottom, represents the confinement of a quanton, otherwise free, to a segment with extremeties which cannot be crossed. The quantized energy levels are indicated on the diagram, and opposite to them, are sketched the wavefunctions of the corresponding stationary states.

For a quanton, the existence of insurmountable walls means that, outside the region of confinement, the probability density of its presence, and hence its wavefunction, are identically zero, whatever might be the state in which the system finds itself. Additionally, we introduce at this point a supplementary hypothesis, namely that of the *continuity* of the wavefunction in x, i.e., of the localization amplitude at all points. Physically, one can very well understand why the localization probability of a quanton (i.e., the square of the modulus of its wavefunction) ought to be continuous. It is a much stronger hypothesis to demand the continuity of the wavefunction itself. However, this is not the first time in this book that we transform a rather natural hypothesis on the probabilities to a much less evident, and a much more restrictive, hypothesis on the amplitudes.

For determining the wavefunction $\psi(x; t)$ of an arbitrary state of the quanton, inside the well, we proceed as in sect. 5.4 and commence, therefore, by calculating the stationary-state wavefunctions.

Within the region $0 \leqslant x \leqslant a$, where the potential is zero, the situation is a priori comparable to that of a free quanton: the relation $E = p^2/2m$ remains true; the momentum spectrum of a stationary state, of energy E, consists of two values $p_E = (2mE)^{1/2}$ and $-p_E$, and the wavefunction of this state is, just as for a free quanton, of the form given by eq. (5.4.16)

$\hbar$
$$\varphi_E(x) = a_+ \, e^{ip_E x} + a_- \, e^{-ip_E x}. \qquad (6.2.1)$$

Evidently, this only holds in the interior of the confinement region, since in the exterior, $\varphi_E(x) = 0$. It is here that the hypothesis on the continuity of the wavefunction comes in: $\varphi_E(x)$ must vanish at $x = 0$ and $x = a$,

$$\begin{cases} \varphi_E(0) = 0, & (6.2.2) \\ \varphi_E(a) = 0. & (6.2.3) \end{cases}$$

The first condition, eq. (6.2.2), implies

$$a_+ + a_- = 0, \qquad (6.2.4)$$

and hence a more restrictive form for the wavefunction

$\hbar$
$$\varphi_E(x) = C \sin p_E x, \qquad (6.2.5)$$

where we have redefined the constant: $C = 2ia_+$. The second condition, eq. (6.2.3), is now written as

ℏ $$\sin p_E a = 0.$$ (6.2.6)

This fixes a discrete set of possible values for p_E,

ℏ $$p_E = n\frac{\pi}{a}, \quad n = 1, 2, 3, \dots .$$ (6.2.7)

Only positive integers appear here, since the value $n = 0$, which makes p_E vanish, and hence also the wavefunction of eq. (6.2.5), does not correspond to any physical situation, while negative integers simply change the sign of $\varphi_E(x)$ – defined in any case only up to a phase – without yielding any new solution. Let us remark in passing that inside the well, the current j is identically zero,

$$j = \frac{p_E}{m}(|a_+|^2 - |a_-|^2) = 0,$$ (6.2.8)

as follows from eq. (6.2.4). This, of course, is only natural: the quanton is confined to the inside of the well and its probability cannot 'flow' out, either from one side or from the other.

To the discrete sequence, eq. (6.2.7), of values for the momentum, there corresponds, as well, a discrete set of proper values of the energy

ℏ $$E_n = n^2 \frac{\pi^2}{2} \frac{1}{ma^2}, \quad n = 1, 2, 3\dots .$$ (6.2.9)

This expression, in its dependence on m and a (and implicitly on ℏ), conforms to what a simple dimensional analysis had already predicted (see ch. 1). At this point we should focus our attention more on the dependence on the integer n, than on the numerical constant $\frac{1}{2}\pi^2$. In particular, the quanton possesses a ground state, of energy

ℏ $$E_1 = \frac{\pi^2}{2ma^2},$$ (6.2.10)

distinct from the minimum energy, $E = 0$, of a classical particle inside the same potential. We have seen (ch. 3) how this difference can heuristically be explained, using the Heisenberg inequalities. This is the second example of a quantization – the discretization of a physical magnitude – that we are encountering, the first being that of the angular momentum (ch. 2). We remark, that in both cases, this discretization results from a *confinement constraint*: the angular variable was, by definition, limited to the interval $[0, 2\pi]$, while here the special circumstances constrain the particle to the interior of the interval $[0, a]$. This is a general result. The integer n, which 'labels' the energy levels, and in a certain way, expresses the quantization of the energy, is an example of a quantity which, according to current usage, is called a 'quantum number'.

The time-independent wavefunctions of the stationary states are, therefore, written as

$$\varphi_n(x) = C \sin\left(n\pi\frac{x}{a}\right), \quad n = 1, 2, 3, \dots . \tag{6.2.11}$$

These can be normalized in a manner which would ensure that the probability of being somewhere inside the entire segment $[0, a]$ would equal unity [condition (5.2.3)]. We immediately obtain

$$C = \sqrt{2/a}, \tag{6.2.12}$$

and hence

$$\varphi_n(x) = \sqrt{2/a} \sin\left(n\pi\frac{x}{a}\right). \tag{6.2.13}$$

The existence of spatial boundaries eliminates here the difficulties with the normalization, encountered in the case of the free quanton. Figure 6.3 represents the shape of the wavefunctions for the first few stationary states of the infinite flat well. We remark, in particular – and this fact is of general validity – that the nth order wavefunction possesses $(n-1)$ zeroes (in addition to the zeroes at the boundaries of the well, which are specific to the situation).

A complete knowledge of the stationary states, i.e. of their energies [eq. (6.2.9)] and their wavefunctions [eq. (6.2.13)], enables us to express, using eq. (5.4.10), the time evolution of any state.

As a simple example, let us consider a state with initial wavefunction

$$\psi(x; 0) = \frac{4}{\sqrt{5a}} \sin^3\left(\pi\frac{x}{a}\right). \tag{6.2.14}$$

It can be verified that this wavefunction is normalized. It clearly satisfies the boundary conditions [eqs. (6.2.2) and (6.2.3)]. To find out what would be the wavefunction $\psi(x; t)$ at an arbitrary time t, we begin by noting that the formula

$$\sin^3 \alpha = \tfrac{3}{4} \sin \alpha - \tfrac{1}{4} \sin 3\alpha, \qquad (6.2.15)$$

enables us to expand $\psi(x; 0)$ in terms of the stationary wavefunctions given by eq. (6.2.13),

$$\psi(x; 0) = \frac{3}{\sqrt{10}} \varphi_1(x) - \frac{1}{\sqrt{10}} \varphi_3(x). \qquad (6.2.16)$$

The expansion of $\psi(x; 0)$, in this particular case, contains only two terms, because of the especially simple choice [eq. (6.2.14)] that has been made. In view of eq. (5.4.10), it follows that at an arbitrary time t,

ℏ

$$\psi(x; t) = \frac{3}{\sqrt{10}} \varphi_1(x) \, e^{-iE_1 t} - \frac{1}{\sqrt{10}} \varphi_3(x) \, e^{-iE_3 t}, \qquad (6.2.17)$$

where the wavefunctions φ_1 and φ_3 and the energy proper values E_1 and E_3 are known. It is instructive to follow the evolution in time of the shape of the wavefunction given by eq. (6.2.17) of a non-stationary state! (fig. 6.4, exercise 6.1).

Let us add that the model of the infinite flat well, as simple as it might seem, can nevertheless serve as the basis for the approximate evaluation of the energy levels for more complicated potentials (exercise 6.2).

The wavefunction of a quanton inside an infinite flat well, at the time $t = 0$, is given by

$$\psi(x; 0) = C \sin^3 \pi x / a.$$

The successive graphs illustrate the evolution of the probability density $\rho(x; t) = |\psi(x; t)|^2$ at the times $t = \tfrac{1}{6}\tau, \tfrac{1}{4}\tau, \tfrac{1}{3}\tau, \tfrac{1}{2}\tau$, where $\tau = ma^2/2\pi$. The evolution continues by passing through the same phases backwards (at the times $t = 2\tau/3, 3\tau/4, 5\tau/6$) and returning to the initial wavefunction configuration at the time $t = \tau$.

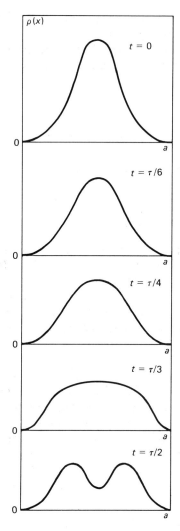

Fig. 6.4. A non-stationary wavefunction.

6.2.2. In three dimensions

Let us now consider an infinite flat well in three dimensions, i.e., a region of zero potential with boundaries which cannot be crossed – in other words, a box within which the quanton is trapped. We are interested in the time-independent wavefunction $\varphi(r)$ of a stationary state of energy E. It has to satisfy the boundary condition $\varphi(r) = 0$ at the periphery (on the 'walls' of the well). Let us consider a rectangular box with sides having lengths a, b, c, where the potential,

$$V = 0 \quad \text{if } 0 < x < a, \qquad 0 < y < b, \qquad 0 < z < c,$$

$$= \infty \quad \text{otherwise} \tag{6.2.18}$$

The boundary conditions are now written as,

(a) $\varphi(x = 0, y, z) = 0, \qquad \varphi(x = a, y, z) = 0;$

(b) $\varphi(x, y = 0, z) = 0, \qquad \varphi(x, y = b, z) = 0;$

(c) $\varphi(x, y, z = 0) = 0, \qquad \varphi(x, y, z = c) = 0. \tag{6.2.19}$

We realize that, separately for each one of the coordinates x, y and z, the wavefunction obeys the same conditions as for a one-dimensional infinite flat well of width a, b and c, respectively. Thus, for example, for x we ought to have the previously obtained dependence in eqs. (6.2.5)–(6.2.7). Thus, we may write,

ħ $$\varphi(x, y, z) = \sin(p_x x) \, \eta(x, y), \tag{6.2.20}$$

with $p_x = n_x \pi / a$, n_x a positive integer. The same reasoning, extended to the other two coordinates, yields the wavefunction in the factorized form

ħ $$\varphi(x, y, z) = A \sin(p_x x) \sin(p_y y) \sin(p_z z), \tag{6.2.21}$$

with

ħ $$p_x = n_x \frac{\pi}{a}, \qquad p_y = n_y \frac{\pi}{b}, \qquad p_z = n_z \frac{\pi}{c}, \quad n_x, n_y, n_z \ \text{integers} > 0. \tag{6.2.22}$$

As for the energy $E = (p_x^2 + p_y^2 + p_z^2)/2m$ of the stationary state, for which eq. (6.2.21) is the wavefunction, it is fixed by the integers n_x, n_y, n_z and has the value,

ħ $$E_{n_x, n_y, n_z} = \frac{\pi^2}{2m} \left(\frac{n_x^2}{a^2} + \frac{n_y^2}{b^2} + \frac{n_z^2}{c^2} \right). \tag{6.2.23}$$

Once again, we encounter here a quantization of the energy, which now depends on the three 'quantum numbers' (n_x, n_y, n_z) – as these integral indices are often called. However, a new phenomenon has appeared – the degeneracy of the energy proper values, namely, the coincidence of the energy levels belonging to different states. Indeed, several sets of values of (n_x, n_y, n_z) could, in principle, lead to the same proper value E_{n_x, n_y, n_z}. This is especially the case when the box is a cube $(a = b = c)$, for which

$$\hbar \qquad E_{n_x, n_y, n_z} = \frac{\pi^2}{2ma^2}(n_x^2 + n_y^2 + n_z^2). \tag{6.2.24}$$

It is clear in this case that each energy level, characterized by three different quantum numbers (n_1, n_2, n_3) is six-fold degenerate, since there are six ways of distributing these three different values between the three numbers (n_x, n_y, n_z). If two of these numbers are equal, the degeneracy is three-fold (one also says that the degree of the degeneracy is equal to three). A level characterized by three equal quantum numbers n is not degenerate – in general, since we could add to these 'normal' degeneracies, due to the symmetry of the problem, also the 'accidental' degeneracies, due to certain numerical coincidences. Thus, e.g.

$$E_{1,1,5} = E_{3,3,3} \qquad \text{since } 1^2 + 1^2 + 5^2 = 3^2 + 3^2 + 3^2 = 27$$

$$E_{1,4,4} = E_{2,2,5} \qquad \text{since } 1^2 + 4^2 + 4^2 = 2^2 + 2^2 + 5^2 = 33$$

$$E_{1,2,7} = E_{3,3,6} = E_{2,5,5} \quad \text{since } 1^2 + 2^2 + 7^2 = 3^2 + 3^2 + 6^2$$
$$= 2^2 + 5^2 + 5^2 = 54.$$

In any event, we see that the energy spectrum of a quanton inside an infinite well, as simple and as symmetric as the system might be, is of a fair degree of complexity (fig. 6.5b). It no longer has the regularity – actually rather misleading – of the one-dimensional infinite well (fig. 6.5a). It is not surprising, therefore, that more realistic physical systems (such as atomic nuclei) display energy spectra of considerable richness and apparent irregularity (fig. 6.5c and exercise 6.3). In spite of its simplicity, the problem of the infinite well in three dimensions furnishes us with a model which allows us to understand certain complex physical phenomena, such as the colour of certain crystals (exercise 6.4).

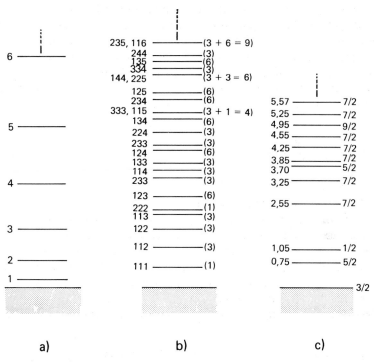

Fig. 6.5. *Energy levels*

(a) The levels of an infinite flat well in one dimension.

(b) The levels of a cubic, infinite flat well in three dimensions (on the left are shown the quantum numbers which mark the level, on the right is indicated its degeneracy).

(c) The levels of the nucleus of nickel, $^{57}_{28}$Ni (to the left of each level, is indicated its energy in MeV above the ground state and to the right the spin s, which gives the degree of degeneracy $g_s = 2s + 1$).

6.2.3. The density of states of a free quanton

An examination of eq. (6.2.24) reveals, that as the dimensions of the box increase, the separation between successive energy levels – of the order of $\pi^2/2ma^2$ – is clearly reduced. The discrete energy levels become more and more closely spaced and the spectrum more and more dense. In the limit where the box grows indefinitely, we recover the continuous spectrum of the free quanton. Conversely, it is sometimes convenient to consider a free quanton as though it were confined to a box – cubic, for example, of volume $V = a^3$. If this volume is sufficiently large, compared to the volume inside which the physical phenomena under investigation take place (that of an atom, for example, or even the entire laboratory), the theoretical computations cannot depend on V, as we have already emphasized, while discussing the problem of the normalization of the wavefunctions of free quantons. Thus, by introducing a large, but finite, volume V, we replace the continuous spectrum of a free quanton by a discrete spectrum which approximates it. This trick enables us to answer an important question, concerning an arbitrary continuous spectrum, namely how would we generalize the notion of *degeneracy*? For a discrete spectrum, the degree of degeneracy $g(E)$ of an energy level E is the number of distinct, stationary quantum states having this common energy proper value. For a continuous spectrum, since the levels cannot be separated, we replace the notion of the *degree of degeneracy $g(E)$* by that of the *density of states*

$$\rho(E) = \lim_{\Delta E \to 0} \frac{\Delta N(E)}{\Delta E}, \tag{6.2.25}$$

where $\Delta N(E)$ is the number of stationary states of energy lying between E and $E + \Delta E$. Let us compute in this way the density of states of a free quanton, by considering it as being enclosed in a cubic box of volume V. Each stationary state of a free quanton is characterized by three positive integers (n_x, n_y, n_z), giving the moduli of the components of the momentum [see eq. (6.2.22)]. In this way, we can mark a state in the 'p-space' by a point, having as coordinates three positive integers in units of π/a. In other words, to each state is associated a small cube of side π/a, centered at the corresponding lattice point. We might say, that to each state is associated an elementary volume

$$h \qquad \tau = (\pi/a)^3 = \pi^3/V, \tag{6.2.26}$$

in the 'p-space' (fig. 6.6). Let us now consider all the energy states contained between E and $E + \Delta E$. They correspond, in p-space, to the points contained between the spheres of radii $p(E)$ and $p(E + \Delta E)$ where

$$p = \sqrt{2mE}, \qquad \Delta p = \sqrt{2m}\,\frac{\Delta E}{2\sqrt{E}}. \tag{6.2.27}$$

The volume of this sector of the spherical shell (let us not forget that we are only considering an octant of the p-space – the one corresponding to positive

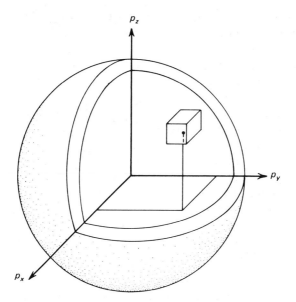

Fig. 6.6. *The 'p-space'*
In this space, each stationary state of a quanton of mass m, inside a cubic box of side a, is represented by a point having (positive) integral coordinates, in units of π/a. In this way, an elementary volume $(\pi/a)^3 = \pi^3/V$ can be attributed to each state. The states having energies lying between E and $E + \Delta E$ correspond to points within the positive octant of the spherical shell of radius $p(E)$ and thickness $\Delta p = (\mathrm{d}p/\mathrm{d}E)\,\Delta E$.

values) is

$$\Delta W = \tfrac{1}{8} 4\pi p^2 \, \Delta p, \tag{6.2.28}$$

and the number of states is obtained by dividing by the volume τ, of each individual state, given by eq. (6.2.26),

$$\hbar \qquad \Delta N = \frac{1}{\tau} \Delta W = \frac{V}{2\pi^2} p^2 \, \Delta p. \tag{6.2.29}$$

Finally, using eq. (6.2.27), this becomes

$$\hbar \qquad \rho(E) = \frac{\Delta N}{\Delta E} = \frac{V}{2\pi^2} \sqrt{2m^3 E}. \tag{6.2.30}$$

The quantity of interest is in fact the number of states per unit interval of energy and unit (volume) density of states:

ħ
$$\frac{1}{V}\rho(E) = \frac{1}{2\pi^2}\sqrt{2m^3 E}.$$ (6.2.31)

Just as it should be, this quantity does *not* depend on V (the density of states is spatially uniform), and increases with E, showing the existence of an increasing number of accessible states of the quanton as its energy is increased. This result is characteristic of a *free* quanton in *three* dimensions (see exercise 6.5). Finally, we should not forget that the notion of the density of states is specifically quantal, since it generalizes the notion of degeneracy which is itself related to the quantization of the energy. It is a good idea to convince oneself of this, by carrying out a dimensional analysis of eq. (6.2.31), and reintroducing its dependence on the quantum constant ħ.

6.3. A potential step

6.3.1. Continuity conditions

We consider now a potential step, i.e. a discontinuity in the potential which, without compromising the generality of our conclusions, we may place at the origin of coordinates. Similarly, let us take the value of the potential to the left as the origin of the energy (fig. 6.7). Thus, we have

$$V = 0 \quad x < 0,$$
$$ = V_0 \quad x > 0.$$ (6.3.1)

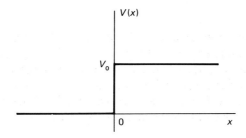

Fig. 6.7.   The potential step.

In each one of the two regions of constant potential, the wavefunction φ_E of a stationary state of the quanton has the general form as given by eq. (5.4.16), with the appropriate value of the momentum and the a priori unknown constants,

$\hbar$

$$\begin{cases} x < 0: & \varphi_E(x) = a_+ \, e^{ip_E x} + a_- \, e^{-ip_E x}, \quad p_E = (2mE)^{1/2} \\ x > 0: & \varphi_E(x) = a'_+ \, e^{ip'_E x} + a'_- \, e^{-ip'_E x}, \quad p'_E = [2m(E - V_0)]^{1/2}. \end{cases} \quad (6.3.2)$$

It is now natural to impose the condition that the discontinuity of the potential does not entail a discontinuity in the probability density $\rho(x) = |\varphi_E(x)|^2$. The continuity condition for this density, at the point of discontinuity $x = 0$, is written as

$$|a_+ + a_-|^2 = |a'_+ + a'_-|^2. \quad (6.3.3)$$

Similarly, it is natural to require that, for each stationary state, the probability current be the same on both sides of the discontinuity of the potential. There is nothing at all unusual here, dealing as we are, with a stationary state. Just think of the case of the electric current: in a stationary situation, the current is the same at each point of the circuit; in particular, the current which 'enters through the − pole' of a generator has the same value as that which comes out ('of the + pole'). A stationary situation in electrokinetics is characterized precisely by the fact that the discontinuities of the potential do not correspond to any charge accumulations. In the same way, here, the quantons 'do not accumulate' at the potential step, so that the probability current is the same on each side of the step:

$\hbar$

$$\frac{p_E}{m}(|a_+|^2 - |a_-|^2) = \frac{p'_E}{m}(|a'_+|^2 - |a'_-|^2). \quad (6.3.4)$$

The two 'physical' conditions [eqs. (6.3.3) and (6.3.4)] affect only the modulii of the four constants $(a_\pm, a'_\pm)$. They are not sufficient, therefore, to relate them unambiguously and to fix the wavefunction completely. In reality, as we shall rigorously establish later, it is necessary to require *the continuity of the wavefunction and its derivative* at a discontinuity of the potential. In the present situation, this double requirement is written as

$\hbar$

$$a_+ + a_- = a'_+ + a'_-, \qquad p_E(a_+ + a_-) = p'_E(a'_+ - a'_-). \quad (6.3.5)$$

These conditions, which are more restrictive than conditions (6.3.3) and (6.3.4) – since they affect both the modulus and the phase of the quantities $(a_{\pm}, a'_{\pm})$ – imply both the continuity of the density and the uniformity of the current, as is easily verified. The case of an infinite potential step (such as the edges of the infinite well in sect. 6.2) is exceptional, with regard to these continuity conditions – precisely because of the infinite value of the potential. In this case, the derivative is not continuous! But then, we had imposed the continuity of the function with a given (zero) function, i.e., a constraint just as strong. It can be verified, by carrying out a study of wavefunctions for a potential step with increasing heights, that the double continuity condition, on the function and its derivative, goes over properly, in the limit, to just the vanishing of the wavefunction at the edges of the infinite wall, justifying our preceding computations (exercise 6.7).

6.3.2. Transmission and reflection

Let us pursue the study of a quanton, subject to this step-like potential, in a special case of great interest. Suppose that there exists a source of quantons, for example an accelerator, situated at infinity to the left $(x = -\infty)$ and just there. Under these conditions, the probability of a quanton arriving from the right, inside the region $x > 0$, is certainly zero, if we suppose that nothing in this region, (neither a source, nor a wall) can send in or send back quantons in this direction. The physical situation is expressed by the condition

$$a'_- = 0. \tag{6.3.6}$$

The equations of continuity now suffice to express the three remaining constants in terms of one of them,

$$a_- = a_+ \frac{p_E - p'_E}{p_E + p'_E}, \qquad a'_+ = a_+ \frac{2p_E}{p_E + p'_E}. \tag{6.3.7}$$

The essential result is that $a_- \neq 0$. We have here a specifically quantum effect. Indeed, suppose we have a *classical* particle approaching a potential step. If $E > V_0$, it would only have its kinetic energy changed from E to $E - V_0$, and would continue its uniform motion with its new velocity. Otherwise, if $E < V_0$, it would 'rebound' from the step. By contrast, for a quanton there exists a non-zero reflection probability even when $E > V_0$. Indeed, $|a_-|^2$ gives the probability density of the quanton to move from the right to the left, inside the region $x < 0$, as though it had 'rebounded' off the wall of the step – even though its energy is greater than the height of the wall! The interesting quantity is the reflection coefficient

$$\mathscr{R} = \frac{|a_-|^2}{|a_+|^2} = \left(\frac{p_E - p'_E}{p_E + p'_E}\right)^2, \tag{6.3.8}$$

which is the ratio of the reflected current to the incident current, or the absolute probability of reflection ($\mathscr{R} = |a_-|^2$ if $|a_+|^2 = 1$). In the same manner, we obtain the transmission coefficient

$$\mathscr{T} = \frac{p'_E |a'_+|^2}{p_E |a_+|^2} = \frac{4 p_E p'_E}{(p_E + p'_E)^2}, \tag{6.3.9}$$

which is the ratio of the transmitted current to the incident current (attention! $\mathscr{T}$ is not equal to $|a'_+|^2/|a_-|^2$). Evidently, we have

$$\mathscr{R} + \mathscr{T} = 1. \tag{6.3.10}$$

Naturally, the coefficients of reflection and of transmission depend upon the energy (fig. 6.8). As we might expect, these tend to their classical values, $\mathscr{R} = 0$ and $\mathscr{T} = 1$, as the energy of the quanton becomes very large compared to the height of the potential step.

In view of eqs. (6.3.2) and (6.3.7), the knowledge of the stationary states enables us to study the motion of a quanton in the most realistic situation in

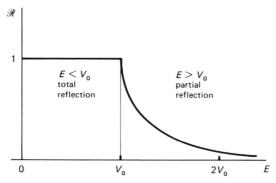

Fig. 6.8. *Reflection off a potential step*
The graph gives the variation of the reflection coefficient, as a function of the energy E of the incident quanton, for a step of height V_0.

which it is represented by an incident wave packet of average energy E_0. The corresponding wavefunction has the form [see eq. (5.4.30)]

$ħ$ $$\psi(x; t) = A \int \hat{\psi}(p)\, e^{i(px - E_p t)}\, dp. \tag{6.3.11}$$

For times well before the arrival of the wave packet at the step, the amplitude in momentum $\hat{\psi}(p)$ displays a peak centered at the average value

$$p_0 = (2mE_0)^{1/2}. \tag{6.3.12}$$

Assuming that the wave packet encounters the discontinuity of the potential around the time $t = 0$, it can be shown by analyzing the probability amplitude for times $t > 0$, that it splits into two: a 'transmitted' wave packet, beyond the step $(x > 0)$, and a 'reflected' wave packet which travels backwards on this side of the step $(x < 0)$ (fig. 6.9). We also see that, in this description of the quanton by a wave packet, which is more realistic than that by a plane wave, the localization has an appreciable probability only within restricted regions. In particular, there is no interference between the incident wave and the reflected wave, except during the reflection, in the neighbourhood of the potential step!

6.3.3. Bouncing back

We had implicitly assumed in the preceding discussion that the energy of the quanton was greater than the value of the height of the potential, $E > V_0$, corresponding to the case where a classical particle would have passed the step without hindrance. Let us now consider the situation in which $0 < E < V_0$ (fig. 6.10). In this case, a classical particle would simply rebound off the wall of the step and move backwards with the same kinetic energy. How does a quanton behave under these circumstances? Its wavefunction in the region $x < 0$ is certainly always of the form

$ħ$ $$x < 0: \quad \varphi_E(x) = a_+ \, e^{ip_E x} + a_- \, e^{-ip_E x}. \tag{6.3.13}$$

One might believe that in the region $x > 0$, the wavefunction would simply vanish. But it is impossible to impose the condition of continuity on the wavefunction, without reverting to the exceptional case of the infinite potential, which does not at all correspond to the present physical situation. Thus, the wavefunction is not zero for $x > 0$. But then, what is its form? We

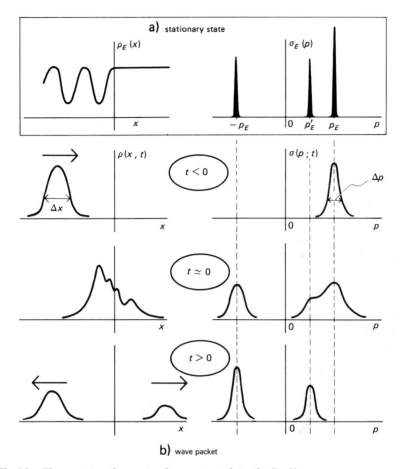

Fig. 6.9. *The scattering of a quanton from a potential step for* $E > V_0$
(a) The probability densities in position $\rho_E(x) = |\hat{\varphi}(x)|^2$ and in momentum $\sigma_E(p) = |\hat{\varphi}(p)|^2$ for a stationary state having (time-independent) wavefunction φ_E.
(b) The probability densities in position and momentum for a (non-stationary) state characterized by an initial wave packet, before, during and after its encounter with the discontinuity of the potential.

shall see here an illustration of the power of the formalism already introduced. Let us have 'blind' faith in eq. (6.3.2), even though for $x > 0$ it gives imaginary values to the momentum,

$$p'_E = \pm [2m(E - V_0)]^{1/2} = \pm i\kappa'_E, \tag{6.3.14}$$

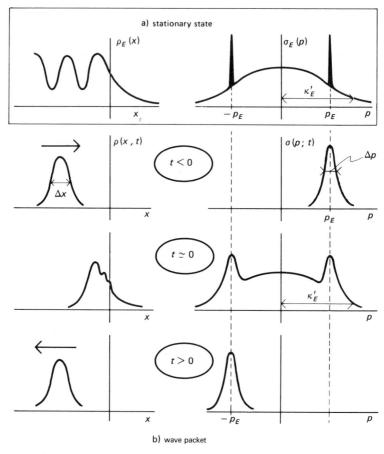

Fig. 6.10. *The reflection of a quanton by a potential step for* $0 < E < V_0$.
(a) The probability densities in position $\rho_E(x) = |\varphi_E(x)|^2$ and in momentum $\sigma_E(p) = |\hat{\varphi}_E(x)|^2$ for a stationary state having (time-independent) wavefunction φ_E.
(b) The probability density in position and in momentum for a (non-stationary) state, characterized by an initial wave packet, before, during and after its encounter with the discontinuity of the potential.

where $\kappa'_E = [2m(E - V_0)]^{1/2}$ is real. To these imaginary values, there corresponds a negative kinetic energy, which clearly does not make sense in a classical theory. But in the quantum theory, after all, these values only appear in the expression for the wavefunction. Using eq. (6.3.14), these latter can be written as

ℏ $$x > 0: \quad \varphi_E(x) = a'_+ \, e^{-\kappa'_E x} + a'_- \, e^{\kappa'_E x}, \tag{6.3.15}$$

which is a combination of two *real* exponentials. The second term, an increasing exponential, cannot be given any satisfactory physical inter- pretation, since it leads to an indefinitely growing probability density, as one penetrates further into the classically forbidden region. We might be willing to overlook, once again, the condition that the wavefunction be normalizable, but we at least do have to assume that it be bounded. Thus, it is necessary in the present situation to impose the condition that $a'_- = 0$. The remaining expression

ℏ $$x > 0: \quad \varphi_E(x) = a'_+ \, e^{-\kappa'_E x}, \tag{6.3.16}$$

yields an entirely acceptable wavefunction. It is exponentially attenuating, in fact very rapidly, in the region $x > 0$. There is, therefore, a certain probability for the quanton to penetrate 'inside' the wall of the potential, since the probability density is not zero. However, this penetration is only appreciable over a distance of the order of $(\kappa'_E)^{-1}$. It is now possible to ensure the continuity, at $x = 0$, of the wavefunction given by eqs. (6.3.13)–(6.3.16), *and* of its derivative. We obtain

ℏ $$a_+ + a_- = a'_+, \quad \mathrm{i}p_E(a_+ - a_-) = -\kappa'_E a'_+. \tag{6.3.17}$$

Hence,

$$a_- = a_+ \frac{p_E - \mathrm{i}\kappa'_E}{p_E + \mathrm{i}\kappa'_E}, \qquad a'_+ = a_+ \frac{2p_E}{p_E + \mathrm{i}\kappa'_E} \tag{6.3.18}$$

(these are evidently the same as eqs. (6.3.7) where p'_E has been replaced by $\mathrm{i}\kappa'_E$). We see, immediately, that $|a_-| = |a_+|$ and consequently, that the reflection is total:

$$\mathscr{R} = \frac{|a_-|^2}{|a_+|^2} = 1. \tag{6.3.19}$$

There is no paradox here: of course, the quanton *always* rebounds off the potential step. The difference with a classical particle lies only in the existence of a certain probability for the quanton to be in the classically forbidden region – in complete conformity with the non-local nature of the quanton (exercise 6.7).

One could analyze the stationary state, represented by the wavefunction, eqs. (6.3.13), (6.3.16) and (6.3.18), more precisely by looking for its amplitude in momentum, rather than in position. Let us recall, for the sake of comparison, that the wavefunction given by (5.4.16) of a stationary state of a free quanton corresponds to a probability amplitude in momentum which shows two peaks for just the two values $p = \pm p_E$, and that the same is true for the probability density (fig. 5.14). In other words, for a free quanton, the only possible values for the momentum are p_E and $-p_E$ (with the relative probabilities $|a_+|^2$ and $|a_-|^2$). If we are only interested in the region $x < 0$, we see that the wavefunction being considered here, for a quanton which is not free but moves under the influence of a potential step, coincides with that of a free quanton [see eq. (6.3.13)]. We expect from this, therefore, that the corresponding probability density in momentum would exhibit the same peak for $p = \pm p_E$. However, it is also necessary to take into account the exponential 'tail' [eqs. (6.3.16)] of the wavefunction. This gives rise to an additional momentum spectrum. Indeed, this part of the wavefunction is expressed as a superposition of plane waves, following the general representation in eq. (5.2.26):

$\hbar$

$$\varphi_E(x) = A \int e^{ipx} \, \hat{\varphi}_E(p) \, \mathrm{d}p, \tag{6.3.20}$$

where $\hat{\varphi}_E(p)$ represents – in conformity with relation (5.2.25) – the transition amplitude from the stationary state of energy E of the *non-free quanton* to the state having wavefunction e^{ipx}, in other words, to the momentum proper state of the *free quanton* characterized by the proper value p. This probability amplitude is calculated by inverting the Fourier integral given by eq. (6.3.20) (see exercise 5.10). We obtain

$$\varphi_E(p) = \frac{\text{const.}}{p - i\kappa'_E}, \qquad \kappa'_E = [2m(V_0 - E)]^{1/2}. \tag{6.3.21}$$

From this, the probability density of the state in momentum is found to be

$$\sigma_E(p) = \frac{|\text{const.}|^2}{p^2 + \kappa'^2_E} \tag{6.3.22}$$

(the constant is of little interest to us here). This defines a distribution, called a 'Lorentzian' distribution, which, although it only assumes appreciable values in an interval $\kappa'_E = [2m(V_0 - E)]^{1/2}$, does nevertheless extend from $-\infty$ to ∞. In other words, although the quanton has a well-defined energy (proper) value, it can in principle, as long as it is under the influence of a potential step, assume *all* possible values of the momentum, and hence all possible values of the energy of a free quanton. The depth of penetration inside the step, $\Delta x \cong (\kappa'_E)^{-1}$, and the width of this distribution in momentum, $\Delta p \cong \kappa'_E$ are clearly related by a Heisenberg relation, $\Delta p \, \Delta x \cong 1$. The same is true for the width in energy and the duration of the interaction. It is the presence of the distribution, given by eq. (6.3.22), in the momentum spectrum, which distinguishes the state of a quanton, subject to a discontinuity in the potential, from the state of a free quanton. Or else, stated differently, a discontinuity of amount V_0 in the potential, which is otherwise constant, has the effect of broadening the momentum distribution of a quanton of energy $E < V_0$. If we now consider the realistic description of the quanton by means of a wave packet, we see, of course, that this broadening of the momentum distribution only takes place within a limited time interval, during which the wave packet interacts with the potential barrier (fig. 6.10b).

6.3.4. The forbidden energies

It remains to examine one last case, the one in which the energy is lower than the bottom of the potential, $E < 0$. Following our previous development of the theory, we are led to believe that the wavefunctions in the two regions (both classically forbidden now) would be of the nature of a real exponential. Indeed, eqs. (6.3.2) become

$\hbar$
$$\begin{cases} x < 0: & \varphi_E(x) = a_+ \, e^{-\kappa_E x} + a_- \, e^{\kappa_E x}, \quad \kappa_E = (-2mE)^{1/2}, \\ x > 0: & \varphi_E(x) = a'_+ \, e^{-\kappa_E x} + a'_- \, e^{\kappa_E x}, \quad \kappa'_E = [2m(V_0 + |E|)]^{1/2}. \end{cases} \quad (6.3.23)$$

For the same reasons as above, we discard the terms growing exponentially to infinity: for $x < 0$, hence $a_+ = 0$, and whence for $x > 0$, $a'_- = 0$. There remain, therefore,

$\hbar$
$$\begin{cases} x < 0: & \varphi_E(x) = a_- \, e^{\kappa_E x}, \\ x > 0: & \varphi_E(x) = a'_+ \, e^{-\kappa_E x}. \end{cases} \quad (6.3.24)$$

It is easy to see that it is not possible to ensure the continuity, at $x = 0$, of both the wavefunction and its derivative. There does not exist an admissible wavefunction for this value of the energy, and hence correspondingly, there is no state either. *A quanton cannot have an energy lower than the minimum value of the potential.* This result has general validity, regardless of the form (step-like or not) of the potential. The situation here is the same as in the classical theory. We may assert, therefore, that the quantum theory is more tolerant than the classical theory, in that it allows the quanton to be in certain regions where, classically, the kinetic energy would be negative and the presence of a particle impossible. However, this tolerance is local and does not extend to the entire space: a quanton cannot exist in a state in which the classical kinetic energy is *everywhere* negative. The behaviour of a quanton subject to a step-like potential, depending on its energy, is summarized in fig. 6.11.

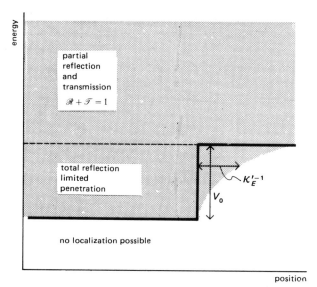

Fig. 6.11. *The quanton and the potential step*
Three situations appear, depending on the value of the energy. The grey region indicates the allowable domains in energy and position.

6.4. The finite flat well

The experience acquired in the preceding section, enables us to treat *any problem with a piecewise flat potential*. To find the wavefunctions of the stationary states, it is sufficient to write down, in each region where the potential is a constant, a linear combination of exponential functions of real or imaginary arguments, depending on whether the energy is smaller or larger than the potential, and then to match the wavefunction from one region to the next by imposing continuity conditions on the function and its derivative. Let us illustrate this technique by looking at the case of the 'finite flat well' or, a well with a flat bottom and of finite depth. This model of a potential well is a little more realistic than the infinite flat well already treated. The potential well, of depth V_0 and width a, is described by

$$V(x) = 0 \qquad |x| > \tfrac{1}{2}a,$$
$$\qquad = -V_0 \quad |x| < \tfrac{1}{2}a. \tag{6.4.1}$$

We have chosen the origin of energies in a manner such that the potential at infinity is zero, following a standard convention, and placed the origin of the abscissas ($x = 0$) at the middle of the well, which facilitates the very fruitful exploitation of its symmetry, as we shall see (fig. 6.12).

In each one of the regions of constant potential, the wavefunction φ_E of a stationary state of the quanton assumes the general form, eq. (5.4.16), with the appropriate value of the momentum,

$$\begin{cases} x \leqslant -\tfrac{1}{2}a: & \varphi_E(x) = b_+ \, e^{ip_E x} + b_- \, e^{-ip_E x}, \\ |x| \leqslant \tfrac{1}{2}a: & \varphi_E(x) = c_+ \, e^{ip'_E x} + c_- \, e^{-ip'_E x}, \\ x \geqslant \tfrac{1}{2}a: & \varphi_E(x) = d_+ \, e^{ip_E x} + d_- \, e^{-ip_E x}, \end{cases} \tag{6.4.2}$$

where

$$p_E = (2mE)^{1/2}, \qquad p'_E = [2m(E + V_0)]^{1/2}.$$

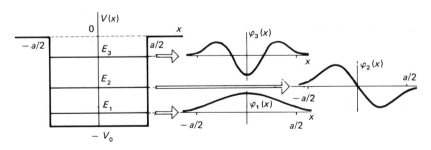

Fig. 6.12. *The finite flat well*
Compare the stationary state wavefunctions to those of the infinite flat well (fig. 6.3).

These expressions are valid – as we now know – for arbitrary values of $E >$ $- V_0$ (since as we just have shown, there do not exist any stationary states of energy less than the minimum value of the potential). In particular, if $- V_0 < E < 0$, eqs. (6.4.2) remain valid, the value of p_E being pure imaginary. It will suffice to set, therefore,

$\hbar$

$$E < 0: \quad p_E = i\kappa_E, \quad \kappa_E = (-2mE)^{1/2} \quad \text{real.} \tag{6.4.3}$$

In this way, there appear, in the expressions for $\varphi_E(x)$, real exponential quantities analogous to eq. (6.3.15). However, although the formal computations are analogous, we shall see now that the wavefunction, eq. (6.4.2), describes physical situations which are quite different depending on whether $E < 0$ or $E > 0$. This is not surprising if we think of what happens to a classical particle in an analogous situation. A particle the energy of which is smaller than the highest value of the potential, finds itself trapped in the well (it is now 'bound'), bouncing back and forth between the walls. On the other hand, if the energy is greater than the potential outside the well, its motion is a sequence of three uniform motions: the first is a velocity $v = (2E/m)^{1/2}$, from $-\infty$ to the edge $x = -\frac{1}{2}a$ of the well, where it undergoes a sudden instantaneous acceleration. This has the effect of imparting to it a larger velocity, which it retains up to the edge $x = \frac{1}{2}a$. There a sudden instantaneous deceleration restores the velocity to this initial value, with which it proceeds indefinitely on its way.

6.4.1. The bound states

(a) *Parity.* Let us begin by studying the bound stationary states, of energy $- V_0 < E < 0$ of a quanton inside the well eq. (6.4.1), and its wavefunctions $\varphi_E(x)$.

Having placed the origin at the middle of the well, we see immediately that the physical conditions are invariant under the operation of a 'reflection in space' which transforms x into $-x$. The same must, therefore, be true of the probability density of localization $\rho(x) = |\varphi_E(x)|^2$, and the modulus of the wavefunction $\varphi_E(x)$ must therefore be an even function in x,

$$|\varphi_E(x)| = |\varphi_E(-x)|. \tag{6.4.4}$$

This imposes on $\varphi_E(x)$ the condition

$$\varphi_E(-x) = e^{i\alpha} \varphi_E(x). \tag{6.4.5}$$

Denoting by Π the operation of spatial reflection, we may also write

$$\varphi_E(-x) = \Pi\varphi_E(x) = e^{i\alpha}\,\varphi_E(x). \tag{6.4.6}$$

On the other hand, the spatial reflection and spatial identity together form a group with two elements: two repeated applications of Π on $\varphi_E(x)$ should give back $\varphi_E(x)$ itself. Hence, $\varphi_E(x)$ must obey the symmetry condition

$$\Pi\,\Pi\varphi_E(x) = \Pi\varphi_E(-x) = \varphi_E(x), \tag{6.4.7}$$

so that

$$e^{2i\alpha}\,\varphi_E(x) = \varphi_E(x). \tag{6.4.8}$$

This fixes the possible values of $e^{i\alpha}$,

$$e^{i\alpha} = \pm 1, \tag{6.4.9}$$

(see also exercise 6.9 for a subtle refinement of this argument). Finally, therefore, there are two classes of possible wavefunctions (6.4.2), the even functions ($e^{i\alpha} = 1$) and the odd functions ($e^{i\alpha} = -1$). This imposes, on the coefficients appearing in eq. (6.4.2) the conditions

$$b_- = \pm d_+, \qquad c_- = \pm c_+, \qquad d_- = \pm b_+, \tag{6.4.10}$$

where the $+$ (respectively $-$) sign corresponds to even (respectively odd) wavefunctions.

(b) *The energy proper values.* In both cases, it is enough, therefore, to determine $\varphi_E(x)$ on the half-line $x > 0$. From this the wavefunction for $x < 0$, can be deduced by symmetry. Now, in the region $x > 0$, the situation physically is that of a quanton encountering a potential step of height greater than its energy proper value (since $-V_0 < E < 0$). In conformity with what has been demonstrated in the preceding section, the wavefunction is, therefore, exponentially decreasing in the interior of the step. Setting

$$p_E = i\kappa_E, \qquad \kappa_E = (-2mE)^{1/2}, \tag{6.4.11}$$

there follows, for $x > \tfrac{1}{2}a$,

$\hbar$ $\varphi_E(x) = d_+\,e^{-\kappa_E x} \quad (d_- = 0). \tag{6.4.12}$

From this we deduce the general forms of the two classes of wavefunctions.

Even functions:

$\hbar$
$$\begin{cases} x < -\tfrac{1}{2}a: & \varphi_E(x) = D\,e^{\kappa_E x} \\ |x| < \tfrac{1}{2}a: & \varphi_E(x) = C\cos p'_E x \\ x > \tfrac{1}{2}a: & \varphi_E(x) = D\,e^{-\kappa_E x}, \end{cases}$$
(6.4.13)

Odd functions:

$\hbar$
$$\begin{cases} x < -\tfrac{1}{2}a: & \varphi_E(x) = -D\,e^{\kappa_E x} \\ |x| < \tfrac{1}{2}a: & \varphi_E(x) = C\sin p'_E x \\ x > \tfrac{1}{2}a: & \varphi_E(x) = D\,e^{-\kappa_E x}, \end{cases}$$
(6.4.14)

where the coefficient C has been defined using c_+. In order to simplify the notation, we have set $d_+ = D$. C and D are determined by requiring that the wavefunction and its derivative be continuous at the point $x = \tfrac{1}{2}a$ (the continuity at $x = -\tfrac{1}{2}a$ is now automatically ensured). For the even functions we get

$\hbar$
$$\begin{cases} C\cos(p'_E a/2) = D\,e^{-\kappa_E a/2} \\ -Cp'_E \sin(p'_E a/2) = -\kappa_E D\,e^{-\kappa_E a/2}. \end{cases}$$
(6.4.15)

Upon taking the quotient (we could just as well have written down directly the condition for the continuity of the logarithmic derivative φ'/φ of the wavefunction), we get

$\hbar$
$$\begin{cases} p'_E \tan(p'_E a/2) = \kappa_E, & \text{for even functions,} \\ p'_E \cotan(p'_E a/2) = -\kappa_E, & \text{for odd functions,} \end{cases}$$
(6.4.16a)
(6.4.16b)

as can easily be established.

The two conditions [eqs. (6.4.16a,b)] can moreover, be condensed into a single one, which is sometimes more convenient to use (as we shall see in sect. 6.7), ℏ $\qquad \cos(p'_E a) + \eta(E) \sin(p'_E a) = 0,$ $\qquad\qquad\qquad$ (6.4.17) .

with

ℏ $\qquad \eta(E) = \dfrac{\kappa_E^2 - p'^2_E}{2\kappa_E p'_E}.$

These conditions are to be interpreted as being the conditions for the quantization of the energy: in effect, they are only satisfied for certain values of E – the proper values, precisely. Of course, by now we are sufficiently familiar with the quantum theory, so as not to be surprised by the fact that a quanton constrained to remain localized inside a well would have a quantized energy.

The allowed values – the proper values – are easily obtained by solving eqs. (6.4.16a,b) graphically (fig. 6.13). We can verify that the spectrum of the quanton consists of a *finite* sequence of discrete values (exercise 6.10). Similarly, we can verify that even for small values of V_0, there always exists at least one bound state. This feature is particular to a one-dimensional problem and is absent in the three-dimensional case. For higher values of V_0, after this first even solution, we obtain a succession of alternately odd and even solutions, bounded above by the value forming the edge of the well. These two properties are completely general.

The wavefunctions of the bound states, unlike those in the (exceptional) case of the infinite well, do not vanish outside a finite well (see fig. 6.12), but display now an 'exponential tail'. In other words, there exists a non-zero probability for finding the quanton *outside* the well, in which it is, nevertheless, ... essentially ... localized (exercise 6.12). We could also say that everything happens as though, because of the non-local nature of the quanton, the potential well had an effective width which was larger than its actual width a (exercise 6.13).

(c) *The deuteron.* An instructive application of these computations is to the deuteron, the simplest of the compound nuclei, which is a bound state of a proton and a neutron. It is known that the nuclear force has a short range and contains an attractive part (clearly, since it can bind the nucleons inside the nucleus!) However, at very short distances, it must become repulsive: as indicated by the almost constant nuclear density, the nucleons cannot approach each other closer than a certain distance. We say that the nuclear force displays a 'hard core'. Thus, we schematize this force by means of an attractive square well potential bounded by an infinitely high barrier (fig. 6.14a)

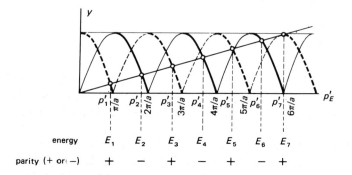

Fig. 6.13. *Graphical determination of the bound state energies inside a finite flat well (of width a and depth V_0)*

Depending on the parity of the wavefunction, the quantization condition is either

$$p'_E \tan(p'_E a/2) = \kappa_E \qquad [\text{eq. (6.4.16a)}]$$

or

$$p'_E \cotan(p'_E a/2) = -\kappa_E \qquad [\text{eq. (6.4.16b)}]$$

These relations are easily transformed into conditions on p'_E alone. For this it is sufficient to square each relation and introduce the quantity

$$q = \sqrt{2mV_0},$$

which is independent of E and hence a characteristic of the well. Thus, the possible values of p'_E are those which satisfy either

$$|\cos p'_E a/2| = p'_E/q,$$

or

$$|\sin p'_E a/2| = p'_E/q.$$

These are the abscissas of the points of intersection of the line $y = p'_E/q$ with one or the other of the two graphs $y = |\cos p'_E a/2|$ (broken curve) and $y = |\sin p'_E a/2|$ (continuous curve). However, the only solutions which are to be retained are those which satisfy $\tan p'_E a/2 > 0$, for the broken curve, and $\cotan p'_E a/2 < 0$, for the continuous curve. There is a *finite* number of bound states – but there is always at least one (exercise 6.10). Note that for $V_0 \to \infty$, i.e., for $q \to \infty$, one obtains the values $p'_n = n\pi/a$, and recovers in this way the energy proper values of the infinite flat well.

$$V(r) = \begin{cases} 0 & r > R, \\ -V_0 & 0 < r < R_0, \\ \infty & r < 0, \end{cases} \qquad (6.4.18)$$

where r is the distance between the two nucleons. R_0 is the 'range' of the force and V_0 the depth of the well, representing the strength of the interaction. As in classical mechanics, the relative motion of the two bodies, having masses M_1 and M_2, interacting via a potential $V(|\mathbf{r}_2 - \mathbf{r}_1|)$ reduces to a motion under the influence of a potential $V(r)$, of a fictive body of 'reduced mass' $m = M_1 M_2/(M_1 + M_2)$. We are thus led to studying the bound states of a quanton of mass $m = \frac{1}{2}M$ (M is the mass of the nucleon) inside the potential $V(r)$. Now, this potential may be regarded as being 'the half' of a flat potential

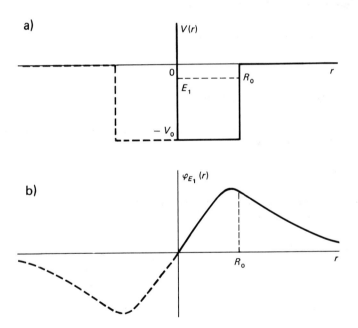

Fig. 6.14. *Potential and wavefunction of the deuteron*
(a) The flat potential well, with a hard core, may be considered as being 'the half' of a flat well with the condition that the wavefunctions vanish at the origin.
(b) The wavefunction of a bound state, for $r \geqslant 0$, is identified, therefore, with that of an odd bound state of a flat well.
Since $|E_1| \ll V_0$, the exponential tail of the wavefunction is particularly long (exercise 6.14).

well, which we have already considered – more precisely, as being the half of a flat, finite potential well with the same V_0 and double the width, $a = 2R_0$, symmetric with respect to the origin. The wavefunctions of the hard-core flat well are, therefore, also the wavefunctions of this flat well. However, they have to satisfy, in addition, the boundary condition

$$\varphi_E(0) = 0, \tag{6.4.19}$$

as a consequence of the hard core (infinite step). This condition selects out, from among the wavefunctions of the double, finite, flat well, those which vanish at the origin, i.e., the odd wavefunctions in eq. (6.4.14) (fig. 6.14b). The energy levels of the nuclear potential, schematized by eq. (6.4.18) are thus the same as the odd levels of a square well of width $2R_0$. We begin now with the remark that the deuteron is the *only* observed bound state of the proton–neutron system. Its observed binding energy is $-E_1 = 2.2$ MeV. Besides, we know that the range of the nuclear forces is of the order of a fermi; we thus get $R_0 \simeq 1$ F. This will enable us to estimate the depth of the well, eq. (6.4.18) – i.e., the strength of the nuclear forces. It amounts to estimating V_0, knowing that the equation

ħ$\qquad \sin(p'a/2) = p'/q, \qquad p' = [2m(V_0 + E)]^{1/2}, \qquad q = (2mV_0)^{1/2}, \tag{6.4.20}$

with

$$a = 2R_0 \simeq 2 \text{ F}, \qquad m = \tfrac{1}{2}M = 0.8 \times 10^{-27} \text{ kg}, \tag{6.4.21}$$

has only one solution

$$E_1 = -2.2 \text{ MeV}. \tag{6.4.22}$$

It is easy to convince oneself, graphically, that the existence of a *single* bound state implies that for this state p'/q should be very close to unity, which in turn means that $V_0 \gg |E_1|$. Under these conditions, we have the approximations

ħ$\qquad \sin(p'a/2) \simeq 1, \quad \text{whence } \tfrac{1}{2}p'a \simeq \tfrac{1}{2}\pi, \quad \text{with } p' \simeq (2mV_0)^{1/2}. \tag{6.4.23}$

Thus finally,

ħ$\qquad V_0 \simeq \dfrac{\pi^2}{2ma^2} = \dfrac{\pi^2}{4MR_0^2}. \tag{6.4.24}$

The numerical computation leads to

ℏ
$$V_0 \simeq 100 \text{ MeV}, \tag{6.4.25}$$

a not too accurate value, in view of the rather rough approximations that we have made, but one which correctly indicates that nuclear interaction potentials measure a few tens of MeVs. Moreover, this estimate can be improved. We also see, in conformity with the actual situation, that the deuteron is a 'barely' bound system ($|E_1| \ll V_0$) (exercise 6.14). Up to now we have implicitly assumed that the neutron–proton system is in a spherically symmetric state, with zero angular momentum, $l = 0$. We can study its states of higher angular momentum by adding to the nuclear potential of eq. (6.4.18) the centrifugal term

ℏ
$$V_{\text{centr}}^{(l)} = \frac{l(l+1)}{2mr^2}. \tag{6.4.26}$$

This additional repulsion forbids the existence of bound states. Thus, there is only one neutron–proton bound state, the deuteron, of orbital angular momentum $l = 0$, and no bound states for $l > 0$ (exercise 6.14).

6.4.2. The scattering states

(a) *Transmission and reflection.* We shall now be interested in the stationary states of energy $E > 0$. The physical situation is that of a quanton emitted at infinity – e.g., to the left of the well, where there is no influence of any potential – by an accelerator which imparts to it its kinetic energy E. Inside the region $[-\frac{1}{2}a, \frac{1}{2}a]$, this quanton is subjected to an attractive potential $-V_0$. Hence, the experiment at hand is that of 'scattering' (in one dimension) by an attractive potential of range a. For this reason, states which are not bound are often given the name of *scattering states*.

The wavefunction of such a state is given by eq. (6.4.2) where, as in the case of the potential step, we set $d_- = 0$ [see eq. (6.3.6)] to signify that there are no sources of quantons to the right. The values of the other coefficients are fixed by the continuity conditions, imposed upon the wavefunction and its derivative at the points $x = \pm \frac{1}{2}a$ of discontinuity of the potential. Thus,

ℏ
$$\begin{cases} b_+ e^{-ip_E a/2} + b_- e^{ip_E a/2} = c_+ e^{-ip'_E a/2} + c_- e^{ip'_E a/2}, \\ p_E b_+ e^{-ip_E a/2} - p_E b_- e^{ip_E a/2} = p'_E c_+ e^{-ip'_E a/2} - p'_E c_- e^{ip'_E a/2}, \\ c_+ e^{ip'_E a/2} + c_- e^{-ip'_E a/2} = d_+ e^{ip_E a/2}, \\ p'_E c_+ e^{ip'_E a/2} + p'_E c_- e^{-ip'_E a/2} = p_E d_+ e^{ip_E a/2}. \end{cases} \tag{6.4.27}$$

We thus have a system of four linear, homogeneous equations with five unknowns, which enables us to express four of the unknown quantities as functions of the fifth, for arbitrary values of the numerical coefficients of the system – in other words, for arbitrary energy proper values E. Unlike the energies of bound states, those of the scattering states are not quantized.... . This is normal, since the quanton is not confined to any region of space.

Of particular interest are the coefficients b_- and d_+ of the reflected and transmitted waves. Since these are proportional to b_+, it is convenient to define the transmission and reflection factors, A_t and A_r respectively, of the amplitudes by

$$b_- = A_r b_+, \qquad d_+ = A_t b_+. \tag{6.4.28}$$

A computation (exercise 6.15) yields

$$\hbar \quad \begin{cases} A_r = \dfrac{i\dfrac{p'^2 - p^2}{2pp'} \sin p'a}{\cos p'a - i\dfrac{p'^2 + p^2}{2pp'} \sin p'a} e^{-ipa}, \\[4ex] A_t = \dfrac{1}{\cos p'a - i\dfrac{p'^2 + p^2}{2pp'} \sin p'a} e^{-ipa}, \end{cases} \tag{6.4.29}$$

hence the reflection and transmission coefficients,

$$\hbar \quad \begin{cases} \mathcal{R} = |A_r|^2 = \dfrac{\left(\dfrac{q^2}{2pp'}\right)^2 \sin^2 p'a}{1 + \left(\dfrac{q^2}{2pp'}\right)^2 \sin^2 p'a}, \\[4ex] \mathcal{T} = |A_t|^2 = \dfrac{1}{1 + \left(\dfrac{q^2}{2pp'}\right)^2 \sin^2 p'a}. \end{cases} \tag{6.4.30}$$

These coefficients depend on the energy, since as we recall,

$$\begin{cases} p^2 = 2mE, \\ p'^2 = 2m(E + V_0), \\ q^2 = 2mV_0 \end{cases} \tag{6.4.31}$$

(we shall henceforth drop the subscript 'E' in order to simplify the expressions).

We verify that $\mathcal{R} + \mathcal{T} = 1$. Let us concentrate our attention on $\mathcal{R}$, which characterizes the effectiveness of the scattering by the potential better than $\mathcal{T}$. Indeed, the outgoing 'forward' wave, of amplitude d_+ can be considered as being a sum of the *incident* (of amplitude b_+) and forward *scattered* waves. The coefficient $\mathcal{T}$, which measures the proportion of the quantons transmitted, simultaneously involves the intensities of the incident and scattered waves and an interference term. It is, therefore, impossible to distinguish, in the forward direction, the quantons 'genuinely' scattered by the well from those which might come from the incident beam, without having interacted with the potential [even though, formally, we can still define effective cross-sections and verify an analogue of the optical theorem in three dimensions (exercise 6.16)]. By contrast, the backward scattering, the extent of which is measured by the reflection coefficient $\mathcal{R}$, is completely due to the action of the potential well on the quantons (if $V_0 = 0$, $\mathcal{R} = 0$). The variation of $\mathcal{R}$ with the energy (fig. 6.15) enables us to distinguish between two particular types of physical situations.

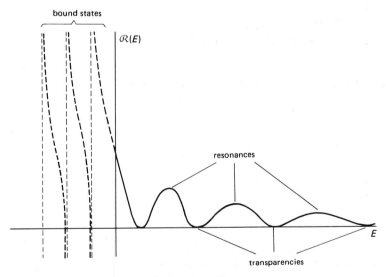

Fig. 6.15. *The reflection coefficient of a finite flat well*
The coefficient $\mathcal{R}(E)$ displays characteristic oscillations:
  maxima: resonances (broader and broader),
  minima: transparencies.
The function $\mathcal{R}(E)$, formally extended to negative values of the energy, exhibits poles (infinite maxima) for the bound states.

(b) *Transparencies.* We note, first of all, that the coefficient of reflection can become zero,

h $p'a = n\pi,$ and hence, $\mathscr{R} = 0,$ $\mathscr{T} = 1.$ (6.4.32)

In other words, there exist particular values of the energy for which the potential does not scatter the quantons backwards. They all go past ($\mathscr{T} = 1$) the interaction region: the scattering is totally ineffective as though the well were 'transparent'. The condition given by eq. (6.4.32) shows that the width of the interaction region corresponds, in this case, to exactly an integral multiple

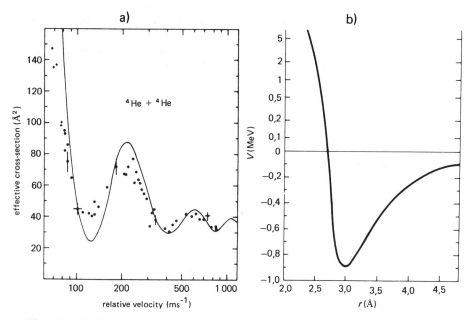

Fig. 6.16.   *The Ramsauer–Townsend effect*
(a) The effective scattering cross-section of two helium atoms at low energies displays deep minima. The first of these is explained with the help of the Ramsauer–Townsend effect. But beware! The succeeding ones, despite their similarity with the curve giving the reflection coefficient for a flat well, are *not* related to this effect (exercise 6.17), and have a completely different explanation (see ch. 7, exercise 7.7) [R. S. Grace et al., Phys. Rev. A 14 (1976) 1006].
(b) The interaction potential between two helium atoms displays the existence of abrupt walls, such as those of a flat bottomed well, which explains its relative transparency for certain values of the energy [H. G. Bennewitz et al., Z. Phys. 253 (1972) 435].

of the interior half-wavelength. This condition is analogous to that of impedance matching inside a classical wave guide. Such a situation had been encountered at the very beginning of experimental quantum physics. In the course of the measurement of the scattering of electrons by the atoms of a noble gas, it was observed, around 1921/22, that the effective cross-section became very small for a certain value of the energy, of the order of an electron volt (fig. 6.16). This behaviour, named the 'Ramsauer–Townsend effect', after the two experimentalists who discovered it, remained incomprehensible until, somewhat later, Bohr made the connection with the just discovered wave aspect of the electron. Once again, this is an effect analogous to one that we have described for the square well, but now in a three-dimensional situation, where 'the impedance of the atom' for the 'electronic wave' vanishes (or almost so) for an appropriate value of the energy. Nowhere is this effect more clearly manifested as for the atoms of rare gases. Their chemical inertness results from their electronic structure of complete shells which gives rise to an interaction potential of limited range and abrupt boundaries, similar to a finite, flat well (in three dimensions!) (exercise 6.17).

(c) *Resonances.* Another type of a particularly interesting situation corresponds to the maxima of the reflection coefficient $\mathcal{R}$. They correspond very nearly to those energy values for which

$$\hbar \qquad p'a \simeq (n + \tfrac{1}{2})\pi, \quad \text{and hence,} \quad \mathcal{R} \text{ maximum.} \qquad (6.4.33)$$

Here, the potential acts maximally on the quanton. Again, in analogy with classical wave physics, we say that a *resonance* has occurred. It can be shown (exercise 6.18) that, under these conditions, the quanton is in a way 'trapped' by the well, in the sense that it takes the maximum amount of time to cross the interaction region, before coming out again. For this reason, resonances are also called 'quasi-bound states'! Moreover, the resonance condition – $\mathcal{R}$ maximum – is very close to the condition which fixes the bound states. These latter are characterized by the fact that there is a non-zero probability of their presence inside the interaction region, even though there is no incident wave. Thus, in the preceding notation, $b_+ = 0$. Indeed, for a bound state, in the region exterior to the potential well, the wavefunction has the character of a real exponential. However, exponentials which increase at infinity ought to be excluded – i.e., the terms involving b_+ and d_- in the wavefunction. In other words, for bound states everything turns out to be as though one had $|b_-/b_+|^2 = \mathcal{R} = \infty$ (and hence a maximum!). In fact, the energy proper values of the bound states are given by the *poles* of the scattering factors A_l and A_r, eq. (6.4.29). The condition

$$\cos p'a - i\frac{p'^2 + p^2}{2pp'} \sin p'a = 0, \tag{6.4.34}$$

requires for its fulfilment that p be pure imaginary, i.e., $E < 0$:

$$p = i\kappa, \qquad \kappa = (-2mE)^{1/2}. \tag{6.4.35}$$

Condition (6.4.34) leads exactly, therefore, to the quantization condition, eq. (6.4.17). If we consider the scattering amplitude as a function of the energy – even for negative energies – we may write (both for A_t and A_r) therefore,

$$A \propto \frac{1}{E - E_k}, \quad \text{for} \quad E \simeq E_k < 0, \tag{6.4.36}$$

where E_k denotes the energy proper value of the bound state of order k. Now, denoting by E_n^* the resonance energy of order n, defined by eq. (6.4.33), it can be shown that the scattering factors of the amplitudes acquire the form

$$A \propto \frac{1}{E - E_n^* + \frac{1}{2}i\Gamma_n} \quad \text{for} \quad E \simeq E_n^* > 0, \tag{6.4.37}$$

(exercise 6.18). In other words, the resonance energies are the positive real parts of the *complex* poles of the scattering amplitude, just as the binding energies are the (negative) positions of the real poles (fig. 6.17).

The imaginary part $\frac{1}{2}\Gamma_n$ of a resonance pole has an important physical significance. Indeed, let us look for the form of a wave packet, initially given by

$$\varphi(x; t) = \int \hat{\varphi}(p) \, e^{ipx - iEt} \, dp, \tag{6.4.38}$$

after it is scattered by the potential. Each component $\hat{\varphi}(p)$ of the initial wave packet is reflected or transmitted with the factor $A_r(p)$ or $A_t(p)$, respectively. Thus, the wave packet splits into two wave packets, one transmitted, the other reflected,

$$\begin{cases} \varphi_t(x; t) = \int \hat{\varphi}(p) \, A_t(p) \, e^{ipx - iEt} \, dp, \\[2ex] \varphi_r(x; t) = \int \hat{\varphi}(p) \, A_r(p) \, e^{-ipx - iEt} \, dp. \end{cases} \tag{6.4.39}$$

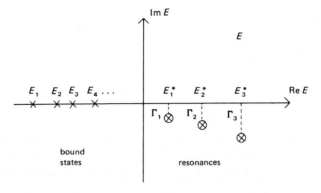

Fig. 6.17. *Bound states and resonances: poles of the scattering amplitude*
We consider the scattering amplitude as a function a complex energy variable (only the values of the function on the real axis enter into the computation of the coefficients of transmission or reflection). It is verified that this function admits two types of poles:
(a) real poles for $E < 0$, giving the positions of the bound states;
(b) complex poles, with Re $E > 0$ and Im $E < 0$, giving the positions, $E^* =$ Re E, and widths, $\Gamma = -2$ Im E, of the resonances.

Let us suppose that the average energy E of the wave packet is close to a resonance energy E^*. For either one of the two scattered wave packets, $\varphi_d = \varphi_t$ or φ_r, we may write, therefore,

$ħ$
$$\varphi_d(x; t) \simeq \int \hat{\varphi}(p) \frac{\text{const.}}{E - E^* + \frac{1}{2}i\Gamma} e^{\pm ipx - iEt} \, dp. \qquad (6.4.40)$$

If, in addition, the dispersion in energy, or what amounts to the same thing, the width of the amplitude $\hat{\varphi}(p)$, is not too narrow (compared to Γ), the variation in the integrand is essentially due to the denominator of the resonance term, and we may write

$ħ$
$$\varphi_d(x; t) \simeq \int_{-\infty}^{\infty} \frac{\text{const.}}{E - E^* + \frac{1}{2}i\Gamma} e^{\pm ipx - iEt} \, dE, \qquad (6.4.41)$$

where we have changed the variable of integration from p to E, and extended the region of integration from $-\infty$ to $+\infty$, in place of from 0 to $+\infty$, taking advantage of the fact that the integrand diminishes very rapidly for $|E - E^*|$ sufficiently large. The integral (6.4.41) is computed by completing the integration along the real line with a semi-circle in the complex plane (fig. 6.18), chosen so that e^{-Et} decreases exponentially on it with E and hence does not contribute to the integral. Thus, it is necessary to choose Im $E > 0$ for $t < 0$ and Im $E < 0$ for $t > 0$. In the first case, the integration, being that along a closed contour of a function which is analytic in the half-plane $E > 0$, gives a null result. Hence,

$$\varphi_d(x; t) \simeq 0, \quad \text{for} \quad t < 0. \tag{6.4.42}$$

This result is very pleasing. It signifies that there is no scattering so long as the quanton (the 'wave packet') does not reach the scattering region. On the other hand, in the second case the contour encloses the pole $E = E^* - \frac{1}{2}i\Gamma$, and the method of residues gives

$\hbar$

$$\varphi_d(x; t) \propto e^{\pm ip_E x - iEt}\big|_{E = E^* - i\Gamma/2}$$

$$\propto e^{\pm ip_E x - iE^*t - \Gamma t/2}, \quad \text{for} \quad t > 0. \tag{6.4.43}$$

The scattered wavefunction displays an exponential decrease with time, just

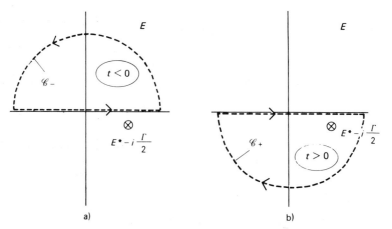

Fig. 6.18. *The energy integral*
(a) For $t < 0$, the term e^{-iEt} decreases exponentially in energy if Im $E > 0$. The integral along the closed contour $\mathscr{C}_-$ is equal to the integral along the real line, in the limit as the semi-circle is pushed to infinity. The function integrated being analytic in the half-plane Im $E > 0$, the integral is zero [eq. (6.4.42)].
(b) For $t > 0$, the term e^{-iEt} decreases exponentially in energy if Im $E < 0$. The integral along the contour $\mathscr{C}_+$ is identical to the integral along the real line. Since $\mathscr{C}_+$ encloses the resonance pole, the integral is given by the residue at this pole [eq. (6.4.43)].

as the probability density of its localization,

$$\hbar \qquad \rho_d(x; t) = |\varphi_d(x; t)|^2 \propto e^{-\Gamma t}. \qquad (6.4.44)$$

We thus have an exponentially decaying state, of 'half-life'

$$\hbar \qquad \tau = 1/\Gamma. \qquad (6.4.45)$$

In other words, at the resonance, the quanton interacts for a long time with the potential and comes out only by means of an exponential 'leak' (fig. 6.19). Finally, let us add that the energy spectrum of a resonance, i.e., the probability density in energy, in the vicinity of a resonance, can be written,

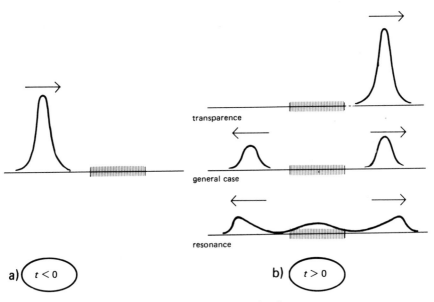

Fig. 6.19. *The scattering of a wave packet by a potential well*
(a) The incident wave packet for $t < 0$ (the shaded region is that of the well, where the interaction takes place at $t \simeq 0$).
(b) The scattered wave packet(s) for $t > 0$. Depending upon the value of the energy, there could be:
  transparency: complete transmission, without reflection;
  general case: reflection and transmission;
  resonance: the quanton remains trapped for a long time inside the region of the well and it escapes from this region by means of a slow leakage of probability.

using eq. (6.4.41), as

$$\sigma(E) \simeq \text{const.} \left| \frac{1}{E - E^* + \frac{1}{2} i \Gamma} \right|^2$$

$$= \frac{\text{const.}}{(E - E^*)^2 + \frac{1}{2} \Gamma^2} . \qquad (6.4.46)$$

This Lorentzian function – also called the Breit–Wigner function – is a bell-shaped curve centered at E^* and has width Γ (fig. 6.20). We could also say that a resonance is a metastable state, of energy E^* which is defined to an accuracy ΔE, where

$$\Delta E = \Gamma. \qquad (6.4.47)$$

We recover again the Heisenberg type of relation between the half-life, eq. (6.4.45), and the energy width ΔE, eq. (6.4.47),

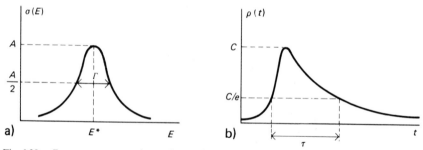

Fig. 6.20. *Energy spectrum and time decay of a resonance*
(a) The Breit–Wigner energy spectrum for a resonance of energy E^* and width Γ.
(b) The exponential decay with half-life $\tau = 1/\Gamma$ of the probability of the presence of the quanton at an arbitrary point (outside the region of the potential).

Fig. 6.21. *Resonances*
(a) The total effective cross-section for the scattering of neutrons by the nucleus of indium [V. F. Sailor and L. B. Borst, Phys. Rev. 87 (1952) 161].
(b) The effective cross-section for the elastic scattering of π^+ and π^- mesons by the proton. [Following S. Gasiorowicz, Elementary Particle Physics (J. Wiley, NY).]
One observes, in these two graphs, well-defined peaks corresponding to different resonances. These are just two examples among many others (note the enormous difference of scale, in the energies and effective cross-sections, between these two phenomena). Compare these curves to those in figs. (6.15) and (6.20a).

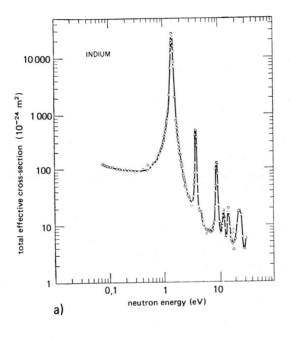

a)

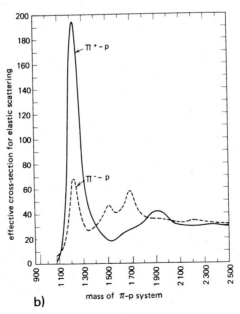

b)

$\hbar$ $\qquad\tau\,\Delta E \simeq (1/\Gamma)\Gamma = 1$ $\qquad\qquad\qquad\qquad\qquad$ (6.4.48)

(see ch. 3, sect. 3.2.1).

We have devoted so much attention to the notion of a resonance, because this notion has turned out to be particularly important for the whole of the physics of atomic, nuclear and particle scattering (fig. 6.21). It enables us to understand the formation, in scattering experiments, of these quasi-bound states, which are definitely non-stationary, but which have 'almost' well-defined energies. The properties of such quasi-bound states (for example, the half-life) depend very little on the conditions of their production (nature and energies of the initial particles) – and this is what enables us to attribute to them an intrinsic identity. Our analysis also indicates that the decay in time ('disintegration') of these states is exponential, with a half-life related to the width of the energy spectrum, which moreover, has the canonical Breit–Wigner form. In particle physics, one does not make a fundamental

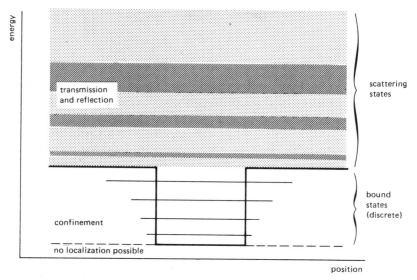

Fig. 6.22. *The quanton in a finite square well potential*
The grey area indicates the allowed region in energy and position:
– confinement 'inside' the well, with discrete energy levels and partial penetration through the 'walls' of the well;
– scattering 'above' the well, with privileged regions of energy (quasi-bound states, of non-zero energy widths).

distinction between stable systems (bound states) and metastable systems (resonances). In certain classification schemes, these latter are considered to be just as much (or as little) elementary as the former. Moreover, the notion of stability is relative (the neutron is stable for the biologist, who uses it as a probe, but it only has a half-life of fifteen or so minutes for the physicist working in weak interactions), and the concept changes in the course of history (the immortality of the proton has been questioned in the context of the recent theories of the unification of interactions).

Figure 6.22 summarizes, in a synthetic overview, our discussion on the different stationary states of a quanton subjected to a flat well potential.

6.5. The 'tunnel' effect

Consider now a potential barrier – a succession of two potential steps, the reverse of the preceding situation – described by (fig. 6.23),

$$V(x) = 0 \qquad |x| > \tfrac{1}{2}a,$$
$$= V_0 > 0 \quad |x| < \tfrac{1}{2}a. \tag{6.5.1}$$

Classically, the movement of a corpuscle in this potential is extremely simple. If its energy is greater than the height V_0 of the barrier, it crosses it. Its motion simply becomes slower, during the crossing of the barrier, than before or after (since during the crossing its energy is diminished by the amount V_0). If, on the other hand, its energy is lower than the height of the barrier, the particle can no longer cross it and is reflected by it, leaving with a uniform motion, with a velocity which is opposite to that of its initial motion.

6.5.1. The transmission coefficient

The quantum theory does not require any new computations in the present situation, the general expressions [eqs. (6.4.28) and (6.4.29)] remain valid, the only change being in the sign of V_0. We note, however, that there are no bound states now, since the potential is not of an attractive nature.

As in the classical theory, we distinguish between two cases, depending on the value of E relative to V_0. In the first case, where $E > V_0$, we recover, with no surprise, the same conclusions as before: the barrier 'scatters' the quanton, and the coefficients of reflection and transmission are still given by

$\hbar$
$$
\left\{
\begin{aligned}
\mathcal{R} &= \frac{\left(\dfrac{q^2}{2qq'}\right)^2 \sin^2 p'a}{1 + \left(\dfrac{q^2}{2qq'}\right)^2 \sin^2 p'a}, \\[2em]
\mathcal{T} &= \frac{1}{1 + \left(\dfrac{q^2}{2qq'}\right)^2 \sin^2 p'a},
\end{aligned}
\right.
\tag{6.5.2}
$$

but now with

$$
\begin{cases}
p^2 = 2mE, \\
p'^2 = 2m(E - V_0), \\
q^2 = 2mV_0.
\end{cases}
\tag{6.5.3}
$$

The fact that the attractive potential (of the well) has changed to a repulsive one (of the barrier), has hardly changed the problem.

In the case where $0 < E < V_0$, on the other hand, the quantum result is more unexpected: there still exists a non-zero probability of transmission, 'across' the barrier, in this case! Indeed, in conformity with what has been said in sect. 7.3, a quanton which comes knocking against the step, of height V_0 and situated at $x = -\tfrac{1}{2}a$, from the left, is reflected by it. However, the quanton has a certain spatial extension, and it 'seeps' into the interior of the step up to a distance of the order of

$\hbar$
$$
(\kappa')^{-1} = [2m(V_0 - E)]^{-1/2}.
\tag{6.5.4}
$$

As long as the barrier under consideration has a width $a \lesssim 1/\kappa'$, the region in which the probability density $|\varphi(x)|^2$ assumes appreciable values extends right up to the other extremity ($x = \tfrac{1}{2}a$) of the barrier. The continuity of the wavefunction across this second discontinuity in the potential now requires that $\varphi(x)$ be also non-zero beyond it. In the region $x > \tfrac{1}{2}a$ the quanton again behaves as a free quanton: its probability density remains constant. Summarizing, the quanton has passed 'right through' the barrier.

This simple argument needs to be examined in detail. The wavefunction

$\varphi(x)$ of the stationary state of energy E is written as

$$\hbar \quad \begin{cases} x < -\tfrac{1}{2}a: & \varphi(x) = b_+\, e^{ipx} + b_-\, e^{-ipx}, \\[2mm] -\tfrac{1}{2}a < x < \tfrac{1}{2}a: & \varphi(x) = c_+\, e^{-\kappa'x} + c_-\, e^{\kappa'x}, \\[2mm] x > \tfrac{1}{2}a: & \varphi(x) = d_+\, e^{ipx}, \end{cases} \quad (6.5.5)$$

where

$$p = (2mE)^{1/2}, \qquad \kappa' = [2m(V_0 - E)]^{1/2}.$$

We have, straightaway, suppressed the $d_-\, e^{-ipx}$ solution, since there is no source of quantons to the right. On the other hand, it is necessary for us to keep the exponentially decreasing solution in the intermediate region, physically admissible here, since in any case it remains bounded.

Thus, imposing the usual continuity conditions, on $\varphi(x)$ and its derivative, at the points $x = \pm\tfrac{1}{2}a$ of discontinuity of the potential, we obtain a system of four linear, homogeneous equations in five unknowns, which always (i.e., for any given value of E) allows us to determine four of them as functions of the fifth. Generally, one keeps b_+, which characterizes the incident 'wave'. Thus, the spectrum of the scattering of a quanton by a potential barrier is continuous, *unquantized* – exactly as the spectrum of the scattering by a potential well.

The expression obtained for the ratio d_+/b_+, namely,

$$\hbar \quad A_t = \frac{d_+}{b_+} = \frac{1}{\cosh \kappa'a - i\dfrac{p^2 - \kappa'^2}{2p\kappa'}\sinh \kappa'a}\, e^{-ipa}, \quad (6.5.6)$$

is, as expected here, identical to the expression obtained from eq. (6.4.29) by setting $p' = i\kappa'$. The transmission coefficient, in turn, is now written as

$$\hbar \quad \mathscr{T} = |A_t|^2 = \frac{1}{1 + \left(\dfrac{q^2}{2p\kappa'}\right)^2 \sinh^2 \kappa'a}. \quad (6.5.7)$$

This quantity never vanishes: thus, the quanton always has the possibility of 'coming out' of the other side of the barrier (figs. 6.23 and 6.24). It can be verified that $\mathscr{T}$ increases with E, in conformity with intuition: the larger the energy of the quanton, the greater is the chance that it would go through the barrier.

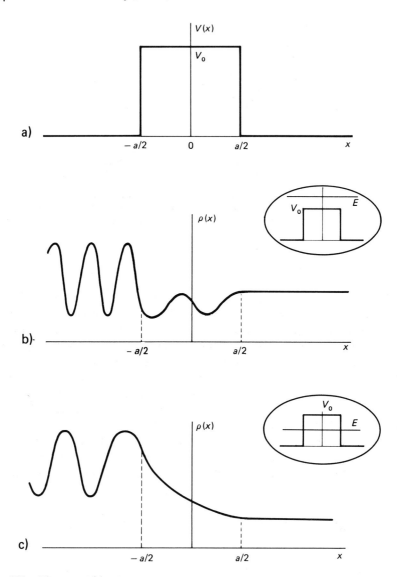

Fig. 6.23. *The potential barrier*
(a) The graph of the potential function.
(b) The probability density of localization, for a stationary state of energy $E > V_0$ (quanton arriving from the left).
(c) The probability density of localization, for a stationary state of energy $E < V_0$ (tunnel effect for a quanton arriving from the left).

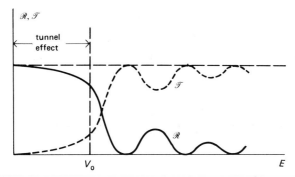

Fig. 6.24.   *The reflection and transmission coefficients of a potential barrier*
In the energy region $0 < E < V_0$, the reflection coefficient is smaller than unity, because of the tunnel effect. For $E > V_0$, the same phenomena (transparency, resonance) occur as for the square well.

6.5.2. Tunnelling or leap-frogging?

This specifically quantum phenomenon appears, in general, under the name of the 'tunnel effect' in the literature. This nomenclature is derived from the picture of a quanton bumping into a 'mountain', of height V_0 and unsurmountable to it – since it finds itself at an 'altitude' $E < V_0$ – and burrowing a tunnel through the mountain to come out of the other side, still at the 'altitude' E. To tell the truth, this terminology is rather poorly chosen since, as we shall demonstrate, the quanton crosses the barrier by actually 'leaping over' it (fig. 6.25). Developing this metaphor, it would be better to speak of the 'leap-frog' effect.

We observe that in this case, as with the simple potential step (sect. 7.3), the quanton, when in a proper state of its total energy, is not in a proper state of kinetic energy, since its potential energy is not constant. Its momentum

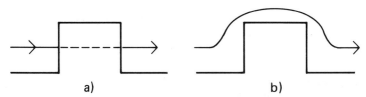

Fig. 6.25.   *Tunnel effect or leap-frogging?*
If a quanton goes past a potential barrier higher than its energy, it is so much passing 'through' (tunnelling, a) as passing 'over' (leap-frogging, b) it: since its stationary state is not identical with that of a free quanton, its kinetic energy has in fact a certain dispersion.

spectrum consists of a distribution of width $\Delta p \simeq \kappa'$, due to the presence of the exponential terms $e^{\pm \kappa' x}$ in the wavefunction (see fig. 6.10). The corresponding kinetic energy spectrum includes components of (kinetic) energy greater than V_0, revealing the existence of non-zero probability amplitudes, for which the quanton has enough energy to go 'over' the top of the barrier.

We could also consider, rather than a stationary (non-localized) state, a quasi-stationary wave packet. The energy of the quanton does not have a well-defined proper value. During the interaction of the quanton with the barrier, it undergoes a fluctuation of order ΔE, something that is possible in a non-stationary state, over a sufficiently small interval of time Δt, in a manner such that $\Delta t \, \Delta E \lesssim \hbar$. If the fluctuation is sufficiently pronounced, so that $E + \Delta E \gtrsim V_0$, the quanton would be able to pass over the barrier, with a velocity of the order of $v \simeq [2(E + \Delta E - V_0)/m]^{1/2}$. In order for it to 'leap' squarely over the barrier, it is necessary that there be enough time, during the interval $\Delta t \simeq \hbar/\Delta E$ to cross the entire width of the barrier with the velocity v, and hence that $v \, \Delta t \gtrsim a$, implying

$$\left(\frac{2(E + \Delta E - V_0)}{m} \right)^{1/2} \frac{\hbar}{\Delta E} \gtrsim a. \tag{6.5.8}$$

If we study the left-hand member of this inequality, as a function of the fluctuation ΔE, we verify that it passes through a maximum, of the order of $\hbar/[2m(V_0 - E)]^{1/2}$, for $\Delta E = 2(V_0 - E)$. Thus, this inequality can only be satisfied, up to a numerical factor, under the condition

$$[2m(V_0 - E)]^{1/2} a = \kappa' a < \hbar. \tag{6.5.9}$$

More precisely, condition (6.5.9) is the condition for which the transmission coefficient would have an appreciable value – of the order of unity. Conversely, the larger the parameter $\kappa' a/\hbar$ is, compared to unity, i.e., the thicker the barrier (for a given height!), the smaller the coefficient of transmission will be. We have temporarily introduced the constant $\hbar$ explicitly into condition (6.5.9), in order to draw attention to its specifically quantum nature. The inverse condition, $\kappa' a \gg \hbar$, expresses the fact that the characteristic action is much larger than the Planck constant. In this case we immediately recover the classical result, with no 'tunnel effect'.

6.5.3. The thick-barrier approximation

The case in which the barrier is sufficiently thick, so that the tunnel effect is small but non-negligible, i.e., the case in which

$ħ$ $$1 \ll \kappa' a = [2m(V_0 - E)a^2]^{1/2} \qquad (6.5.10)$$

(we once again set $ħ = 1$) is of great physical interest. The transmission amplitude A_t [eq. (6.5.6)] assumes the approximate form

$ħ$ $$A_t = \frac{d_+}{b_+} = \frac{4 i p \kappa'}{(p + i \kappa')^2} e^{-\kappa' a} e^{-i p a}, \qquad (6.5.11)$$

and its physical interpretation is very simple: the coefficient $4 i p \kappa'/(p + i \kappa')^2$ can, by reintroducing the imaginary momentum $p' = i \kappa'$, be written as

$$\frac{4 i p \kappa'}{(p + i \kappa')^2} = \frac{2p}{p + p'} \frac{2p'}{p + p'}. \qquad (6.5.12)$$

We recognize in $2p/(p + p')$, expression (6.3.7) for the fraction of the incident amplitude, transmitted by the first discontinuity ('ascending' step), from 0 to V_0, at $x = -\frac{1}{2}a$ [see also eq. (6.3.18)].

In the same manner, the second factor, $2p'/(p + p')$, in eq. (6.5.12) represents the fraction of the amplitude transmitted from the left to the right by the second discontinuity ('descending' step), from V_0 to 0. The structure of the transmitted amplitude away from the barrier, $\varphi(\frac{1}{2}a)$, now appears clearly:

$ħ$ $$\varphi(\tfrac{1}{2}a) = \varphi(-\tfrac{1}{2}a) \frac{2p}{p + p'} e^{-\kappa' a} \frac{2p'}{p + p'}, \qquad (6.5.13)$$

where $\varphi(-\frac{1}{2}a)$ represents the amplitude incident upon the barrier, $b_+ e^{-i p a/2}$. Therefore, during the crossing of the barrier, the incident amplitude undergoes, in turn:
- the effect of the first discontinuity, accounted for by the coefficient $2p/(p + p')$;
- an attenuation inside the tunnel: $e^{-\kappa' a}$ (or rather a loss of amplitude during the leap-frogging!);
- the effect of the second discontinuity: $2p'/(p + p')$.

One verifies that in this analysis, it is only the vicissitudes suffered by $\varphi(x)$, in its variation from the left to the right, that figure: the effects of reflection (from right to left) have been neglected completely. This is a consequence of our original assumption that $\kappa' a \gg 1$. More precisely, this hypothesis allows us to neglect the reflection from the second discontinuity (the reflection from the first discontinuity being already taken into consideration by the attenuation $e^{-\kappa'a}$): the amplitude at the second discontinuity, having been greatly attenuated by the factor $e^{-\kappa'a}$ ($\ll 1$), we are allowed to neglect the reflected portion by comparison with the incident wave.

From the point of view of experiment, it is $\mathcal{T}$, the transmission coefficient of the barrier, that is measured. Again, with the hypothesis $\kappa'a \gg 1$, we get

$\hbar$
$$\mathcal{T} = \frac{4E(V_0 - E)}{V_0^2}\, e^{-2\kappa'a}. \tag{6.5.14}$$

Only a fraction $16E(V_0 - E)/V_0^2$ of the incident current, attenuated by an exponential factor $e^{-2\kappa'a}$, is transmitted. Expression (6.5.14) holds for all energies (except, of course, for $E = V_0$; but in that case, the hypothesis $\kappa'a \gg 1$ does not make sense). Studying the variation of the ratio $16E(V_0 - E)/V_0^2$, as a function of the energy E, we verify that it has a maximum, equal to 4, for the value $E = \frac{1}{2}V_0$, and that it is of the order of unity except at the end of the interval $0 < E < V_0$. In other words, for 'average' values of E, the effects of the two discontinuities are small and almost compensate – only the effect of the attenuation survives, so that one may use the approximate expression

$\hbar$
$$\mathcal{T} = e^{-2\kappa'a}. \tag{6.5.15}$$

6.5.4. Barrier of arbitrary shape

This last remark will allow us to treat, in an approximate fashion, the case of a non-square barrier $V(x)$, of arbitrary shape. Let us break up the barrier being considered into a succession of elementary flat barriers, of the same width Δx, and each having for its height the average value of $V(x)$ in the interval Δx in question (fig. 6.26). Let x_1 and x_2 be the abscissas of the points for which $E = V(x_1) = V(x_2)$ and let us try to determine the coefficient of transmission, $\mathcal{T}$, of this barrier for a quanton of energy E, emitted to the left on the x-axis. For simplicity, we assume that the elementary barriers, the heights of which are definitely lower than E, have practically no effect on the quanton. Here we are only interested in the 'true' elementary barriers, the ones for which the quanton does not have sufficient (classical!) energy.

Let $\mathcal{P}(x)$ be the probability for the quanton to be present at the point x ($x_1 < x < x_2$). What, then, is the probability, $\mathcal{P}(x + \Delta x)$, of its being present at the point $x + \Delta x$? This is the product of the probability $\mathcal{P}(x)$ for the quan-

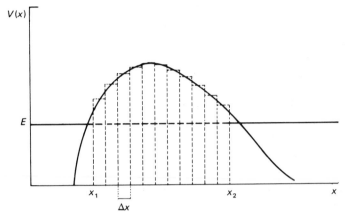

Fig. 6.26.   *The tunnel effect for an arbitrary barrier*
By considering the barrier as a succession of infinitely thin flat barriers, one can obtain an
approximate expression, eq. (6.5.21), for the transmission coefficient.

ton to arrive at x, multiplied by the probability that it would pass through
the elementary square barrier that appears between x and $x + \Delta x$. Assuming
that this barrier only has an attenuating effect, we get

$\hbar$ $\qquad \mathcal{P}(x + \Delta x) = \mathcal{P}(x) \exp[-2\kappa(x)\,\Delta x],$ $\qquad\qquad$ (6.5.16)

with

$$\kappa(x) = [2m(V(x) - E)]^{1/2}. \qquad\qquad (6.5.17)$$

This in turn yields, upon replacing the finite intervals Δx by the differential
elements dx,

$\hbar$ $\qquad \mathcal{P}(x + dx) = \mathcal{P}(x)\,[1 - 2\kappa(x)\,dx], \qquad\qquad$ (6.6.18)

$\hbar$ $\qquad \dfrac{d\mathcal{P}(x)}{dx} = -2\kappa(x)\,\mathcal{P}(x). \qquad\qquad$ (6.6.19)

This differential equation is easily integrated,

$$\mathcal{P}(x) = \text{const.} \exp\left[-2\int \kappa(x)\,dx\right]. \qquad\qquad (6.5.20)$$

From this we obtain, for the transmission coefficient, the expression:

$$\hbar \qquad \mathcal{T} = \frac{\mathcal{P}(x_2)}{\mathcal{P}(x_1)} \exp\left[-2 \int_{x_1}^{x_2} \kappa(x)\, dx \right], \qquad (6.5.21)$$

an expression which is very commonly used in numerous applications of the theory (see, e.g., exercises 6.21 and 6.22).

The 'proof' given here makes no claim to be rigorous. In particular, it is clear that the 'singular' points x_1 and x_2 pose problems, since precisely formula (6.5.15), on which the above reasoning is based, is definitely not valid at these points (called the 'turning points' – a reference to the classical situation).

6.5.5. *Applications: α-decay*

One cannot overestimate the importance of the tunnel effect in the applications of the quantum theory to different physical situations. It is scarcely an exaggeration to say that *every* specifically quantum phenomenon can be interpreted in terms of the tunnel effect. It is essentially due to the tunnel effect that the possibility exists for a quanton to penetrate and go through regions, which otherwise are forbidden classically: the tunnel effect, explains just those phenomena which are forbidden in the classical theory and, hence, are characteristic to the quantum theory. This can very well be seen, moreover, in the fundamental formula, eq. (6.5.21), giving the coefficient of transmission by means of the tunnel effect. Reintroducing the Planck constant, this formula may be written as

$$\mathcal{T} = \exp(-\mathcal{A}/\hbar), \qquad \mathcal{A} \triangleq 2 \int_{x_1}^{x_2} [2m(V(x) - E)]^{1/2}\, dx. \qquad (6.5.22)$$

We note that $\mathcal{A}$ is a typical magnitude of action. The classical approximation, defined by

$$\mathcal{A} \gg \hbar \qquad (6.5.23)$$

implies, therefore,

$$\mathcal{T} \ll 1, \qquad (6.5.24)$$

where we have deliberately indicated by the symbol $\ll$, that the exponential dependence of $\mathscr{A}$ implies an *extremely* rapid decrease as $\mathscr{A}/\hbar$ increases. In particular, we note that $\mathscr{T}$ does not depend analytically on the constant $\hbar$, which shows the impossibility of understanding even a weak ($\mathscr{T} \ll 1$) quantum phenomenon by continuity, or by extension, starting with a classical theory ($\mathscr{T} = 0$).

From a practical point of view, the tunnel effect plays an essential role in modern electronics, in which a number of devices depend on the ability of electrons to pass through classically forbidden regions. We can see that the exponential dependence of the transmission factor, relative to the physical characteristics of these regions, would allow one to control the electronic current sharply. It is only very recently that it has been possible to display the passage of an electronic current across an *empty* region of space, which constitutes, if we wish, one of the more direct demonstrations of the tunnel effect (fig. 6.27 and exercise 6.20).

Among the most important applications of the tunnel effect in the understanding of certain physical phenomena, let us again cite α-decay, which we now proceed to treat briefly.

The explanation of α-decay was a major historic success of the quantum theory. This was the first application (around the year 1930) of the theory to the atomic nucleus, i.e., at scales 10^5 times smaller than atomic and molecular systems, in the context of which alone the theory had been developed until then. The phenomenon to be explained is that of the emission of α-particles (helium nuclei) by a nucleus, with energies lower than the potential barrier, which, in principle, holds them inside the nucleus. Indeed, the interaction potential between an α-particle and the rest of the nucleus consists of an attractive potential well, due to specifically nuclear forces, in which the α-particle is bound, and a repulsive electrostatic part which constitutes the exterior rim of the barrier (fig. 6.28a). We can easily estimate the height of the barrier for a nucleus ${}_{Z}^{A}N$, containing Z protons out of A nucleons. It corresponds to the value of the repulsive Coulomb potential at the edge of the nuclear well – hence on the surface of the nucleus – of radius given by $R \simeq A^{1/3} R_0$ ($R_0 = 1.2$ F). The charge of the α-particle being $2q_e$, that of the residual nucleus $(Z-2)q_e \simeq Zq_e$ (if Z is sufficiently large), the potential, at the summit of the barrier, has the value

$$V_{\mathrm{m}} \simeq \frac{2Ze^2}{R} = 2\frac{Z}{A^{1/3}}\frac{e^2}{R_0} \simeq 3\frac{Z}{A^{1/3}} \text{ MeV.} \qquad (6.5.25)$$

For example, for ${}^{238}\mathrm{U}$ ($A = 238$; $Z = 92$) we have $V_{\mathrm{m}} \simeq 30$ MeV. Now, this nucleus is α-radioactive and emits α-particles of energy 4.3 MeV – thus, much

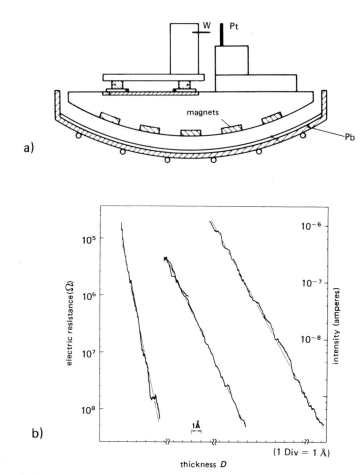

Fig. 6.27. *Tunnel effect across empty space*

(a) Two metallic electrodes, a tungsten point W and a platinum plate Pt, are separated by a region of empty space. Their distance, of a few angstroms, is controlled by a highly sensitive piezoelectric system. The entire set-up is insulated from any mechanical vibration by magnetic levitation above a section of superconducting lead.

(b) The electric current jumps across the empty space between the two electrodes, the electrons crossing by means of the tunnel effect. The resistance of this space is measured as a function of its thickness D. The curves obtained depend upon the state of the surface of the electrodes, which fixes the height of the effective potential barrier. But in all cases, the exponential dependence, characteristic of the tunnel effect, is obtained (see exercise 6.20). [G. Binnig, H. Rohrer, Ch. Gerber, E. Weibel, Appl. Phys. Lett. 40 (2) (1982) 178.]

In the past few years, this cumbersome and delicate device has been improved into a much simpler and sensitive apparatus – the 'tunnel effect microscope' which is now commonly used to explore and observe surfaces with unprecedented accuracy (fractions of an angström).

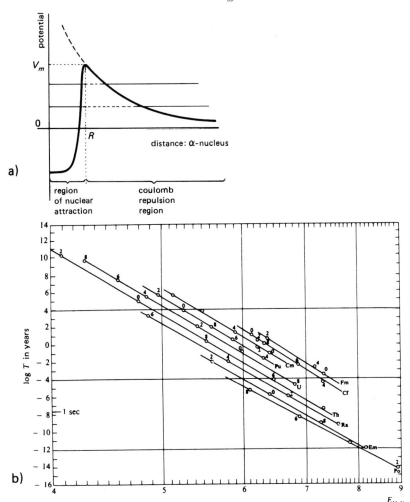

Fig. 6.28. *The Coulomb barrier in radioactivity*

(a) The helium nuclei, emitted by α-radioactive nuclei, have energies lower than the height of the Coulomb potential barrier which ought to hold them in. They 'go through' the barrier by the tunnel effect. The effective thickness of the barrier depends very sensitively on the energy. The same is true for the half-life τ of the α-radioactive nuclei.

(b) For the different isotopes of any given type of nucleus (Z fixed), there exists a linear relation between the logarithm of the half-life and the inverse of the square root of the energy of the emitted α-particles (the Gamow–Condon–Gurney law; see exercise 6.21). For each type of nucleus, the number corresponding to the experimental point is the last digit of the mass number and likewise indicates the corresponding isotope (example: the last point at the bottom right of the diagram corresponds to the ^{212}Po nucleus) [L. Valentin, Physique Subatomique (Hermann, Paris, 1975)].

lower than V_m! The emission is to be interpreted through the tunnel effect. The α-particle, assumed to exist (pre-formed)* in the interior of the nucleus (which is just a convenient picture), has an excitation energy of a few MeV, implying a velocity $v = 10^7$ m s^{-1}, or a little larger. Since the diameter of the nucleus is $2R \simeq 7$ F $= 7 \times 10^{-15}$ m, we see – using a classical picture – that the particle 'collides' with the inner wall of the potential well with a periodicity $\Delta t = 2R/v \simeq 7 \times 10^{-22}$ s. The α-particle has a non-zero probability of being transmitted to the other side of the barrier, instead of just bouncing back and forth and remaining confined to the interior of the well. This probability $\mathscr{T}$ is very small and can be estimated with the help of the general formula, eq. (6.5.22), adapted to the present situation (exercise 6.21). We could also say that the particle only has a $1/\mathscr{T}$ th chance of being transmitted. It will have to make, therefore, an average of τ^{-1} collisions with the wall of the well to come out. The average half-life of the nucleus is thus estimated by

$$\tau = \mathscr{T}^{-1} \Delta t. \tag{6.5.26}$$

Given the extremely small values of $\mathscr{T}$, we can understand how the above half-life could be very large, indeed. Experimentally, in the present case, it is found to be of the order of 10^{10} years. We also realize that the half-lives of α-radioactive nuclei display extreme sensitivity to the disintegration energy, since they range from 10^{10} years for ^{238}U, with $E = 4.3$ MeV, to 10^{-14} years (3×10^{-7} s) for ^{212}Po, with $E = 8.9$ MeV (fig. 6.28b). This is because the transmission probability $\mathscr{T}$ depends exponentially on the thickness a of the barrier, which in its turn increases very rapidly as the energy diminishes (fig. 6.28a) in the case of the Coulomb barrier. An application of formula (6.5.22) leads to a relationship between the energies of the emitted particles and the half-life of the nucleus – a formula which is remarkably well verified, given the simplicity of the model, which ignores, e.g., all details of nuclear structure (exercise 6.21). The tunnel effect in nuclear physics works backwards too, rendering possible the penetration of the potential barrier by charged particles outside the nucleus, whose kinetic energies would otherwise be too small, classically, to allow them to enter the nucleus. It is in this way that thermonuclear fusion reactions are rendered possible, which supply the energies of stars – and H-bombs.

---

* Parentheses added by translator.

6.6. The double well

There exist numerous physical situations where the potential, to which a quanton is subjected, possesses several minima and has the shape of a succession of wells. In the simplest case, we have a double well (fig. 6.29). A standard example is furnished by the molecule of ammonia, NH_3, in the situation where we are interested in the position of the nitrogen atom N, in relation to the plane defined by the three hydrogen atoms H. In general, one says the shape of the molecule is pyramidal (but we shall see by and by the rather subtle sense in which this assertion is to be understood). This means that the N atom is in equilibrium (wtih respect to the molecular binding forces) when it is at a fixed distance from the three H atoms, which form an equilateral triangle (fig. 6.30). Furthermore, this atom could be situated on

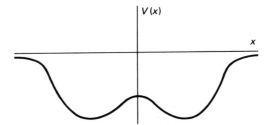

Fig. 6.29.   *A double-well potential.*

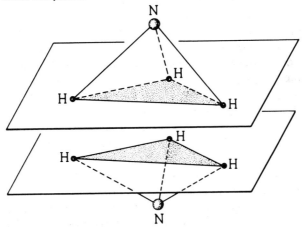

Fig. 6.30.   *The ammonia molecule*
For a given position of the hydrogen atoms (at the vertices of an equilateral triangle), there exist two positions of equilibrium for the nitrogen atom N, at the vertex of a pyramid, on one side or the other of the plane of the hydrogen atoms. The potential of N, as a function of its distance from the plane, has the shape of the curve in fig. 6.29.

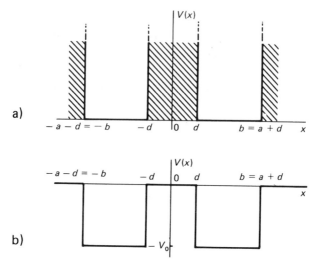

Fig. 6.31.   *Flat double well*
Two flat wells, of width a, are separated by an interval of width $2d$. We consider two cases:
(a) infinitely deep well,
(b) well with finite depth V_0.
(The origins of energies have been chosen differently in the two cases, for the same reasons of convenience as in sect. 6.2 and 6.4 of this chapter.)

either side of the plane of the hydrogen atoms. Thus, there exist *two* positions of equilibrium, symmetric with respect to the plane, and hence two minima for the potential function. The latter, expressed as a function of the distance from the plane, is described well by the shape of fig. 6.29. Let us study the behaviour of a quanton in such a potential with the help of the model of a piecewise flat potential, i.e., in a flat bottomed double well. To begin with, we shall consider the almost trivial case of an infinite double well (fig. 6.31a), then a finite double well (fig. 6.31b).

6.6.1. The infinite double well

The problem is to find wavefunctions, which would be harmonic – as for free quantons – inside the 'free' regions, where the potential is zero, and which would be identically zero on the outside (inside the regions of infinite potential), obeying the continuity condition at the edges of the well. Now, the wavefunctions of each well *separately* satisfy these conditions. Thus, the stationary states of the double well are simply given by the collection of the stationary states of the two wells! Hence, we have a double series of stationary states: those of the left well, having wavefunctions and energy proper values given by (see sect. 6.2)

ℏ
$$\begin{cases} -a-d<x<-d: & \varphi_n^{(\mathrm{L})}(x)=\sqrt{2/a}\,\sin\left(n\pi\dfrac{x+d}{a}\right), \\[2mm] \text{elsewhere:} & \varphi_n^{(\mathrm{L})}(x)=0, \end{cases}$$

$$E_n^{(\mathrm{L})}=n^2\frac{\pi^2}{2ma^2}, \quad n=1,2,\dots, \tag{6.6.1}$$

and those of the right well, with the corresponding expressions

ℏ
$$\begin{cases} d<x<a+d: & \varphi_n^{(\mathrm{R})}(x)=\sqrt{2/a}\,\sin\left(n\pi\dfrac{x-d}{a}\right), \\[2mm] \text{elsewhere:} & \varphi_n^{(\mathrm{R})}(x)=0, \end{cases}$$

$$E_n^{(\mathrm{R})}=n^2\frac{\pi^2}{2ma^2}, \quad n=1,2,\dots. \tag{6.6.2}$$

The energy spectrum of the double well is the same as that of the single well, except *for a degeneracy of degree two*, since

$$E_n^{(\mathrm{L})}=E_n^{(\mathrm{R})}. \tag{6.6.3}$$

In other words, every level is a *double* level.

We make at this point the following fundamental observation: since $\varphi_n^{(\mathrm{R})}$ and $\varphi_n^{(\mathrm{L})}$ are wavefunctions corresponding to two stationary states *of the same energy*, any linear combination of these two functions again defines a stationary state! Suppose indeed that we have to find a time-dependent wavefunction $\psi(x;t)$ for which, at the instant $t=0$, we have

$$\psi(x;0)=\alpha\varphi_n^{(\mathrm{L})}(x)+\beta\varphi_n^{(\mathrm{R})}(x). \tag{6.6.4}$$

Its temporal evolution is immediately obtained by inserting in this expansion the appropriate harmonic evolution factor $\mathrm{e}^{-\mathrm{i}Et}$, before each stationary state function, i.e.,

ℏ
$$\psi(x;t)=\alpha\,\mathrm{e}^{-\mathrm{i}E_n^{(\mathrm{L})}t}\,\varphi_n^{(\mathrm{L})}(x)+\beta\,\mathrm{e}^{-\mathrm{i}E_n^{(\mathrm{R})}t}\,\varphi_n^{(\mathrm{R})}(x). \tag{6.6.5}$$

However, in virtue of the degeneracy given by eq. (6.6.3), this temporal factor is common and we may write

$$\hbar \quad \begin{cases} \psi(x; t) = e^{-iE_n t}[\alpha\varphi_n^{(L)}(x) + \beta\varphi_n^{(R)}(x)] \\ \quad = e^{-iE_n t}\psi(x; 0). \end{cases} \tag{6.6.6}$$

The wavefunction $\psi(x; 0)$ has, therefore, a harmonic evolution and describes a stationary state of the same energy as well. We have here a general property. Any linear combination of wavefunctions belonging to a degenerate energy level is again a wavefunction for this level. In other words, *the wavefunctions of a degenerate energy level form a vector space of dimension equal to the degree of degeneracy.*

This remark enables us to understand the relationship between the wavefunctions $\varphi_n^{(L)}$ and $\varphi_n^{(R)}$ and two other wavefunctions, which we can introduce, exploiting the spatial symmetry of the problem. This symmetry

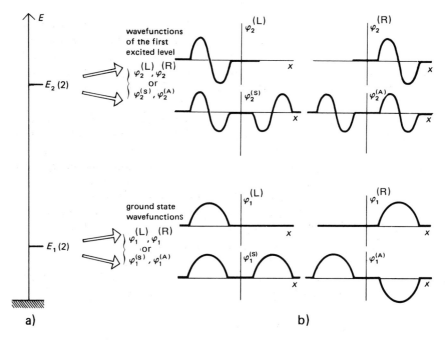

Fig. 6.32. *Stationary states of the infinite double well*
(a) The energy spectrum (showing the degeneracy of order 2).
(b) The wavefunctions of the first two levels.

actually leads us to consider the stationary wavefunctions of fixed parity (see sect. 4.1). These are, respectively, the symmetric and antisymmetric functions,

$$
\left\{
\begin{array}{l}
\varphi_n^{(S)}(x) = \dfrac{1}{\sqrt{2}} [\varphi_n^{(L)}(x) - \varphi_n^{(R)}(x)], \\[4mm]
\varphi_n^{(A)}(x) = \dfrac{1}{\sqrt{2}} [\varphi_n^{(L)}(x) + \varphi_n^{(R)}(x)],
\end{array}
\right.
\tag{6.6.7}
$$

which are both stationary state wavefunctions of the energy E_n. We observe that $\varphi_n^{(L)}$ and $\varphi_n^{(R)}$ on the one hand, and $\varphi_n^{(S)}$ and $\varphi_n^{(A)}$ on the other, form two bases for the two-dimensional vector space of the wavefunctions corresponding to the level E_n (fig. 6.32): each wavefunction for this level is a linear combination of $\varphi_n^{(L)}$ and $\varphi_n^{(R)}$ or, equally, of $\varphi_n^{(S)}$ and $\varphi_n^{(A)}$.

6.6.2. The finite double well

The situation is more complicated now; the solution to the problem does not reduce to a simple union of the results obtained for each individual finite well. The reason for this is mathematically evident: the wavefunctions of the bound states of one of the wells (the one on the left, say) are *not* admissible since they do not have the correct behaviour in the region to the right, where they display a real exponential dependence, instead of the required harmonic dependence. The physical explanation of the phenomenon is fundamental: a quanton cannot remain confined to one of the wells *because of the tunnel effect*! The barrier which separates the two wells is quantically not impermeable, and a quanton localized initially in one of the wells may later, with a non-zero probability, find itself in the second. Needless to say, it is certainly not in a stationary state. The wavefunctions of the actual stationary states of the double well cannot be obtained directly using those of isolated wells. It is necessary for us to call upon our computational arsenal once again. We can – and should(!) – moreover, exploit the symmetry of the problem, made clear by our choice of axes (fig. 6.31b). First, let us consider the stationary wavefunctions of the symmetric states. Using a henceforth systematic procedure, we can immediately write these down in the form

$$
\left\{
\begin{array}{ll}
0 < x < d: & \varphi^{(S)}(x) = A \cosh \kappa x, \\[2mm]
d < x < b: & \varphi^{(S)}(x) = B \sin p(b-x) + C \cos p(b-x), \\[2mm]
b < x: & \varphi^{(S)}(x) = D\, e^{-\kappa x},
\end{array}
\right.
\tag{6.6.8}
$$

$\hbar$

with $b = a + d$, and where we have, as usual, set

$$\kappa^2 = -2mE, \qquad p^2 = 2m(V_0 + E) \quad (-V_0 < E < 0). \tag{6.6.9}$$

The form 'cosh' of the central wavefunction displays its symmetry explicitly. The expression for the wavefunction for $x < -d$ is deduced by symmetry from the expression for $x > d$. The combination of the complex exponentials e^{ipx} and e^{-ipx}, in the region inside the well has been put into a particularly convenient (general) form. It remains to impose the continuity conditions on $\varphi^{(S)}$ and its derivatives at $x = d$ and $x = b$. These lead, following the standard procedure, to the relation

$$\tan pa = -\frac{p}{\kappa} \frac{1 + \tanh \kappa d}{\tanh \kappa d - p^2/\kappa^2} \quad \text{(symmetric)}, \tag{6.6.10}$$

which implicitly defines the energy proper values, or in other words, is the quantization condition. The computation of the antisymmetric wavefunctions is immediate. It is enough to set

$$\hbar \qquad \begin{cases} 0 < x < d: & \varphi^{(A)}(x) = A \sinh \kappa x, \\ d < x < b: & \varphi^{(A)}(x) = B \sin p(b - x) + C \cos p(b - x), \\ b < x: & \varphi^{(A)}(x) = D\, e^{-\kappa x}, \end{cases} \tag{6.6.11}$$

in other words, to replace 'cosh' by 'sinh', hence 'tanh' by 'cotanh', which leads to

$$\hbar \qquad \tan pa = -\frac{p}{\kappa} \frac{1 + \cotanh \kappa d}{\cotanh \kappa d - p^2/\kappa^2} \quad \text{(antisymmetric)}. \tag{6.6.12}$$

It is instructive to consider conditions (6.6.10) and (6.6.12) in the limit where the barrier separating the two wells is infinitely thick. As $d \to \infty$, $\tanh \kappa d \to 1$ and *both* conditions reduce to the quantization condition, eq. (6.4.17), for an isolated well! In other words, the two series of energy levels merge: there is degeneracy of order two. Now, the barrier can no longer be crossed by means of the tunnel effect and we recover a situation analogous to that of the infinite wells (except that now it is the thickness, and not the height, of the barrier which, being infinite, renders it opaque).

In the general case, the quantization condition inside an isolated well is replaced by the two conditions given by eqs. (6.6.10) and (6.6.12). In other words, each level of the isolated well, of energy E_n, is replaced by two levels, of energies $E_n^{(A)}$ and $E_n^{(S)}$, corresponding, respectively, to the 'antisymmetric' stationary state, having wavefunction $\varphi_n^{(A)}$, and the 'symmetric' stationary state, having wavefunction $\varphi_n^{(S)}$. The degeneracy, of order two, observed in the very special case in which the two wells are unable to communicate with each other (infinitely thick barrier), now loses its validity because of the tunnel effect (we say that the tunnel effects 'lifts' the degeneracy).

If the barrier is very thick, so that $\kappa d \gg 1$, we get the approximate expression $\tanh \kappa d \simeq 1 - 2\,e^{-\kappa d}$. Thus, the two conditions given by eqs. (6.6.10) and (6.6.12) differ from condition (6.4.17) by the terms involving $e^{-2\kappa d}$ and the energy levels $E_n^{(A)}$ and $E_n^{(S)}$ differ from E_n by a correction of order $e^{-2\kappa d}$. In the present case, explicit computation shows that the lifting of the degeneracy is accomplished by a symmetrical splitting of E_n (exercise 6.23). In other words, we have

$$\begin{cases} E_n^{(A)} = E_n + \tfrac{1}{2}\delta E_n, \\ E_n^{(S)} = E_n - \tfrac{1}{2}\delta E_n, \end{cases} \tag{6.6.13}$$

the difference δE_n between the two levels being given, in order of magnitude, by

ℏ
$$\delta E_n \simeq \varepsilon\,e^{-2\kappa d}, \tag{6.6.14}$$

where ε is an energy related to the characteristics of the well and to the level. Evidently, it is the exponential factor in eq. (6.6.14), related to the tunnel effect, which is decisive.

Figure 6.33 summarizes the results obtained so far. It should be compared to fig. 6.32. Once again we obtain results of a general nature.

– an alternation of levels corresponding to symmetric and antisymmetric ('even' and 'odd' levels) wavefunctions;

– even (symmetric) nature of the ground state;

– existence of k zeroes for the wavefunction of the kth excited level ($k = 0$ corresponding to the ground level).

Finally, let us consider the wavefunctions

$$\begin{cases} \varphi_n^{(L)}(x) = \dfrac{1}{\sqrt{2}}[\varphi_n^{(S)}(x) + \varphi_n^{(A)}(x)], \\[2mm] \varphi_n^{(R)}(x) = \dfrac{1}{\sqrt{2}}[\varphi_n^{(S)}(x) - \varphi_n^{(A)}(x)], \end{cases} \tag{6.6.15}$$

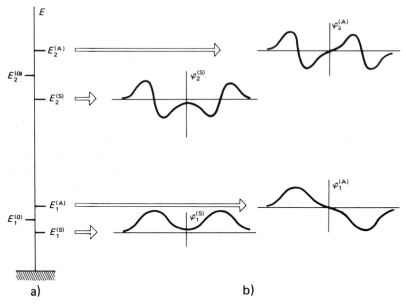

Fig. 6.33. *The stationary states of the finite double well*
(a) The energy spectrum. The levels $E_n^{(0)}$ ($n = 1, 2$) are those of the isolated, finite flat well. The levels $E_n^{(S)}$ and $E_n^{(A)}$ correspond to the stationary states of the symmetric and antisymmetric wavefunctions.
(b) The wavefunction of the first two stationary states.
Compare with fig. 6.32.

linear combinations of the wavefunctions $\varphi_n^{(S)}$ and $\varphi_n^{(A)}$ of the stationary states of energies $E_n^{(S)}$ and $E_n^{(A)}$. These wavefunctions, $\varphi_n^{(R)}$ and $\varphi_n^{(L)}$, are very close to the wavefunctions which would correspond to the stationary states of energy $E_n^{(0)}$ of each one of the two wells (left and right), considered in isolation (fig. 6.34). They describe, therefore, the states in which the quanton is essentially well-localized inside the left or right well. These states are *not* stationary. Indeed, let $\psi(x; t)$ be the time-dependent wavefunction, such that at the initial instant of time one has

$$\hbar \qquad \psi(x; 0) = \varphi_n^{(L)}(x) = \frac{1}{\sqrt{2}} [\varphi_n^{(S)}(x) + \varphi_n^{(A)}(x)]. \qquad (6.6.16)$$

Its time evolution, using the general theory, is given by

$$\hbar \qquad \psi(x; t) = \frac{1}{\sqrt{2}} [e^{-iE_n^{(S)}t} \varphi_n^{(S)}(x) + e^{-iE_n^{(A)}t} \varphi_n^{(A)}(x)], \qquad (6.6.17)$$

which in turn, by eq. (6.6.13), can be written as

$\hbar$

$$\psi(x;t) = \frac{1}{\sqrt{2}} e^{-iE_n^{(0)}t} [e^{i\delta E_n t/2} \varphi_n^{(S)}(x) + e^{-i\delta E_n t/2} \varphi_n^{(A)}(x)]$$

$$= e^{-iE_n^{(0)}t} [\cos(\delta E_n t/2) \varphi_n^{(L)}(x) + i \sin(\delta E_n t/2) \varphi_n^{(R)}(x)]. \qquad (6.6.18)$$

We see that ψ oscillates between $\varphi_n^{(L)}$ and $\varphi_n^{(R)}$, inducing an oscillation in the probability of the presence of the quanton inside the left and the right well, with a pulsation δE_n. Of course, we get here once more a quantum beat phenomenon – a particular case of the general situation considered in ch. 5, sect. 5.1.3. In other words, returning to our physical system, the geometrical form of NH_3 is determined only up to a symmetry with respect to the plane of the hydrogen atoms! In its ground state, of energy $E_1^{(S)}$, the NH_3 molecule does not have a definite shape (a 'proper' shape), but rather, is in a state of superposition of the two symmetric configurations of fig. 6.30.

Expression (6.6.18) of the time-dependent wavefunction $\psi(x;t)$ enables us, in addition, to evaluate generally the difference δE_n between the two energy levels, making clear its connection with the tunnel effect. First of all, let us give a meaning to the coefficients $\cos(\delta E_n t/2)$ and $i \sin(\delta E_n t/2)$ in eq. (6.6.18).

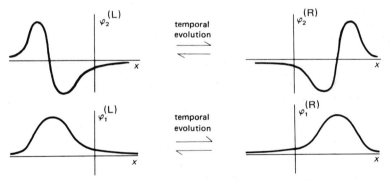

Fig. 6.34. *The localized states of the finite double well*
The wavefunctions

$$\varphi_n^{(L),(R)} = \frac{1}{\sqrt{2}} [\varphi_n^{(S)} \pm \varphi_n^{(A)}],$$

of the finite double well, for $n = 1, 2$. These wavefunctions correspond to the non-stationary states, and they both evolve with a periodicity T_n related to the difference of the corresponding energy levels, $\delta E_n = E_n^{(A)} - E_n^{(S)}$, by $T_n = 2\pi/\delta E_n$.

Very simply, they are the probability amplitudes for finding the quanton to the left or the right, without making precise the exact point. To convince ourselves of this, it is enough to look at eq. (6.6.18), in conformity with the general rules for the computation of amplitudes given in ch. 4, in the form

$\hbar$ $\langle$quanton localized at $x | \psi \rangle_t$

$$= \underbrace{\langle\text{quanton at } x | \text{quanton to the left}\rangle}_{e^{-iE_n t}\, \varphi_n^{(L)}(x)} \underbrace{\langle\text{quanton to the left} | \psi\rangle_t}_{\cos(\delta E_n t/2)}$$

$$+ \underbrace{\langle\text{quanton at } x | \text{quanton to the right}\rangle}_{e^{-iE_n t}\, \varphi_n^{(R)}(x)} \underbrace{\langle\text{quanton to the right} | \psi\rangle_t}_{i \sin(\delta E_n t/2)}.$$

(6.6.19)

Thus, the amplitude $\langle$quanton to the right$| \psi \rangle_t$ of localization to the right, i.e., the term $i \sin(\delta E_n t/2)$, increases linearly with time, starting at $t = 0$, as long as $\delta E_n t \ll 1$. Indeed,

$\hbar$ $\langle$quanton to the right$| \psi \rangle_t \simeq \frac{1}{2}\delta E_n t$ (t small),   (6.6.20)

up to a phase factor. Next, in the quasi-stationary state having wavefunction $\varphi_n^{(L)}(x)$, let us consider the quanton inside the left well. Classically, a particle so bound would have an oscillatory motion inside the well, bouncing back and forth between the walls. Quantically, we may consider an analogous motion – that of a wave packet – constructed from a superposition of several stationary wavefunctions, close to the one for the level E_n being considered [this would amount to a generalization of the considerations of example (6.2.14) and exercise 6.1]. In such a state, the quanton oscillates inside the well, with a frequency given by the energy dispersion of the wave packet. This dispersion is of the same order of magnitude as the average difference between the energy levels around the mean energy E_n of the wave packet (the wave packet is formed by the superposition of the wavefunctions of these levels). Let ΔE_n be this average difference. Then, the quanton 'encounters' the barrier with this same frequency ΔE_n. At each one of these encounters, a small proportion of the amplitude is transmitted, via the tunnel effect to the well on the right. This proportion is of order $|A_t| = \sqrt{\mathcal{T}}$, where A_t and $\mathcal{T}$ are, respectively, the amplitude and the coefficient of transmission through the barrier. After a lapse of time t, sufficiently long compared to the period $2\pi/\Delta E_n$, there will have been $2 \Delta E_n t/2\pi$ encounters, and the amplitude of localization to the right would, therefore, be of order

$\hbar$ $\qquad$ $\langle \text{quanton to the right} | \psi \rangle_t \simeq \dfrac{1}{2\pi} \Delta E_n t | A_t |.$ $\qquad$ (6.6.21)

Comparing expressions (6.6.20) and (6.6.21), as is permissible for times t for which $1/\Delta E_n \ll t \ll 1/\delta E_n$, we obtain

$$\frac{\delta E_n}{\Delta E_n} \simeq \frac{1}{\pi} | A_t | = \frac{1}{\pi} \sqrt{\mathscr{T}}, \qquad (6.6.22)$$

which demonstrates the direct connection between the splitting of the levels δE_n and the transmission factor due to the tunnel effect $\mathscr{T}$. Taking account of the approximate expression, eq. (6.5.15), for the transmission factor, we recover exactly the estimate (6.6.14) [remember that here the width of the barrier is $a = 2d$ and that the factor of π in eq. (6.6.22) does not modify the orders of magnitude]. Expression (6.6.22) is, however, much more general, and does not depend on the specific form of the barrier through which the tunnelling proceeds. Thus, it allows us to estimate the separation between the doubly split levels of an arbitrary double well, e.g., by evaluating the transmission factor $\mathscr{T}$ using the approximate general formula given by (6.5.22).

6.6.3. Orders of magnitude. Localization of quantons

It is necessary once more, in order to appreciate the physical importance of these considerations, to have an idea of the order of magnitude of the separation δE_n of the energy levels, induced by the existence of a finite barrier between the two potential wells. If the time $\tau = 1/\delta E_n$, characterizing the rate with which the configuration evolves, is very large – a situation which corresponds to a very narrow separation δE_n, and hence to a very thick barrier – we may, for all practical purposes, consider the localized states $\varphi^{(L)}$ and $\varphi^{(R)}$ as being stationary: a quanton localized in one of the wells remains there – approximately. This is not the case for the NH_3 molecule, where the separation of the levels δE_1 and the characteristic times have the approximate values

$$NH_3: \quad \delta E_1 \cong 10^{-4} \text{ eV} = 24\,000 \text{ MHz}, \quad \tau \cong 4 \times 10^{-11} \text{ s}.$$

(Note that the two levels $E_1^{(A)}$ and $E_1^{(S)}$ are the levels on which the ammonia maser operates.) But a closely related molecule, AsH_3, leads to the values

$$AsH_3: \quad \delta E_1 \cong 0.8 \times 10^{-22} \, eV = 2 \times 10^{-8} \, Hz, \quad \tau \cong 5 \times 10^7 \, s \cong 2 \text{ years}.$$

We encounter here the extreme sensitivity of the phenomenon of the tunnel effect to tiny variations in the physical parameters – a sensitivity due to the presence of the exponential factor in eq. (6.6.14) giving the separation δE_n between the levels (see exercise 6.25).

Another example is furnished by similar considerations, explaining the effective stability of optical isomers. This is the name given to molecules which can exist in two symmetric configurations having opposite optical activities – one form being levorotatory and the other dextrorotatory. The molecules 'with asymmetric carbon atoms', such as those of CHClFBr, provide us with an example of such optical isomers (fig. 6.35). Exactly as with the NH_3 molecule, these two configurations correspond to the two symmetric minima of the potential energy of the molecule, separated by a potential barrier. *Strictu senso*, neither one of these configurations corresponds to a stationary state and neither is stable: an initially levorotatory state should, in principle, transform itself into a dextrorotatory state, then return to its initial form, etc., with the periodicity $2\pi/\delta E$, related to the separation between the symmetric and antisymmetric levels. However, in the case in which we are interested, the barrier is so thick, δE so small and the period so long, that the optical isomers may be considered as being stable.

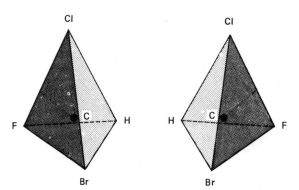

Fig. 6.35. *Optical isomers*
The CHClFBr molecule, containing an asymmetric carbon, possesses two configurations of minimum potential, symmetric with respect to one another. The corresponding states are, in principle, not stationary, although practically they are almost so.

More generally, these ideas enable us to discuss the question of the localization of a quanton in any physical situation in which there are several sites (regions of minimum potential), separated by classically inaccessible regions (potential barriers). As an example of great current interest, we might wish to know if it would be permissible to consider the electrons in a given material as being bound to particular atoms, or if, on the contrary, they have a diffuse distribution throughout it (delocalized electrons). In the light of what has been said before, it is clear that the greatest possible delocalization exists in the very stationary states of a complex system. But the crucial question is that of the transit time from one site to the other, the value of which – large or small – determines the nature – whether approximately localized or not – of the actual states. This characteristic time can be evaluated using expression (6.6.22) for the splitting of the levels, since the difference δE induced by the tunnel effect governs, as we have seen [eq. (6.6.20)], the rapidity of the passage from one well to the other. Here, the average difference ΔE between the levels of a Coulomb well is found to be of the same order as the binding energy E_0 inside this well [see eq. (3.3.23)]. Likewise, the potential barrier may be roughly approximated (see fig. 6.36) by a flat barrier of height E_0, beyond the levels being considered, and having a width of the same order as the distance R between the sites. Thus, using expression (6.5.15) for the transmission factor, where $\kappa' = (2mE_0)^{1/2}$, we obtain

$\hbar$

$$\delta E \simeq E_0\, e^{-(2mE_0)^{1/2}R}. \qquad (6.6.23)$$

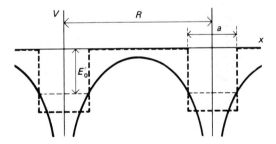

Fig. 6.36. *The double Coulomb well*
In order to study the tunnel effect between two atomic levels, we replace the double Coulomb well by a double flat well, the height of the barrier above the level considered being equal to the binding energy E_0.

Introducing a characteristic dimension a, corresponding to the energy E_0,

$\hbar$
$$E_0 = \frac{1}{2ma^2}, \tag{6.6.24}$$

we may also write

$$\delta E \simeq E_0 \, e^{-R/a}, \tag{6.6.25}$$

and, of course, for the characteristic transit time,

$$\tau = \hbar/\delta E. \tag{6.6.26}$$

Let us apply now these estimates to the following three cases:

(1) *Valence electrons in molecules or metals:* We take $E_0 \simeq 4\,\text{eV}$, so that $a \simeq 1\,\text{Å}$ and $R \simeq 2\,\text{Å}$ (interatomic distance). Then, $\delta E \simeq$ a few tens of eVs, hence $\tau \simeq 10^{-15}$ to 10^{-14} s. The characteristic time scale is just that of atomic and molecular times, which is so short that the tunnel effect has full play, and hence the valence electrons in a molecule or a metal cannot be considered as being localized to any given atomic site. Quite the contrary, their delocalization is the very key to molecular binding and the properties of metals, as we shall soon see.

(2) *Electrons in the deep atomic shells*, again in molecules or metals: We take $E_0 \simeq 10^4\,\text{eV}$ and $a \simeq 10^{-2}\,\text{Å}$ (remember that for the deeper shells of an atom, of atomic number Z, the energy varies as Z^2 and the radius as Z^{-1}), again with $R \simeq 2\,\text{Å}$. Thus we get, $\delta E \simeq 10^{-83}\,\text{eV}$, hence $\tau \simeq 10^{68}$ s which is 10^{60} years.... At this scale, we may therefore consider the electrons of the deep atomic shells as being completely *intra*-atomic, bound to 'their' molecules and localized to definite sites. Only the valence electrons need, therefore, be considered in *inter*-atomic processes, i.e., those of the physics of solids (see following section), of chemistry and of biochemistry.

(3) *Electrons of the atoms (or molecules) of a gas.* Let us take the surface electrons, for which [as in (1)], $E_0 \simeq 4\,\text{eV}$, $a \simeq 1\,\text{Å}$, but with $R \simeq 30\,\text{Å}$ – the average distance between two molecules of a gas under normal conditions $[R^3 = 22.4\,\text{litres}/(6 \times 10^{23})]$. We now get, $\delta E \simeq 10^{-12}\,\text{eV}$, hence $\tau \simeq 10^{-3}$ s, a time much too long on the scale of molecular collisions in the gas (of the order of 10^{-10} s). This amounts to saying that the electrons, even the most

superficial (and *a fortiori* the deepest) ones, in a gas, may be considered as being firmly attached to their respective atoms or molecules. Here, it is the thickness of the potential barriers (and not their heights, unlike in the preceding case), which renders them opaque and allows us to consider the electrons as though they were localized.

6.6.4. The quantum binding: exchange forces

We can summarize the essential results, obtained in our study of the double well, as follows: the ground state of a quanton inside a double well has an energy lower than that of its ground state when inside one or the other of the wells in isolation. This lowering of the energy, δE, from the level $E^{(0)}$, of the isolated well to the symmetric level $E^{(S)}$ of the double well, results from the tunnel effect. It allows us to consider the quanton, because of its delocalization, as though it were 'simultaneously' inside both wells. Thus, it wins... or rather loses... on both counts (of energy). The closer the two wells, the lower is, therefore, the energy of the total system consisting of the quanton and the wells. In other words, the energy difference δE plays the role of the potential energy of an interaction between the two wells, or, in more physical terms, between the two physical entities in interaction with the quanton. The 'exchange' as it is often called, or more correctly, the sharing of the quanton, through the tunnel effect, brings about in this manner an interaction between the two entities – an interaction which is collectively mediated by the quanton. This effect is often described by the epithet of the 'exchange force'. Let us give a few examples.

(a) *Chemical bonding.* Consider two protons, a distance R apart, and an electron. This electron feels the effect of a double Coulomb potential, typically of the form illustrated by fig. 6.36. If the distance R is very large (compared to the Bohr radius), we may consider the states in which the electron is approximately localized to one of the protons, as being quasi-stationary: hence the system consists of a hydrogen atom and an isolated proton – in two possible ways. These states have an energy $E^{(0)} = -13.6\,\text{eV}$, the binding energy of the hydrogen atom. But, as we have seen, a true stationary state corresponds to an unlocalized electron and its sharing between the two protons, within a symmetric state of the wavefunction. The lowering of the energy, $\delta E(R)$, which grows exponentially as R diminishes, corresponds to an attractive potential between the two protons, which becomes appreciable as R approaches a few Å, as we have seen above. Thus, there develops an attraction between the protons through the sharing of the electron, and the formation of a bound system: this is the molecular ion H_2^+, the simplest of polyatomic systems and the prototype of molecules (fig. 6.37). However, as R decreases further, the Coulomb repulsion between the protons manifests itself and ends up dominating the exchange attraction (fig. 6.38). Thus, there exists a minimum of the potential, which corresponds to the stable configuration of the molecular ion H_2^+. Evidently, the sharing of the electron in the antisymmetric state raises the energy and leads to a repulsive interaction. A similar reasoning can be made for more complex molecules –

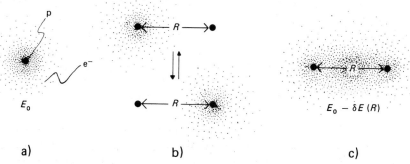

a)   b)   c)

Fig. 6.37.   *The H_2^+ molecular ion*
(a) An isolated hydrogen ion, of energy $E_0 = -13.6$ eV.
(b) When two protons are at a finite distance R, the two states, formed from an electron bound to one of the protons (that is, atom + isolated proton) are non-stationary states of energy E_0.
(c) The ground state of the system is the state in which the electron is symmetrically shared by the two protons. Its energy is $E_0 - \delta E(R)$.

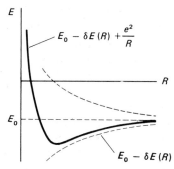

Fig. 6.38.   *The molecular interaction potential*
The energy of the system formed by the electron and the two protons as a function of their distance R, contains an attractive exchange term $\delta E(R)$ and the Coulomb repulsion term e^2/R.

with important modifications due to the presence of several electrons. Let us remember the fundamental idea that chemical bonding is due to the sharing of electrons between atoms, which brings about, through the effect of the delocalization, a lowering of the potential energy and hence an attraction.

This explanation of chemical bonding also helps us understand an important point, namely, the order of magnitude of the energies which give rise to it. Indeed, it is not uncommon for certain types of molecular or metallic bondings to have binding energies much lower – of the order of an electron volt or even a few tens of eVs – than what might be expected on purely dimensional grounds (see ch. 1, sect. 1.4.1). The point is that the energy differences, corresponding to the binding of two atomic or molecular structures by the tunnel effect, are given by expressions such as eq. (6.6.23), where the critical exponential factor could easily reach the value 10^{-2} or lower, thus giving binding energies which could approach tens or hundreds of electron volts [see case (1) in sect. 6.6.3]. In this way we understand the full richness of atomic physics, encompassing at once the rather strong (intra-atomic) as well as the fairly weak (interatomic) bindings.

(b) *Nuclear forces.* Exactly the same mechanism explains the origin of nuclear forces, responsible for the binding between nucleons (neutrons and protons) and, hence, for the existence of atomic nuclei. The exchange, or rather the sharing, is now of various types of mesons, and in the first instance, of the lightest one among them – the π meson or pion. The potential barrier which the pion has to cross to go from one nucleon to another is just its own mass energy! As a matter of fact, the conservation of energy forbids the creation of a 'real' pion *ex nihilo* from a solitary nucleon. An additional amount of energy, $V_0 = mc^2$ (at least) is necessary to create such a pion. But we may consider this energy as being the height of a barrier, which separates the two nucleons, and then invoke the tunnel effect to enable the pion to go through it, with a total energy equal to zero, i.e., with a negative kinetic energy $-V_0 \simeq mc^2$ (fig. 6.39). Its momentum p, which has now to be computed relativistically, and given by $(p^2 c^2 + m^2 c^4)^{1/2} = 0$, is, therefore, pure imaginary, $p = i\kappa$ with $\kappa = me$. The potential barrier, being of the rectangular type, may be treated using the elementary theory given above. For a distance R between the two nucleons (width of the barrier), the attractive potential due to the pion exchange would be $\delta E(R) \propto e^{-\kappa R}$, if the system were spatially one dimensional. In reality, we have to take into account the three-dimensionality of the space. For this we need to replace, in the theoretical analysis, the unidimensional waves e^{ipx} by the 'spherical waves', propagating away from an emitting nucleon, and having amplitude

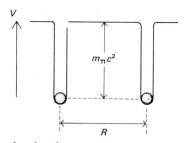

Fig. 6.39. *Pion exchange and nuclear forces*
By supplying an amount of energy $m_\pi c^2$ a π-meson can be extracted from a nucleon. In other
words, for a π-meson a nucleon is a potential well of depth $m_\pi c^2$. Two nucleons constitute a
double well, for which the ground state of the pion corresponds to its being shared, and leads to
an attraction between the two nucleons.

e^{ipr}/r, where r is the radial distance from the nucleon (see ch. 5, sect. 5.2.4).
Thus, lowering of the energy due to the sharing of the pion is now of the form
$\delta E(R) \simeq e^{-\kappa R}/R$. Again we say that there exists an attractive force between
the two nucleons, due to the 'exchange' of the pion and characterized by a
potential $V_Y(R) \simeq e^{-\kappa R}/R$. This potential is called the Yukawa potential, after
the Japanese physicist who, in fact, in 1935, had predicted the existence of the
pion on the basis of the properties of nuclear forces. The 'range' of this
potential, which is the characteristic distance beyond which it becomes
negligible, is determined by its exponential decrease. It is of the order $a \simeq 1/\kappa$
which, upon reintroducing the dimensional constants, yields $a \simeq h/mc$, in
other words, the Compton wavelength of the pion. We see how the present
approach makes more precise something that we had done earlier (ch. 3, sect.
3.3.1), based upon the Heisenberg inequalities, to describe the same phenomenon. In particular, we see clearly that if the mediating quanton has zero
mass, so that $\kappa = 0$, the interaction potential is simply $V_C(R) \propto -1/R$, and we
understand how the Coulomb potential is related to the mass of the photon
being zero. Finally, let us indicate that, in general, the Yukawa potential is
written in the form $V_Y(R) = g^2 e^{-\kappa R}/R$, the constant g^2 playing here the role
of the constant e^2 in the case of the Coulomb potential, $V_C(R) = -e^2/R$. Just
as e^2 is the 'coupling constant' of the electromagnetic interaction, characterizing the strength of this interaction, g^2 is a 'coupling constant' of the nuclear
interactions (see ch. 1, sect. 1.4.3, and exercise 3.12).

(c) *Virtual quantons.* This exchange mechanism, via the tunnel effect, in
which certain quantons are the mediators, or vectors, for the forces of

interaction between the other quantons, is very general. These intermediate quantons are often called 'virtual', in view of their 'abnormal' energy properties, as opposed to 'real' quantons, which can be observed and manipulated experimentally. This distinction is somewhat formal: once again, it is a classical habit to consider these 'virtual' quantons as having properties which are 'not real'. The quantum theory clearly demonstrates the possibility of the existence of these non-classical states. Moreover, from the empirical point of view itself, the distinction is hard to sustain. From the exchange of a pion between two nucleons, the 'virtual' existence of which lasts 10^{-24} s, imparting to it a large negative kinetic energy – of the order of its mass – to the exchange of an electron between two protons in the H_2^+ ion, with negative kinetic energies of a few electron volts, right up to quantons considered as being 'real' which, to all accounts, have no more than a transitory existence between the accelerator which produces and the counter which absorbs them, there is no absolute distinction possible.

Let us conclude by emphasizing the specifically quantum nature of the 'exchange forces'. The crucial term in the exchange potential is the exponential factor

$$V_{\text{exch}}(R) \simeq e^{-\kappa R/\hbar}, \tag{6.6.27}$$

where we have explicitly displayed the Planck's constant to indicate, not just its presence, but also the highly singular dependence of the exchange potential on it: V_{exch} is a non-analytic function of $\hbar$ at '$\hbar = 0$'. In other words, the classical theory is a *singular* limit of the quantum theory. It should come as no surprise that the quantum theory should display characteristics that seem so strange compared to the classical theory!

6.7. The battlement-like potential. Theory of band structure

Let us consider now the more complicated case of a potential formed by a regular alternation of finite flat wells, or – of barriers. We shall call this the 'battlement potential' (or a potential of 'crenelles and merlons'). The barriers are considered to be identical, of width a and height V_0. They are separated by equal distances, of length b and zero potential. The battlement potential is, therefore, a periodic potential, of period

$$l = a + b, \tag{6.7.1}$$

and it is, in this respect, that its study has more than just academic interest for us. In fact, the main result which we are about to obtain – namely, the existence of energy *bands* for which the quanton propagates as though it were actually free – holds for any *periodic* potential and does not depend on the detailed form of the potential within each period. Now, it is just this periodic nature which characterizes the potential acting on an electron inside a crystal or at least (since we shall limit ourselves to a one-dimensional problem) along a chain of atoms. The battlement potential of fig. 6.40 should be considered as being a schematic representation of the potential which acts on an electron inside a solid, made up of a single species of atoms. The regularly spaced ions in the chain define as many potential wells, into which the electron might 'drop' (to form a neutral atom), but out of which it might also escape... by means of the tunnel effect.

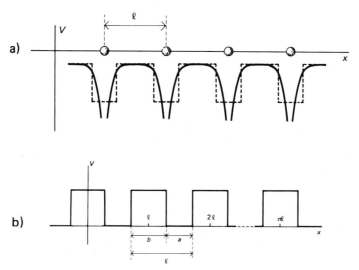

Fig. 6.40. *A periodic potential*
(a) A Coulomb type potential, to which an electron, moving along a chain of atoms is subjected, can very roughly be schematized by a battlement potential. The main results of the theory are consequences of the periodicity of the potential and do not depend in principle on the exact form of the potential.
(b) A potential of crenelles and merlons.

6.7.1. *The iteration matrix*

Our problem is thus to study the stationary states of a quanton, of energy E (assuming that $0 < E < V_0$), inside the potential $V(x)$ of fig. 6.40b, in which the origin of the x-axis has been chosen to coincide with the centre of a barrier, so that the nth barrier is centered at the point $x = nl$ on the abscissa. In order to extend the results of sect. 6.5 (for an isolated barrier, centered at the origin) to the chain of barriers of the battlement potential (centered at different points along the x-axis), it is useful to introduce a slight modification into the notation of the stationary-state wavefunction $\varphi_E(x)$, of energy E. Let us agree to denote by $\varphi_n(x)$ the wavefunction which coincides with this wavefunction inside the crenelle *to the left* of the nth barrier and to write $\varphi_n(x)$ in the form

$\hbar$
$$\varphi_n(x) = A_n\, e^{ip(x-nl)} + B_n\, e^{-ip(x-nl)}, \tag{6.7.2}$$

for

$$(n-1)l + \tfrac{1}{2}a \leqslant x \leqslant nl - \tfrac{1}{2}a, \tag{6.7.3}$$

where we recall that

$$p = (2mE)^{1/2}. \tag{6.7.4}$$

Here, the real coefficients A_n and B_n are to the nth barrier what the coefficients b_+ and b_-, used in eq. (6.5.5), were to the isolated barrier.

Similarly, within the interval

$$nl + \tfrac{1}{2}a \leqslant x \leqslant (n+1)l - \tfrac{1}{2}a, \tag{6.7.5}$$

i.e., on the other side of the barrier, within the crenelle to the right of the nth barrier and hence to the left of the $(n+1)$th barrier, $\varphi_E(x)$ coincides with

$\hbar$
$$\varphi_{n+1}(x) = A_{n+1}\, e^{ip[x-(n+1)l]} + B_{n+1}\, e^{-ip[x-(n+1)l]}, \tag{6.7.6}$$

or,

$\hbar$
$$\varphi_{n+1}(x) = A_{n+1}\, e^{-ipl}\, e^{ip(x-nl)} + B_{n+1}\, e^{ipl}\, e^{-ip(x-nl)}, \tag{6.7.7}$$

which is to be compared with expression (6.7.2) for $\varphi_n(x)$. It appears, therefore, that it is the quantities $A_{n+1}\, e^{-ipl}$ and $B_{n+1}\, e^{ipl}$ (and not A_{n+1} and B_{n+1}) which for the nth barrier play the role of the coefficients d_+ and d_-, introduced for an isolated barrier. But now, unlike in the case of an isolated barrier, for which there was no reflected amplitude to the right of the barrier (forcing us to set $d_- = 0$), the existence of the $(n + 1)$st barrier (and succeeding ones) means that the coefficient B_{n+1} could possibly be non-zero.

A method of solution using matrices seems to suggest itself. Indeed, the coefficients (A_{n+1}, B_{n+1}) for the index $(n + 1)$ are manifestly obtained from the coefficients (A_n, B_n) by a linear transformation (just as d_+, d_- are related linearly to b_+, b_-). Furthermore, by successive iterations of this transformation from one end of the chain to the other, (A_n, B_n) can be determined in terms of (A_0, B_0).

Hence, let **M** be the matrix defined by

$$\hbar \qquad \begin{pmatrix} A_n \\ B_n \end{pmatrix} = \mathbf{M} \begin{pmatrix} A_{n+1}\, e^{ipl} \\ B_{n+1}\, e^{ipl} \end{pmatrix}. \qquad (6.7.8)$$

This matrix depends on p, and hence on E, but does not depend on n, since the chain is assumed to be sufficiently long, so that the system is invariant under translation. Defined in this way, the matrix **M** is precisely the one which relates the coefficients (d_+, d_-) to (b_+, b_-) introduced in sect. 6.4 [eq. (6.4.2)],

$$\begin{pmatrix} b_+ \\ b_- \end{pmatrix} = \mathbf{M} \begin{pmatrix} d_+ \\ d_- \end{pmatrix} = \begin{pmatrix} M_1 & M_2 \\ M_3 & M_4 \end{pmatrix} \begin{pmatrix} d_+ \\ d_- \end{pmatrix}. \qquad (6.7.9)$$

Let us express the coefficients of **M** with the help of the reflection and transmission coefficients, $A_r = b_-/b_+$ and $A_t = d_+/b_+$, of an isolated barrier. The coefficients M_1 and M_2 are easily obtained by setting $d_- = 0$. Taking account of the analytic form of A_r and A_t, given by eq. (6.4.28), we get

$$\hbar \qquad \begin{cases} M_1 = 1/A_t = (\cosh \kappa' a + i\eta \sinh \kappa' a)\, e^{ipa}, \\ M_3 = A_r/A_t = -i\varepsilon \sinh \kappa' a, \end{cases} \qquad (6.7.10)$$

where

$$\begin{cases} p = (2mE)^{1/2}, \qquad \kappa' = [2m(V_0 - E)]^{1/2}, \\[2mm] \eta = \dfrac{\kappa'^2 - p^2}{2\kappa' p} = \dfrac{V_0 - 2E}{2[E(V_0 - E)]^{1/2}}, \\[2mm] \varepsilon = \dfrac{\kappa'^2 + p^2}{2\kappa' p} = \dfrac{V_0}{2[E(V_0 - E)]^{1/2}}. \end{cases} \qquad (6.7.11)$$

The coefficients M_2 and M_4 can be determined by symmetry considerations alone. Changing p to $-p$, i.e., changing the orientation of the system, amounts to interchanging the roles of d_- and b_+ and of d_+ and b_-. At the same time $\mathbf{M}(p)$ is transformed into $\mathbf{M}(-p)$ (exercise 6.26). This reasoning allows us to express the matrix $\mathbf{M}$ in the form

$$\mathbf{M} = \begin{pmatrix} M_1 & \bar{M}_3 \\ M_3 & \bar{M}_1 \end{pmatrix}, \tag{6.7.12}$$

where M_1 and M_3 are given by eq. (6.7.10). The coefficients (A_n, B_n) are then related to the coefficients (A_{n+1}, B_{n+1}) by means of a new matrix $\check{\mathbf{M}}$

$$\begin{pmatrix} A_n \\ B_n \end{pmatrix} = \check{\mathbf{M}} \begin{pmatrix} A_{n+1} \\ B_{n+1} \end{pmatrix}, \tag{6.7.13}$$

$\check{\mathbf{M}}$ is simply related to $\mathbf{M}$. Indeed, from eq. (6.7.8),

$$\check{\mathbf{M}} = \begin{pmatrix} M_1\,e^{-ipl} & \bar{M}_3\,e^{ipl} \\ M_3\,e^{-ipl} & \bar{M}_1\,e^{ipl} \end{pmatrix}. \tag{6.7.14}$$

6.7.2. The quantization inequality

By iterating relation (6.7.13) we obtain

$$\begin{pmatrix} A_0 \\ B_0 \end{pmatrix} = \check{\mathbf{M}}^n \begin{pmatrix} A_n \\ B_n \end{pmatrix}. \tag{6.7.15}$$

To be physically acceptable, the amplitudes (A_n, B_n) must remain finite: the wavefunction must maintain a finite value no matter how far we move down the chain. Hence, we have to require that the matrix $\check{\mathbf{M}}^n$ remain bounded as $n \to \pm\infty$. This condition is very simply expressed by writing $\check{\mathbf{M}}$ in its diagonal form. It then amounts to requiring that the nth power of the proper values of $\mathbf{M}$ remains bounded as $n \to \pm\infty$, and hence that all the proper values have moduli less than or equal to one.

It can be shown (exercise 6.27) that this implies imposing the condition

$$|\mathrm{Re}\,\check{M}_1| \leqslant 1, \tag{6.7.16}$$

in which case the two proper values m' and m'' of $\check{\mathbf{M}}$ are complex conjugates, and of modulus one,

$$
\begin{cases}
m' = e^{i\alpha} \\
m'' = e^{-i\alpha}
\end{cases}, \quad \text{with } \cos \alpha = \text{Re } \check{M}_1.
\tag{6.7.17}
$$

In view of the expression for $\check{\mathbf{M}}$ and of eq. (6.7.10), condition (6.7.16) bears upon the function

ℏ
$$
\text{Re } \check{M}_1 = \cosh \kappa' a \cos pb + \eta \sinh \kappa' a \sin pb.
\tag{6.7.18}
$$

For a given form of the battlement potential (a, b, V_0 fixed), this condition, applied to E, determines the only possible energies that a quanton placed in the potential under consideration can have. [Relation (6.7.18) has been established here for the case $0 < E < V_0$. It can be shown that an analogous relation exists for energies $E > V_0$ (see exercise 6.27).] This condition, while similar to the various other quantization conditions mentioned in the previous sections, does however differ from them, since this one is an *inequality* and not an equality. It restricts the values of the energy by excluding the intervals in which condition (6.7.16) is not satisfied, but allows all other values. Thus, we do not have a discretization of the energy into distinct levels here, but rather a weaker 'quantization' – if the term may at all be used – which selects out certain intervals and eliminates the rest. Thus we speak of 'allowed' and 'forbidden' energy 'bands', respectively.

Do the bound state energies of a *single*, isolated flat well, of width b and depth V_0, form part of an 'allowed band' for the infinite chain? For the specific values p_n of p, satisfying the quantization condition given by eq. (6.4.17) for the bound states of a single well, the relation $\eta(p_n) \sin p_n b = -\cos p_n b$ enables us to write:

ℏ
$$
\text{Re } \check{M}_1(p_n) = (\cosh \kappa'_n a - \sinh \kappa'_n a) \cos p_n b
$$

$$
= e^{-\kappa'_n a} \cos p_n b.
\tag{6.7.19}
$$

Clearly, therefore, inequality (6.7.16) is satisfied and the levels of the isolated well belong also to the allowed bands of a chain of wells.

Let us turn our attention now to the forbidden energies, to note first of all ... that they definitely do exist. Indeed, so long as $\cosh \kappa' a$ is greater than unity, condition (6.7.16) can certainly not be satisfied when $|\cos pb| = 1$, $\sin pb = 0$, i.e., for all the values $p_n = \eta\pi/b$. Hence, the corresponding energies,

$E_n = n^2\pi^2/2mb^2$, are strictly excluded. Figure 6.41 represents the variation of Re $\check{M}_1$ as a function of E. We see from it that the forbidden energies are sandwiched between the energies for which Re $\check{M}_1$ vanishes and which, at least in the case $\kappa'a \gg 1$, coincide with the bound state energies of the single well. Similarly, we also notice that $E_n = n^2\pi^2/2mb^2$ are not the only forbidden energies: the continuous function Re $\check{M}_1$, which is strictly greater than one in absolute magnitude for the values E_n, cannot suddenly fall below unity. Thus, around each value E_n, there exists a continuum of forbidden energies and, similarly, around each bound state energy of the isolated well, there are other allowed values of the energy (Re $\check{M}_1$ cannot suddenly pass from a value which is practically zero to one which is greater than unity).

In conclusion, the energy spectrum of a quanton, placed in a periodic battlement potential, displays an *alternation* of *allowed bands* and *forbidden bands*.

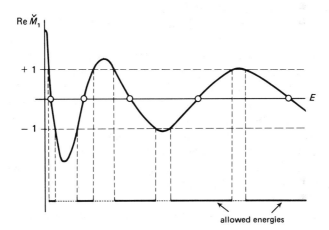

Fig. 6.41. *Band structure*
The alternating structure, of allowed and forbidden bands, of the energy spectrum of a quanton in a battlement potential. The allowed bands are the bands for which the values of Re $\check{M}_1$, as a function of E [see eq. (6.7.18)], lie between -1 and $+1$.

6.7.3. Band spectrum and delocalization

Let us summarize the preceding discussion by considering a situation in which, starting with an ensemble of N individual wells, of width b and height V_0 (thought of as representing N isolated atoms, or rather a model of the Coulomb potential for the attraction of an electron by the nuclei), and extending indefinitely, we approximate them in such a way as to form a regular succession of wells, separated by a distance l. In short, we have an imaginary situation in which we 'construct' a crystal by bringing together its individual atoms.

When the atoms are infinitely far apart ($l = \infty$), the individual bound states of an electron within each atom are all equally stationary. Moreover, the quantization inequality, eq. (6.7.16), reduces to equality (6.7.19). These 'atomic' states are N in number. They are both *stationary* and *localized* (up to a translation of the atomic lattice). The corresponding energy levels form a *discrete* spectrum. Each level is *N-fold degenerate* since it corresponds to N states localized at different sites.

Bringing the atoms closer to one another has the effect of reducing a (and hence l), while holding the 'atomic' characteristics, b and V_0, constant. Thus, for each one of the discrete levels of the isolated atom, the quantity $e^{\kappa' a} = \exp[2m(V_0 - E)]^{1/2} a$ becomes smaller: the barrier, separating two successive atoms for an electron, becomes permeable. The electron, initially bound to a specific nucleus, now has the possibility of 'jumping' via the tunnel effect (or leap-frogging) from one well to another. Its stationary states are no longer localized and, evidently, the least localized are the ones with the highest energies, for which the quantity $V_0 - E$ (and consequently $e^{\kappa' a}$) is the smallest.

At the same time, the degeneracy of the original levels is lifted. Just as the levels of the double potential will appear in pairs, corresponding to each energy level of the bound states of the simple well, so also, over here, each level of the single well splits into N levels, which, in the limit of large N, form a pseudo-continuum around the level which gives rise to them (fig. 6.42). Thus, it is by the close packing of the discrete levels, which become more and more dense as N grows larger, that the allowed energy 'bands' are formed. Let us note, that in the model treated here, each band contains as many discrete levels as there are atoms in the chain representing the crystal. Unlike many of the other results obtained in this section, which are of general validity, this last property is specific to crystals having only a single atom per 'link' of length l.

As we have already noted, in connection with the double well (see sect. 6.6), the tunnel effect, inside a crystalline solid, is fully operative only for the external electrons, for which the barrier is the lowest. While the 'valence electrons' are delocalized and 'belong' to the widest bands, the internal electrons themselves remain strongly attached to their own atoms, and the deeper they are, the thinner their allowed energy bands happen to be. It is the delocalization of the outermost electrons of atoms and the band structure of the energy spectrum which give crystalline solids their specific properties,

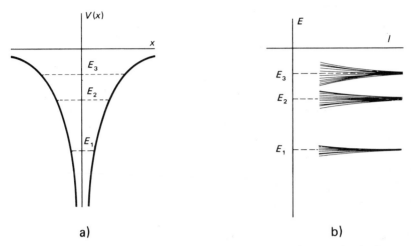

Fig. 6.42. *Broadening of the electronic levels of an isolated atom into the energy bands of a crystal*
(a) Usual schematic representation of the energy levels of a peripheral (called valence) electron
of an *isolated atom*. The electron is assumed to be placed in the (Coulomb) field of the nucleus
and the other electrons of the atom. In the figure, this potential has been represented using a one-
dimensional model.
(b) Energy spectrum of a valence electron inside a *crystal* (chain of N atoms separated by a
distance l). Each electronic level is N-fold degenerate when $l = +\infty$. As the atoms are brought
closer together, this degeneracy is lifted due to the tunnel effect. Each level ($n = 1, 2, ...$) gives rise
to N levels which, for $N \gg 1$, have the appearance of a continuum – hence the name, 'energy
bands'.

particularly the cohesion which typifies them as being solids. It can be
checked from fig. 6.42 that the lowest energy of a band is always smaller than
the energy of the corresponding isolated level (a phenomenon already verified
in connection with the double-well potential). However, this lowering of the
energy is not sufficient to explain the cohesiveness of solids. It remains to be
seen what happens when *all* the electrons of a crystal are considered which,
contrary to all expectations (at least from our present stand-point) cannot all
occupy this level (see ch. 7).

6.7.4. The quasi-momentum

The stationary states of a quanton, subject to a periodic potential
(delocalized and having a spectrum weakly quantized into energy bands), are
intermediate between its bound states inside a well (localized and having a
discrete spectrum) and its free states (delocalized and with unquantized

spectrum). Indeed, the physical situation under consideration is intermediate between that of an isolated well, where one discrete region of space happens to be privileged, and that of a constant potential, invariant under an arbitrary spatial translation. This situation can be described as being 'partially invariant', in the sense that the translations which leave it invariant constitute a restriction from the set of all continuous translations of space, to the discrete subset of those translations which are integral multiples of the period of the potential. From this we expect a priori analogies (the extent of which remains to be specified) between the properties of the states of these quantons with those of free quantons.

We recall that for a free quanton, complete invariance under spatial translations leads to the existence of stationary states which are also proper states of an important physical magnitude, the momentum. More precisely, for the same value E of the energy, there exist two special stationary states – one characterized by the proper value $p = \sqrt{2mE}$ of momentum and the other by the proper value $-p$. For each one of these (stationary) states, the momentum, which in the course of its temporal evolution maintains the same value, is a conserved magnitude. The corresponding wavefunctions $\varphi_p(x)$ and $\varphi_{-p}(x)$ are (see ch. 5) spatially harmonic functions

$$\varphi_{\pm p}(x) = e^{\pm ipx}, \tag{6.7.20}$$

which, for an arbitrary spatial translation of length ξ, get multiplied by a simple phase factor

$$\varphi_{\pm p}(x + \xi) = e^{\pm ip\xi}\, \varphi_{\pm p}(x). \tag{6.7.21}$$

In the case of a more restricted invariance, such as the one which is of interest to us now, do there similarly exist stationary states which are invariant, not under arbitrary spatial translations, but rather under the discrete set of translations which leave the potential invariant? More precisely, do there exist, among the stationary, wavefunctions $\varphi_E(x)$, of the energy E, functions which, on being subjected to a translation through an integer multiple of the period l, get multiplied by a simple phase factor $e^{i\delta}$? If such functions exist, they have to satisfy the relation

$$\varphi_E(x + l) = e^{i\delta}\, \varphi_E(x). \tag{6.7.22}$$

Considering the general form, eq. (6.7.2), of the functions $\varphi_E(x)$, this, in turn, gives

$$\hbar \qquad \begin{pmatrix} A_n \\ B_n \end{pmatrix} = e^{-i\delta} \begin{pmatrix} A_{n+1} \\ B_{n+1} \end{pmatrix}. \qquad (6.7.23)$$

For each allowed value of E, there exist in effect two functions satisfying these conditions: as can be seen, by comparing eq. (6.7.23) to eq. (6.7.13), these are the functions for which the coefficients (A_n, B_n) are proper states of the matrix $\check{\mathbf{M}}$. One of them, denoted $\varphi_E^+(x)$ is associated to the proper value $m' = e^{i\alpha}$ [see eq. (6.7.17)] and the other, $\varphi_E^-(x)$, to the value $m'' = e^{-i\alpha}$. We have, therefore,

$$e^{i\delta} = e^{\pm i\alpha}. \qquad (6.7.24)$$

These functions, which play over here a role analogous to that of the harmonic function $\varphi_p(x)$, of the free quanton, may be described as being *quasi-harmonic*. Note the close resemblance between the restriction of the energy proper values to certain (allowed) bands and the quasi-harmonicity of the corresponding wavefunctions: for the forbidden values of the energy, the matrix $\check{\mathbf{M}}$ has real proper values, and there do not exist any wavefunctions such that their translations lead only to simple phase shifts.

In order to make this notion of quasi-harmonicity precise, let us set

$$\hbar \qquad \alpha = \not{p}l, \qquad (6.7.25)$$

so as to give the relation (6.7.22) a form comparable to (6.7.21),

$$\hbar \qquad \varphi_E^\pm(x+l) = e^{\pm i\not{p}l} \varphi_E(x). \qquad (6.7.26)$$

The magnitude $\not{p}$ introduced via eq. (6.7.25), characterizes the function $\varphi_E^\pm(x)$ in the same way as δ or α. It plays, vis-à-vis these functions, a role comparable to that played by the momentum p in $\varphi_p(x)$. We shall agree to call it the 'quasi-momentum'.

The analogy between $\not{p}$ and p is, however, limited. Thus, while the values p and $-p$, characterizing the two functions $\varphi_p(x)$ are, for each value of E, uniquely defined by the dispersion relation $E = p^2/2m$, the same is not true of $\not{p}$. Indeed, the dispersion relation connecting $\not{p}$ to E, obtained by combining eqs. (6.7.25), (6.7.17) and (6.7.18), can be written as,

$$\hbar \qquad \cos \not{p}l = \cosh(\kappa'a) \cos(pb) + \eta \sinh(\kappa'a) \sin(pb), \qquad (6.7.27)$$

where κ', p and η are functions of E [see eq. (6.7.11)]. While relation (6.7.27) determines $\cos pl$ unambiguously, it allows for a multiplicity of values – up to $2\pi/l$ – of p. In other words the characteristic quasi-momentum of the states $\varphi_E^\pm(x)$ is only defined up to $2\pi/l$. This quantity p, characteristic of stationary states, is conserved in time – but its conservation, just like its definition, holds only up to $2\pi/l$. Thus, to the spatial periodicity of the potential – of period l – there corresponds a so-called 'reciprocal' periodicity in the momentum – of period $2\pi/l$. The energy of a stationary state is a function $E(p)$ of its quasi-momentum, which displays this periodicity.

Very often the convention is made to define p uniquely, by restricting its variation to the interval $[-\pi/l, +\pi/l]$, called the 'Brillouin zone'. It should not be forgotten, however, that in this 'reduced' representation, the extremities of the interval, that is, the values $-\pi/l$ and $+\pi/l$ have to be identified. It is easy, in this representation, to understand the form of the graph giving the energy as a function of the quasi-momentum. Indeed, let us assume that the crenelles of the periodic potential are very shallow, i.e., the depth of the crenelles is small compared to the energy of the quanton ($V_0 \ll E$). To a first approximation the quasi-momentum is, therefore, equal to the (true) momentum – accurate to the extent that it is only defined up to $2\pi/l$. We may say that the effect of the periodicity of the battlement potential continues to be felt no matter how weak it is. In other words,

$\hbar$ $p \simeq p$,   modulo $2\pi/l$.   (6.7.28)

[This moreover, is what we would deduce from the relation analogous to eq. (6.7.27) in the case where V_0 is small – see exercises 6.28 and 6.29.] The energy, in view of the weakness of the potential, is evidently given by a relation similar to the one which holds for a free quanton,

$\hbar$ $E \simeq \dfrac{p^2}{2m}$.   (6.7.29)

From here the energy, as a function of the *quasi*-momentum can be obtained by translating the parabolic relation, eq. (6.7.29), through multiples of $2\pi/l$, which impose on it the desired periodicity (fig. 6.43). We may concentrate our attention on the reduced zone $[-\pi/l, +\pi/l]$ in which the parabola given by eq. (6.7.29) is seen to 'fold over' to make the different energy bands appear. These bands are contiguous and form a continuum in the special case of a free quanton. However, if the periodic potential is not constant ($V_0 \neq 0$), the function $E(p)$ is only approximately given by eqs. (6.7.28) and (6.7.29). It

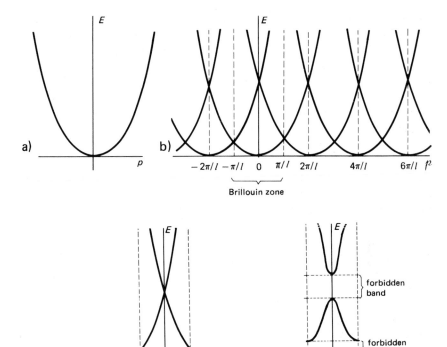

Fig. 6.43. *Energy and quasi-momentum*

(a) For a free quanton, the energy depends quadratically on the momentum, $E = p^2/2m$.

(b) In a weak battlement potential, the energy is a periodic function of the quasi-momentum p, which practically gets tangled up with the preceding 'periodicized' parabola.

(c) Within the Brillouin zone $(-\pi/l, +\pi/l)$ the graph of $E(p)$ is now obtained, from the parabolic graph of (a), by 'folding' along the edges.

(d) When the effect of the potential is no longer negligible, it deforms the curve $E(p)$, most notably by lifting the degeneracies at the edges and in the middle of the Brillouin zone.

In summary, while the energy of a free quanton is unambiguously defined by the value of its momentum p, which assumes all possible real values (from $-\infty$ to $+\infty$), the energy of a quanton inside a periodic potential is multi-valued: it is defined by the value of its quasi-momentum – which may be restricted to the first Brillouin zone (from $-\pi/l$ to $+\pi/l$) – and by the specification of one particular branch, corresponding to one of the allowed bands.

would then differ from the folded-over parabola, most particularly in the neighbourhood of $p = -\pi/l$, 0, $+\pi/l$. Indeed, to these abscissa points correspond the intersections of the two branches of the multiple parabola, i.e., in fact, to degenerate situations, in which two different stationary states (corresponding to the two branches) have the same energy. As in the case of the double well, the change in the potential has the effect of 'lifting' this degeneracy, bringing about, in other words, the existence of two *different* values of the energy. In this way, the intersections of the branches of the graph disappear and there appear forbidden bands between the allowed bands.

6.7.5. *Bragg reflection*

The fact that it is not possible for a quanton, placed inside a periodic potential, to be in a proper state of the (true) momentum, is easily understood. It is again a case of the manifestation of the neither corpuscular, nor undulatory, nature of the quanton. A stationary and harmonic wave-function in x, of the type $e^{-i(Et-px)}$ (respectively $e^{-i(Et+px)}$), describes the uniform propagation of a quanton to the right (respectively to the left). This type of progressive movement in 'one direction' is only possible if the quanton does not encounter any obstacles. Now, a potential barrier, of any height, is always an obstacle for a quanton. In this case, it is always both scattered forward and reflected (retro-scattered) backwards. Its wavefunction is always a 'mixture' of waves progressing in one direction (e^{-ipx}, for example) and the other (e^{ipx}). This is why, in the case of a periodic potential, the wavefunction of the quanton can at best be of the 'quasi-progressive' type, $\varphi_E(x)$ – a linear combination of free progressive waves.

The quasi-harmonic (or rather quasi-progressive) nature of the wavefunctions $\varphi_E(x)$ manifests itself even better, if we try to isolate a harmonic factor from their analytic expression, by writing them in the form

$ħ$ $$\varphi_E^\pm(x) = e^{\pm i\not px} u_{\pm\not p}(x). \tag{6.7.30}$$

Had the wavefunctions been real harmonic, $u_{\pm\not p}(x)$ would have reduced to a simple constant, independent of x, and $\varphi_E(x)$ would have represented the propagation to the right or the left, of a constant probability density $|u_{\pm\not p}|^2$. As a matter of fact, relation (6.7.26) forces $u_{\pm\not p}(x)$ to be periodic, with period l,

$$u_{\pm\not p}(x+l) = u_{\pm\not p}(x). \tag{6.7.31}$$

Put into this form, the functions $\varphi_E(x)$ are given the name, 'Bloch (wave-) functions'. We see that the periodic distribution $|u_{\pm\not p}(x)|^2$ is in a way 'carried' (in the same sense as we speak of carrier waves in classical wave theory) by the harmonic waves $e^{\pm i\not px}$, with a phase shift $e^{\pm i\not pl}$ for each spatial period.

However, the quanton (or at least the probability density of its presence) ceases to be 'transported' as soon as the periodicity of the carrier wave coincides with the period l. Now the transport effected by the waves $e^{\pm i \not{p} x}$ only reproduces identically the periodic distribution itself. We have here a stationary situation (in the sense in which we understand this term in wave theory): the quanton does not progress in any sense.

The physical reason for this phenomenon is easy to understand. At each barrier, the quanton can just as well be reflected backwards as it can be scattered forwards. The functions $\varphi_n(x)$ mix, in equal parts, the terms in $e^{i \not{p} x}$ and $e^{-i \not{p} x}$, as can easily be verified (exercise 6.32). In order to emphasize the collective nature of this reflection by all the barriers, one often says that the quanton is reflected by the battlement potential. Clearly, this reflection phenomenon always exists, whatever might be the value of $\not{p}$. But it is only when there is agreement between the quasi-undulation $\not{p}$, of the carrier waves $e^{\pm i \not{p} x}$, and the period l that the phenomenon becomes important; i.e., when

$$\hbar \qquad e^{\pm i \not{p} l} = \pm 1. \tag{6.7.32}$$

Under these conditions, there is no longer any phase shift from one site to another, and hence no more progression. If we choose to restrict $\not{p}$ to the Brillouin zone – the interior of the interval $[-\pi/l, +\pi/l]$, the matching condition, eq. (6.7.32), for the phase becomes

$$\hbar \qquad \not{p} = 0 \quad \text{and} \quad \not{p} = \pm \pi/l. \tag{6.7.33}$$

The condition $\not{p} = \pm \pi/l$ is (formally, since $\not{p}$ is only a quasi-momentum) analogous to the Bragg condition, governing the reflection of X-rays by the atomic planes of a one-dimensional crystal, at distances l apart. As a matter of fact, in the one-dimensional case, where reflection corresponds to backward scattering $2\theta = \pi$, the Bragg condition, $2l \sin \theta = \lambda$, reduces to $2l = \lambda$, implying $p = 2\pi/\lambda = \pi/l$. That is why this inability of the quantons to propagate, under certain conditions, in a periodic potential, is also called *Bragg reflection*. This name is fully justified since, after all, X-ray photons are also governed by the same general laws as other quantons. Moreover, it was this very Bragg reflection which had been observed by Davisson and Germer, in 1927, in a historic experiment on the reflection of electrons by a crystal – an experiment which was to establish, experimentally, the 'wave nature' of electrons, two years after de Broglie had proposed this principle (exercise 4.21).

It has probably been noticed that, when condition (6.7.33) for Bragg reflection is satisfied, the proper values $\alpha = \not{p}l$ assume the values $\alpha = 0, \pm \pi$, and the function Re $\check{M}_1 = \cos \alpha$ [eq. (6.7.18)] attains the values ± 1, which mark the limits of the allowed energy bands, eq. (6.7.16) (fig. 6.41): the Bragg reflection and the existence of forbidden energies are intimately related.

As a matter of fact, as long as the value of $\not{p}$, characterizing a stationary state, defines a true progressive wavefunction, the corresponding motion of a quanton is a displacement along the atomic chain. On this displacement, the battlement potential has a reduced effect on the average. The situation is completely different inside and at the boundaries of the Brillouin zone, where the wavefunction is stationary. The quanton now actually 'stays put', enabling the potential to make its effect strongly felt, maximally modifying thereby the values of the energy relative to those of a free quanton. In a sense, the Bragg reflection traps the quanton.

6.7.6. Electrical conduction in metals

Surprising as it may seem, it was necessary to wait until the advent of the quantum theory to obtain a satisfactory explanation for electrical conduction in metals. To say that a metal conducts electricity is to say that its valence electrons perform, under the effect of a uniform electric field $\mathscr{E}$, a collective motion forming a charge current.

In a classical model, in which the electrons are considered as being particles, the motion of the electrons is obtained by applying to them the basic dynamical law

$$F = -q_e \mathscr{E} = \frac{\mathrm{d}p}{\mathrm{d}t}.$$
(6.7.34)

If we assume that $\mathscr{E}$ is uniform and does not vary with time, the electrons acquire an acceleration which allows them to circulate across the ions of the crystal. However, this motion does not carry them too far, since very quickly, the electrons begin to undergo collisions with the ions of the crystal and lose the acquired momentum. This classical model, which attributes to the conduction electrons a mean free path of the order of the interatomic distance, leads unfortunately to a much too high value for the electrical resistance. We have to find out, therefore, why the mean free path of an electron inside a metal could be much larger than the interatomic distances.

The problem was believed to have been solved in the thirties, using the phenomenon of the tunnel effect. The latter allows the electrons to go across the potential barriers separating the atomic sites, guaranteeing them a quasi-propagation, and hence endowing them with a mean free path larger than that obtained from simple classical considerations. In this model, the electric current appears as a manifestation of the Bloch waves. The basic dynamical law is modified in order to adapt it to the case of a partial translation invariance: the applied force $-q_e \mathscr{E}$ determines the temporal variations of the quasi-momentum

$$-q_e \mathscr{E} = \frac{\mathrm{d}\not p}{\mathrm{d}t}. \qquad (6.7.35)$$

A potential difference, applied to the ends of a metallic wire changes $\not p$, and the state of the electrons is described as a succession of states of increasing or decreasing $\not p$, depending on the sign of $\mathscr{E}$.

One could think, therefore, that the electrons would propagate without meeting any obstacle inside a crystal, their quasi-momentum increasing indefinitely under the effect of the electric field $\mathscr{E}$. However, that is not what happens, since there is an essential difference between the 'quasi-propagation' of an electron and free propagation. Indeed, for values of the quasi-momentum at the edges of the Brillouin zone ($\not p = \pm \pi/l$), the Bloch wavefunctions do not correspond to a true propagatory motion, but rather to a spatially stationary situation. Under the effect of the electric field, $\not p$ increases up to the value $\not p = \pi/l$: the motion stops, in order to reverse itself subsequently, since the values $\not p = \pi/l$ and $\not p = -\pi/l$ are to be identified and since, once this critical value is attained, the quasi-momentum (in the reduced zone) becomes negative. This is seen more clearly by considering the velocity of quasi-propagation, given by the usual expression $v = \mathrm{d}E/\mathrm{d}\not p$ which vanishes inside or at the edges of the zone (exercise 6.29), and thus undergoes a periodic change – alternately positive and negative – under the effect of the field. If the initial velocity is zero, this oscillatory variation in the velocity implies an oscillatory motion of the electron itself, and not a monotonic displacement.

This apparently strange phenomenon is just the manifestation of the Bragg reflection. The interplay of the transmission and the reflection of plane waves in the periodic potential gives rise to the Bloch waves – a certain proportion of the plane waves progressing in one direction, the rest in the other. At the edges of the Brillouin zone, as in the middle, the two contributions are equal: there is maximal reflection under Bragg conditions, which leads to the establishment of a stationary, rather than a progressive, phenomenon.

It would seem, finally that the electron, subjected as it is to a zig-zag motion, unable to move a distance greater than the average distance of these oscillations, could not give rise to a charge transport. Let us estimate the length D of the oscillations of the electron. The energy stored in the electron during the first part of its oscillatory motion (as $\rlap{/}p$ goes from 0 to π/l), due to the work done by the electric force $q_e\mathscr{E}$ over the distance of oscillation D – of amount $q_e\mathscr{E}D$ – is equal to the width of the allowed band – typically, a few eV. If we take a value of 10 V m^{-1} for $\mathscr{E}$ (typically 100 volts over 10 metres), we obtain $D \simeq 0.5$ m for the average distance of oscillation. Had the crystals been perfect, the currents could not have moved over distances larger than 0.5 m and our 'electrified' civilization would have been impossible. The quantum theory of conduction within a perfect crystal, in spite of its initial appearance, does not turn out to be better than the classical theory. But perfect crystals are only a figment of the imagination: an electron always encounters an impurity before it travels a distance of 0.5 m. Its mean free path never in fact exceeds 1 mm. Thus, it is in the imperfections of the crystal that the definite key to its conductivity lies.

At the end of this analysis, therefore, we make no other claim than to have only laid the foundation of a genuine explanation. In spite of its complexity, the 'theory of bands' (of which the present section is a simplified version) is just a prelude to the study of a phenomenon as commonplace as electrical conduction in metals. As Feynman once said: 'Physics does not explain any simple phenomenon simply'.

Exercises

6.1. A quanton inside an infinite, flat well of width a possesses, at time $t = 0$, a wavefunction

$$\psi(x; 0) = C \sin^3(\pi x/a).$$

(a) Calculate the normalization constant C.

(b) Expand $\psi(x; 0)$ in terms of the stationary wavefunction $\varphi_n(x)$ and, hence, deduce the wavefunction $\psi(x; t)$ at time t.

(c) Illustrate graphically the *time evolution* of the probability density. Re-derive the results of fig. 6.4.

(d) Calculate the quadratic dispersion Δx as a function of time.

(e) Repeat the same questions for the wavefunction

$$\psi(x; 0) = C \sin(\pi x/a) \cos^3(\pi x/a).$$

6.2. We wish to determine, approximately, the energy levels of a quanton of mass m, subject to a potential of the form $V(x) = \frac{1}{2}kx^2$, called a *harmonic potential* (because of the harmonic nature of the classical motion), of pulsation $\omega = \sqrt{k/m}$, in such a potential. More conveniently, we write

$$V(x) = \tfrac{1}{2} m \omega^2 x^2.$$

The approximation used consists in replacing the harmonic potential by a cleverly chosen infinite flat well – the levels of which are known. Let us consider the level E_n and let $(-R_n, R_n)$ be the amplitude of the classical motion for the same energy, defined by $E_n = V(R_n)$.

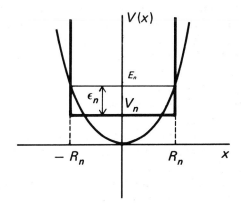

The harmonic potential is replaced by an infinite flat well,
– of width $2R_n$,
– of depth (to the bottom of the well) V_n equal to the average value, over the interval $(-R_n, +R_n)$, of the actual potential $V(x)$,
– such that its nth level coincides with that of the harmonic potential: $E_n = V_n + \varepsilon_n(2R_n)$, where $\varepsilon_n(a)$ is the energy of the nth level of an infinite, flat well, counting from the bottom of the well.

(a) Determine V_n by taking the average of $V(x)$ with respect to a uniform probability density over $(-R_n, R_n)$. We obtain in this way, $V_n = \tfrac{1}{6} m \omega^2 R_n^2$.

(b) Using this value of V_n and expression (6.2.9) for $\varepsilon_n(a)$, determine the values of E_n and R_n. Compare this approximate value of E_n to the exact value,

$$E_n = n \hbar \omega - \tfrac{1}{2} \hbar \omega \quad (n = 1, 2, \ldots).$$

(c) It would be more appropriate to calculate the average value V_n, using the actual probability density of a quanton inside the flat well, i.e.,

$$V_n = \int_{-R_n}^{R_n} V(x) \, |\varphi_n(x)|^2 \, dx,$$

where φ_n is the wavefunction of the nth stationary state of the infinite, flat well [see eq. (6.2.13)]. Carry out this computation and show that it changes the results of (b) only very slightly.

6.3. Determine the first twenty or thirty energy levels, and their degeneracies, for a three-dimensional infinite, flat well, shaped as a rectangular parallelepiped, the sides of which are in the proportions:

(a) $a = \sqrt{2}\, b = 2c.$

(b) $a = \sqrt{3}\, b = \sqrt{3}\, c.$

6.4. *Colour centres of ionic crystals*

(a) Consider a cubic box of side a, inside which there is a constant potential, conventionally taken to be zero, and the potential outside is infinite ('infinitely deep, flat, cubic well' in three dimensions). A quanton of mass m is enclosed in it. What are the energies of the first two levels E_1 (ground state) and E_2 (for the first excited state). What are the degeneracies of these levels?

(b) [Independent of (a)] When an anion is ejected from inside a cubic crystal, such as NaCl (e.g., under the effect of X-ray radiation), it leaves behind a highly electronegative lacuna and is apt to capture an electron. Such a lacuna, occupied by an electron is called an 'F centre' or 'colour centre' (from the German 'Farbenzentrum'). The objective of this problem is to understand this nomenclature. When the crystal is illuminated, the presence of a luminous *absorption* line is observed. The energy ε of the absorbed photons is measured as a function of the size of the lattice spacing a, for various alkyl halides. Figure 1 gives the experimental results. Show that they may be represented fairly well in the form

$\varepsilon = Ka^n$ (Mollwo–Ivey law)

and calculate the experimental values of the constants K and n.

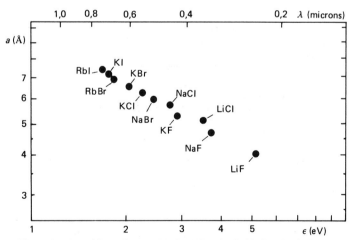

[From *Physics of Color Centers*, W. Beall Fowler (ed.), Academic Press, New York (1968)]

Figure 1

(c) Normally, the alkyl halides are colourless and appear as whitish crystals (... such as NaCl). Three packets, containing NaCl, KI and KCl crystals, respectively, have unfortunately lost their labels and now appear to be identical. In order to identify them, they are irradiated using X-rays. The creation of F centres colours them by *absorption* (attention!). If we get in this way a yellow, a green and a mauve sachet, what is the content of each. [Recall that the colours of the visible spectrum correspond roughly to the following photon energies: red (1.65 to 2.0 eV), orange (2.0 to 2.1 eV), yellow (2.1 to 2.3 eV), green (2.3 to 2.55 eV), blue (2.55 to 2.65 eV), violet (2.65 to 3.1 eV).]

(d) To a first approximation, the lacuna of a centre F can be considered to be a cubic box of the type studied in (a), and having sides equal to a, the lattice spacing (fig. 2). An electron placed inside this box could occupy the calculated levels E_1 and E_2. We consider the absorption of the light to be due to the excitation of the electrons, of the 'F centres', from their ground state to the first excited state. Show that this interpretation enables us to understand the Mollwo–Ivey law. What are the theoretical values of K and n? It can be verified that $n_{\text{theor}} \simeq n_{\text{exp}}$, but that $K_{\text{exp}} \neq K_{\text{theor}}$. Can you explain the disagreement between theory and experiment in the values of K? (Hint: In the actual situation, what is the order of magnitude of the depth of the well? What would you conclude on the spacing of the levels?)

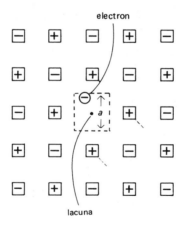

electron

lacuna

Figure 2

(e) Upon absorption of the light, the F centres are de-excited, emitting a photon. However, the emission lines (luminescence) are displaced towards the higher wavelengths (or, towards lower energies). Figure 3 shows an example of this 'Stokes displacement'. Does this emission render the theory in (c), on the colour of crystals, invalid?

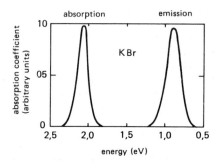

[From W. Gebbart, A. Kuhnert, Phys. Stat. Solidi, *14* (1966) 157]

Figure 3

(f) The Stokes displacement is due to a deformation of the centre F, which loses its cubic symmetry and is transformed into a parallelepiped, having a square base of side b and height c, while keeping its volume constant $(b^2 c = a^3)$ (fig. 4). The deformation is characterized by the parameter $\delta = (b/c)^{1/3}$.

(1) Express b and c as functions of a and δ.

(2) Calculate the effect of the deformation on the energy levels E_1 and E_2: express, as functions of a and δ, the energies of the ground state and the first few excited states of the electron inside the box, now shaped as a parallelepiped. Show that the degeneracy of E_2 is lifted, and that for $\delta \neq 1$, there are two excited levels: E'_2, of degeneracy 1 and E''_2, of degeneracy 2.

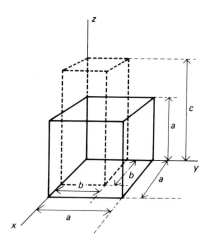

Figure 4

(3) Plot the curves giving the energies $E_1(\delta)$, $E'_2(\delta)$ and $E''_2(\delta)$. In particular, calculate the positions of the minima of these curves. What is the lowest energy attainable by simple deformation, starting from the level $E_2(\delta = 1)$? Does this minimum correspond to the level E'_2 or to E''_2? Let δ_0 be the corresponding parameter. Is the unit cell flattened or stretched for $\delta = \delta_0$?

(g) The luminescent emission is now explained by assuming that after the absorption, the F centre gets deformed until its energy reaches a minimum, which is $E'_2(\delta_0)$. Then it emits a photon which de-excites it to the energy $E_1(\delta_0)$, before regaining its initial form and its ground state energy $E(\delta = 1)$ (fig. 5).

(1) What happens to the energy lost during the deformation phases (2) and (4)?

(2) Calculate the energy of the photon emitted during the de-excitation. Calculate the ratio of the energies of the absorbed and the emitted photons, $\varepsilon_{ab}/\varepsilon_{em}$ and compare with the experimental results of fig. 3.

(3) Among the different crystals in fig. 1, for which one would the emission line be in the visible region?

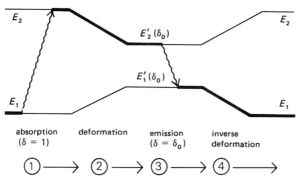

Figure 5

6.5. (a) For a quanton of mass m in two dimensions, free to move on a surface of area S, establish the expression,

$$\rho_2(E) = \frac{S}{2\pi} m,$$

for the density of states.

(b) In the same way, establish the expression

$$\rho_1(E) = \frac{2L}{\pi}\left(\frac{m}{2E}\right)^{1/2},$$

for the density of states of a quanton of mass m in one dimension, along a segment of length L.

(c) Establish the expression

$$P_\gamma(E) = \frac{V}{\pi^2 c^3} E^2,$$

for the density of states of a photon inside a box of volume V. (One takes into account the fact that, for each direction of propagation of a photon, there exist two possible states of polarization – see ch. 4.)

(d) Calculate the density of states of a free 'Einsteinian relativistic' quanton inside a volume V. Verify the agreement of the formula so obtained with that obtained in (c), for a photon.

(e) In each case, reintroduce into the formulae the quantum constant $\hbar$, using dimensional analysis.

6.6. Consider a free quanton of mass m in one dimension. In order to avoid problems with the continuous spectrum, arising from the infinite nature of the space, we may replace the entire line by a ('large') segment of length l, and confine the quanton to it by means of 'infinite potential barriers'. This is the technique that has been used up to now. Another trick, which we shall study in this problem, involves considering a segment of length l, so that $0 \leqslant x \leqslant l$, and imposing on the wavefunction $\varphi(x)$ the (so-called Born–von Karman) 'cyclic conditions':

$$\varphi(0) = \varphi(l), \qquad \varphi'(0) = \varphi'(l).$$

(a) By considering the most general (time-independent) wavefunction of a stationary state of energy $E = p^2/2m$, and imposing the cyclic conditions on it, show that the energy is quantized.

(1) Find the proper values and their degeneracies.

(2) Write down the stationary state wavefunctions, suitably normalized.

(b) Compute the number $N(E)$ of states of energy lower than E, and compare it with the value obtained by the method of quantization by confinement.

(c) Consider the wavefunction $u(x) = A \sin^2(\pi x/l)$.

(1) Show that it satisfies the cyclic conditions.

(2) Write down its expansion in the basis consisting of the stationary state wavefunctions.

(3) Calculate the time-dependent wavefunction $\psi(x; t)$ which satisfies the initial condition $\psi(x; 0) = u(x)$.

(4) Calculate the probability density of the localization of the quanton, $\rho(x; t) = |\psi(x; t)|^2$, and sketch the graph of $\rho(x; t)$ as a function of x, for different values of the time t.

6.7. Consider the wavefunction of a stationary state of a quanton, in the presence of a *potential step* of height V_0, for the case where $E < V_0$. Study the behaviour of the wavefunction and its derivative in the neighbourhood of the discontinuity as $V_0 \to \infty$ (E remaining fixed). Show that the wavefunction remains continuous, but not its derivative. Recover the boundary conditions imposed at the edges of an *infinite* step (sect. 6.2).

6.8. For studying certain properties of neutrons, it would be convenient to be able to work with large quantities of them, and hence to stock-pile them at the exit to nuclear reactors which produce them. We shall see that it is possible to enclose effectively the neutrons in '*neutron bottles*', appropriate for the purpose, but that this storage is bound to be temporary.

A 'neutron bottle' is a box, which we take to be cubic, for convenience, of side $l =$ 20 cm, having thick walls made of copper, for example. Let $M = 63.6$ g be the atomic mass of copper and $\mu = 9.0$ g cm^{-3} its mass density.

(a) Calculate the density $\mathcal{N}$ of the copper nuclei in the metal. What is the dimension and the value of the magnitude $d = \mathcal{N}^{-1/3}$? What is its physical significance?

The nuclei of copper scatter the neutrons. At low energies, this scattering is characterized by a scattering length $a + ia'$, where $a = 7.7$ F and $a' = 1.3 \times 10^{-3}$ F.

(b) The neutrons in copper are subjected to a constant potential. Show that it can be written as

$$V = 2\pi \frac{a}{md^3},$$

in a system of units in which $\hbar = 1$ (used in the entire problem). Calculate V in electron volts.

(c) Hence deduce that the 'bottle' behaves like a flat bottomed, cubic potential well, and that the neutrons remain confined to it if each one of the three components of their velocities is smaller than a certain critical velocity v_e (calculate it numerically). Why are these neutrons called 'ultracold'?

(d) It will be convenient, in the following, to define the magnitude,

$$\lambda_c = (\pi d^3/a)^{1/2}.$$

What is its dimension, its value and its physical significance? What is the value of the wavelength of a neutron of kinetic energy $E = \frac{1}{2}V$?

(e) For neutrons confined to the bottle, the states of accessible energy are discrete. Show that *if* the box were an infinitely deep potential well, the number S of states of energy $E \ll V$ can be expressed in the form $S = K(l/\lambda_c)^3$, where K is a certain numerical constant. Compute S. Hence deduce that one can justifiably approximate the spectrum of the accessible energy levels of the neutrons by the continuum $0 \leqslant E \leqslant V$. Henceforth, we shall simplify the problem and only consider neutrons which propagate in the direction Ox, between the two walls of the box, parallel to the plane yOz, having abscissas $x = 0$ and $x = l$.

(f) Suppose that we have such a confined neutron, of energy E. In the following we shall set $\kappa = [2m(V - E)]^{1/2}$ and $p = (2mE)^{1/2}$. What is the form of its wavefunction for $x \leqslant 0$? Calculate the average penetration depth δ, of a neutron inside a wall. In the typical case in which $E = \frac{1}{2}V$, compare δ to λ_c and find the numerical value of δ.

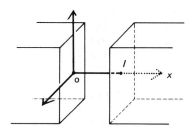

(g) The nuclei of copper are able to *absorb* neutrons. Show that the effective cross-section of absorption of a nucleus, for a neutron of energy E, is roughly $\sigma_a = 4\pi a'/p$. Compute it numerically in the case where $E = \frac{1}{2}V$.

(h) Calculate the *average* number n of copper nuclei, per unit surface area of the wall, with which the neutrons can interact as they penetrate the wall. Show that $n = \mathcal{N}\delta$. (Hint: How many nuclei are there within a slice of copper of unit surface area, thickness dx at depth x, and what is the probability that a neutron, arriving at the surface, would be localized inside this slice?)

(i) Hence deduce the probability for a neutron to be absorbed in the wall as a result of a *single* collision with it, $P = \mathcal{N}\sigma_a\delta$. Show that, for the case where again $E = \frac{1}{2}V$, we simply get $P = Ca'/a$, where C is a numerical coefficient. Compute the probability P.

(j) Show that the probability of finding a neutron (of energy $E = \frac{1}{2}V$) inside the box, at an instant of time t following its introduction, is of the form $\mathscr{P}(t) = \exp(-t/\tau)$, where τ is the characteristic time: $\tau = l/Pv$. Calculate τ and compare it to the intrinsic half-life of the neutron, $T \simeq 1000$ s. Is the neutron bottle a useful device?

6.9. The *invariance under spatial reflection* of the probability density of localization, for a wavefunction $\varphi(x)$, is written as

$$|\varphi(-x)|^2 = |\varphi(x)|^2.$$

This requires, therefore, that $\varphi(-x) = e^{i\alpha(x)}\varphi(x)$, where the phase $\alpha(x)$ could, in principle, depend on x, contrary to what we had implicitly assumed in eq. (6.4.5).

Show that the function $\alpha(x)$ must obey the condition $\exp i[\alpha(x) + \alpha(-x)] = 1$.

We define a wavefunction, physically equivalent to φ (since it only differs from it in phase): $\Phi(x) = \varphi(x) e^{i\alpha(x)/2}$. Show that the new wavefunction Φ is either even or odd, justifying more rigorously, thereby, the assertions made at the beginning of sect. 6.4.1.

6.10. Show that a finite, flat well, of depth V_0 and width a, possesses exactly N *bound states*, where N is the integer defined by

$$N - 1 = \text{integral part of } (2mV_0 a^2/\pi^2)^{1/2}.$$

6.11. Consider a 'δ-*well*', i.e., a flat well in the limit in which it becomes indefinitely narrower and deeper, so that $V_0 \to \infty$ and $a \to 0$ while $V_0 a = U = \text{const}$. Examine the results of sect. 6.4 in the limit of $V_0 \to \infty$, under these conditions.

(a) What happens to the bound states, their energies and their wavefunctions?

(b) What happens to the expressions for the amplitudes and the coefficients of transmission and reflection?

6.12. Consider the wavefunctions of the bound states in a finite, flat well [eqs. (6.4.13) and (6.4.14)].

(a) Write down the *normalization* condition for the wavefunctions and hence deduce the value of the constant C:

$$|C|^2 = \frac{\kappa_E}{1 + \frac{1}{2}a\kappa_E}$$

(treat the cases of odd and even functions separately).

(b) Calculate the probability $\mathscr{P}_{\text{ext}}$ for the quanton to be localized outside the classically allowed region $[-\frac{1}{2}a, \frac{1}{2}a]$. Discuss the variation of $\mathscr{P}_{\text{ext}}$ as a function of the energy E

$(-V_0 < E < 0)$ of the bound state, and more particularly, the cases of the levels close to the bottom of the well $(E \gtrsim -V_0)$ and to its surface $(E \lesssim 0)$.

(c) Calculate the average values of the potential energy $\langle V \rangle$, and the kinetic energy $\langle T \rangle$ of the quanton.

Discuss their variations as functions of the energy E, comparing the quantum results with the classical values.

6.13. Consider the energy quantization condition, eq. (6.4.17), for the finite flat well, in the case of the lowest bound states of a deep well, that is when $V_0 - E \ll V_0$, so that $p' \ll \kappa_E$. Show that the quantization condition can approximately be written as $p' a_{\text{eff}} = n\pi$, with $a_{\text{eff}} = a + 2\kappa_E^{-1}$. (The finite, flat well is, therefore, roughly equivalent to an infinite, flat well of width a_{eff}, which results in the penetration of the quanton inside the classically forbidden region, just up to a distance $l = \kappa_E^{-1}$.)

6.14. Consider the *deuteron*, a neutron–proton bound state, of energy $-E_1$, assuming a hard-core [eq. (6.4.18)], flat potential.

(a) Show that the wavefunction is of the form

$$\varphi_{E_1}(r) = C \sin p' r, \quad \text{for } 0 < r \leqslant R_0,$$
$$= D e^{-\kappa r}, \quad r > R_0,$$

with $p' = [2m(V_0 - |E_1|)]^{1/2}$ and $\kappa = (2m|E_1|)^{1/2}$, $(m = \frac{1}{2}M)$. Use the approximate value $V_0 \simeq \pi^2/8mR_0^2 \gg E_1$ [see eq. (6.4.24)].

(b) Show, graphically, the shape of $\varphi_{E_1}(r)$. Prove that $\kappa R_0 \ll 1$ and that the 'exponential tail' of $\varphi_{E_1}(r)$ extends well beyond $r = R_0$.

(c) Write down the equations of continuity and hence deduce the approximate relation between the coefficients C and D: $|C| \simeq |D|$.

(d) Compute the probabilities $\mathcal{P}_{\text{int}}$ and $\mathcal{P}_{\text{ext}}$ for the proton–neutron distance to be, respectively, smaller than or greater than R_0 (i.e., for the deuteron to be inside or outside the classically admissible region). Show that $\mathcal{P}_{\text{ext}}/\mathcal{P}_{\text{int}} \simeq (\kappa R_0)^{-1} \gg 1$. Thus, the deuteron is a 'loose' system, having a size much larger than the range of the binding forces.

(e) Confirm this result by computing the average size $\langle r \rangle$ of the deuteron, and by showing that $\langle r \rangle > R_0$.

(f) Consider the states of the neutron–proton system, of angular momentum $l \neq 0$, subject to the potential $V(r) + V_{\text{centr}}^{(l)}(r)$ [see eqs. (6.4.18) and (6.4.26)]. Prove, by studying this function for $l = 1, 2, \ldots$, that for $l \geqslant 2$, this potential does not contain an attractive part and hence cannot give rise to any bound state. For $l = 1$, there exists a 'small' attractive potential well. Representing this potential schematically by means of a flat, hard-core, potential, show that it is also incapable of ensuring the existence of a bound state.

6.15. Solve the system of eqs. (6.4.27) to obtain b_- and d_+ as functions of b_+ [eqs. (6.4.28) and the *reflection and transmission coefficients* $\mathcal{R}$ and $\mathcal{T}$, eq. (6.4.30)]. Study the function $\mathcal{R}(E)$ and trace its graph (fig. 6.15). The manipulations can be greatly simplified by setting

$$\alpha = pa, \qquad \beta = p'a \quad \text{and}$$
$$\sinh \lambda = \frac{p'^2 - p^2}{2pp'} = \frac{q^2}{2pp'}, \qquad \cosh \lambda = \frac{p'^2 + p^2}{2pp'}.$$

6.16. Consider the *scattering* of a quanton, by a one-dimensional, finite, flat well, governed by eqs. (6.4.27)–(6.4.29).

(a) Writing the total amplitude in the form $\psi = \psi_{inc} + \psi_{scat}$, and by considering the backward and forward scattered waves ψ_{scat}, show that one can define a forward 'scattering amplitude'

$$f_{for} = \frac{d_+ - b_+}{b_+} = A_t - 1,$$

and a backward 'scattering amplitude'

$$f_{back} = \frac{b_-}{b_+} = A_r,$$

then the forward and backward 'scattering coefficients' (one-dimensional analogues of the three-dimensional differential effective cross-section), $\chi_{for} = |f_{for}|^2$ and $\chi_{back} = |f_{back}|^2 = \mathscr{R}$ and a 'total scattering coefficient' (analogue of the total effective cross-section), $\sigma = \chi_{for} + \chi_{back}$.

(b) Verify that the relation

$$\sigma = |f_{for}|^2 + |f_{back}|^2 = -2 \operatorname{Re} f_{for},$$

is the one-dimensional analogue of the optical theorem. (It would be useful to adopt the notation of exercise 6.15.)

(c) Show that at very low energy (for $E \ll V_0$, i.e., $p \ll p'$), we have, approximately, $f_{for} \simeq -1$ and $f_{back} \simeq e^{-ipa}$. Hence deduce that $\chi_{for} \simeq \chi_{back}$ (isotropic scattering).

(d) Study the variations of χ_{for} and χ_{back} with energy. Show that at the resonance energies, one gets $\chi_{for} \simeq \chi_{back}$ (the 'decay' of the resonance does not depend on the conditions – e.g., the initial direction – of its creation).

6.17. The *Ramsauer–Townsend effect*

We consider the effective cross-section for the scattering of two low-energy helium atoms (fig. 6.16a). We identify the interaction potential between the two atoms (fig. 6.16b) with a finite, flat well of width a and depth V_0.

(a) Interpreting the *first* minimum as being due to the Ramsauer–Townsend effect, calculate, using the conditions (6.4.32) for transparency, $p'a = n\pi$ with $n = 1$, the depth V_0 taking $a = 1$ (remember that the theory describes the relative motion of two atoms and that the mass to be considered is the reduced mass $M = M_{He}/2$).

(b) Calculate the kinetic energies and the relative velocities corresponding to transparencies of higher orders ($n = 2, 3$) than envisaged by the elementary theory. Hence deduce that the second and third minima of the effective cross-section (fig. 6.16a) *cannot* be explained by the Ramsauer–Townsend effect (see exercise 7.7).

6.18. Consider a one-dimensional potential well

$$V(x) = 0 \quad \text{if} \quad |x| > a,$$
$$= -V_0 \quad \text{if} \quad |x| < a, \quad (V_0 > 0),$$

and study the resonant scattering states (see sect. 6.4.2c) of a quanton of mass m in this potential. Denote by $E(>0)$ the energy of the stationary states of the quanton and adopt the following notation:

$$E = p^2/2m, \qquad E + V_0 = p'^2/2m, \qquad V_0 = q^2/2m.$$

Assume that the quanton is emitted at $x = -\infty$.

(a) Write the reflection amplitude in the form

$$A_r = |A_r|\, e^{i\delta}\, e^{-ipa}.$$

Express $|A_r|$ and δ as functions of p and p'. It is useful to set

$$g(E) = \frac{p'^2 + p^2}{2p'p}.$$

Under what conditions on A_r and δ is there resonance?

(b) In order to study the variation of δ as a function of E, first prove the following mathematical lemma.

Let x and y be two quantities related by the following functional relation

$$\tan y = \alpha \tan x \quad (\alpha > 1).$$

Show that, if we choose for y the functional dependence which has the value 0 when $x = 0$, then y and x simultaneously assume the values 0, π, 2π,... . What is the value of the derivative dy/dx at these points? Show that y and x also assume the values $\pi/2$, $3\pi/2$, $5\pi/2$,... , but with a different derivative. Compute this derivative. Trace the curve of the function $y(x)$.

(c) Using the results proved in (b), study the variation of δ with E. [Note that, except for small values of E, $g(E)$ varies slowly with E.] Show that the resonance energies E_n^* are the energies for which $\delta(E_n^*) = n\pi$. Trace, one below the other, the curves of the variations of $|A_r|$ and δ as functions of E. Compare these curves with the curves giving the modulus and the phase of the impedance Z of an LCR circuit in the neighbourhood of the resonance pulsation $\omega = (LC)^{-1/2}$.

(d) Write down the equation of motion of the centre of a wave packet, of the form given by eq. (6.4.38), in front of and beyond the well. Express, as a function of δ, the time taken by the wave packet to cross the region of the potential well. Show that, in general, a quanton spends less time inside the well than a classical particle, of velocity $v_0 = (2E/m)^{1/2}$, would do, except when there is a resonance, in which case it stays there much longer.

(e) By making an approximate expansion of the quantity $g \tan pa$, in a neighbourhood of $E = E_n^*$, show that A_r can be put into the form

$$A_r \propto \frac{1}{E - E_n^* + \tfrac{1}{2}i\Gamma_n}.$$

Express Γ_n as a function of m, a and E_n^*. Show that Γ_n characterizes the width ΔE_n^* of the resonance. Can one derive a Heisenberg-type relation here between ΔE_n^* and τ? What happens in the limit of high energies?

6.19. In this exercise, we show how it is possible to establish, *without computing*, the nature of the local variation of the probability density for the stationary states of a quanton subject to a succession of flat potentials.

(a) Show that in each flat potential region, of value V_0, the wavefunction of a stationary state of energy $E > V_0$ has the form

$$\varphi_E(x) = e^{i\alpha}(A\,e^{ipx} + B\,e^{i\beta}\,e^{-ipx}),$$

where p, α, β, A and B are positive, real quantities.

(b) Determine the shape of the function $\rho(x)$, the probability density, in the interior of a region of constant potential. Show that the continuity conditions imposed upon $\varphi_E(x)$, imply that while crossing a discontinuity of the potential, the curve of $\rho(x)$ shows neither a discontinuity nor a 'jump in the slope'.

(c) Establish the relation,

$$j^2 = v^2[\rho(x)]_{\max}[\rho(x)]_{\min} = v^2 \rho_{\max}\rho_{\min},$$

connecting the probability current j to the maximal and minimal values of the probability density (with $v = p/m$). Under what circumstances would there be no 'evolution' of the probability. How would one recognize a progressive 'wave'?

(d) Consider a beam of quantons, emitted from the left at infinity, along the axis, with an energy $E(>0)$ and falling upon an 'ascending' potential step of height $V_0 < E$. Draw the curve of the variation of ρ. Determine how it joins up at the step. Taking the value of the probability density to the right of the step as the standard, determine the limits between which $\rho(x)$ varies to the left. What happens to these results in the case of a 'descending' potential step?

(e) Consider now a rectangular barrier of height $V_0 > 0$, which we look upon as being a succession of an ascending step and descending step. As in (d), the quantons are emitted from the left with an energy $E > V_0$. Proceeding from right to left, draw the shape of the variations of ρ in the three regions of flat potential, determining in each case how they join up, and the limits between which ρ varies. Recover, in this way, the resonance condition, eq. (6.4.33).

Repeat the same problems for a potential 'well' of depth V_0.

(f) We would now like to extend the method to the case of a barrier of height $V_0 > E$ (tunnel effect). Let us set, as in the text,

$$\kappa = [2m(V_0 - E)]^{1/2}, \quad \text{real} > 0.$$

Show that $\rho(x)$ can be joined, at the points of discontinuity of the potential, without any jump in the slope, only if

$$A = B \neq 0 \quad \text{and} \quad \beta \neq 0.$$

Hence deduce that the wavefunction, inside the barrier, cannot be reduced to a single exponential component (increasing or decreasing). Taking into account the fact that j has to be conserved on either side of the barrier, show that $\rho(x)$ can, however, only decrease inside the barrier. Trace the curve of $\rho(x)$.

(g) Study qualitatively, and without going into details, how the results in (e) and (f) are modified when the barrier (or the well) is asymmetric.

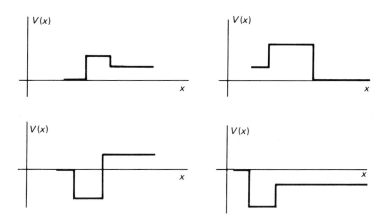

6.20. *Tunnel effect in vacuum*

The empty space between two metallic electrodes may be considered as being a flat potential barrier, for electrons, of height equal to the threshold potential V_0 of the metal, and of width equal to the distance between the electrodes. The passage of electrons across the barrier via the tunnel effect enables an electric current to pass across the void! The resistance of the space between the two electrodes is evidently proportional to the transmission factor, and hence depends exponentially on the width of the barrier. Deduce from the experimental results (fig. 6.27), the height V_0 of the potential barrier in the different cases. (The kinetic energy E of the electrons may be neglected, compared to the height V_0 of the barrier.)

6.21. *α-decay*

The function

$$V(x) = Z_1 Z_2 \, e^2/x \quad (Z_1, Z_2 > 0),$$

may be considered to be a one-dimensional representation of the Coulomb potential, which prevents a particle of charge $Z_1 q_e > 0$ (e.g., an α-particle) from penetrating into the interior of a nucleus of charge $Z_2 q_e$. Suppose that the nucleus is placed at $x = 0$ and that the α-particle arrives from $x = +\infty$, to the right (see fig. 6.28).

(a) Trace the curve of $V(x)$ and compute the probability for the particle to penetrate, up to $x = 0$, inside the Coulomb barrier. To do this, use formula (6.5.21) and determine the abscissa x_2 of the classical turning point. One obtains $x_2 = R$ (the nuclear radius). Application: take $Z_1 = 2$, $Z_2 = 90$, $R = 7$ F.

(b) Is the (so-called Gamow) transmission coefficient calculated in (a), equal to that which appears in the treatment of radioactive α-decay?

(c) Hence deduce the Gamow–Condon–Gurney law

$$\log T = a + bE^{-1/2},$$

which connects the half-life T of an α-radioactive nucleus to the energy of the α-particles which it emits. Compare with fig. 6.28b. From the theoretical expression obtained for the coefficient b, estimate the radius of the nuclei in question.

6.22. It is known that in order to extract electrons from a metal, it is necessary to supply the most energetic electrons with a (threshold) energy energy Φ, characteristic of the metal (see the photoelectric effect). The 'release' of the electrons can also be facilitated by the application, to the metallic sample under consideration, of an electric field $\mathscr{E}$, perpendicular to one of its faces. Discuss how exactly!

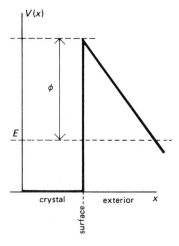

This effect is known as *field emission*. The electrons of the metal now find themselves under the influence of a potential $V = -\mathscr{E}x + \text{const.}$, which along (the direction x of) the field $\mathscr{E}$ has the form shown in the figure. Establish the formula $\mathscr{T} = \exp[-\tfrac{4}{3}(2m\Phi^3)^{1/2}/\mathscr{E}]$ for the transmission coefficient of this barrier. For what values of the electric field $\mathscr{E}$ does this coefficient become appreciable?

6.23. Show that in the *thick* barrier approximation, conditions (6.6.10) and (6.6.12) can be written as

$$\tan pa = -\frac{2p/\kappa}{1 - p^2/\kappa^2}\left(1 \pm \frac{1 + p^2/\kappa^2}{1 - p^2/\kappa^2}e^{-2\kappa d}\right).$$

Hence, comparing with the levels of the isolated well given by eq. (6.4.17), deduce that the levels $E_n^{(A)}$ and $E_n^{(S)}$ are given by

$$E_n^{(A),(S)} = E_n\left(1 \pm \frac{4\kappa}{q^2 a}e^{-2\kappa d}\right).$$

6.24. A quanton of mass m is subjected to a one-dimensional *double-well* potential $V(x)$.

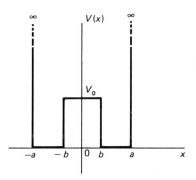

$$V(x) = +\infty \qquad a < |x|,$$
$$= 0 \qquad b < |x| < a,$$
$$= V_0 > 0 \qquad |x| < b.$$

We are interested in the stationary states of the quanton for energy proper values $E < V_0$. We allow these stationary state wavefunctions to be either even or odd, in view of the symmetry of the potential. Set

$$p^2 = 2mE, \qquad \kappa^2 = 2m(V_0 - E), \qquad q^2 = 2mV_0.$$

(a) Find the forms of the even and odd stationary state wavefunctions, in the five regions in which the potential is constant.

(b) Imposing the relevant matching conditions, show that the energy is quantized by the condition

$$\pm p \cotan p(a - b) = \kappa (\tanh \kappa b)^{\pm 1},$$

(the exponent ± 1 corresponding, respectively, to even and odd functions).

(c) What happens to this double condition in the limit in which $b \to 0$ and $V_0 \to \infty$, simultaneously in a way such that $q^2 b = \text{const.}$ ('δ-function' barrier)?

(d) First, in the case of the odd functions, show that one obtains in this way half the energy levels of an infinite well of width $2a$, *without* a central barrier, and explain the result.

(e) In the case of even functions show, using a graphical solution, that at least for $\lambda = \kappa a \gg 1$, the successive levels $E_n^{(+)}$ differ slightly from the levels $E_n^{(-)}$ calculated in (d). Compute, approximately, the difference $\Delta E_n = E_n^{(+)} - E_n^{(-)}$ of the 'doublet' of levels $E_n^{(\pm)}$.

(f) What happens to the energy levels $E_n^{(+)}$, in the case of $\lambda \to 0$ [the opposite of (e)]? Explain.

6.25. *Doubling by inversion of the ground level of ammonia.*
We wish to compare the doubling δE, due to the tunnel effect, of the ground level for the

three ammonia molecules, corresponding to the three different isotopes of hydrogen H and D (deuterium). We write

$$Am_1 = {}^{14}NH_3, \qquad Am_2 = {}^{14}NH_2D, \qquad Am_3 = {}^{14}NHD_2.$$

(a) Show that the mass which ought to enter into the calculation of the tunnel effect factor is the reduced mass of the four atoms, namely $\mu = mM/(m + M)$, where m is the mass of the three hydrogen atoms and M the mass of the nitrogen atom. Compute the reduced masses μ_i ($i = 1, 2, 3$) for the three molecules being studied.

(b) The atomic potentials, and, in particular, the double well governing the inversion of the molecule, are assumed to be independent of the isotopic constitution. Show, using formulae (6.6.14), that the difference δE_k for the molecule of the kth species is related to the corresponding reduced mass μ_k by the relation

$$\ln \delta E_k = A + B\mu_k^{1/2},$$

where A and B are constants which are the same for all the molecules. Compare with the experimental results: $v_1 = 23\,786$ MHz, $v_2 = 12\,182$ MHz, $v_3 = 5160$ MHz.

(c) Compute, using v_2/v_1, the order of magnitude of the average height of the barrier separating the two minima of the double well, knowing that the distance $D = 0.8$ Å.

6.26. (a) Show that for the battlement potential (sect. 6.7), the invariance under *spatial reflection* implies the property $\mathbf{M}(-p) = \mathbf{M}^{-1}(p)$ for the matrix $\mathbf{M}$ defined in eq. (6.7.9).

(b) Hence deduce that $M_4(p) = \bar{M}_1(p)$ and $M_2(p) = \bar{M}_3(p)$.

(c) One can obtain the same results in another way: show that if $\varphi_E(x)$ is the wavefunction of a stationary state of energy E, the same is necessarily true for the complex conjugated function $\overline{\varphi_E(x)}$. Hence recover the results of (b).

(d) Define the matrix $\mathbf{S}$ connecting

$$\begin{pmatrix} b_- \\ d_+ \end{pmatrix} \quad \text{to} \quad \begin{pmatrix} b_+ \\ d_- \end{pmatrix}.$$

How do you interpret it? Compute its matrix elements using those of $\mathbf{M}$. Show that it is unitary.

6.27. Consider a quanton subject to a battlement potential $V(x)$ of the form shown in fig. 6.40b. Let $\mathbf{M}$ be the *transfer matrix* defined through relation (6.7.9) in the text.

(a) Find the expression for the coefficients of $\mathbf{M}$ as a function of E (energy of the quanton assumed to be in a stationary state), V_0 (height of the barriers), a (width of each barrier). Why does the periodicity l of the potential $V(x)$ not figure in the expression for the coefficients of $\mathbf{M}$?

(b) Define $\check{\mathbf{M}}$, using $\mathbf{M}$, via relation (6.7.14) of the text. Calculate the proper values m' and m'' of $\check{\mathbf{M}}$. Show that their product is equal to one. Hence deduce that if one of them is real, then so also is the other, that in this case one of them has modulus greater than and the other less than one, and that wavefunction $\varphi_E(x)$ cannot remain bounded at the end of the chain.

(c) Hence deduce that, in order for $\varphi_E(x)$ to remain bounded, it is necessary that

$$|\operatorname{Re} \check{M}_1| \leqslant 1,$$

and that the two proper values are now complex conjugates of each other, of unit modulus.

6.28. Consider the stationary states of a quanton of energy E in a battlement potential of height V_0, for the case $E \gg V_0$.

(a) Demonstrate the existence of alternately allowed and forbidden *energy bands*, by establishing an inequality of the type of eqs. (6.7.16)–(6.7.18), suitably modified.

(b) Study the function $E(\mu)$, and compare it with the formula $E = p^2/2m$ which holds for a free quanton, in the limit $E \gg V_0$.

6.29. Show that at the edges of the Brillouin zone and at its centre, the curve $E(\mu)$ admits horizontal tangents. [Perform a finite expansion of the 'dispersion relation', eq. (6.7.27), around the values of μ under consideration.] Hence deduce the form of the function $v(\mu)$ giving the *velocity of propagation*, $v = \mathrm{d}E/\mathrm{d}\mu$ as a function of the quasi-momentum.

6.30. In this exercise we shall attempt to make precise the results of the preceding exercise. We restrict ourselves to the stationary states of the quanton of energy E, where $0 < E < V_0$ (the geometry of the potential and the notation are shown in fig. 6.40).

(a) Let $E_n (0 < E_n < V_0)$ be the energy of the nth bound state of the single well. Assume that $\kappa'_n a \gg 1$, $\kappa'_n = [2m(V_0 - E_n)]^{1/2}$. Establish the dispersion relation $E(\mu)$ in the neighbourhood of $E = E_n$, in the form

$$2\,\mathrm{e}^{-\kappa'_n a} \cos \mu l = (\kappa' - \kappa'_n) F'(\kappa'_n),$$

where

$$F(\kappa'_n) = (\cos pb + \eta \sin pb)_{E=E_n}.$$

Hence deduce that in the neighbourhood of $E = E_n$, the form of the curve $E(\mu)$ is given by the relation

$$E = E_n - 4E_n\,\mathrm{e}^{-\kappa'_n a} \cos \mu l \, [\kappa'_n F'(\kappa'_n)]^{-1}.$$

Hence deduce the expression for the width ΔE_n of the band originating from E_n. Recover the result that the narrower bands originate from the deeper levels.
Show that the lower edge of the band corresponds either to $\mu = 0$ or to $\mu = \pm \pi/l$ (in the reduced-zone diagram).

(b) Show that at the centre of the reduced-zone diagram (along the vertical $\mu = 0$ and its immediate neighbourhood) the *dispersion curve* can be identified with a succession of parabolas alternately concave upwards and downwards. Set up the equation of these parabolas and show that they become gradually more and more pointed as E increases.

6.31. Consider a crystal composed of identical atoms placed at the nodes of a three-dimensional lattice. Let a, b, c be the dimensions of a unit cell, formed along the axes Ox, Oy, Oz.

Denote by R the vector designating the position of an atom. Each atom behaves as a point scattering centre with respect to a beam of X-rays. Let p be the characteristic momentum of the incident beam (parallel, monochromatic light).

(a) Show that the scattering directions in the interior of the crystal are determined by the condition $e^{i(p'-p)\cdot R} = 1$, for arbitrary R, where p' denotes the momentum of the photons after elastic scattering.

(b) Show that this condition is equivalent to the Bragg condition, as traditionally written, $2d \sin \theta = k\lambda$.

(c) In p-space, construct a lattice of points, the coordinates p_x, p_y, p_z of which are integral multiples of the quantities $2\pi/a, 2\pi/b, 2\pi/c$, respectively ('*reciprocal lattice*'). Show that the condition written down in (a) demands that the momentum of the photon be 'conserved up to a vector of the reciprocal lattice'. Would it not be better, under these conditions to speak of the quasi-momentum?

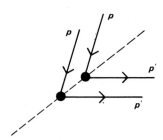

6.32. (a) Express the components of the proper vector of the matrix $\check{\mathsf{M}}$ [eq. (6.7.14)] as functions of the important parameters p, κ' and η [all depending on E, see eq. (6.7.11)].

(b) Show that at *Bragg reflection* ($e^{\pm i\mu l} = 1$), the forward scattering amplitude from an (arbitrary) barrier is equal, in absolute value, to the backward scattering amplitude.

6.33. A low-energy ($\simeq$ a few tens of eV), parallel, monoenergetic beam of electrons is passed through a crystal. The crystal is cut in a way such that the face perpendicular to the electrons is parallel to one of its family of lattice planes, for which the distance between successive planes is d. We are interested in the *reflection of electrons from the crystal*.

The interaction of an electron with the crystal is represented schematically by a *one-dimensional battlement potential* (placed along the normal z to the lattice planes). The wells of this potential coincide with the lattice planes and have a depth of about 10 eV.

(a) Taking into account the fact that the only source of electrons is to the left of the crystal, set up the matching conditions for the wavefunctions at the crystal/vacuum interface, $z = 0$, in the form

$$p = \pm \not\!p,$$

where p denotes the modulus of the momentum of the incident electrons in vacuum and $\not\!p$ their quasi-momentum in the interior of the crystal.

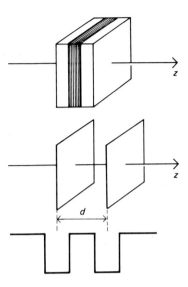

Figure 1

(b) Figure 2 shows, facing each other, two band diagrams, one displaying the reduced zone of a one-dimensional crystal and the other the curve of the variation of the reflectivity of the crystal, in a given direction, as a function of the energy E. How would you interpret the existence of the reflectivity peaks?

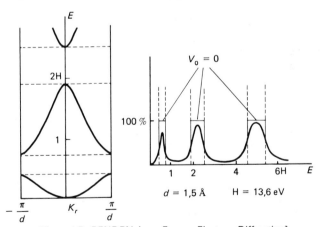

[From J.B. PENDRY *Low Energy Electron Diffraction*]

Figure 2

(c) Why are the peaks of the reflectivity curve not in fact equal to 100%? Hence deduce that the absorption, which is always present in a real crystal, has the effect of reducing the width of the forbidden bands. Explain how.

7

Collective behaviour: Identical quantons

7.1. Composite systems

7.1.1. The composite factorization principle

Let us consider two particular quantum transitions effected by two different quantons. Quanton no. 1, for example an electron, of initial state $p^{(1)}$, is projected onto a final state $r^{(1)}$, while quanton no. 2, a photon if we wish, of initial state $q^{(2)}$, is projected onto the final state $s^{(2)}$. If we are interested in the composite system, consisting of the two quantons, we may say that it has undergone a transition from an initial state, characterized by the couple of individual states $(p^{(1)}, q^{(2)})$, to the final state, characterized by the couple of individual states $(r^{(1)}, s^{(2)})$: This transition is described by a probability amplitude $\langle r^{(1)}, s^{(2)} | p^{(1)}, q^{(2)} \rangle$, from which we derive the probability

$$\mathscr{P}^{(1,2)} = |\langle r^{(1)}, s^{(2)} | p^{(1)}, q^{(2)} \rangle|^2. \tag{7.1.1}$$

However, the transition of the composite system consists of two individual transitions, which, since they modify different physical objects, should be considered as independent events. Each one of these transitions occurs with a certain probability, derived from the corresponding amplitude for the respective quanton

$$\mathscr{P}^{(1)} = |\langle r^{(1)} | p^{(1)} \rangle|^2, \qquad \mathscr{P}^{(2)} = |\langle s^{(2)} | q^{(2)} \rangle|^2. \tag{7.1.2}$$

The probability of the joint event, i.e., the transition of the composite system, consisting of the two independent transitions of the individual quantons, is, according to the classical theory of probability, given by the product of the individual probabilities,

$$\mathscr{P}^{(1,2)} = \mathscr{P}^{(1)} \mathscr{P}^{(2)}. \tag{7.1.3}$$

This factorization of the probability requires, in view of eqs. (7.1.1) and (7.1.2), the factorization of the modulus of the amplitude of the composite system in terms of the modulii of the individual amplitudes. Indeed, a basic principle of quantum theory, in complete accord with the rules of combination of amplitudes, already enunciated in ch. 4, extends this factorization to the amplitudes themselves, including the phase. We may thus state the

---

Composite factorization principle

For a composite quantum system, undergoing a transition between two states characterized by the individual states of its components, the probability amplitude of the transition is the product of the probability amplitudes of each one of the subsystems,

$$\langle r^{(1)}, s^{(2)} | p^{(1)}, q^{(2)} \rangle = \langle r^{(1)} | p^{(1)} \rangle \langle s^{(2)} | q^{(2)} \rangle. \qquad (7.1.4)$$

In this statement, the principle has been extended from the example considered, in order to make it applicable to arbitrary subsystems and not just to individual quantons. So long as these subsystems are independent, the reasoning evidently remains valid. Of course, it can be applied not just to the case of two subsystems, but also to an arbitrary number of them. The composite factorization principle should be carefully distinguished from the sequential factorization principle (see sect. 4.4.3).

7.1.2. Limit of validity of the composite factorization principle

The preceding discussion might seem highly trivial, were it not for the fact that the fundamental principles of quantum theory set limits on its validity, in a manner which is perhaps unexpected but is, nevertheless, physically essential. Let us consider again the transition of a system of two quantons, but suppose that this time we have two electrons, two photons or two iron atoms – i.e., two *identical* quantons. A major difference now appears from the case with two different quantons. If we characterize the initial state of the composite system by two individual states (p, q), the identity of the two quantons prevents us from labelling the states in a manner which refers to one or the other quanton: there is no distinguishing mark to enable us to differentiate quanton 'no. 1' from quanton 'no. 2', and hence to attribute the states p and q to one or the other separately.

As a concrete (and deliberately extreme!) example, suppose that p is a state localized in Paris and q a state localized in Quebec. For two distinct quantons, such as an electron (no. 1) and a photon (no. 2), the state $(p^{(1)}, q^{(2)})$ describes an electron in Paris and a photon in Quebec, which is quite different from the state $(p^{(2)}, q^{(1)})$ describing a photon in Paris and an electron in Quebec. However, for two identical quantons, e.g., two electrons, there is only one state (p, q): one electron in Paris and one in Quebec. Thus, the symbols p and q refer to the individual states, and not to the quantons which occupy them and which cannot be individualized. From this it follows that we can no longer apply the composite factorization principle, to calculate the transition amplitude $\langle r, s | p, q \rangle$ of a system composed of identical quantons. For distinct quantons, the same properties which differentiate between them also enable us to determine which quanton has made the transition from the state p to the state r, and which one from q to s. Going back to the case of an electron and a photon, e.g., if the initial state p and the final state r are occupied by an electrically charged quanton, then it must be the electron which has made the transition $r \leftarrow p$, hence the possibility of labelling the transitions: $r^{(1)} \leftarrow p^{(1)}$ and $s^{(2)} \leftarrow q^{(2)}$. For identical quantons, it is impossible to analyze the transition $(r, s) \leftarrow (p, q)$ as the conjunction of two individual transitions, $r \leftarrow p$ and $s \leftarrow q$ or, for that matter, of $r \leftarrow q$ and $s \leftarrow p$ (fig. 7.1). Just as in a Young's double slit experiment, the quanton does not pass through either one *or* the other of the two slits (there is only one transition), similarly, we have here just one transition $(r, s) \leftarrow (p, q)$ and not the two

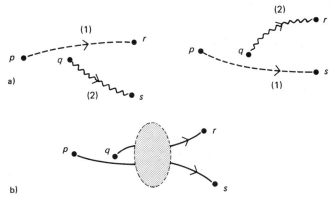

Fig. 7.1. *Two-quanton transitions*
(a) For two distinct quantons, for the same initial state (quanton 1 in p, quanton 2 in q), there exist two final states, and hence two quantum transitions, depending on whether the final state of quanton 1 is r and that of quanton 2 is s, or the other way around.
(b) If the two quantons are identical, it is not possible to differentiate between them so as to be able to either label them or to follow their trajectories. Thus, there is now only one final state and only one transition.

possibilities $(r \leftarrow p; s \leftarrow q)$ or $(r \leftarrow q; s \leftarrow p)$. The identity – sometimes called 'indistinguishability', although this term is highly subjective – of the quantons, prevents us from separating a collective transition into a set of independent, distinct, individual transitions. Even when the (identical) quantons do not interact, or influence each other through specific forces, their collective behaviour cannot be understood in terms of individual behaviour. We may say, in a deliberately figurative way, that *identical quantons, even when they are free, are not independent.*

In classical theory, identical particles can be individually marked, in view of the possibility of localizing them: it suffices to follow their trajectories. Moreover, one could also say that it is this very localizability which prevents us from considering them as being 'truly' identical (each having its own history). However, in the present case, the spatial extension of the quantons destroys the very notion of a trajectory, and renders it impossible to follow, from within an ensemble of identical quantons, the individual destiny of each one separately. We have here another demonstration of the unusual nature of quantons, studied in ch. 2: while it is true that within a collective system, classical particles may be denumerated and individualized, when at the same time classical fields can be neither denumerated nor individualized, quantons can, however, be denumerated but not individualized. More precisely, it ought to be stated that quantons may be denumerated and not individualized, but that, under certain circumstances, ensembles of quantons may be denumerated and practically individualized – these then are called classical 'particles' – while under other circumstances, they seem to be neither denumerable nor individualizable – what are then called classical 'fields' (see sect. 7.6).

Systems composed of identical quantons possess, therefore, collective properties which distinguish them radically from systems of classical particles. It is the notion of the probability amplitude, suitably extended, which enables us to understand these fundamental aspects of quantum physics. Let us add that we also understand why, under certain conditions, even identical quantons may, to an excellent approximation, be treated as being independent. Our example of electrons in Paris and Quebec was, in this respect, deliberately exaggerated: one *can*, in practice, deal only with the Parisian electrons and consider their quantum transitions, forgetting all about their Québecois cousins.

7.1.3. Collective description of composite systems

Before moving on to the study of systems of identical quantons, it would be appropriate to return to the very description of composite systems. So far we have specified the state of such a system in terms of the individual states, of the type (r, s) for a two-quanton system. In this description of the states r and s one state could, for example, be a proper state of localization at a point A, and the other a proper state of localization at a point B. Or otherwise, r and s could be proper states of momentum, having respectively proper values p_α and p_β. The two individual states r and s could just as well be of a different nature: for example, one a proper state of localization at c and the other a proper state of the momentum p_γ.

However, we could just as well choose another specification of the states of a composite system, in which we consider mutually compatible collective magnitudes of the system – e.g., for a two-quanton system, the position of the centre of mass or the relative momentum.

Thus, for a two-quanton system, we might consider a simultaneous proper state of the position of the centre of mass of the two quantons and the relative momentum of the two quantons, or e.g., a simultaneous proper state of the total momentum and the relative angular momentum of the two quantons. The crucial point now is that such a state is not in any way identical to any state of the type (r, s), specified by two individual states. Indeed, a collective magnitude (such as the position of the centre of mass) has no reason to be compatible with the set of individual physical magnitudes referring only to one or the other quanton. A proper state of such collective magnitudes cannot, therefore, be a proper state of the individual magnitudes and cannot be characterized by a pair of individual states. It is, therefore, impossible, in general, to describe the state of a composite quantum system in terms of the individual states of its components. This is in stark contrast to the description of classical systems, which always allows a labelling of the individual states. We say, therefore, that we have, in the quantum theory, a fundamental 'non-separability', which moreover is remarkably illustrated by the behaviour of composite systems of identical quantons (but which is not limited to these systems alone). In any case, let us remember that states of the type (r, s), specified by the individual states, constitute only a very particular (although extremely important!) class of states of a composite system.

7.2. Systems of two identical quantons

7.2.1. Amplitudes of two identical quantons

(a) *Symmetry and antisymmetry.* Let Ψ be a state of a system of two identical quantons, the exact specification of which is not important here. We ask for the probability of finding in it the two quantons, in well-defined, individual states, which we call r and s, respectively. Hence, we are interested in the probability of the transition $(r, s) \leftarrow \Psi$, from the state Ψ to a state (r, s), specified by two individual states. According to the basic formulation of quantum theory, this probability is calculated by means of a certain transition amplitude $\langle r, s | \Psi \rangle$,

$$\mathscr{P}_{r,s \leftarrow \Psi} = |\langle r, s | \Psi \rangle|^2. \tag{7.2.1}$$

When the quantons are distinct, it is naturally appropriate that the first state symbol (here r) be applied to one of the quantons (call it 1), and the second (s) to the other, which we call (2). The state, which we may denote more precisely by $(r^{(1)}, s^{(2)})$ differs, as we have seen, from the state $(s^{(1)}, r^{(2)})$. The probability of finding 1 in the state s and 2 in the state r is, therefore, in general not the same as that of finding 1 in the state r and 2 in the state s! Thus, for a hydrogen atom, in the course of a measurement of the velocities of its two constituents, the probability of obtaining the velocities 2×10^6 m s^{-1} and 10^4 m s^{-1} for the electron and the proton, respectively, is much greater than the probability of obtaining 2×10^6 m s^{-1} for the proton and 10^4 m s^{-1} for the electron (exercise 1.14):

$$\mathscr{P}_{r^{(1)}, s^{(2)} \leftarrow \Psi} \neq \mathscr{P}_{s^{(1)}, r^{(2)} \leftarrow \Psi}, \quad \text{for distinct quantons.} \tag{7.2.2}$$

On the other hand, as soon as we consider the two quantons to be identical, the situation becomes symmetrical. The states defined by (r, s) and (s, r) are the same, and we must obtain only one probability, which does not depend, therefore, on the order of the arguments (r, s):

$$\mathscr{P}_{r,s \leftarrow \Psi} = \mathscr{P}_{s,r \leftarrow \Psi}, \quad \text{for identical quantons.} \tag{7.2.3}$$

Attention, however – and we repeat once more – one *cannot* say now that, 'the probability is the same of finding 1 in r and 2 in s, as of finding 1 in s and 2 in r'! The very identity of the quantons forbids labelling them and asking 'which one is in which given state'... .

The transition probability, of a system of two identical quantons, from an arbitrary state Ψ onto a state characterized by two individual states is, therefore, a symmetric function, following eq. (7.2.3), of the individual states. Thus, the corresponding transition amplitude, defined by (7.2.1), is symmetric *in modulus*,

$$|\langle r, s | \Psi \rangle| = |\langle s, r | \Psi \rangle|. \tag{7.2.4}$$

(from now on we shall consider identical quantons and hence not indicate the fact next to the formulae). The 'interchange' of the arguments (r, s) in the function $\langle r, s | \Psi \rangle$ thus only gives rise to a phase shift

$$\langle s, r | \Psi \rangle = e^{i\alpha} \langle r, s | \Psi \rangle. \tag{7.2.5}$$

However, a second interchange, by introducing a second phase factor, must lead back to the initial amplitude

$$\langle r, s | \Psi \rangle = e^{i\alpha} \langle s, r | \Psi \rangle = e^{2i\alpha} \langle r, s | \Psi \rangle. \tag{7.2.6}$$

Thus, we have

$$e^{2i\alpha} = 1, \quad \text{so that} \quad e^{i\alpha} = \pm 1, \tag{7.2.7}$$

implying that two cases arise for the two-quanton amplitudes

symmetry: $\quad \langle r, s | \Psi \rangle = + \langle s, r | \Psi \rangle, \tag{7.2.8}$

antisymmetry: $\quad \langle r, s | \Psi \rangle = - \langle s, r | \Psi \rangle. \tag{7.2.9}$

The above is a restriction of considerable importance, the physical implications of which can hardly be overestimated: it determines in an essential way the properties of systems with a large number of quantons – among others, the macroscopic world in which we live, as this chapter will show.

The symmetry or antisymmetry of the amplitudes is not a particular property of the transition considered, which would depend, e.g., on the choice of the states Ψ, r and s, for the two given quantons. In fact, it depends only on the quantons in question and we have, either *always* eq. (7.2.8) or *always* eq. (7.2.9) for each species of quantons. This fact is easily proved. Indeed, let us assume that, for the same two-quanton system, we could have two amplitudes corresponding to the same final state (r, s) – one symmetric, given by a state Φ_1, the other antisymmetric, given by a state Φ_2. Thus,

$$\langle r, s | \Phi_1 \rangle = \langle s, r | \Phi_1 \rangle,$$
$$\langle r, s | \Phi_2 \rangle = - \langle s, r | \Phi_2 \rangle. \tag{7.2.10}$$

Let us now consider the transition to (r, s) from an arbitrary state Ψ, under experimental conditions which allow just the two intermediate states Φ_1 and

Φ_2. The rules for the addition and factorization of amplitudes enable us to write

$$\langle r, s | \Psi \rangle = \langle r, s | \Phi_1 \rangle \langle \Phi_1 | \Psi \rangle + \langle r, s | \Phi_2 \rangle \langle \Phi_2 | \Psi \rangle. \qquad (7.2.11)$$

We see that this amplitude is, in general, neither symmetric nor anti-symmetric, and hence is inadmissible. If, on the other hand, $\langle \Phi_2 | \Psi \rangle = 0$ (or, for that matter, $\langle \Phi_1 | \Psi \rangle = 0$) for all states Ψ, there is no difficulty any more. But then, since the transition $\Phi_2 \leftarrow \Psi$ is not possible, Φ_2 is *not*, in fact, a state of the system! There exist, therefore, two major classes of quantons. Those having (always!) symmetric amplitudes are called *bosons*, because of the so-called 'Bose–Einstein' statistics, which describes their behaviour when in large aggregates. Those having (always!) antisymmetric amplitudes are called *fermions*, in view of the so-called 'Fermi–Dirac' statistics.

> symmetry : bosons
>
> antisymmetry : fermions

This dichotomous property of quantons is called 'statistics', for historical reasons, even in situations where one is dealing with only a few quantons (2 or 3,...) and where the statistical viewpoint, in its proper sense, has no relevance. A stricter terminology would be preferable: one could, e.g., speak of the *permutability* of quantons, and distinguish between quantons of even permutability, or bosons, and odd permutability, or fermions.

The symmetry or antisymmetry of a two-quanton amplitude is a general property, which holds just as well when the final state is not specified in terms of the individual states. Indeed, for an arbitrary state Φ, we can always define a state Φ_{Exch}, which results from it through an interchange of all the arguments related to the two quantons: if Φ is defined as a proper state of the collective physical magnitudes $\mathscr{L}(1, 2)$, $\mathscr{M}(1, 2)$, etc., Φ_{Exch} would be the proper state, with the same proper values of the permuted magnitudes $\mathscr{L}(2, 1)$, $\mathscr{M}(2, 1)$, etc. In the example cited above, of a proper state Φ of the position of the centre of mass $X = \frac{1}{2}(X_1 + X_2)$ of two quantons, and their relative momentum $p = p_2 - p_1$, the state Φ_{Exch} would be the proper state of $\frac{1}{2}(X_2 + X_1) = X$ and $p_1 - p_2 = -p$. This is the same as saying that if Φ has the proper values X_0 and p_0, Φ_{Exch} has the proper values X_0 and $-p_0$. The above reasoning, dealing with the interchange of individual states, extends without difficulty to an exchange operation carried out on an arbitrary state of two quantons. We may write the general result in the form

$$\langle \Phi_{Exch} | \Psi \rangle = \pm \langle \Phi | \Psi \rangle. \tag{7.2.12}$$

Finally, the symmetry or antisymmetry of two-quanton amplitudes holds just as well for the initial state of a transition as for the final state, as proved immediately by complex conjugation (see sect. 4.4.5)

$$\langle \Phi | \Psi_{Exch} \rangle = \overline{\langle \Psi_{Exch} | \Phi \rangle} = \pm \overline{\langle \Psi | \Phi \rangle} = \pm \langle \Phi | \Psi \rangle. \tag{7.2.13}$$

On a general level, it is important to note the fundamental role that is played here by an invariance principle, namely the invariance of the physical laws under permutations of (identical) quantons. These permutations form a *group* – the simplest one possible, since it only consists of two elements: the identity operation and the operation of exchange of the quantons. The existence of such invariance groups is certainly not specific to quantum theory (the same invariances hold, in general, in classical theory too), but here they lead to deeper consequences.

It ought to be admitted that the foregoing 'proof', leading us to the properties of symmetry or antisymmetry, eqs. (7.2.8) or (7.2.9), is a gross over-simplification, lacking in rigour. The reader is invited to find out for himself the gaps in the reasoning, before reading the more correct version which follows – through which one might wish to go quickly, on first reading, knowing the validity of the results obtained.

(b) *A matter of phase.* It was from eqs. (7.2.5) and (7.2.6) that we had jumped too quickly. It is true that the interchange of the arguments (r, s) in the amplitude $\langle r, s | \Psi \rangle$ 'leads to a simple phase shift'. But this latter depends a priori on the states in question. In place of eq. (7.2.5), it is necessary, therefore, to write more precisely

$$\langle s, r | \Psi \rangle = e^{i\alpha(s,r)} \langle r, s | \Psi \rangle. \tag{7.2.14}$$

From this we get, in place of eq. (7.2.6),

$$\langle r, s | \Psi \rangle = e^{i\alpha(r,s)} \langle s, r | \Psi \rangle = e^{i\alpha(r,s) + i\alpha(s,r)} \langle r, s | \Psi \rangle, \tag{7.2.15}$$

and there is no reason to believe that the phase α is a symmetric function. A priori, we have

$$\alpha(r, s) \neq \alpha(s, r), \tag{7.2.16}$$

which destroys the conclusions in eq. (7.2.7). However, we shall see that they can be recovered. Let us adopt a more compact notation setting

$$\langle r, s | \Psi \rangle \triangleq A(r, s) = A \quad \alpha(r, s) \triangleq \alpha,$$

$$\langle s, r | \Psi \rangle \triangleq A(s, r) = A' \quad \alpha(s, r) \triangleq \alpha'. \tag{7.2.17}$$

Thus, we have the relations

$$A = e^{i\alpha} A', \qquad A' = e^{i\alpha'} A, \tag{7.2.18}$$

hence, evidently,

$$e^{i(\alpha + \alpha')} = 1. \tag{7.2.19}$$

Let us define a new amplitude

$$B(r, s) \triangleq e^{-i\alpha(r,s)/2} A(r, s). \tag{7.2.20}$$

Since B only differs from A by a phase factor, it is physically equivalent to A. What, now, is the effect on the amplitude B of the interchange of the states (r, s)? We have

$$\begin{aligned}
B(s, r) &= e^{-i\alpha(s,r)/2} A(s, r) \quad &\text{by eq. (7.2.20),} \\
&= e^{-i\alpha'/2} A' \quad &\text{by eq. (7.2.17),} \\
&= e^{-i\alpha'/2} e^{i\alpha'} A \quad &\text{by eq. (7.2.18),} \\
&= e^{i\alpha'/2} e^{i\alpha/2} B(r, s) \quad &\text{by eq. (7.2.20).}
\end{aligned}$$

Since, by eq. (7.2.19),

$$e^{i(\alpha + \alpha')/2} = \pm 1, \tag{7.2.21}$$

we again get one of the two relations

$$B(s, r) = \pm B(r, s). \tag{7.2.22}$$

In other words, it is not necessarily the amplitude A which is (anti)symmetric, but rather an equivalent amplitude B. More precisely, among the set of physically equivalent transition amplitudes (i.e., differing only in their phases), there exists one which is (anti)symmetric. Clearly, it is this privileged amplitude which is always chosen in order to characterize the transition under consideration [actually, there exists a whole class of (anti)symmetric amplitudes, differing from one another by their phase factors which are themselves symmetric].

(c) *Collective amplitudes and individual states.* This analysis can be summarized in a special case of great practical importance – one in which the transition proceeds from an initial state, which is itself marked by a pair of individual states, (p, q) say, to the final state (r, s). We shall see an illustration of this in the scattering of two quantons, marked, for example, by their respective momenta before and after their collision. If the quantons are distinct, there is no ambiguity: a transition amplitude $\langle r, s|p, q\rangle$, which we may denote more precisely by $\langle r^{(1)}, s^{(2)}|p^{(1)}, q^{(2)}\rangle$, enables us to compute the probability of the quanton M_1 to be projected from the state p onto the state r and the quanton M_2 from q to s. The interchange of the final states r and s leads to a completely different process, in which M_1 passes from p to s and M_2 from q to r, characterized by an amplitude $\langle s, r|p, q\rangle$, distinct from $\langle r, s|p, q\rangle$ (see fig. 7.1). The probability of obtaining the two final states r and s, without worrying about the identity of the quantons occupying the states, is obtained in the usual way, by adding the probabilities corresponding to the two independent processes:

$$\mathscr{P}^{\neq} = |\langle s, r|p, q\rangle|^2 + |\langle r, s|p, q\rangle|^2$$
$$= \mathscr{P}_{s,r \leftarrow p,q} + \mathscr{P}_{r,s \leftarrow p,q}. \tag{7.2.23}$$

Now suppose, that the quantons are identical. Then there is only one final state, since it is impossible to label the quantons by M_1 or M_2 and to know whether the quanton, which after the transition is in the state r, was previously in the state p or in the state q. Thus, we have here a process which can be realized following two indistinguishable modes (fig. 7.2). We know that this is characterized by an overall amplitude, obtained by adding the amplitudes of the individual processes,

$$A^{\equiv} = A_{s,r \leftarrow p,q} + A_{r,s \leftarrow p,q}. \tag{7.2.24}$$

We have intentionally avoided the notation $\langle \ldots | \ldots \rangle$ for it tends to conceal an essential phenomenon: the arbitrariness in the choice of the phases of the amplitudes. Indeed, $A_{r,s \leftarrow p,q}$ is *one* of the amplitudes equivalent to the

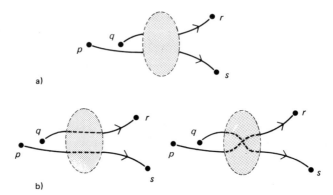

Fig. 7.2. *Transition of two identical quantons*
The single transition (a) can be characterized by the two modes (b).

amplitude $\langle r, s | p, q \rangle$, written previously, and $A_{s,r \leftarrow p,q}$ *one of the amplitudes* equivalent to $\langle s, r | p, q \rangle$. Thus, in general we have

$$A_{r,s \leftarrow p,q} = e^{i\beta} \langle r, s | p, q \rangle,$$
$$A_{s,r \leftarrow p,q} = e^{i\gamma} \langle s, r | p, q \rangle. \tag{7.2.25}$$

The phases β and γ are of no importance *as long as the quantons are distinct*, as shown by eq. (7.2.23). However, in the superposition, eq. (7.2.24), these phases – or more correctly, the relative phase $(\beta - \gamma)$ – enter in a crucial manner. The preceding argument now holds, and the overall amplitude $A^{\equiv}$ has to be either symmetric or antisymmetric under the interchange $r \leftrightarrow s$. It is easy to see that this forces the phase factors $e^{i\beta}$ and $e^{i\gamma}$ to be either identical or opposite. We may, therefore, choose the phase of the total amplitude $A^{\equiv}$ in such a way as to have – returning to the notation $\langle \dots | \dots \rangle$ – one of the two cases,

symmetry (*bosons*):

$$\langle r, s | p, q \rangle_{\mathrm{b}} = \langle r, s | p, q \rangle + \langle s, r | p, q \rangle, \tag{7.2.26}$$

antisymmetry (*fermions*):

$$\langle r, s | p, q \rangle_{\mathrm{f}} = \langle r, s | p, q \rangle - \langle s, r | p, q \rangle. \tag{7.2.27}$$

The explicative (and provisional) indices 'b' and 'f' are meant to remind us that the amplitudes on the left refer to identical quantons – bosons or fermions – while those on the right are computed 'as though' the quantons were different. Thus, we may apply to the latter the composite factorization principle, eq. (7.1.4), here written as

$$\langle r, s | p, q \rangle^{\neq} = \langle r | p \rangle \langle s | q \rangle, \tag{7.2.28}$$

and obtain finally,

$$\langle r, s | p, q \rangle_{b} = \langle r | p \rangle \langle s | q \rangle + \langle r | q \rangle \langle s | p \rangle \tag{7.2.29}$$

$$\langle r, s | p, q \rangle_{f} = \langle r | p \rangle \langle s | q \rangle - \langle r | q \rangle \langle s | p \rangle. \tag{7.2.30}$$

In these expressions, the symmetry or the antisymmetry is evident, whether for the initial or for the final states,

$$\langle s, r | p, q \rangle_{b,f} = \pm \langle r, s | p, q \rangle_{b,f}, \tag{7.2.31}$$

$$\langle r, s | q, p \rangle_{b,f} = \pm \langle r, s | p, q \rangle_{b,f}. \tag{7.2.32}$$

It is essential to realize that, despite appearances, the fermionic case is in accord with the superposition principle: in eq. (7.2.27), as in eq. (7.2.26), *we add* the amplitudes, the difference being that due to the relative phase. One could be pedantic (however, avoiding any possible confusion thereby) and write

$$\langle r, s | p, q \rangle_{f} = \langle r, s | p, q \rangle + (-1) \langle s, r | p, q \rangle, \tag{7.2.33}$$

in place of eq. (7.2.27).

It is time now to show, in an important example – that of the mutual scattering of two quantons – the far-reaching experimental effects of the symmetry or antisymmetry of two-quanton amplitudes (see, already, exercises 7.1 and 7.2).

7.2.2. Scattering of two quantons

Consider a two-quanton scattering experiment. We have seen in ch. 5 that the phenomenon is described by a certain transition amplitude, from which one calculates the effective scattering cross-section. We shall study the scattering in the reference frame of the centre of mass of the quantons. In this frame the scattering looks particularly simple, since now a single parameter, the scattering angle θ, characterizes the final state, as observed with respect to the initial state. The scattering amplitude is, therefore, a function $f(\theta)$ of this parameter and the differential effective cross-section is given by

$$\chi(\theta) = |f(\theta)|^2. \tag{7.2.34}$$

Recall that this quantity measures the probability of *each* quanton being deflected into a direction making an angle θ with its initial direction – in other words, the probability that the detector D_1 would register the quantum M_1 (fig. 7.3a). The final state is unambiguously determined, since the counter D_1 is sensitive specifically to quantons of type M_1. A detector D_2, which registers quantons of type M_2, placed at the same spot, would show a different final state – one in which the two quantons are deviated through an angle $(\pi - \theta)$, having thus 'exchanged' their final state (fig. 7.3b). Of course, we might not be interested in the effective difference between M_1 and M_2, and use a non-discriminating detector D, which is equally sensitive to both M_1 and M_2 (fig. 7.3c). Nevertheless, there still do exist two possible final states, corresponding to two distinct processes, depending on whether M_1 or M_2 is recorded at D, and hence on whether the scattering angle is θ or $(\pi - \theta)$. The overall probability is then obtained by adding the probabilities of the individual processes. In other words, the differential effective cross-sections of the processes, being proportional to the probabilities, add to yield the global differential effective cross-section $\chi_{gl}^{\neq}$

$$\chi_{gl}^{\neq}(\theta) = \chi(\theta) + \chi(\pi - \theta). \tag{7.2.35}$$

The symbol '$\neq$' connotes that we are dealing with different quantons. In terms of the scattering amplitude, this becomes

$$\chi_{gl}^{\neq}(\theta) = |f(\theta)|^2 + |f(\pi - \theta)|^2. \tag{7.2.36}$$

The situation is completely different for two identical quantons. Now, it is no longer possible to distinguish between the two processes, since it is impossible to label the particles and the question of being able to tell which one (M or M?) is recorded at A, loses its meaning (fig. 7.4). There exists just a single final state. Thus, it is necessary to add the amplitudes corresponding to an interchange of the labels of the individual states, i.e., the interchange $\theta \leftrightarrow (\pi - \theta)$. Of course, one must take into account their relative phase. The general case has been treated earlier, and we have seen that the symmetry of the phenomenon allows only for two values of this relative phase. Thus, there are two ways of expressing the differential effective cross-section as a function

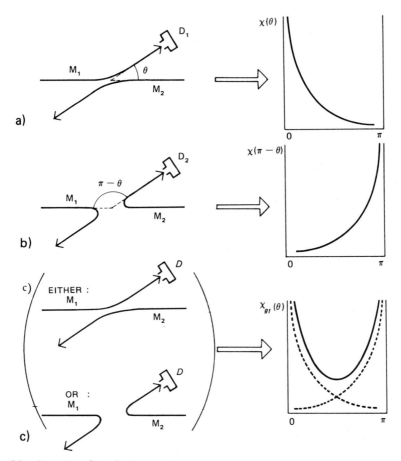

Fig. 7.3. *Scattering of two distinct quantons*

(a) The detector D_1 is sensitive only to the quanton M_1. The two quantons are deflected through the angle θ. The probability is measured by the effective cross-section $\chi(\theta)$.

(b) The detector D_2 is sensitive only to the quanton M_2. The two quantons are deflected through the angle $(\pi - \theta)$. The probability is measured by the function $\chi(\pi - \theta)$.

(c) The detector D records both M_1 and M_2 equally. The quantons are deflected through θ or through $(\pi - \theta)$. The overall probability is measured by the global effective cross-section, $\chi_{gl}(\theta) = \chi(\theta) + \chi(\pi - \theta)$.

of the scattering amplitude, depending on whether we are dealing with bosons or fermions,

$$\text{bosons:} \quad \chi_b(\theta) = |f(\theta) + f(\pi - \theta)|^2, \tag{7.2.37}$$

$$\text{fermions:} \quad \chi_f(\theta) = |f(\theta) - f(\pi - \theta)|^2. \tag{7.2.38}$$

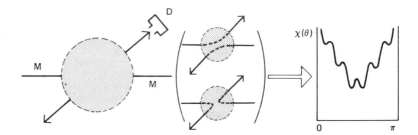

Fig. 7.4. *Scattering of two identical quantons*
There exists a single final state and thus, a single transition. Its amplitude is obtained by adding the scattering amplitudes corresponding to the two modes of transition (with an appropriate phase factor, depending on whether we are dealing with fermions or bosons). The effective cross-section displays characteristic oscillations of the interference between the two amplitude terms.

These two expressions differ significantly from one another and from formula (7.2.36) which holds for different quantons. We can further write

$$\chi_{b,f}(\theta) = \chi(\theta) + \chi(\pi - \theta) \pm 2 \, \mathrm{Re} \, \overline{f(\theta)} \, f(\pi - \theta). \tag{7.2.39}$$

The third term on the right-hand side, absent in eq. (7.2.35), is a characteristic interference term, which now depends on the phase of the amplitude $f(\theta)$ and not just on its modulus.

In a general way, the difference between the three cases (distinct quantons, bosons, fermions) can be best illustrated by a consideration of the differential effective cross-sections – for the same amplitude $f(\theta)$ – in the median direction, corresponding to $\theta = \frac{1}{2}\pi$. In this case, we have from eqs. (7.2.36)–(7.2.38)

$$\chi_{gl}^{\neq}(\tfrac{1}{2}\pi) = 2|f(\tfrac{1}{2}\pi)|^2 \qquad = 2\chi(\tfrac{1}{2}\pi),$$
$$\chi_b(\tfrac{1}{2}\pi) = |2f(\tfrac{1}{2}\pi)|^2 \qquad = 4\chi(\tfrac{1}{2}\pi),$$
$$\chi_f(\tfrac{1}{2}\pi) = |f(\tfrac{1}{2}\pi) - f(\tfrac{1}{2}\pi)|^2 = 0. \tag{7.2.40}$$

Thus, the probability of observing two bosons, scattered at right angles to their initial direction, is double that of the same probability for two distinct quantons, and is identically zero in the case of two fermions (see exercise 7.3).

Of course, it is in general difficult to compare the expressions (7.2.35) and (7.2.39), since the amplitudes depend specifically on the two quantons considered and one cannot, in general, find a pair of bosons, a pair of fermions and a pair of distinct quantons, all three pairs having the same scattering amplitude! There does exist, however, a case in which comparison is possible: this is the case of Coulomb scattering for which, as we have seen (sect. 5.3.2) the scattering amplitude is given by eqs. (5.3.12)–(5.3.14)

$$f_c(\theta) = \frac{Z_1 Z_2 e^2}{4E} \frac{e^{-i\zeta(\theta)}}{\sin^2 \frac{1}{2}\theta}. \tag{7.2.41}$$

The theoretical effective cross-section for two distinct quantons, which agrees perfectly with the experimental results (fig. 7.5a), is given by the same expression as in the classical theory,

$$\chi_c(\theta) = \left(\frac{Z_1 Z_2 e^2}{4E}\right)^2 \frac{1}{\sin^4 \frac{1}{2}\theta}. \tag{7.2.42}$$

Its angular dependence is the same, irrespective of the masses, the charges and the relative energy of the quantons, which enables a comparison to be made between Coulomb scattering processes involving different types of quantons. The global effective cross-section for the scattering of two distinct quantons corresponding to a detection, in the direction θ, of either one of the two quantons, would be given by

$$\chi_{gl}^{\neq}(\theta) = \left(\frac{Z_1 Z_2 e^2}{4E}\right)^2 \left(\frac{1}{\sin^4 \frac{1}{2}\theta} + \frac{1}{\cos^4 \frac{1}{2}\theta}\right). \tag{7.2.43}$$

It is the sum of the classical cross-section and its symmetric counterpart, with respect to the direction $\theta = \frac{1}{2}\pi$. The graph of this function is as regular as it could possibly be, and shows a minimum at $\theta = \frac{1}{2}\pi$. Now, the scattering experiment carried out with identical nuclei yields very different results (figs. 7.5b and c). In this case the differential effective cross-section exhibits definite oscillations around the function $\chi_{gl}^{\neq}$ – oscillations which are clearly due to the interference between the two amplitudes $f(\theta)$ and $f(\pi - \theta)$ in eqs. (7.2.39), and brought about by the specifically quantum phase of the Coulomb amplitude [see eqs. (5.3.13) and (5.3.14)]. Moreover, an approximate evaluation of this phase is possible, allowing us to interpret these oscillations (exercise 7.4). Figure 7.5b clearly shows, for $\theta = \frac{1}{2}\pi$, the expected factor of 2, in the effective bosonic cross-section, as in eq. (7.2.40). It also enables us to understand how eq. (7.2.39), in which the third term is typically quantum in nature, can be approximated, under certain conditions by the classically

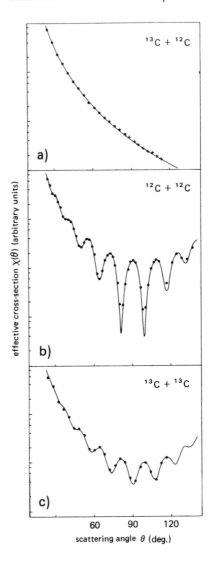

Fig. 7.5.   *Coulomb effective cross-sections*

Comparison between the Coulomb differential effective cross-sections of ^{12}C and ^{13}C carbon isotopes – hence of nearly equal masses – each at energy 2 MeV.

(a) ^{12}C + ^{13}C: distinct nuclei
(b) ^{12}C + ^{12}C: identical bosons
(c) ^{13}C + ^{13}C: identical fermions.

The dots correspond to experimental measurements. The continuous curves correspond to the expected expression eq. (7.2.35). For this it is sufficient that the scale of the characteristic oscillations of this interference term be smaller than the resolving power of the detector. The latter then measures a cross-section averaged out over many oscillations (see fig. 7.6), making them disappear, and the classical result is recovered. It can be shown that the criterion for experimentally detecting the oscillations corresponds to the criterion for having to use a quantum theory, and vice versa.

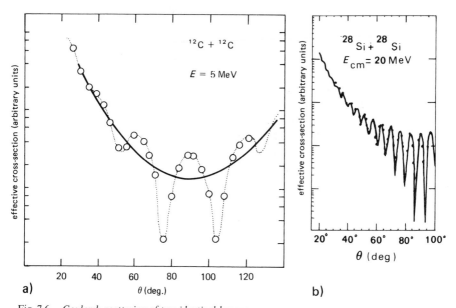

Fig. 7.6. *Coulomb scattering of two identical bosons*
(a) Scattering of two ^{12}C nuclei of kinetic energy 5 MeV [D. A. Bromley, J. A. Kuehner and E. Almquist, Phys. Rev. Lett. 4 (1960) 365]. The continuous curve indicates the symmetrized Rutherford cross-section eq. (7.2.43) which would have been obtained for distinct quantons having the same charge and energy.
(b) Scattering of two ^{28}Si nuclei having kinetic energy 20 MeV [Ferguson, 1971, in: Nuclear Spectroscopy and Reactions, p. 236 (Academic Press, NY 1974)]. One verifies, by comparing with (a), the narrowing of the oscillations with the increase of energy (despite the difference in scales!).

---

theoretical formulae (7.2.36), (7.2.37), (7.2.49) respectively, where eq. (7.2.41) has been taken into account. The quantum interference effects, coming from the fact that the quantons are identical, are evident in (b) and (c) and the agreement of the experimental results with the theory is excellent. [G. R. Plattner and I. Sick, Eur. J. Phys. 2 (1981) 109.]

It would be nice to have a clean experimental result, such as in fig. 7.5b, for the Coulomb scattering of two identical *fermions*. However, fig. 7.5c does not correspond to formulae (7.2.39) and (7.2.40) for the fermionic case. In particular, we see that $\chi(\frac{1}{2}\pi) \neq 0$. The fact is that the situation is now more complicated, and its analysis enables us to make the notion of identity and its quantum effects more precise. In order that there be symmetry or anti-symmetry, of the amplitudes, it is necessary that the two quantons being considered be truly identical. That is, they should neither differ in any of their internal properties nor carry any 'labels' which distinguish between them. Thus, the Coulomb scattering of the two ^{12}C nuclei would not display the effects of symmetry if one of them is in an excited state of a different energy, or if it is replaced by its isotope ^{13}C (fig. 7.5a), which would make it identifiable. Now, some of the quantons possess a property which marks them out: their spin, whenever it is non-zero. We have seen in fact, that a quanton of spin s has $2s + 1$ different spin states, depending on the proper value $m\hbar$ (m assuming the values $s, s - 1, \dots, -s$) of one of the components of its spin vector. Two quantons of spin s can be considered to be identical only if their spin state is the same. Otherwise, the proper value of this spin component would serve as a label and would allow us to distinguish between the two quantons – or rather to distinguish between the different processes. Now, fermions necessarily have a non-zero spin, as we shall see later (in sect. 7.3.3), and this fact complicates the scattering experiments involving them, by forcing us to take the different spin states into account. Naturally, similar complications would arise for bosons with non-zero spin also. However, there do exist spin-zero bosons and it is this case that we have considered so far.

Let us consider, therefore, the Coulomb scattering of two spin-$\frac{1}{2}$ fermions (fig. 7.7). This is the simplest case, in which each quanton has two possible spin states, corresponding to the proper values $\pm\frac{1}{2}\hbar$ of one component of its spin vector. Let us denote these two states by $\uparrow$ (up) and $\downarrow$ (down), following a common convention (which amounts implicitly to choosing the vertical component of spin for defining the states). The Coulomb scattering does not depend on the spin state (which would not be the case if one were to take the magnetic interactions into account – but these latter are negligible, at low energies, compared to the electrostatic Coulomb interaction) and does not change this state: the two-fermion final spin states are the same as their initial states. Moreover, this condition is indispensable, in order for us to be able to use the spin state to 'label' the quantons. Whenever this condition is not satisfied, so that the interaction modifies the spin state, the description of the scattering process becomes more complicated (see exercise 7.5). There are four possible cases (fig. 7.7):

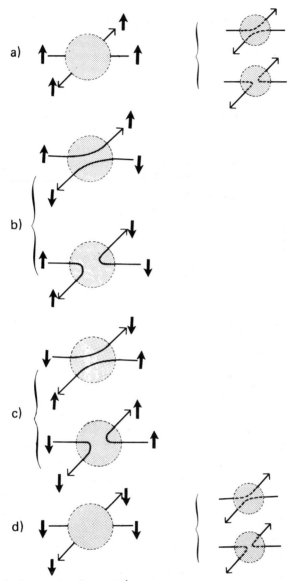

Fig. 7.7.   *Coulomb scattering of two spin-$\frac{1}{2}$ quantons*
Depending on the spin of two quantons, there exist one or two final states for a scattering in the
direction θ. If the two quantons have the same individual spin state [cases (a) and (d)], they are
identical quantons – there is just one final state and hence one single transition. If the individual
spin states are different, there are two possible final states [cases (b) and (c)].

(a) $\uparrow\uparrow$, the symbol indicating that both quantons are in the spin 'up' state. The quantons are now surely identical and the effective cross-section may be written as

$$\chi_C^{\uparrow\uparrow}(\theta) = |f_C(\theta) - f_C(\pi - \theta)|^2, \tag{7.2.44}$$

where f_C is the same Coulomb amplitude as in eq. (7.2.41).

(b) $\uparrow\downarrow$, one quanton in the 'up' and the other in the 'down' state. The two quantons should be considered as being different since if the counter is able to detect the spin state, there would be two different final states, depending on whether the $\uparrow$ state or the $\downarrow$ state is detected by one of the counters. If we are not interested in the final spin, we simply add the probabilities and, thus, also the effective cross-sections,

$$\chi_C^{\uparrow\downarrow}(\theta) = |f_C(\theta)|^2 + |f_C(\pi - \theta)|^2; \tag{7.2.45}$$

(c) $\downarrow\uparrow$: as in (b),

$$\chi_C^{\downarrow\uparrow}(\theta) = |f_C(\theta)|^2 + |f_C(\pi - \theta)|^2; \tag{7.2.46}$$

(d) $\downarrow\downarrow$: as in (a)

$$\chi_C^{\downarrow\downarrow}(\theta) = |f_C(\theta)|^2 + |f_C(\pi - \theta)|^2; \tag{7.2.47}$$

Now a serious experimental problem makes a separate measurement of these cross-sections very difficult. It is in general hard to prepare spin-$\frac{1}{2}$ quantons in a definite state of spin, i.e., to 'polarize' them. A priori, the ensemble of quantons in the beam coming out of an accelerator, and even more so the ensemble of quantons in a stationary target (i.e., the nuclei in a sample of matter), is unpolarized. Half of these quantons are in the spin-up and half in the spin-down state. The experimental measurement of the scattering probability, carried out by collecting a large number of elementary scattering events, yields therefore, an average over all possible states of quanton pairs. Under the conditions in which neither the beam nor the target is polarized, one-quarter of the collisions correspond to quantons in case (a) above, i.e., in the state $\uparrow\uparrow$, a quarter to quantons in case (b), i.e., $\uparrow\downarrow$, etc. Thus finally, we observe the average probability

$$\chi_C^{\text{unpol}}(\theta) = \tfrac{1}{4}[\chi_C^{\uparrow\uparrow}(\theta) + \chi_C^{\uparrow\downarrow}(\theta) + \chi_C^{\downarrow\uparrow}(\theta) + \chi_C^{\downarrow\downarrow}(\theta)], \tag{7.2.48}$$

implying,

$$\chi_C^{unpol}(\theta) = |f_C(\theta)|^2 + |f_C(\pi - \theta)|^2 - \text{Re } \overline{f_C(\theta)} \, f_C(\pi - \theta).$$ (7.2.49)

It is interesting to compare this with the result

$$\chi_C^{\uparrow\uparrow}(\theta) = |f_C(\theta)|^2 + |f_C(\pi - \theta)|^2 - 2 \text{ Re } \overline{f_C(\theta)} \, f_C(\pi - \theta),$$ (7.2.50)

which would have been obtained in the case of complete polarization, i.e., for effectively identical fermions. Equation (7.2.49) agrees well with experimental observations (figs. 7.5c and 7.8). In particular, for $\theta = \frac{1}{2}\pi$, we immediately verify that

$$\chi_C^{unpol}(\tfrac{1}{2}\pi) = |f_C(\tfrac{1}{2}\pi)|^2 = \tfrac{1}{2}\chi_C^{\neq}(\tfrac{1}{2}\pi)$$ (7.2.51)

[compare with eq. (7.2.40)].

We see that being unpolarized reduces the interference effects arising from the identity of the particles. This phenomenon becomes more pronounced for higher spin values, both for fermions and for bosons. Indeed, the scattering of

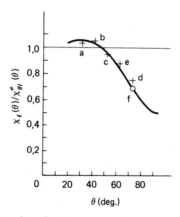

Fig. 7.8. *Coulomb scattering of two fermions*
Scattering of two protons having a kinetic energy of 30 keV [Chr. Gerthsen, Ann. Phys. 5 (1931) 769]. The (experimental) points and the (theoretical) curve, correspond to the ratio of the cross-section, eq. (7.2.48), for unpolarized spin-$\frac{1}{2}$ fermions to the symmetrized cross-section, eq. (7.2.36), for distinct quantons. The reduction by a factor of $\frac{1}{2}$ for $\theta = \frac{1}{2}\pi$ [see eq. (7.2.51)] is clearly seen.

bosons having (integral) spin, calls for an analogous treatment, and leads to expressions in which the interference term is reduced, just as in eq. (7.2.49). Here again, in a general way, is another reason for the validity of the classical approximation, in which the effects coming from the identity of the quantons become negligible (exercise 7.5).

7.3. Systems of N identical quantons

7.3.1. N-quanton amplitudes. Composite quantons

Let us consider a system with an arbitrary number, N, of identical quantons. The analysis given in sect. 7.2.1 can be generalized readily. Indeed, the transition amplitude $\langle r_1, r_2, \dots r_N | \Psi \rangle$, from an arbitrary state Ψ to a collective state, specified by the N individual states $(r_1, r_2, \dots, r_N)$, can be considered, in turn, for each pair of states. Following the argument given above, the identity of the quantons requires that the amplitude be either symmetric or antisymmetric in the interchange of any two states, so that

$$\langle r_1 r_2 \cdots r_j \cdots r_i \cdots r_N | \Psi \rangle_{b,f} = \pm \langle r_1 r_2 \cdots r_i \cdots r_j \cdots r_N | \Psi \rangle_{b,f}. \quad (7.3.1)$$

In general, an arbitrary permutation P of the N states $\{r_1 \cdots r_N\}$ can be expressed as a product of interchanges. The permutation is called 'even' or 'odd', depending on whether it contains an even or an odd number of factors. Thus, we may write

bosons:

$$\langle P(r_1 \cdots r_N) | \Psi \rangle_b = \langle r_1 \cdots r_N | \Psi \rangle_b, \quad (7.3.2)$$

fermions:

$$\langle P(r_1 \cdots r_N) | \Psi \rangle_f = \varepsilon_P \langle r_1 \cdots r_N | \Psi \rangle_f, \quad (7.3.3)$$

where $\varepsilon_P = \pm 1$ is the parity of the permutation P. We say that the bosonic amplitudes are 'completely symmetric' and the fermionic amplitudes are 'completely antisymmetric'. The symmetry, eq. (7.3.2), or antisymmetry, eq. (7.3.3), properties have the effect of imposing on the function $\langle r_1 \cdots r_N | \Psi \rangle$ of N variables as many conditions as there are independent permutations of the N arguments. It is more or less obvious that in the case of N distinct quantons the effect of the constraints becomes greater as N increases. Hence, the larger the ensemble of identical quantons, the stronger are the effects of the 'statistics' (of permutability).

Up to now, we have implicitly assumed that our quantons are elementary, i.e., constituting the fundamental units taking part in the physical phenomena being considered. However, this notion of elementarity is completely relative: atomic nuclei, which are 'elementary' from the point of view of atomic physics, where their internal structure does not have to be taken into consideration, have to be considered as being composed of nucleons (neutrons and protons) in nuclear physics. In turn, the nucleons lose their apparent elementarity in elementary particle physics. Thus, there ought to exist a relationship between the 'statistics' of a composite quanton (such as a nucleus) and that of its components (in this case, the nucleons). Let us examine the simplest case, that of a quanton composed of two 'elementary' quantons. To fix ideas, consider the deuteron, or the deuterium nucleus, consisting of a proton and a neutron. Of course, its 'statistics' only shows up in systems having many deuterons. Here again, we opt for simplicity: a two-deuteron system. Let Ψ be an arbitrary state of the system and R, S two states of a deuteron. We would like to know if the deuteron is a boson or a fermion, i.e., whether the amplitude $\langle R, S | \Psi \rangle$, or what amounts to the same thing – being also more convenient for our purposes here – whether the amplitude $\langle \Psi | R, S \rangle$ is symmetric or antisymmetric with respect to an interchange of the states R and S. As we have seen, this property cannot depend on the particular state Ψ chosen. Hence, let Ψ be a state specified by the individual states (r, r', s, s') of four nucleons. More precisely, let r, r' be the states of a proton and s, s' those of a neutron. Thus, the amplitude $\langle rr'ss' | RS \rangle$ describes a transition from a two-deuteron state, occupying the individual states R and S, onto a state of four nucleons with two protons in the states r and r' and two neutrons in the states s and s'. Let us suppose for a moment that the two protons are distinct as also are the two neutrons. The deuterons, which are made up of these, would now also be distinct. For example, if one of the protons were 'white' (say, the one in the state r) and the other 'black' (r') and similarly if the neutrons were also 'white' (s) and 'black' (s'), then one of the deuterons would be 'white and white' and the other 'black and black' or else one could be 'white and black' and the other 'black and white'. Suppose we are in the first case: then the white–white deuteron, in the state R, is composed of the white nucleons (r and s) (fig. 7.9). The only transition from a white deuteron (R) to a two-nucleon state pair is $\langle r, s | R \rangle$, and the other amplitudes are zero,

$$\langle rs' | R \rangle = \langle r's | R \rangle = \langle r's' | R \rangle = 0.$$

Similarly, the other deuteron (S) couples only to the black nucleons (r' and

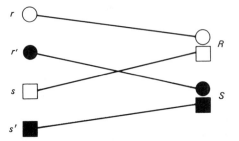

Fig. 7.9. *'Black' and 'white' nucleons and deuterons*
If the protons and neutrons were of the two types, 'black' and 'white' (protons ● and ○, neutrons ■ and □), one could have made 'black' and 'white' deuterons (●, ■ and ○, □). The transition amplitude, from a state of two deuterons – one black (S), the other white (R), to a four-nucleon state would then factorize into two amplitudes, corresponding to the independent transitions of the black and white quantons.

s'), so that $\langle r's' | S \rangle \neq 0$, but

$$\langle rs | S \rangle = \langle rs' | S \rangle = \langle r's | S \rangle = 0.$$

Thus, the factorization rule for amplitudes applies and one can write

$$\langle rr'ss' | RS \rangle = \langle rs | R \rangle \langle r's' | S \rangle \quad \text{(distinct quantons)}. \tag{7.3.4}$$

If we wish to have a less artificial example than the example with coloured nucleons, we might replace one of the deuterons by an anti-deuteron and two of the nucleons by anti-nucleons. The deuteron and anti-deuteron, the proton and anti-proton and the neutron and anti-neutron are distinct (having different baryon numbers) and the preceding argument applies.

Let us return to our identical deuterons. It is now no longer possible to set up a correspondence between the states of a nucleon and one of the states of a deuteron: we cannot say that the proton in the state r belongs to the deuteron in the state R, or that it belongs to the other.... . Stated differently, there are four indistinguishable transition modes (fig. 7.10):

$$(r, s) \leftarrow R \quad \text{and} \quad (r', s') \leftarrow S,$$

$$(r, s') \leftarrow R \quad \text{and} \quad (r', s) \leftarrow S,$$

$$(r', s) \leftarrow R \quad \text{and} \quad (r, s') \leftarrow S,$$

$$(r', s') \leftarrow R \quad \text{and} \quad (r, s) \leftarrow S.$$

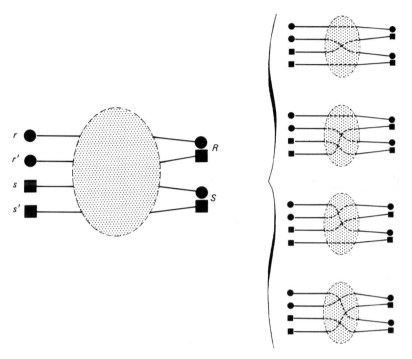

Fig. 7.10. *Nucleons and deuterons*
Since, like the nucleons, the deuterons are identical, the transition amplitude does not factorize, but rather appears as a sum of the amplitudes corresponding to the four transition modes.

Hence, we obtain the total amplitude by adding the four corresponding amplitudes which respectively factorize, as in eq. (7.3.4). Finally, in view of a principle, discussed many times before, we must take into account any phase factors that might appear. Thus, we may write

$$
\begin{aligned}
\langle rr'ss' | RS \rangle = &\, \eta_1 \langle rs | R \rangle \langle r's' | S \rangle \\
&+ \eta_2 \langle rs' | R \rangle \langle r's | S \rangle \\
&+ \eta_3 \langle r's | R \rangle \langle rs' | S \rangle \\
&+ \eta_4 \langle r's' | R \rangle \langle rs | S \rangle,
\end{aligned} \tag{7.3.5}
$$

where $\eta_1, \eta_2, \eta_3, \eta_4$ are phase factors. The fermionic nature of the protons and the neutrons demands that this amplitude be antisymmetric in the interchange of the proton states (r, r') on the one hand, and the neutron states (s, s') on the other,

$$\langle rr'ss' | RS \rangle = - \langle r'rss' | RS \rangle, \tag{7.3.6}$$

$$\langle rr'ss' | RS \rangle = - \langle rr's's | RS \rangle. \tag{7.3.7}$$

Making a simple substitution of the labels in eq. (7.3.5), we see that this implies, respectively,

$$\eta_1 = -\eta_3 \quad \text{and} \quad \eta_2 = -\eta_4, \tag{7.3.8}$$

$$\eta_1 = -\eta_2 \quad \text{and} \quad \eta_3 = -\eta_4. \tag{7.3.9}$$

From this, taking $\eta_1 = 1$ to fix the global phase, we obtain the expression

$$\begin{aligned}
\langle rr'ss' | RS \rangle = {} & \langle rs | R \rangle \langle r's' | S \rangle \\
& - \langle rs' | R \rangle \langle r's | S \rangle \\
& - \langle r's | R \rangle \langle rs' | S \rangle \\
& + \langle r's' | R \rangle \langle rs | S \rangle,
\end{aligned} \tag{7.3.10}$$

which should be compared with eq. (7.3.4). We have now reached the end of our analysis: the amplitude, eq. (7.3.10), possesses either a symmetry or an antisymmetry property, *with respect to the states R and S*, as would be the case for a description of two identical quantons (the deuterons). In fact, we observe that

$$\langle rr'ss' | RS \rangle = \langle rr'ss' | SR \rangle, \tag{7.3.11}$$

as shown by the interchange $(R \leftrightarrow S)$ in eq. (7.3.10). The amplitude is symmetric and the deuteron is a boson. It is obvious that the symmetry is brought about by the double change of sign arising from the two fermionic permutations, eqs. (7.3.6) and (7.3.7): relations (7.3.8) and (7.3.9) imply $\eta_1 = \eta_4$ and $\eta_3 = \eta_4$ and hence eq. (7.3.10). Stated otherwise, the deuteron is a boson because it consists of two fermions. The preceding discussion easily generalizes to two quantons, composed of an arbitrary number of fermions and bosons. The effect of the exchange of the states of two composite quantons would be the same as that of the exchange of all the states of one of the constituent quantons. Thus, *there are as many sign changes as there are fermions.* Hence the simple rule:

A composite quanton is a fermion if and only if the number of its fermionic constituents is odd.

7.3.2. Some examples

(a) *Nitrogen-14 and the neutron.* Up until the 1930s the only known 'massive elementary particles' were the proton and the electron. It was thought (hoped?) that they were sufficient for constructing any composite system. Thus, atomic nuclei were thought of as being made up of protons and electrons: a nucleus having mass number A and atomic number Z would have to contain A protons to give it most of its mass, and $(A - Z)$ electrons to give it a charge of $+Zq_e$. [One could also say that the atom was supposed to contain A electrons, of which Z would be in the external electronic cloud and $(A - Z)$ in the interior of the nucleus.]

$$^A_Z\text{nucleus} = A \text{ protons} + (A - Z) \text{ electrons} = (2A - Z) \text{ fermions.}$$

The statistics of any particular type of nucleus is fixed by the parity of the total number of fermions within it, which in this case is $(2A - Z)$. It would depend, therefore, on the parity of Z only. Under these conditions, the nitrogen-14 nucleus, or $^{14}_7\text{N}$, for example, would have to be a fermion. Now, the spectrum of the N_2 molecule, which depends on the permutational properties of the two nuclei, shows that these nuclei are bosons and hence contain an *even* number of fermions. This remark is one of the arguments which lead us to reject the idea of intranuclear electrons and to substitute it by that of neutrons. We now know that

$$^A_Z\text{nucleus} = Z \text{ protons} + (A - Z) \text{ neutrons} = A \text{ fermions.}$$

The statistics of a nucleus is, therefore, uniquely fixed by the parity of its mass number A. Hence, the nitrogen-14 atom, with A even, *is* a boson.

(b) *The two heliums.* There exist two stable isotopes of the helium nucleus: ^3He or helium-3 (2 protons + 1 neutron), and, ^4He or helium-4 (2 protons + 2 neutrons) which is more frequent. The first is a fermion and the second a boson. There probably does not exist a more spectacular example of the difference between bosons and fermions, than the one obtained by a comparison of the macroscopic properties of the fluids, helium-3 and helium-4, respectively. Indeed, we note that their atomic properties are identical: the common electronic structure of the two types of helium atoms (one saturated shell having 2 electrons) implies the identity of the interactions between atoms. The mass difference between the helium-3 and helium-4 atoms can

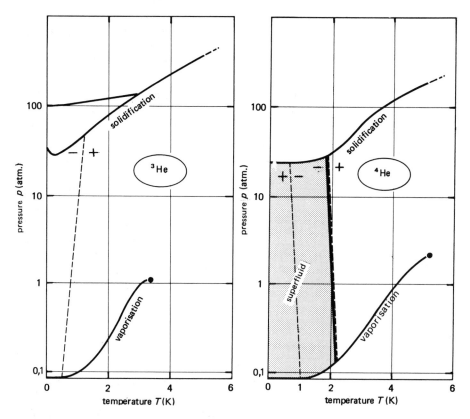

Fig. 7.11. *Phase diagrams for helium*
For the two isotopes ^{3}He and ^{4}He, respectively, the pressure–temperature diagrams show, in the continuous curves, the changes of state liquid↔gas (vaporization) and liquid↔solid (solidific-ation). The broken curves separate the regions in which the coefficient of expansion of the liquid is negative (−) or positive (+). The most obvious difference between the macroscopic properties of the two isotopes is the existence of a superfluid phase for ^{4}He (indicated by shading).

only lead to limited quantitative differences between their properties. However, as shown most convincingly by the phase diagrams of the two fluids, at low temperatures (fig. 7.11) they behave completely differently. The statistics alone – fermionic or bosonic, respectively – is responsible for this difference. In particular, it is the bosonic character of helium-4 which explains its 'superfluidity' as we shall soon see (sect. 7.5.2).

(c) *Quarks.* Today, hadrons are thought of as being made up of elementary (*more* elementary!) quantons, called 'quarks'. The baryons (proton, neutron, etc.), being fermions, can only consist of an odd number of fermions. Indeed, they are made up of three quarks. The mesons, in turn, are made up essentially of a quark–antiquark pair and hence are definitely bosons. The different bound states of a given three-quark system form a family of baryons. One can, in this way, regroup and classify the multitudes of known hadrons – just as the (periodic) table of Mendeleev expresses in simple terms the structures formed by protons and neutrons. However, the quark theory had, for a long time, run into a major obstacle: in order to explain the existence of the most stable hadrons, one was forced to assume that they arose from a quark configuration which violated the antisymmetry principle of fermionic amplitudes. This difficulty, very similar to that encountered for nuclei, half a century earlier, eventually led to the theory of 'coloured' quarks. According to this theory, there exist three kinds of quarks, having three different 'colours', and only quarks of unlike 'colours' can be bound together to form hadrons. Since they are different quarks, the amplitudes are not required to be antisymmetric anymore, and the configurations in question are allowed. The notion of a fermion played, therefore, a major role in the development of 'quantum chromodynamics', or the field theory of 'coloured quarks', which is the present-day theory of strong interactions.

7.3.3. Spin and statistics

The collective behaviour of quantons, their 'statistics' or permutability – i.e., their fermionic or bosonic nature – is related to another fundamental property, the spin. The intrinsic angular momentum magnitude, or spin, of a quanton is quantized (see sect. 2.3.4): it is given by an integer or a half-integer s. More precisely, the square of the modulus of the spin vector has the proper value $s(s + 1)\hbar^2$ and any one of its components has the proper values $[s\hbar, (s - 1)\hbar, ..., -s\hbar]$. Now, the dichotomy between integral and half-integral spins is exactly reproduced by the dichotomy between bosons and fermions.

This fundamental result is known as the

> *Spin–statistics relation*
> integral spin quanton = boson,
> half-integral spin quanton = fermion.

Table 7.1 lists a few examples. Referring to table 1.1, we see that among the 'elementary particles', the photon is a boson and the leptons are fermions, while among the hadrons, the mesons are bosons and the baryons are fermions.

This relation, first established experimentally, presently constitutes a theorem which can be proved within the general framework of quantum theory. Unfortunately, the different existing proofs are all given within the frameworks of elaborate formalisms, which are considerably beyond the scope of this book. Moreover, despite its fundamental nature, there is, up to now, no elementary way to understand this theorem in which physical intuition could substitute for mathematical rigour. We mention, however, as an indication of the measure of the difficulties and complexities inherent in this theorem, that *Einsteinian* relativity plays a crucial role in it. Thus, it is not possible to prove it without appealing to 'quantum field theory' (see ch. 1,

Table 7.1
Fermions and bosons (some examples).

Quantons	Spin	Quantons	Spin
Bosons		**Fermions**	
Elementary		*Elementary*	
Photon γ	1	Leptons: e, μ, ν,...	1/2
Gluons	1	Quarks	1/2
Weak bosons $W^\pm$, Z°	1		
		Composite	
Composite		Baryons:	
Mesons: π, K,...	0	nucleons n, p	1/2
η, ρ, ω,...	1	Λ, Σ,...	1/2
Even-A nuclei		N*, Y*,...	3/2
^{2}H	1	Odd-A nuclei	
^{4}He	0	^{1}H, ^{3}H	1/2
^{14}N	1	^{3}He	1/2
^{40}K	4	^{15}N	1/2
...		^{39}K	3/2
		...	

sect. 1.1) in one form or another. In other words, if we replace Einsteinian relativity by its Galilean approximation, as we have done in this book, the theorem can no longer be proved .... It is the Einsteinian structure of space–time, more restrictive as it is than that of Galilean space–time – in particular, from the stand-point of the notion of causality – which plays a crucial role. Nevertheless, it remains extremely difficult to understand a priori, the very reason for such a connection – without going into the specifics of its nature – between the collective behaviour of quantons and one of their spatial properties, namely, the spin.

However, even though we are unable here to prove the spin–statistics relation, we can at least verify its consistency. In particular, the rule for the combination of the spins of a system composed of several quantons must conform to the rule governing the statistics of such systems. The addition of angular momenta in quantum mechanics is quite complicated. An elaborate formalism is needed, in order to determine the possible proper values of the total spin of a system made up of two quantons of known spins s and s'. However, whether the combined spin is integral or half-integral – and this is all that is of importance to us – is fairly evident. If angular momentum is to retain the same importance in the quantum theory that it has in the classical theory, then it has to be a conserved additive magnitude. The addition of two angular momenta J_1 and J_2 yields a total angular momentum J according to the rules of vector addition. Thus, for example,

$$J_z = J_{1z} + J_{2z},$$
(7.3.12)

and the same holds true for all the other components. In this way, the proper values of J_z appear as the sums of the proper values of J_{1z} and J_{2z}. Hence, in units of $\hbar$, they are integral or half-integral, depending on whether J_{1z} and J_{2z} are both alike in nature (both integral or half-integral) or unlike in nature (one integral and the other half-integral). A composite system of two quantons, having spins S and S' is a little more complicated, since its total angular momentum – *its* spin, if it is regarded as being elementary – is obtained by combining the two spins, as well as the final orbital angular momentum L of the two quantons, in a manner such that

$$J_z = S_z + S_{z'} + L_z.$$
(7.3.13)

But an orbital angular momentum can only assume integral proper values (see ch. 2, sect. 2.3.4), and hence it does not change the integral or half-integral character of the sum of S and S'. If one of the spins is integral, and the other half-integral, the composite spin is half-integral. More generally, a system composed of several (identical)* quantons has a half-integral spin if and only if it consists of an odd number of quantons, each having a half-integral spin. We see that this rule is exactly isomorphic to the one giving the statistics of a composite quanton. A quanton is a fermion (with half-integral spin) if and only if it consists of an odd number of fermions (each with a half-integral spin).

7.4. Fermions

7.4.1. The Pauli exclusion principle

An N-fermion amplitude is 'totally antisymmetric', so that,

$$\langle r_1 r_2 \cdots r_j \cdots r_i \cdots r_N | \Psi \rangle = -\langle r_1 r_2 \cdots r_i \cdots r_j \cdots r_N | \Psi \rangle. \qquad (7.4.1)$$

From this, it immediately follows that if two individual states are identical, for example, $r_i = r_j = s$, the amplitude vanishes

$$\langle r_1 r_2 \cdots s \cdots s \cdots r_N | \Psi \rangle = 0. \qquad (7.4.2)$$

In other words, the transition probability of a system of identical fermions is zero, as soon as two of the individual final states become identical. Hence, the system can never assume such a configuration. This is what is traditionally called the

---

Pauli exclusion principle
A system of fermions can never occupy a configuration of individual states in which two individual states are identical.

This principle (which really is no longer a principle any more, since it can be deduced, as we have just done, from more fundamental principles of quantum theory) is often formulated in a simpler way: 'two fermions can never occupy the same state'. However, this statement is ambiguous, since it encourages the supposition that to each fermion a state could be attributed. But then, the essential characteristic of systems of identical quantons is just

---

* Parentheses added by translator.

the impossibility of attributing a state to each quanton individually! While it is certainly possible to specify the state of the system using a set of individual states, the symmetrization or antisymmetrization requires precisely, so to say, that all the quantons occupy together all the states.

7.4.2. Spatial consequences. The Heisenberg–Pauli inequality

The Pauli principle has a considerable impact on the spatial distribution of a system of identical fermions. To demonstrate this, let us take the localization proper states as the individual states. The Pauli principle now asserts that it is impossible to find two fermions in the same place – provided, of course, that the other state specifications (e.g., the spin variables) are the same. More generally, let us consider the probability that the fermions of a certain system should occupy a given spatial configuration. This is the probability of finding a fermion (although it is impossible to say exactly which one!) at each one of the N points $(A_1, \dots, A_N)$. This probability must tend to zero whenever two points approach each other. Somehow, the antisymmetrization tends to keep the fermions at a distance from one another. This plays the role of a fictitious, although highly effective, mutual repulsion being exerted within the system, irrespective of any other actual forces or interactions (Coulombic, nuclear, etc.) that might be present.

Moreover, it is possible to express the Pauli principle, or more generally the fermionic nature of an N-particle system, using a modified version of the spatial Heisenberg inequalities.

We have already seen, in chs. 2 and 3, that the quantic concept of a state differs considerably from its classical analogue. In particular, the state of a particle is classically defined by giving its position and momentum. But in the quantum theory, these two magnitudes are incompatible and cannot simultaneously assume exact numerical values (or proper values) in the same state. The Heisenberg inequality

$$\Delta p \, \Delta x \gtrsim \hbar, \tag{7.4.3}$$

shows that, for an arbitrary state, position and momentum have spectra with correlated widths. In other words, while classically the state of a particle is represented by a *point* in a space having coordinates (p, x) – called the 'phase space' – the state of a quanton corresponds to a *domain* having characteristic dimensions Δp and Δx. These dimensions are variable, but their domain is at least of order $\hbar$. Thus, we are led to associating a phase-space 'cell' of size (at least) $\hbar$ to each quantum state. This argument, holding in one dimension, is easily generalized to three, where the Heisenberg inequalities

$$\Delta p_x \, \Delta x \gtrsim \hbar, \qquad \Delta p_y \, \Delta y \gtrsim \hbar, \qquad \Delta p_z \, \Delta z \gtrsim \hbar, \qquad (7.4.4)$$

demonstrate the necessity of attributing, to each quantum state, a cell of size $\hbar^3$ in a six-dimensional phase space.

Suppose now, that we have an ensemble of N particles. Classically, the state would be given by an N-point configuration in phase space. Quantically, we shall have N regions, in this same space, each of size $\hbar$ (in one dimension) or $\hbar^3$ (in three dimensions) (fig. 7.12). The collective properties of the system depend upon the overall size of this collection of regions. If together they make up a region, the typical dimensions of which, along the position and momentum axes, are of the order of Δl and Δp, respectively, these would also be the values of the characteristic spatial extension and the characteristic momentum of the system. The Pauli principle enters at this point and prevents two identical states from appearing together in the configuration. In other words, since each individual state corresponds to a particular cell, the total region occupied by an N-fermion system in phase space must consist of N distinct cells. Thus, the total volume of this region is at least of order $N\hbar^3$.

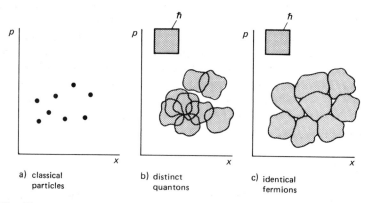

| a) classical particles | b) distinct quantons | c) identical fermions |

Fig. 7.12. *Phase space*
(a) The state of an N-particle classical system is represented by N points in phase space (having coordinates of momentum and position).
(b) The Heisenberg inequality allows us at best to associate a certain region of phase space, of size $\hbar$, to the state of each quanton. For N distinct quantons, there are N independent, and hence possibly overlapping, regions.
(c) For N fermions, the Pauli exclusion principle forces the domains to be disjoint.

In terms of the characteristic size and momentum, Δl and Δp, respectively, we obtain the inequality

$$(\Delta p)^3 (\Delta l)^3 \gtrsim N\hbar^3, \tag{7.4.5}$$

i.e.,

$$\Delta p \, \Delta l \gtrsim N^{1/3}\hbar \quad \text{(Heisenberg–Pauli).} \tag{7.4.6}$$

This inequality, which we shall call the 'Heisenberg–Pauli inequality', expresses the basic property of an N-fermion system. In fact, compared to the standard Heisenberg inequality,

$$\Delta p \, \Delta l \gtrsim \hbar, \tag{7.4.7}$$

we notice that, in this case, the presence of a large number of identical fermions forces the system either to become spatially more extended for a fixed typical momentum dispersion, or to increase its typical momentum dispersion for a fixed typical spatial extension, or more generally, to increase in both these ways. We could also say that for a fermionic system in its ground state, the average energy *per quanton* increases with the density of the system (exercise 7.8). Of course, the exponent $\frac{1}{3}$ in eq. (7.4.6) comes from the spatial dimensionality (it would be 1 in a one-dimensional model). This inequality, confirmed by a rigorous theory, holds for *identical* fermions. The identity in question requires that all physical magnitudes be taken into account. As an example, formula (7.4.6) applies to electrons only if they are all in the same spin state. Otherwise, there could be the possibility of having two different states of spin correspond to each phase space cell. It would then be enough to have $N/2$ such cells in order to characterize N states, and we would end up with the inequality

$$\Delta p \, \Delta l \gtrsim (\tfrac{1}{2}N)^{1/3}\hbar. \tag{7.4.8}$$

But the factor, $2^{-1/3} \simeq 0.79$, being very close to unity, and with inequality (7.4.6) holding in any case only to an order of magnitude – i.e., up to a numerical constant (which hopefully is close to unity) – this refinement, while conceptually important, is for the moment devoid of any quantitative interest.

A picturesque way of interpreting the Heisenberg–Pauli inequality is to compare eq. (7.4.6) with eq. (7.4.7) and to think of the quantity on the right-hand side of it as an 'effective fermionic Planck constant',

$$\hbar^f(N) = N^{1/3}\hbar, \tag{7.4.9}$$

which grows larger with N. We could also say that antisymmetrization, which typifies fermionic amplitudes, amplifies those quantum effects which are affected by the Heisenberg inequality. It does so to a degree which becomes significant if the number N of identical fermions is large.

7.4.3. The Pauli principle and ordinary matter

We now proceed to examine some of the 'everyday' consequences of the fermionic nature of electrons. The examples which follow have been chosen, from among many others, because the notion of temperature is (almost) not used there. A general and complete theory of matter would involve both a quantum theory and a statistical treatment (otherwise called quantum statistics) – obtained by combining the quantum effects that concern us here, and the thermodynamic concepts of temperature and entropy, which we leave aside.

In what follows, we shall understand by 'ordinary matter' a set of nuclei and electrons subject only to electromagnetic forces: the system is not large enough for gravitational forces to be important (as would have been the case if we were studying stellar matter), and neither dense nor hot enough for nuclear forces to play any role other than just to ensure the stability of the nuclei. Thus, ordinary matter will mean for us the matter which makes up our surrounding world.

Under these conditions, the system is mainly governed by the combined interplay of Coulomb forces – attractive between electrons and nuclei, and repulsive between the electrons themselves or the nuclei themselves (magnetic interactions are, in general, secondary). The fermionic nature of electrons plays an absolutely essential role and the Pauli exclusion principle reigns supreme, 'without which things would never be what they are!' Remaining on a qualitative level, we shall describe its effects in the extreme cases of matter 'in detail' (the atom) and of matter 'in bulk', for which we shall examine the most stable, minimal energy ground state.

(a) *Matter 'in detail': atoms.* Had electrons not been fermions, the ground state of an atom of atomic number Z, could have been obtained by minimizing the total energy of the Z electrons, evaluated in terms of their average momentum and the size of the atom, taking only the Heisenberg inequalities into consideration (exercise 3.11). We could also say, that for a nucleus with charge Zq_e, neglecting the interaction between the electrons, each electron has a minimal energy of the order of $-Z^2 E_H$ (the coupling constant e^2 being multiplied by Z, as compared to the case of hydrogen [see eq. (3.3.11)]. Since there are Z electrons, there would thus be an energy

$$E_Z \simeq -Z^3 E_H, \tag{7.4.10}$$

for the entire atom, and its size, compared to that of hydrogen, would be reduced by Z,

$$r_Z \simeq Z^{-1} a_0. \tag{7.4.11}$$

Since taking the Coulomb repulsion between the electrons is not enough to modify qualitatively these estimates (particularly, their dependence on Z), atoms with higher Z would be more and more concentrated, and smaller and smaller, because of the increasing Coulomb attraction. Given that the atomic number varies from 1 to around 100, across the Mendeleev table, the atom of lead or uranium would have to be about a hundred times smaller than that of hydrogen!

However, it is a fact that the angstrom (or rather, the Bohr radius a_0) gives the scale of the size of *all* atoms. We shall see here, how as a consequence of the Pauli principle, the electrons are compelled to maintain their distances. The energy state, eq. (7.4.10), described above, corresponds to the Heisenberg inequality, eq. (7.4.7). But the electrons ought to obey the much stricter Heisenberg–Pauli inequality, eq. (7.4.6); this latter inequality easily allows us to compute the energy and size of an atom with Z electrons. It is enough to replace $\hbar$ by

$$\hbar^f(Z) = Z^{1/3} \hbar$$

in the estimates given by eqs. (7.4.10) and (7.4.11).

Taking into account the expressions for E_H (in $\hbar^{-2}$) and a_0 (in $\hbar^2$), we get

$$E_Z^f \simeq -Z^{7/3} E_H, \tag{7.4.12}$$

$$r_Z^f \simeq Z^{-1/3} a_0. \tag{7.4.13}$$

Actually, because of the Pauli principle, the atomic structure is more complex than what the preceding considerations would tend to indicate. In reality, the first few excited states in Z of a (fictitious) one-electron atom have larger and larger average radii. Under these conditions, the interaction between the electrons, ignored up to now, plays a much more subtle role here than in the non-fermionic case – where they only modify the energy estimate of the unique quanton state, used Z times in order to construct the symmetric collective state. Now, the interelectronic repulsion affects the Z different states, used to construct the collective state, differently. In fact, if the first z states have already been used up, the $(z + 1)$st state would correspond to a situation in which the electric charge Zq_e of the nucleus would appear screened by the amount $-zq_e$, corresponding to the z electrons already used and situated closer to the nucleus. In particular, the Zth and last state to be considered is of an electron which only feels an effective charge, $Zq_e - (Z - 1)q_e = q_e$, and hence is comparable to the ground state of the hydrogen atom. If therefore, as in chemistry, the dimension of the atom has to be taken as being that of the outermost region in which an electron could be situated, this dimension would essentially be that of the hydrogen atom, a_0. Estimate (7.4.13) is an ensemble average, over all electrons. Viewed in this way, it shows that the distribution of electronic charge is far from being uniform, becoming more concentrated with increasing Z. The outer region of this distribution remains hydrogen-like, however, which explains the constancy (in order of magnitude) of atomic sizes and ionization potentials – in other words, of the binding energy of the first electron, which is always a few electron-volts. This analysis of the atomic structure can be sharpened by introducing angular momentum considerations for the different atomic states. It can then be verified that the Pauli principle, governing the electrons, explains the well-known 'shell' structure of the atom as well as its basic chemical properties.

(b) *Matter 'in bulk'.* By matter 'in bulk' we shall understand matter as considered in macroscopic physics, fluid and solid mechanics or thermodynamics. Certain properties of matter, such as its equation of state, are not supposed to depend on the total quantity present. That is, the total number of particles does not figure in the determination of these properties. The boiling temperature of water, at a given pressure, is the same in a cooking pot as inside a boiler. The density of iron, under given conditions, is the same for a kilogram as for a ton. The point is, that the latent heat of vaporization, or the volume, are extensive quantities, i.e., proportional to the number of particles. It is, therefore, necessary to satisfy ourselves of the existence of such quantities, and to begin with, to verify that energy is an extensive quantity. Effectively, the binding energy $E_0(N)$ of a system of N atoms (forming a solid body, for example) increases linearly with N, in order for the properties of matter to be uniform. We must have, therefore,

$$E_0(N) \propto N, \tag{7.4.14}$$

in order that the binding energy *per* particle be independent of N,

$$\varepsilon(N) \triangleq \frac{1}{N}|E_0(N)| \propto \text{const.} \tag{7.4.15}$$

This condition is seen to hold very well for ordinary matter, where a few electron-volts are enough to strip off an electron, whether from a snow-flake or from an iceberg. One says that the Coulomb forces are 'saturated'. On a larger scale, at which gravitational forces come into play, saturation does not occur any more (exercise 7.9), and classical thermodynamics can no longer describe such systems. This saturation, an everyday experimental fact, has to be accounted for. Let us consider the simplest case – a neutral system of N positive particles having charge $+q_e$, and N negative particles, having charge $-q_e$ (protons and electrons, for example). The total potential energy is the result of a certain cancelling out between the Coulomb attractions and repulsions. In fact, each particle tends to be neutralized by the others, of opposite sign, forming a screen at a certain distance. Thus, the effective potential energy is just that of these N, say positive, particles, each interacting with an effective opposite charge, situated at a characteristic distance l, which is simply the typical distance between nearest neighbours (fig. 7.13). Hence,

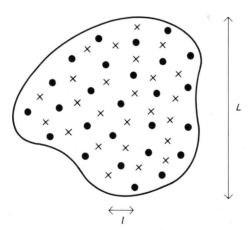

Fig. 7.13. *A Coulomb system*
N positive charges (●) and N negative charges (×) occupy a volume of dimension L. The average distance l between nearest neighbours also characterizes the dimension of the volume per particle.

we may write for this potential energy.

$$U \simeq -N\frac{e^2}{l}. \tag{7.4.16}$$

An expression of this type shows that, classically, since the stability of the system is not ensured, the problem of saturation does not even arise. As in the case of hydrogen, to have $U \to \infty$, it is enough to let $l \to 0$: there is no state of minimal energy (ground state). We can bring quantum theory to the rescue, therefore, thanks to the correlation which it establishes between the kinetic and potential energies, through the Heisenberg inequalities. The kinetic energy can be written as

$$E_{\text{kin}} = N\frac{\tilde{P}}{2M} + N\frac{\tilde{p}^2}{2m}, \tag{7.4.17}$$

where m and $\tilde{p}$ are the mass and the average momentum of, e.g., the negative particles and M and $\tilde{P}$ are those of the positive particles. If, as occurs in nature, $M \gg m$, we may rightfully ignore the kinetic energy of the nuclei by comparison with that of the electrons, and estimate the total energy using

$$E = N\frac{\tilde{p}^2}{2m} - N\frac{e^2}{l}. \tag{7.4.18}$$

Contrary to appearances, the linearity of E in N is not ensured since $\tilde{p}$ and l in eq. (7.4.18) depend on N! Indeed, let L be the characteristic dimension of the system as a whole, i.e., the extension of the spatial region accessible to each one of the particles. From the Heisenberg inequality we must have,

$$\tilde{p}L \gtrsim \hbar. \tag{7.4.19}$$

Moreover, if l is the distance between nearest neighbours, each particle contributes an amount of the order of l^3 to the total volume L^3 (fig. 7.13). Thus, we get

$$L^3 = Nl^3, \tag{7.4.20}$$

i.e.,

$$l = N^{-1/3}L. \tag{7.4.21}$$

Finally, from eqs. (7.4.19) and (7.4.20), we get

$$E \gtrsim N \frac{\hbar^2}{2m} \frac{1}{L^2} - N^{4/3} \frac{e^2}{L}, \qquad (7.4.22)$$

as a function of L only. As with the hydrogen atom, there exists an optimal balance between the kinetic and potential energies. This is obtained for

$$L \simeq N^{-1/3} a_0 \qquad (7.4.23)$$

and the energy of this ground state has the value

$$E_0(N) \simeq -N^{5/3} E_H. \qquad (7.4.24)$$

There is stability (the energy is bounded below), but not saturation. We see that the density of the system

$$\rho = \frac{N}{L^3} \simeq N^2 \frac{1}{a_0^3}, \qquad (7.4.25)$$

increases very rapidly with N, as does the binding energy per particle,

$$\varepsilon = \frac{|E_0(N)|}{N} \simeq N^{2/3} E_H. \qquad (7.4.26)$$

To appreciate how serious the problem is, it is only necessary to consider a gram of hydrogen, so that $N \simeq 10^{24}$. From eq. (7.4.23), this mass, in its ground state, would occupy a region of dimensions 10^{-8} Å (much smaller than a *single nucleus!*), and it would be necessary, in view of eq. (7.4.26), to expend $10^{16} E_H \simeq 10^{-2}$ J to detach a *single* atom. Once again, here it is the Pauli principle which saves the world from collapse by preventing the electrons from forming such a condensed collective state. As we have seen, to observe its effect via the Heisenberg–Pauli inequality, it is enough to replace $\hbar$ by $\hbar^f = N^{1/3} \hbar$. When this substitution is made in eqs. (7.4.23)–(7.4.26), it leads to the expected expressions for the size of the system,

$$L^f \simeq N^{1/3} a_0, \qquad (7.4.27)$$

its density

$$\rho^f \simeq \frac{1}{a_0^3}, \tag{7.4.28}$$

the energy of its ground state

$$E_0^f(N) \simeq -NE_H, \tag{7.4.29}$$

and the binding energy per particle

$$\varepsilon \simeq E_H. \tag{7.4.30}$$

Thus, there is saturation of the Coulomb forces, and eq. (7.4.28) clearly shows that the density of matter does not depend on the total number of particles. It is important to note that this result depends on the fermionic nature of *only one* of the two types of charged particles. Here, just the kinetic energy of the electrons is enough to balance the effective Coulomb attraction, and hence the statistics of the nuclei is of little consequence. This conforms to experience: although the finer details of the behaviour of matter depend on the statistics of nuclei (compare, e.g., ^{3}He and ^{4}He), its existence itself and its qualitative properties (such as density and cohesion) are independent of it.

We should mention, however, that the problem of the quantum saturation of Coulomb forces had for long been overlooked and was not formulated, and resolved, until fairly recently (Dyson and Lenard 1965). This was in spite of the fact that the notion is crucial to our understanding of matter, and that it may be considered as being a major contribution of quantum theory to this understanding. (It ought to be conceded that a rigorous analysis of this problem is of a level of technical difficulty which our heuristic presentation scarcely allows us to imagine.)

(c) *Matter in 'large bulk'*. If we consider a macroscopic block of matter, of increasing size, there comes a point where gravitational forces become decisive. In fact, the gravitational interaction, which at the individual level is much weaker than the Coulomb interaction (see ch. 1, sect. 1.4.3), dominates it at the collective level as a result of its purely attractive nature. In the gravitational case, there is no partial compensation between attraction and repulsion and no screening effect. Thus, a system having characteristic size L, consisting of N atoms of mass M, has a gravitational energy proportional to the number of interacting pairs, namely $\frac{1}{2}N(N-1) \simeq \frac{1}{2}N^2$ (since $N \gg 1$). The average distance between two atoms being of order L, the gravitational potential energy may be approximated using

$$U_{\text{grav}} \simeq -N^2 \frac{GM^2}{L},$$ (7.4.31)

where G is the Newton constant. This expression should be compared to the Coulomb potential energy previously evaluated [see eq. (7.4.22)] taking the screening effect into consideration,

$$U_{\text{el}} \simeq -N^{4/3} \frac{e^2}{L}.$$ (7.4.32)

We see that while U_{grav} is negligible compared to U_{el} for small N, it becomes of comparable magnitude at the transition value

$$N_t \simeq \left(\frac{e^2}{GM^2} \right)^{3/2}.$$ (7.4.33)

For $N > N_t$, the structure of the system is essentially governed by the gravitational interaction and its behaviour becomes vastly different. Indeed, the ground state of the system can now be characterized by minimizing the total energy

$$E \simeq N^{5/3} \frac{\hbar^2}{2mL^2} - N^2 \frac{GM^2}{L},$$ (7.4.34)

in which we have, of course, taken the fermionic nature of the electrons (which, because of their small mass, contribute essentially only through the kinetic energy) into consideration. The minimum of this expression is obtained for

$$L \simeq N^{-1/3} \frac{\hbar^2}{GM^2 m}.$$ (7.4.35)

We see this time that the size of the system diminishes as it mass $\mathcal{M} = NM$ increases! Clearly, here it is the effect of the gravitational attraction which, as shown by eq. (7.4.31), increases *more* rapidly with N than the kinetic energy. Consequently, the size of an object must pass through a maximum for

$N \simeq N_t$, a size which is straightforward to evaluate: eqs. (7.4.35) and (7.4.27) yield, as they must, the same result when the Coulomb and gravitational energies become comparable. We obtain

$$L_t \simeq \left(\frac{e^2}{GM^2}\right)^{1/2} \frac{\hbar^2}{me^2}. \tag{7.4.36}$$

The corresponding mass, $\mathcal{M}_t = N_t M$, has the value

$$\mathcal{M}_t \simeq \left(\frac{e^2}{GM^2}\right)^{3/2} M. \tag{7.4.37}$$

Let us compute these two magnitudes numerically, in the two cases of composite objects – 'light' matter, such as hydrogen ($M = m_p$) and 'heavy' matter, such as iron $M \simeq 60m_p$. We thus obtain for:

light matter:

$$N_t = 1.4 \times 10^{54}, \qquad L_t = 5.9 \times 10^4 \text{ km}, \qquad \mathcal{M}_t = 1.7 \times 10^{27} \text{ kg}$$

heavy matter:

$$N_t = 6.5 \times 10^{48}, \qquad L_t = 9.8 \times 10^2 \text{ km}, \qquad \mathcal{M}_t = 1.1 \times 10^{22} \text{ kg. (7.4.38)}$$

These magnitudes characterize the size of astronomical objects. As a matter of fact, asteroids and planets display masses and sizes well in accord with our description, for $N \leqslant N_t$ or $N \simeq N_t$ (fig. 7.14).

But how is it that stars, such as the sun, have radii much larger than the critical size calculated in eq. (7.4.38) (in fact, $R_\odot \simeq 0.7 \times 10^6$ km $\gg L_t$)? The point is, our analysis applies only to the ground state of the body! In reality, a star is in an extremely excited state of energy, and indeed, that is why it radiates.... . It is necessary, therefore, to take its thermal energy into account. In other words, it is the ordinary thermodynamic pressure, arising from its high temperature, and not the quantum kinetic energy of the electrons, in so far as they are fermions, which holds the star in equilibrium against gravitational collapse. Nevertheless, there exist (relatively!) cold stars which, having exhausted their reservoirs of internal energy, and undergone a drop in pressure with the fall in temperature, have contracted in size to a point where a new stability has been reached, thanks, this time, to the Pauli principle. These stars, called 'white dwarfs', do indeed conform to our analysis in the case $N > N_t$: the larger their mass $\mathcal{M} = NM$, the smaller is the radius, as

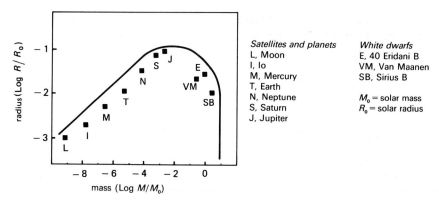

[From W.C. De Marcus, *Handbuch der Physik* L II, 420, S. Flügge (ed.), Springer-Verlag]

Fig. 7.14. *(Cold) matter at large scales*
A system of N atoms is primarily governed by Coulomb forces in its ground state, and its size is proportional to the 1/3rd power of its mass (its density being constant). Above a transitional mass, gravitational forces dominate, and its size decreases with increasing mass, up to a certain critical point (called the Chandrasekhar mass – see exercise 7.11), beyond which the system is no longer stable. Thus, quantum theory qualitatively explains these fundamental aspects, for astronomical objects such as planets or white dwarfs. The points in the graph correspond to these objects, while the curve indicates the theoretical relationship between mass and radius for a system formed of pure hydrogen, and limits the physically accessible region.

predicted by eq. (7.4.35) (see fig. 7.14). Additionally, for masses just a little larger than that of ordinary stars (such as that of the sun), our theory has to be modified to take into account the effects of Einsteinian relativity. We now discover that white dwarfs *cannot* be stable above a critical mass (the so-called the Chandrasekhar limit), corresponding to a number, $N_c = \alpha^{3/2} N_t$, of nuclei (exercise 7.11). Thus, at least in so far as 'cold matter' is concerned, quantum theory, in particular through the Pauli principle, plays a crucial role at the macroscopic level, as evidenced by the presence of the constant $\hbar$ in eq. (7.4.36) for a genuine astronomical magnitude.

7.4.4. Energetic consequences

We now look at the exclusion principle from a different angle, by directly analyzing its consequences in the evaluation of the energy of an N-fermion system.

(a) *The Fermi level.* Let us consider a system of N independent, i.e., mutually non-interacting fermions, and let us concentrate on its ground state. The external forces, which act separately on each fermion, determine the one-quanton stationary states, characterized by a series of energy levels $(\varepsilon_1 \ll \varepsilon_2 \ll \cdots \ll \varepsilon_k \ll \cdots)$. Analyzed in terms of these individual stationary states, a collective state of the system cannot, as a result of the Pauli principle, contain two identical states. It necessarily consists of N different individual states in a configuration. Thus, the ground state is formed by using N individual states, characterized by the lowest energies $(\varepsilon_1, \varepsilon_2, \ldots, \varepsilon_N)$. The energy of this ground state would then be

$$E_{\text{ground}}^{N\,\text{ferm}} = \varepsilon_1 + \varepsilon_2 + \cdots + \varepsilon_N. \tag{7.4.39}$$

This result is in contrast with that for a system of N different quantons (or for N bosons). In this latter case we would have obtained the minimum energy by using N individual states each having the lowest energy (or, in other words, from a single individual state, having occupation number N),

$$E_{\text{ground}}^{N\,\neq} = N\varepsilon_1 \leqslant E_{\text{ground}}^{N\,\text{ferm}}. \tag{7.4.40}$$

We see, therefore, that the effect of the Pauli principle is to raise the energy of a system of quantons. The result in eq. (7.4.39) is often stated by saying that the ground state of a system of fermions is obtained 'by putting the N fermions into the first N individual states'. But it is not true that we now have a fermion in each state: the N fermions collectively occupy the first N states (fig. 7.15).

It is necessary, moreover, to be careful in identifying stationary states with energy levels: to a given level there could correspond several different states. Their number is the degree of degeneracy of the level. Let us consider, e.g., the case of a system of N free electrons. A state of an electron is specified entirely by giving its momentum p and the projection ($\frac{1}{2}$ or $-\frac{1}{2}$) of its spin. However, it is only the value of its momentum which enters into the expression for the energy ε of these states ($\varepsilon = p^2/2m$), meaning that each level has a degeneracy of degree 2 (spin degeneracy). The ground state of the system of N free electrons is obtained, therefore, by 'putting' two electrons (of opposite spins) into each one of the first $\frac{1}{2}N$ individual energy levels (see, e.g., exercise 7.12). Generally, the ground state of a system of N independent fermions of spin s is obtained by filling up, at the rate of $(2s + 1)$ fermions per level, the individual levels one by one.

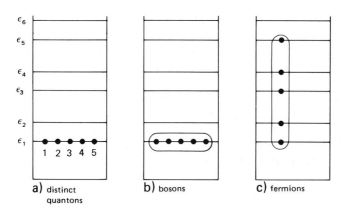

a) distinct quantons   b) bosons   c) fermions

Fig. 7.15. *Ground state of N independent quantons*

A stationary state of a system of N independent quantons can be characterized using individual stationary states, assumed to be non-degenerate here. Let $(\varepsilon_1, \varepsilon_2, \varepsilon_3, ...)$ be the proper energies, or 'energy levels', of these individual states.

(a) N different quantons (marked by numbering them), can all occupy the same individual level. The ground state is then obtained, by putting all N quantons into the lowest state.

(b) N identical bosons can no longer individually be marked for 'each' to occupy a state. Nevertheless, their collective state is characterized by using N individual states which could be identical. The ground state has the same energy as in the previous case.

(c) For N identical fermions, the collective state is specified by using N different individual states. Thus, the ground state energy is the sum of the energies of the first N individual levels.

Thus, there exists an individual level of maximum energy which demarcates the 'occupied' from the 'empty' levels. This is called the *Fermi level*, and is denoted by ε_F. In the ground state of the system, the energy levels below ε_F are completely occupied and the ones with energy higher than ε_F are all empty (this is the zero-temperature configuration, in the absence of thermal excitations).

The position of the Fermi level in the scale of the individual levels is easily obtained by noting that all energy levels below ε_F are occupied at the rate of $(2s + 1)$ per level, except possibly the level ε_F, which could only be partially occupied. When, as is generally the case, we have to deal with a pseudo-continuum of states, the computation is done using the notion of the density of states $\rho(\varepsilon)$ introduced in ch. 6. Thus, ε_F is determined using the relation

$$(2s + 1) \int_0^{\varepsilon_F} \rho(\varepsilon)\, d\varepsilon = N, \qquad (7.4.41)$$

since, as we recall, $\rho(\varepsilon)\,d\varepsilon$ is the number of energy states lying between ε and $\varepsilon + d\varepsilon$.

To end with these generalities, we stress the fact that the notion of the Fermi level only makes sense for a system of N independent fermions, where the collective state may be described by means of the individual levels obtained from the study of a single quanton. This condition may seem so restrictive as to exclude all actual physical systems: after all, nuclei exist only because nucleons are subject to mutual nuclear interactions, and, in atoms, the Coulomb interactions between electrons are surely effective, as we have already observed. Nevertheless, in numerous situations, it is seen that a system of fermions can successfully be described, approximately, by an independent quanton model. Precisely here, again, we find another consequence of the Pauli principle. The stubborn individualism expressed by this principle, keeps the fermions apart, so to say, and forces them into isolated seclusion, where eventually each one is subject to the action of all the others only in the form of a global effect. In other words, it is in general permissible to replace this action by an average external potential. In this way, a system of N fermions interacting with a certain external potential *and* among themselves, may be adequately simulated by a system of N fermions, *without* mutual interaction, but subject to an appropriately modified external potential. This 'independent fermion approximation' has a large domain of applicability (we have already used it in the description of the atom, at the end of sect. 7.4.3a). Clearly, it allows us to enlarge greatly the domain of applicability of the notion of the Fermi level.

Let us now study more closely, the simplest model of a system of independent fermions: a system of free fermions.

(b) *The free fermionic gas.* A system of N free fermions, without any interaction, but just enclosed in a 'box' of volume V, provides us with an elementary and yet one of the most useful of models. It is often given the name of a 'Fermi gas' (even though its density could be quite high). As we showed in ch. 6 [see eq. (6.2.31)], the density of states is now given by

ħ
$$\rho(\varepsilon) = \frac{V}{2\pi^2}(2m^3\varepsilon)^{1/2}. \tag{7.4.42}$$

The Fermi level of this system is computed using eq. (7.4.41). We find, once the integration has been performed,

ħ
$$\varepsilon_{\mathrm{F}} = \frac{1}{2m}\left(\frac{6\pi^2}{2s+1}\frac{N}{V}\right)^{2/3}. \tag{7.4.43}$$

It is important to note that the height of the Fermi level in fact depends on the density, N/V, of the system, the level being higher the higher the *density*. As we have seen, quantum effects only play a role for high densities.

It is interesting to calculate at the same time, the total energy of the system in its ground state, which clearly is

$$E_0 = (2s + 1) \int_0^{\varepsilon_F} \varepsilon \rho(\varepsilon) \, d\varepsilon. \tag{7.4.44}$$

The computation, using eq. (7.4.42), yields

$\hbar$
$$E_0 = (2s + 1) \frac{V}{4\pi^2} (2m)^{3/2} \frac{2}{5} \varepsilon_F^{5/2} = \left(\frac{3^5 \pi^4}{2(2s + 1)^2} \right)^{1/3} \frac{1}{5m} N^{5/3} V^{-2/3}, \tag{7.4.45}$$

which can easily be put into the form

$$E_0 = \tfrac{3}{5} N \varepsilon_F. \tag{7.4.46}$$

We see that the average energy of the fermions, $\tilde{\varepsilon} = E_0/N$, has the value

$$\tilde{\varepsilon} = \tfrac{3}{5} \varepsilon_F. \tag{7.4.47}$$

(compare with the qualitative results of exercise 7.8). Thus, the Fermi energy gives at once the upper limit of the energies of occupied states, and an order of magnitude for the average energy of an occupied state. In the absence of the Pauli principle, the ground state would have been constructed using the individual ground state levels inside the 'box', each having an energy $(3\pi^2/2m) V^{-2/3}$, and we would have had

$\hbar$
$$E_0 = \frac{3\pi^2}{2} \frac{1}{m} N V^{-2/3} \quad \text{(no Pauli principle)}, \tag{7.4.48}$$

which should be compared to eq. (7.4.45). Thus, the effect of the Pauli principle is to give an average energy per particle which depends only on the density of the system.

One can also interpret the foregoing results from a thermodynamic viewpoint. Expression (7.4.45) shows that the energy of the ground state of a system of N fermions increases as the confinement volume of the system is reduced. In other words, there exists an intrinsic *pressure* in the system. Since this is the pressure in the ground state, and hence at zero temperature, it has nothing to do with the usual pressure of a gas which is related to thermal agitation. Its origin lies in the exclusion principle and the repulsion effect which the latter produces. This so-called 'Fermi' pressure is easily computed by considering its contribution to the variation of the internal energy of the system, in the course of an expansion, everything else being held fixed,

$$P_F = -\frac{\partial E_0}{\partial V}.$$
(7.4.49)

We obtain in this way

$$P_F = \left(\frac{6\pi^2}{2s+1}\right)^{2/3} \frac{1}{5m} \left(\frac{N}{V}\right)^{5/3} = \frac{2}{5} \frac{N}{V} \varepsilon_F.$$
(7.4.50)

ℏ

It only depends on the density, i.e., on the ratio N/V, no matter how large V is. This would never have been the case for a system of bosons (exercise 7.13). For a fermionic system at a given temperature, a comparison between the Fermi pressure and the thermal pressure enables us to see whether or not quantum effects play an important role. In these terms we could reformulate the considerations in sect. 7.4.3, on matter 'in large bulk', by saying that cold matter, such as exists in white dwarfs, is maintained in equilibrium against gravitational collapse by the Fermi pressure, and not by thermal pressure which governs the hot matter of ordinary stars.

(c) *Energetic structure of metals.* The foregoing general considerations apply, in particular, to the description of a metallic crystal. Suppose we have a crystal, formed out of N atoms of a metal, having atomic number Z. The problem is one of describing the system of NZ electrons in Coulomb interaction with the N nuclei, placed on a regular lattice, but also undergoing mutual interaction – which tremendously complicates the analysis. Nevertheless, the reasoning developed above allows us to replace this mutual interaction between the electrons by an average effect, and to consider the electrons as being subject to an appropriately modified external potential, but independent of one another. Thus, the system can be described using individual states of *one* electron in a periodic potential. These stationary states follow the analysis carried out in ch. 6, sect. 6.7. They form energy bands, each allowed band being a pseudo-continuum comprising a number of levels, this number being a multiple of N. The ground state of the crystal is obtained now by distributing the NZ electrons, considered independent, among these individual levels, in accordance with the Pauli principle, which means for example, placing two electrons per level into the lowest levels. (Beware! We stress once again that this is an abuse of the usual language. To be completely rigorous, we ought to say that the ground state of the crystal is constructed using the NZ different individual states of the lowest energy, taking the second-order spin degeneracy into consideration.)

To fix ideas, consider the case of lithium ($Z = 3$). Lithium crystallizes in a way such that each mesh of the lattice contains only a single atom, so that each bond of metallic lithium has as many levels as there are atoms in the crystal, namely N. Thus, the problem becomes one of distributing the $3N$ electrons among the lowest levels at the rate of two per level (see fig. 7.16). Thus, we start by placing $2N$ electrons in the N levels of the lowest band (the 1s band coming from the 1s level of the free atom). There remain N additional electrons, which we put into the first $\frac{1}{2}N$ levels of the next higher band (labelled 2s). This band is, therefore, half empty (or half full). The Fermi level splits the last allowed energy band into two. This result is true for all monovalent metals; they always have a band which is half full.... Note, however, that the Fermi level ε_F does not coincide with the middle of this band. Indeed, the individual levels are not squeezed together towards the bottom of the band. In other words, in the continuum limit, the density of states, $\rho(\varepsilon)$, is not constant. In fact, as our discussion in ch. 6, sect. 6.7, pointed out, a conduction band is sufficiently well-described, except near the boundary of the Brillouin zone, by the formula

$$\varepsilon = \frac{1}{2m} \hbar^2, \tag{7.4.51}$$

similar to the one which holds for the energies of a free electron (see fig. 6.4.3). Thus, the density of states is described by the same expression, eq. (7.4.42), and the Fermi level can be evaluated using eq. (7.4.43) as a function of the electronic density inside the metal. We find in this way, that for monovalent metals, the Fermi level is of the order of a few electron volts (exercises 7.14, 7.15). This description, although brief, does nevertheless enable us to understand numerous phenomena in the physics of solids: specific heats, paramagnetism, photoelectric effect, resistance of junctions, etc. (exercises 7.16 to 7.25).

(d) *Electrical properties of solids: insulators and semiconductors.* Among the physical characteristics of a material, the resistivity (or its inverse, the

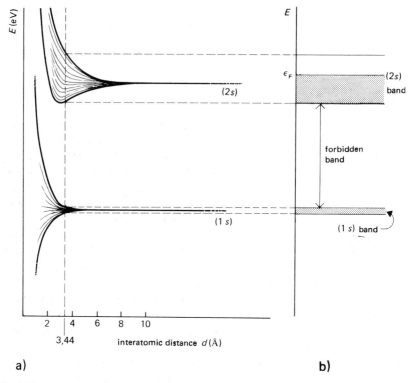

Fig. 7.16. *Energy bands of metallic lithium*
(a) Broadening of the atomic levels of lithium (denoted 1s and 2s, following spectroscopic usage), as one approaches the atoms, leading to crystal formation. In an actual crystal ($d = 3.44$ Å), the 1s level is practically not broadened at all, while the 2s band extends over several eVs.
(b) The electrons of a lithium crystal occupy the first band completely and half of the second band.

conductivity) is definitely the one with the most widely varying numerical value, as one passes from one material to another. For example, there is a ratio of 10^{32} between the resistivity of a good conducting metal ($\simeq 10^{-8}$ Ω m) and that of an insulator 10^{24} Ω m). This enormous variation, the importance of which in our daily life hardly needs demonstrating, is itself a consequence of the fermionic nature of the electrons, together with the band structure of the energy spectrum of the states of individual electrons.

We have seen in ch. 6, sect. 6.7.4 that when an electric field $\mathscr{E}$ is applied to a crystal, the second law of motion, as applied to the whole system (electrons + crystal), is written as

$$-q_e \mathscr{E} = \frac{\mathrm{d}\not{p}}{\mathrm{d}t}. \tag{7.4.52}$$

This relation involves the quasi-momentum of the electrons, i.e., the magnitude defining the stationary electronic states for a given allowed energy band. Hence, eq. (7.4.52) governs the displacement of the point representing the electron state on the curve $E(\not{p})$. We must now take into account the fact that the crystal contains not just one, but rather a large number of electrons. This system of electrons has an energy configuration which we shall now discuss: in the independent-electron approximation, its collective state may be described using individual states, characterized by the value of their quasi-momentum $\not{p}$. The Pauli principle implies that all the states, corresponding to the values $0 \leqslant \not{p} \leqslant \not{p}_F$, are 'occupied'. The effect of the electric field is then to modify the quasi-momentum $\not{p}$ of each one of these states, in accordance with eq. (7.4.52). From the band structure of the energy it follows that only states from the partially filled band can effectively contribute to a change in the collective state. It is often said (though incorrectly) that the electrons from this band alone contribute to the conduction.

Figure 7.17 shows, for a one-dimensional model, the two last, non-empty, alkaline type bands. One of them, labelled I, is completely filled and the next, II, is filled only up to the Fermi level, $\varepsilon_F = \not{p}_F^2/2m$. To say that band I is full, is the same as saying that all the states characterized by the values of $\not{p}$ lying between $-\pi/l$ and $+\pi/l$ are occupied. On the other hand, for band II the spectrum of $\not{p}$-values, corresponding to the occupied states, is more restricted – excluding the interval from $-\not{p}_F$ to $+\not{p}_F$. In order to fix ideas, let us assume that $\not{p}$ increases by an amount $\delta\not{p} > 0$, in the interval of time δt, under the effect of an applied field, following eq. (7.4.52). It is easily seen, from fig. 7.17c, that this increase does not modify the structure of the filled band I: since $\not{p}$ is defined only up to $2\pi/l$ the states which become occupied in the transition from (b) to (c) are exactly the ones which had been vacated. For the unfilled band II, on the other hand, things are different: the distribution of states on the curve $E(\not{p})$ is shifted to the right and the newly occupied states are not the same as the vacant states. The $\not{p}$-distribution, which is centred at $\not{p} = 0$ in the absence of the electric field, is block displaced by $\delta\not{p}$. The average value of $\not{p}$, over the different states has become positive: there are more states characterized by a positive value of $\not{p}$ than those characterized by a negative value. The

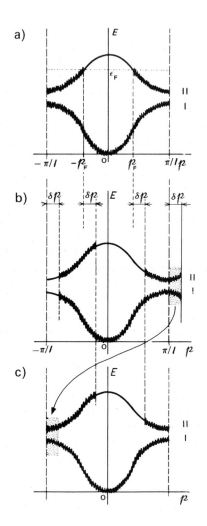

Fig. 7.17. *Electronic conduction inside a crystal*

(a) Band structure of an alkali 'in one dimension' in the reduced-zone scheme. Only the two last non-empty bands, I and II, are shown. The p-spectrum of the occupied states in each band has been emphasized. Since p is defined only up to $2\pi/l$, this spectrum extends from $-\pi/l$ to π/l for band I; it excludes the interval $(-p_F, p_F)$ for band II.

(b) Upon application of the field $\mathscr{E}$ over an interval of time δt, the two distributions of states are block translated by δp.

(c) The same in the reduced-zone scheme. Distribution I is equivalent to the one shown in (a). The same is not true for distribution II.

overall balance of electron transport is positive: the electronic system has acquired a non-zero total quasi-momentum and there is propagation. The application of the electric field $\mathscr{E}$ has the effect of moving the electrons in the direction opposite to $\mathscr{E}$. A current flows.... An alkali is a conductor. At least nothing prevents electricity from being conducted. This does not necessarily mean that it is conducted for, as we saw in ch. 6, sect. 6.7, because of Bragg reflection, the motion of an electron generated by $\mathscr{E}$ is, in fact, oscillatory. Actually, a complete explanation of conduction in metals brings crystalline impurities into play in a fundamental way.

Be that as it may, an alkali is potentially able to conduct electricity. By contrast, materials in which all bands are either completely filled or completely empty, i.e., materials for which the Fermi level is situated at the top of an allowed band, are insulators. Given that the number of available states in the interior of a band is twice the number of electrons per elementary mesh, only materials having an even number of electrons per elementary mesh can be insulators. Let us remember the following result from this analysis: conductors are materials whose band structure contains at least one partially filled band (see also exercise 7.25).

Finally, for certain materials, the highest occupied band is filled and separated from the first empty band by a small gap, of the order of 1 eV or less. This is the case in silicon and germanium, for example. It would seem that these materials would fall into the category of insulators. In fact, that is what they are at zero temperature. However, the gap separating the filled band from the empty band is so thin that it is quite possible for an electron, having acquired a surplus energy due to thermal excitation, to make a transition into one of the states at the bottom of the empty band. This latter band is called the 'conduction band' (as opposed to the last filled band, which is called the 'valence band'), since the electrons which have access to it give rise to conduction in the same manner as described above. Materials which are insulators at 0 K and conductors at non-zero temperatures are called intrinsic semiconductors (usual semiconductors are of a different type).

Figure 7.18 summarizes this whole discussion.

(e) *The Fermi model of the nucleus.* Let us look at one last example of the use of the free-fermion gas model, as applied to the atomic nucleus. In this case, the independence hypothesis might seem absurd or paradoxical, since it is by virtue of the *strong* mutual interactions of the nucleons, that the nucleus can exist at all! Nevertheless, one can even here consider, to a first approximation, that for each nucleon the collective effect of all the others is essentially to confine it to the interior of a finite volume – the volume of the nucleus. In

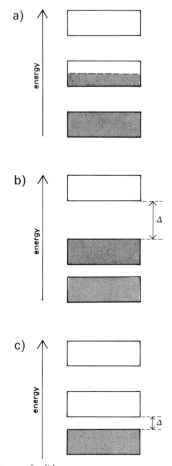

Fig. 7.18.   *The different types of solids*

(a) Conductor: in a monovalent metal, such as an alkali, the highest band is only half filled (it is called the conduction band).

(b) Insulator (carbon type): the filled valence band and the empty conduction band are very far apart.

(c) Semiconductor (silicon type): the (filled) valence band and the (empty) conduction band are separated by a small gap. An electron from the valence band can pass into the conduction band by simple thermal excitation.

other words, we consider each nucleon as having no interaction with the others, but subject to an average nuclear potential having radius R and depth U_0. Independently, these nucleons have individual energy levels which are the energy levels corresponding to the stationary states of the well. The ground state of the nucleus is constructed using these individual states by applying the Pauli principle, separately for protons and neutrons, and by filling up the energy levels of the well up to the Fermi level (fig. 7.19). If the potential well is sufficiently deep, we may consider it to be infinite, and apply the formulae established above for a free-fermionic gas, strictly confined to the volume V. The Fermi level is now computed using eq. (7.4.43).

For light, stable nuclei, the number of neutrons N_n and the number of protons N_p are equal, so that $N_p = N_n = \frac{1}{2}A = Z$, where A is the mass number and Z the atomic number. Thus we get

$$\varepsilon_F = \frac{\hbar^2}{2M} \left(\frac{3\pi^2}{2} \frac{A}{V} \right)^{2/3} . \tag{7.4.53}$$

Furthermore, the volume V is exactly proportional to the mass number:

$$V = \frac{4}{3}\pi R^3 = \frac{4}{3}\pi A r_0^3 \quad \text{(with } r_0 = 1.2 \text{ F).} \tag{7.4.54}$$

Finally,

$$\varepsilon_F = \frac{(9\pi)^{2/3}}{8} \frac{\hbar^2}{M r_0^2} \simeq 33 \text{ MeV.} \tag{7.4.55}$$

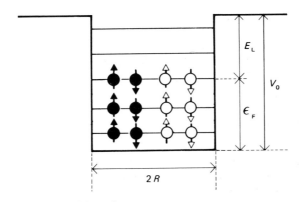

Fig. 7.19. *The Fermi model of the nucleus*
A spherical potential of radius R, having a flat bottom of depth V_0, simulates the average effect on each nucleon, of all the others. The nucleons, considered independent, now occupy the stationary states of this well up to the Fermi level, at the rate of two protons ● (having different spin states) and two neutrons ○ (having different spin states) per level.

Also, we know that the binding energy of a *single* nucleus in an ordinary nucleus is about 8 MeV,

$$E_L \simeq 8 \text{ MeV.} \tag{7.4.56}$$

Obviously, this is the energy which needs to be imparted to a nucleon in order to remove it from the highest-energy stationary state, i.e., from the Fermi level. In other words, the Fermi level is located around 8 MeV below the 'edge' of the well, and this latter has the depth

$$U_0 = \varepsilon_F + E_L \simeq 45 \text{ MeV} \tag{7.4.57}$$

(see fig. 7.19). Finally, let us note that the average (kinetic) energy of a nucleon *inside* the nucleus, in view of eq. (7.4.47) is

$$\tilde{\varepsilon} = \tfrac{3}{5}\varepsilon_F \simeq 22 \text{ MeV.} \tag{7.4.58}$$

We recover in this way the well-known orders of magnitude of nuclear physics.

We can also understand *why* the number of neutrons N_n and the number of protons N_p are nearly equal for a nucleus, at least for small A. Indeed, suppose that $N_p \neq N_n$. The total ground state energy of the nucleus can then be computed with the help of formula (7.4.45) for protons and neutrons, giving

$$E_0 = \left(\frac{3^5 \pi^4}{2^3}\right)^{1/3} \frac{\hbar^2}{5MV^{2/3}} (N_p^{5/3} + N_n^{5/3}). \tag{7.4.59}$$

As usual, set

$$N_p = Z, \qquad N_n = A - Z. \tag{7.4.60}$$

In the case where $N_p = N_n = \tfrac{1}{2}A$, introducing the Fermi level ε_F computed in eq. (7.4.53), expression (7.4.59) assumes the more appealing form

$$E_0(Z = \tfrac{1}{2}A) = \tfrac{3}{5}A\varepsilon_F, \tag{7.4.61}$$

which is the minimum value of the energy $E_0(Z)$ as a function of Z. As a matter of fact, this is the function described by eq. (7.4.59), which to lowest order in $\delta Z = Z - \frac{1}{2}A$ is

$$E_0(Z) = \tfrac{3}{5}A\varepsilon_F + \frac{4}{3A}\varepsilon_F(\delta Z)^2 + \cdots . \tag{7.4.62}$$

We see, therefore, that for a fixed number A of nucleons, the most stable nucleus is the one for which $N_p = N_n$ ($\delta Z = 0$). The fact that this rule is violated for nuclei with $A \gtrsim 30$, is due to another contribution to the total nuclear energy, which has not been taken into account here, namely the positive potential energy due to the Coulomb repulsion between the protons, which obviously increases as the proportion of protons in the nucleus becomes larger. The most stable nucleus (for a given value of A) is the result of a balance between the tendency to achieve proton–neutron symmetry, arising from the Pauli principle [see eq. (7.4.62)], and the tendency to reduce the number of charged protons. This nucleus has an excess of neutrons, which is larger for larger A (about 50% more neutrons than protons for $A \gtrsim 200$). Thus, the fermionic nature of nucleons turns out to be absolutely essential for an understanding of nuclear structure (see also exercises 7.26 and 7.27).

7.5. Bosons

7.5.1. The principle of Panurgean gregarity

Consider a system consisting of N *different* quantons, the state of which is characterized by N individual states $(r_1, r_2, \dots, r_N)$. We shall consider a transition to a final state, also characterized by N individual states $(s_1, s_2, \dots, s_N)$. The composite factorization principle applies here, since the quantons are distinct and hence independent. Its expression in eq. (7.1.4) generalizes immediately, enabling us to write the transition amplitude

$$\langle s_1, s_2, \dots, s_N | r_1, r_2, \dots, r_N \rangle = \langle s_1 | r_1 \rangle \langle s_2 | r_2 \rangle \cdots \langle s_N | r_N \rangle. \tag{7.5.1}$$

The probability, of course, factorizes in the same way as well,

$$\begin{aligned}
\mathscr{P}[(s_1, s_2, \dots, s_N) \leftarrow (r_1, r_2, \dots, r_N)] \\
= \mathscr{P}(s_1 \leftarrow r_1)\,\mathscr{P}(s_2 \leftarrow r_2) \cdots \mathscr{P}(s_N \leftarrow r_N).
\end{aligned} \tag{7.5.2}$$

Here, the final state is specified for *each* individual quanton: s_k is the final state of the specifically marked kth quanton. Suppose that we are not interested in knowing which quanton occupies which individual state. This would be the case if, e.g., the N individual final states were defined by as many counters, sensitive to the presence of the incident quantons, but not to their nature. The only thing that would matter then would be the fact that each individual state s_k is occupied by a certain quanton. Such a situation corresponds to having as many (collective) final states as there are different possible ways of distributing the N quantons among the N individual states $(s_1, s_2, \ldots, s_N)$. The number of such collective final states is $N!$, the number of possible permutations of the N individual states $(s_1, s_2, \ldots, s_N)$ when the latter are considered different. These $N!$ states being distinct, the total probability of reaching the ensemble of states $\{s_1, s_2, \ldots, s_N\}$, regardless of the identity of the quantons which occupy them, and which we denote using curly brackets, is obtained as the sum over the $N!$ probabilities corresponding to each permutation

$$\mathscr{P}[\{s_1, s_2, \ldots, s_N\} \leftarrow \{r_1, r_2, \ldots, r_N\}]$$

$$= \sum_{N!\,\text{permut.}} \mathscr{P}[(s_{i_1}, s_{i_2}, \ldots, s_{i_N}) \leftarrow (r_1, r_2, \ldots, r_N)]$$

$$= \sum_{N!\,\text{permut.}} \mathscr{P}(s_{i_1} \leftarrow r_1)\, \mathscr{P}(s_{i_2} \leftarrow r_2) \cdots \mathscr{P}(s_{i_N} \leftarrow r_N), \qquad (7.5.3)$$

where $(i_1, i_2, \ldots, i_N)$ denotes one of the $N!$ possible permutations of the indices $(1, 2, \ldots, N)$. In the case where some of the final individual states are identical, it is clearly necessary to adjust the computation of the probability by considering the states which have been counted and which should actually have been excluded. When the same state s_k appears n_k times in the list of the N individual states $\{s_1, \ldots, s_N\}$, we often say that this state is 'occupied by n_k quantons' and we call n_k the 'occupation number' of the state s_k. A permutation of these n_k identical states among themselves clearly does not affect the final collective state. For example, if $N = 3$, there are six different final states if s_1, s_2, s_3 are different, but only three if $s_1 = s_2 \neq s_3$, and just one if $s_1 = s_2 = s_3$. Hence, to evaluate the number of collective final states, it is necessary to divide expression (7.5.3) by the product of the factorials of the occupation numbers of the individual states. From now on, we shall be interested exclusively in the case in which all the individual final states are the same, $s_1 = s_2 = \cdots = s_N = s$. Thus, there is only one final state of the ensemble, and the total probability is obtained by dividing eq. (7.5.3) by the factor $N!$,

$$\mathscr{P}[\{ss \cdots s\} \leftarrow \{r_1, r_2, \ldots, r_N\}]$$

$$= \frac{1}{N!} \sum_{\text{permut.}} \mathscr{P}(s \leftarrow r_1)\, \mathscr{P}(s \leftarrow r_2) \cdots \mathscr{P}(s \leftarrow r_N). \qquad (7.5.4)$$

Since the sum obviously contains $N!$ identical terms, we finally obtain

$$\mathscr{P}[\{ss \cdots s\} \leftarrow \{r_1 r_2 \cdots r_N\}] = \mathscr{P}(s \leftarrow r_1)\, \mathscr{P}(s \leftarrow r_2) \cdots \mathscr{P}(s \leftarrow r_N), \quad (7.5.5)$$

which is just the obvious and natural result that could have been obtained directly. However, the detour that we have made will be useful to us in the case of identical bosons that we shall soon encounter. Before that, let us rewrite our result in eq. (7.5.5) more compactly, displaying explicitly that we are dealing with distinct quantons and recalling that the N individual final states are the same,

$$\mathscr{P}^{\neq}(N) = \mathscr{P}_1 \mathscr{P}_2 \cdots \mathscr{P}_N. \quad (7.5.6)$$

Let us now consider the same situation for a set of N identical bosons. This time, the composite factorization principle no longer applies, since now the set of N individual final states $(s_1, s_2, \dots, s_N)$ defines a single collective final state and a single (symmetric) amplitude. Now it is the amplitude itself (and not the probability), which appears as a symmetrized sum of $N!$ terms, corresponding to all the possible permutations. Thus, one obtains, following the model of the two-boson expression in eq. (7.2.29),

$$\langle s_1, s_2, \dots, s_N | r_1, r_2, \dots, r_N \rangle_b = \langle s_1 | r_1 \rangle \langle s_2 | r_2 \rangle \cdots \langle s_N | r_N \rangle$$
$$+ \langle s_2 | r_1 \rangle \langle s_1 | r_2 \rangle \cdots \langle s_N | r_N \rangle$$
$$+ \cdots, \quad (7.5.7)$$

or, with the notation already introduced,

$$\langle s_1, s_2, \dots, s_N | r_1, r_2, \dots, r_N \rangle_b = \sum_{\text{permut.}} \langle s_{i_1} | r_1 \rangle \langle s_{i_2} | r_2 \rangle \cdots \langle s_{i_N} | r_N \rangle. \quad (7.5.8)$$

This amplitude allows us to compute the transition probability

$$\mathscr{P}^b[\{s_1, s_2, \dots, s_N\} \leftarrow \{r_1, r_2, \dots, r_N\}]$$
$$= |\langle s_1, s_2, \dots, s_N | r_1, r_2, \dots, r_N \rangle_b|^2, \quad (7.5.9)$$

where now, the result is independent of the order of the states $(s_1, s_2, \dots, s_N)$, by hypothesis and because of the symmetry of the amplitude. For an ensemble of N bosons, it is no longer possible to associate a definite state to each boson. The expression in eq. (7.5.9), like its analogue for distinct quantons, eq. (7.5.3), holds when the N individual final states are all different. Otherwise, as before, it is necessary to divide by the product of the factorials of the occupation numbers of the individual states, so as not to count the same state in the ensemble more than once. In the case of particular interest to us, in which the N individual final states are all the same, so that $s_1 = s_2 = \cdots = s_N = s$, the expression in eq. (7.5.9) ought to be corrected by a factor of $1/N!$ Moreover, in this case, the amplitude of eq. (7.5.8) is a sum of $N!$ equal terms,

$$\langle ss \cdots s | r_1, r_2, \ldots, r_N \rangle = N! \langle s|r_1 \rangle \langle s|r_2 \rangle \cdots \langle s|r_N \rangle. \tag{7.5.10}$$

Thus, we obtain finally,

$$\mathscr{P}^b[\{ss \cdots s\} \leftarrow \{r_1 r_2 \cdots r_N\}]$$

$$= \frac{1}{N!}(N!)^2 |\langle s|r_1 \rangle|^2 |\langle s|r_2 \rangle|^2 \cdots |\langle s|r_N \rangle|^2. \tag{7.5.11}$$

On the right-hand side we recognize the individual transition probabilities

$$\mathscr{P}_k = |\langle s|r_k \rangle|^2. \tag{7.5.12}$$

Using a notation similar to that used in eq. (7.5.6), we write the total probability in terms of the individual probabilities,

$$\mathscr{P}^b(N) = N! \mathscr{P}_1 \mathscr{P}_2 \cdots \mathscr{P}_N. \tag{7.5.13}$$

The essential result here is the presence of an additional factor of $N!$, in the case of bosons in eq. (7.5.13) – the 'bosonic factorial', as we shall call it – as compared to that for distinct quantons. This result is full of consequences. The preceding discussion can be schematized as follows:

$$\text{Classical probability} = \frac{1}{N!} \sum_1^{N!} \text{probabilities}$$

$$\rightarrow \text{a factor } \frac{N!}{N!} = 1;$$

$$\text{Bosonic quantum probability} = \frac{1}{N!} \left| \sum_1^{N!} \text{amplitude} \right|^2$$

$$\rightarrow \text{a factor } \frac{(N!)^2}{N!} = N!.$$

The physical significance of this result is very simple: the probability for obtaining a system of N bosons, all in the same individual state, is $N!$ times larger than the analogous probability for N distinct quantons. Thus, collective states in which all the bosons are in the same individual state are overwhelmingly preferred: bosons 'like' being identical. ...While fermions display a stubborn aloofness – true lonely wolves – bosons exhibit a herd mentality, like the Panurgean sheep in Rabelais. Conversely, the probability that a system of bosons, thus occupying identical individual states, would undergo a transition, in which *one* of them changes states, is relatively much smaller than in the case of distinct quantons (or in the case of fermions). Bosonic systems display very strong coherence. It is difficult to change their state.

A little later (sect. 7.5.3) we shall see another derivation of the bosonic factorial. However, first let us show what its implications might be.

7.5.2. Some examples

(a) *Superfluidity and quantum fluids.* It is the coherence arising from the bosonic factorial which explains the superfluidity of liquid helium – more precisely, helium-4 (^{4}He). This is a highly special state of a fluid, reached at a temperature lower than about 2 K, in which the liquid no longer displays *any* viscosity. This brings about a horde of effects, such as the tendency of superfluid helium to come out, all by itself, of the container holding it, creeping over container walls by capillarity, with no friction holding it back, and other, even stranger, thermomechanical properties. Superfluidity is a perfect example of the kind of rigidity which exists in a bosonic system, which is what helium-4 atoms really are. They can only move all together. Thus flowing liquid helium cannot lose its energy bit by bit, like an ordinary viscous liquid, in which interactions with the walls slow down the individual atoms and diminish the kinetic energy of the entire fluid. But how is it that superfluidity shows up only at low temperatures? The gregarious nature of bosons, characterized by the bosonic factorial, becomes evident only when the collective state consists of a large number of identical individual states. This is the case at zero temperature, where, by definition, the system is in its ground state. For mutually non-interacting bosons, this state is obtained using N states, each one of which is identical to the individual ground state, i.e., using N individual stationary states of minimum energy. Indeed, as soon as the temperature of the system falls below a certain critical temperature T_b, there is, among the N individual states making up the collective state, a significant proportion of states which are just the (individual) ground state.

This phenomenon, of a macroscopic occupation of the individual state of minimum energy, is called 'Bose–Einstein condensation'. Thus, a veritable phase transition takes place below a critical temperature, which is low enough so that almost none of the individual states appearing in the collective state is an excited state with an energy higher than that of the ground state. This notion can be more explicitly stated, once again, with the help of the Heisenberg inequalities. In the helium fluid, having density μ, each atom, of mass m, occupies an average volume

$$\mathscr{V} = \frac{m}{\mu}, \tag{7.5.14}$$

and may be thought of as being confined (unlike in the case of a perfect gas) to a region having linear dimensions a, such that

$$\mathscr{V} \simeq a^3. \tag{7.5.15}$$

Now, a quanton constrained to occupy a finite spatial volume undergoes a quantization of its energy levels. The ground state possesses an average kinetic energy

$$\varepsilon_0 = \frac{\tilde{p}^2}{2m}, \tag{7.5.16}$$

where $\tilde{p}$ is the average momentum, related to the dimension a of the confinement region, through the saturated Heisenberg inequality

$$\tilde{p}a \simeq \hbar. \tag{7.5.17}$$

This value of the energy, namely,

$$\varepsilon_0 \simeq \frac{\hbar^2}{2ma^2} \simeq \frac{\hbar^2}{2m}\left(\frac{\mu}{m}\right)^{2/3}, \tag{7.5.18}$$

also gives the scale of energies of the various excited states of the quanton, and in particular, the difference between the ground state and the first excited state. To make sure that not too many of the occupied states are excited, it is necessary that the average thermal energy kT, where T is the temperature, be lower than ε_0/k. We could also say that, roughly ε_0 gives the temperature T_b, below which bosonic coherence effects appear, so that

$$kT_b \simeq \frac{\hbar^2}{2m}\left(\frac{\mu}{m}\right)^{2/3}. \tag{7.5.19}$$

Of course, we recover in this way the value already obtained in ch. 1, sect. 1.3.1, using purely dimensional arguments – except that now we have gained, in the meantime, more physical insight. Expression (7.5.19) also gives a correct order of magnitude: $T_b \simeq 3.1$ K, to be compared with the value $T_\lambda = 2.2$ K of the superfluid transition in helium. However, an exact explanation of superfluidity demands a much more detailed analysis. In particular, it requires the existence of interactions between the bosons: a fluid of free bosons is not superfluid even though it displays some quantum effects. More generally, the temperature T_b is the temperature below which, in a fluid, macroscopic quantum properties manifest themselves. In principle, these appear during a phase transition, showing up as Bose–Einstein condensation, superfluidity or other phenomena.

One last argument should, *a contrario*, be given to justify the connection established between superfluidity and the bosonic nature of helium-4 atoms: it lies in the fact that their behaviour is so completely different from that of helium-3 atoms, as shown already in fig. 7.11. To recall, the interaction forces arising from the single electrons in the atoms are the same for helium-3 and for helium-4 atoms. It is only the nuclei of these atoms which are different, but the nuclei do not figure directly in the macroscopic properties in question. However, this intimate difference is enough, through its strictly quantum effect on the 'statistics' of the atoms, to lead to remarkable results.

One might justifiably ask, why of all simple bosonic matter, it is helium-4 alone, which displays the property of superfluidity. The answer depends on a double characteristic of this atom: it is both light and inert, because of its saturated electronic structure. This has the consequence that at the quantum temperature T_b, its average kinetic energy ε is still much greater than the weak attractive potential between two atoms. For most other materials, the atomic mass m is much too large, so that ε_0 is very much smaller: the attractive potential dominates, the atom is now bound and the material becomes solid! True, the hydrogen molecule H_2 is even lighter than helium, but nevertheless the strong interaction between hydrogen molecules again dominates. Thus, helium is the only material which is still liquid at a temperature at which quantum effects in a fluid can manifest themselves. Let us note, however, that some current research deals with 'polarized' fluids, in which the atomic spins are made to assume the same orientation. In polarized fluid hydrogen, for example, all the atoms are in the same spin state. As we have seen, in this case, the Pauli principle implies a kind of additional repulsion, and the interaction between the hydrogen atoms is considerably weaker than in the non-polarized fluid. For this reason, polarized hydrogen – which we denote by H↑ – remains in the atomic state, rather than forming H_2 molecules, and from this it is expected that it would manifest extremely interesting quantum properties: given the weakness of the interactions, it would be very close to being an ideal free-boson gas and could display an almost 'pure' Bose–Einstein condensation. In general, the study of quantum fluids is, at the moment, a highly active area (see exercise 7.28).

There are other bosonic 'fluids' which also display a behaviour analogous to superfluidity. This is the case of the conduction electrons in certain metals which, interacting via the vibrations of the crystalline lattice, form pairs of bound states – the 'Cooper pairs'. These electron pairs may be considered as being bosons and their 'superfluidity' – analogous to that of helium atoms – gives to the metal the property of 'superconductivity': below a certain critical temperature, its resistance vanishes completely, just as does the viscosity of a superfluid. The complete explanation of superconductivity has certainly been one of the greatest triumphs of the quantum theory of condensed matter. There are other examples of the pairing off mechanism, which transforms a system of interacting fermions into a system of pseudo-bosons. Thus, helium-3 itself becomes superfluid in this way, when two of its fermionic atoms form a bosonic pair. Of course, this phenomenon has nothing to do with the superfluidity of helium-4, as borne out by the extremely low and very different temperature at which helium-3 becomes superfluid ($T \simeq 10^{-3}$ K).

(b) *Induced transitions and lasers.* There is another way of writing and interpreting the probabilities given by eqs. (7.5.6) and (7.5.13), the former corresponding to the case where N different quantons and the latter to the case where N bosons make a transition to a state consisting of N identical individual states. To see this, let us compare, in the two cases, the probabilities for $N + 1$ and N quantons. In the case of distinct quantons, we have

$$\mathscr{P}^{\neq}(N + 1) = \mathscr{P}^{\neq}(N)\,\mathscr{P}, \tag{7.5.20}$$

where $\mathscr{P}$ is the individual transition probability of the last quanton. In other words, the probability of a final state, in which $(N + 1)$ quantons are in the same state, is given by the product of the probability of a state in which N quantons are in this state and the probability that the $(N + 1)$st quanton is in there too – normal factorization for independent quantons. But for bosons, as a result of the bosonic factorial, the analogous expression, obtained using eq. (7.5.13), is

$$\mathscr{P}^{\mathrm{b}}(N + 1) = (N + 1)\mathscr{P}^{\mathrm{b}}(N)\,\mathscr{P}. \tag{7.5.21}$$

In other words, the probability of obtaining the $(N + 1)$ bosons in the same state is multiplied by $(N + 1)$, as compared to the case of distinct quantons, eq. (7.5.20). Yet another way of expressing this result is by introducing the transition probability of the $(N + 1)$ bosons to the same final state, *given that N* of them are already there, to be denoted by $\mathscr{P}^{\mathrm{b}}(N + 1 \leftarrow N)$. By the very definition of such a conditional probability, we have

$$\mathscr{P}^{\mathrm{b}}(N + 1) = \mathscr{P}^{\mathrm{b}}(N + 1 \leftarrow N)\,\mathscr{P}^{\mathrm{b}}(N), \tag{7.5.22}$$

and, consequently, comparing with eq. (7.5.21),

$$\mathscr{P}^{\mathrm{b}}(N + 1 \leftarrow N) = (N + 1)\mathscr{P}. \tag{7.5.23}$$

Thus, the probability of obtaining $(N + 1)$ bosons in the same state, given that N of them are already in it, is $(N + 1)$ times larger than the transition probability of a single boson [the bosons are *not* independent, and eq. (7.5.23) is *not* the transition probability of 'the $(N + 1)$st boson'!]. Writing this transition probability in the form

$$\mathscr{P}^{\mathrm{b}}(N + 1 \leftarrow N) = \mathscr{P} + N\mathscr{P}, \tag{7.5.24}$$

we see that it is the sum of two terms, the first of which is the term which exists only for independent quantons and the second term is proportional to the number of bosons already present. The first term corresponds in a way to a *spontaneous* natural transition, which occurs 'in any case' while the second term describes a transition *induced* by the presence of the N bosons, there.

These ideas are applicable to all bosonic transitions, in particular to photon emission. For example, the phenomenon of induced emission plays a crucial role in the functioning of lasers. In fact, in a laser, an ensemble of excited atoms, occupying metastable energy states, is made to lose its excitation energy in the form of a coherent monochromatic radiation. The first few photons, emitted spontaneously, *induce* the emission of the next few, gradually releasing in this way a veritable avalanche of photons. The photons emitted by a laser, unlike those coming from an ordinary source of light (e.g., an incandescent lamp) are, therefore, not independent. Rather, they form a collective system of bosons, having well-defined phase relationships, thus giving the laser, besides its power and monochromaticity, an exceptionally high degree of coherence.

(c) *Black-body radiation.* A 'black body' is a body which is in thermodynamic equilibrium with the electromagnetic radiation which it emits and absorbs. The aim of the theory of black-body radiation is to calculate the density of the electromagnetic energy of this radiation at a given temperature T, and its spectral composition: empirically, it is known that a 'black body' becomes... red, then white, as its temperature rises (in other words, it becomes 'white hot'), implying that the average wavelength of the emitted radiation decreases. The classical theory stumbled and fell against this phenomenon, which historically by the way, has been one of the origins of the quantum theory. We shall now establish the theoretical expression for black-body radiation, based on the foregoing considerations on bosonic systems (we have already indicated a qualitative quantum approach in exercise 1.7). Thus, consider an enclosure filled with radiation – i.e., with photons – in equilibrium at a temperature T with the atoms of the walls. Being in equilibrium, the number of photons emitted per unit time by these atoms is equal to the number of photons absorbed. Consider a fixed pulsation ω of the radiation and let us try to find the average number $n(\omega)$ of photons in the enclosure, having the corresponding energy $\hbar\omega$. Let us use a particular model, assuming that the enclosure is made up of atoms having two energy levels, the ground state with energy E_g, and an excited state with energy E_e. The levels are assumed to be separated by an energy difference of exactly $\hbar\omega$, so that atoms in the energy state E_e can emit and those in the energy state E_g can absorb a photon (fig. 7.20a). Thus,

$$E_e - E_g = \hbar\omega. \tag{7.5.25}$$

Let $\mathcal{N}_g$ and $\mathcal{N}_e$ be the numbers of atoms in the ground state and the excited state, respectively. At equilibrium, these two numbers must be in a ratio given by the Boltzmann factor,

$$\frac{\mathcal{N}_e}{\mathcal{N}_g} = \frac{e^{-E_e/kT}}{e^{-E_g/kT}} = e^{-\hbar\omega/kT}. \tag{7.5.26}$$

This will be our only appeal to thermodynamics! The average number of photons emitted is now obtained by multiplying the probability of emission

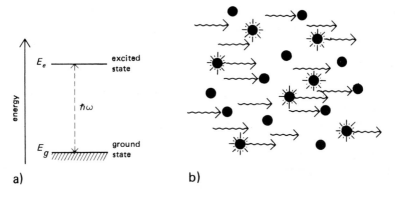

Fig. 7.20. *Photons in thermodynamic equilibrium*
(a) We use a simplified model of atoms, having two energy levels E_g (ground state) and E_e (excited state, separated by an energy difference equal to the energy $\hbar\omega$ of the photons being considered.
(b) At thermodynamic equilibrium, the number of atoms in the ground state (shown as ●) is greater than that of the excited atoms (shown as ⚡●⚡). But the emission induced by the presence of the photons makes emission by the excited atoms more probable than absorption by atoms in the ground state. Equilibrium is established when the total number of photons absorbed is equal to the number of photons emitted.

of a photon by a *single* atom, by the number of excited atoms $\mathcal{N}_e$. Since, by hypothesis, there are already n atoms present, this probability is the probability $\mathcal{P}(n + 1 \leftarrow n)$, introduced previously in eq. (7.5.23). Thus, we may write

$$\text{Average number of emitted photons} = \mathcal{N}_e(n + 1)\mathcal{P}, \qquad (7.5.27)$$

where $\mathcal{P}$ is the probability of emission of a single photon by an atom. Quite symmetrically, to calculate the average number of photons absorbed, we multiply the number of atoms in the ground state $\mathcal{N}_g$ by the probability of absorption of a photon, given that n photons are already there, that is, by $\mathcal{P}(n - 1 \leftarrow n)$ [see eq. (7.5.24)]. To compute this last probability we appeal to the symmetry of probabilities, established in ch. 4 [see eq. (4.4.36)]. Using this rule we get simply,

$$\mathcal{P}(n - 1 \leftarrow n) = \mathcal{P}(n \leftarrow n - 1) = n\mathcal{P}, \qquad (7.5.28)$$

by eq. (7.5.23) (note the consistency of this result: the probability of absorption of a photon is zero when there are no photons present...). From this we may write

Average number of photons absorbed $= \mathcal{N}_\mathrm{g} n \mathcal{P}.$ (7.5.29)

At equilibrium, the average numbers of photons emitted, eq. (7.5.27), and absorbed, eq. (7.5.29), are equal (fig. 7.20b), so that

$$\mathcal{N}_\mathrm{g} n \mathcal{P} = \mathcal{N}_\mathrm{e}(n+1)\mathcal{P}.$$ (7.5.30)

From this we easily obtain

$$n = \frac{1}{(\mathcal{N}_\mathrm{g}/\mathcal{N}_\mathrm{e}) - 1},$$ (7.5.31)

and finally, by eq. (7.5.26),

$$n(\omega) = \frac{1}{e^{\hbar\omega/kT} - 1},$$ (7.5.32)

where the notation $n(\omega)$ reiterates the fact that we are dealing with the average number of photons of energy $\hbar\omega$. Although we established eq. (7.5.32) for a specific model of absorbing–emitting atoms, it necessarily has a more general validity since, at thermodynamic equilibrium, the radiation characteristics can only depend on the temperature and not on the specific properties of the material. From this we derive the energy content of the radiation, for the pulsation ω, by multiplying the number of photons $n(\omega)$ by their energy $\hbar\omega$,

$$\varepsilon(\omega) = \frac{\hbar\omega}{e^{\hbar\omega/kT} - 1}.$$ (7.5.33)

[It is interesting and satisfying to verify that, in the region of validity of the classical approximation, i.e., for $kT \gg \hbar\omega$ (see exercise 1.7), this expression tends to the classical formula $\varepsilon = kT$, and the equipartion of energy is recovered.] We have, however, neglected up to now an essential aspect of the situation, by having argued as though the pulsation ω were sufficient to define the state of a photon. Of course, that is not true, since the radiation occupies a certain volume V, the momentum of a photon can have different orientations and finally, everything else being equal, a photon can still have two states of polarization. Thus, we ought to introduce the number of different states available to a photon, having energy ω and lying within $d\omega$, inside the volume V. This computation is exactly that of the density of states, introduced in sect. 6.2.3. In the particular case of photons, having zero mass, we obtain (exercise 6.5) for the number of states

$$\rho(\omega)\,d\omega = \frac{V}{\pi^2 c^3}\omega^2\,d\omega. \qquad (7.5.34)$$

Since, for each one of these states, the energy is $\varepsilon(\omega)$ as given by eq. (7.5.33), we finally obtain the total radiation energy within the volume V for pulsations in the range ω to $\omega + d\omega$:

$$dW = \varepsilon(\omega)\,\rho(\omega)\,d\omega = \frac{V}{\pi^2 c^3}\,\frac{\hbar\omega}{e^{\hbar\omega/kT}-1}\,\omega^2\,d\omega. \qquad (7.5.35)$$

In other words, the spectral and volume density of the energy is

$$u(\omega) \triangleq \frac{1}{V}\frac{dW}{d\omega} = \frac{1}{\pi^2 c^3}\,\frac{\hbar\omega^3}{e^{\hbar\omega/kT}-1}. \qquad (7.5.36)$$

This is the famous Planck formula (see fig. 7.21), the quantum and more specifically, the bosonic nature of which is apparent. Indeed, induced

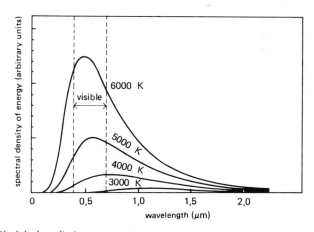

Fig. 7.21. *Black-body radiation*
The graphs show the spectral density of black-body radiation as a function of the wavelength at various temperatures, as given by Planck's law.

emission – a phenomenon characteristic of the gregariousness of bosons – plays a fundamental role in matter/radiation thermodynamic equilibrium, since in combination with spontaneous emission, it compensates for the small number of emitting atoms, compared to the absorbing atoms, as shown clearly by the equilibrium criterion, eq. (7.5.30) (see also exercises 7.29–7.31).

7.5.3. *Back to the bosonic factorial*

Another derivation might be useful in helping us understand better the origin of the bosonic factorial. As we have seen, the latter comes into play when we try to find amplitudes for transitions to a collective final state, defined by N identical individual states. For concreteness, consider the situation in which the bosons of a certain system are detected within a counter of area S. We are interested in the probability that, in a given period of time, the counter would register k quantons. This probability clearly depends on the size of the counter: a counter with twice the surface area would have doubled probability for registering one impact, quadrupled for two, eightfold more for three, etc. – provided the area of the counter is small enough so that the spatial amplitude does not vary appreciably over its surface. Stated differently, the probability of recording k quantons is proportional to S^k. We denote this by $\mathscr{P}(k)S^k$, where $\mathscr{P}(k)$ is the probability of recording k bosons *per unit area*. It is related, of course, to a probability amplitude $\mathscr{A}(k)$

$$\mathscr{P}(k) = |\mathscr{A}(k)|^2. \tag{7.5.37}$$

We proceed to compute $\mathscr{P}(k)$ in terms of the individual probability $\mathscr{P} = \mathscr{P}(1)$, by considering the case of different quantons and that of bosons. We denote by $\mathscr{P}^{\neq}(k)$ and $\mathscr{P}^b(k)$, respectively, the corresponding probabilities. Let us begin with the case $k = 2$. Divide the surface of the detector D into two equal parts D′ and D″ of area $\frac{1}{2}S$ each. The probability of recording two impacts in D can always be written in terms of the probabilities of the three disjoint events (see fig. 7.22), which are
  (i) two impacts in D′;
  (ii) one impact in D′ and the other in D″;
  (iii) both impacts in D″.
In the case where the quantons are distinct, the event 'impacts in D′ and D″', subdivides, in its turn, into either 'quanton 1 in D′ and quanton 2 in D″', or '2 in D′ and 1 in D″'. Thus we get the relation

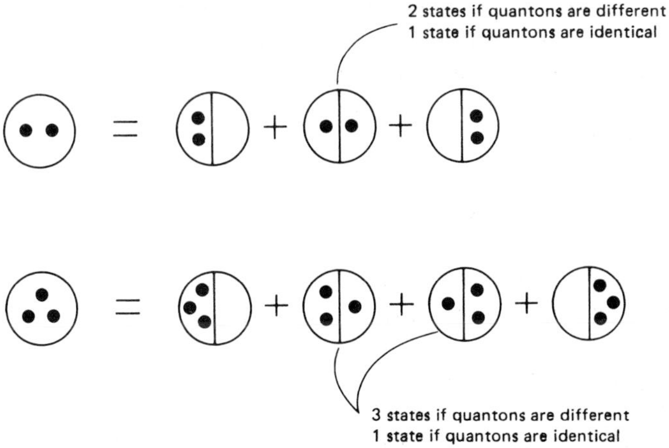

Fig. 7.22. *Detection probability*

A detector having surface area S is considered as being cut into two parts, each of area $\frac{1}{2}S$. To the detection of 2 quantons correspond three events, depending on whether 2, 1 or 0 are recorded by each half-surface. From this we derive a recurrence relation for the probabilities. The detection of a quanton on each surface corresponds to 2 distinct states when the quantons are different (probabilities add), but to only one state when the quantons are identical (amplitudes add). The idea generalizes to the detection of 3, 4, etc., quantons.

$$\text{Prob}(1 \text{ and } 2 \text{ in D}) = \text{Prob}(1 \text{ and } 2 \text{ in D}')$$
$$+ \text{Prob}(1 \text{ in D}' \text{ and } 2 \text{ in D}'')$$
$$+ \text{Prob}(2 \text{ in D}' \text{ and } 1 \text{ in D}'')$$
$$+ \text{Prob}(1 \text{ and } 2 \text{ in D}''). \tag{7.5.38}$$

It is now easy to express these probabilities in terms of the $\mathscr{P}^{\neq}(k)$. Equation (7.5.38) can be written, therefore, as

$$\mathscr{P}^{\neq}(2) \, S^2 = \mathscr{P}^{\neq}(2) \, (\tfrac{1}{2}S)^2 + 2[\mathscr{P}(1)\tfrac{1}{2}S][\mathscr{P}(1)\tfrac{1}{2}S]$$
$$+ \mathscr{P}^{\neq}(2) \, (\tfrac{1}{2}S)^2. \tag{7.5.39}$$

We finally obtain

$$\mathscr{P}^{\neq}(2) = [\mathscr{P}(1)]^2 = \mathscr{P}^2, \tag{7.5.40}$$

as we should. The bosonic case differs in that the two bosons, one in D' and one in D″, correspond to only one final state. Thus, we now have to write,

$$\text{Prob}(1 \text{ and } 2 \text{ in D}) = \text{Prob}(1 \text{ and } 2 \text{ in D}')$$
$$+ \text{Prob}(1 \text{ and } 2 \text{ in D}' \text{ and D}'')$$
$$+ \text{Prob}(1 \text{ and } 2 \text{ in D}''), \tag{7.5.41}$$

and,

$$\text{Prob}(1 \text{ and } 2 \text{ in D}' \text{ and D}'') = |\text{Ampl}(1 \text{ in D}') \times \text{Ampl}(2 \text{ in D}'')$$
$$+ \text{Ampl}(2 \text{ in D}') \times \text{Ampl}(1 \text{ in D}'')|^2,$$

$$\tag{7.5.42}$$

according to the now well-known quantum rules – applied to the bosonic case. With the notation introduced earlier, eqs. (7.5.41) and (7.5.42) allow us to write

$$\mathscr{P}^{\mathrm{b}}(2) \, S^2 = \mathscr{P}^{\mathrm{b}}(2) \, (\tfrac{1}{2}S)^2 + |2\mathscr{A}^2(1)|^2 (\tfrac{1}{2}S)^2 + \mathscr{P}^{\mathrm{b}}(2) \, (\tfrac{1}{2}S)^2, \tag{7.5.43}$$

which, since $|\mathscr{A}(1)|^2 = \mathscr{P}^{\mathrm{b}}(1)$, becomes

$$\mathscr{P}^{\mathrm{b}}(2) = 2[\mathscr{P}^{\mathrm{b}}(1)]^2 = 2\mathscr{P}^2, \tag{7.5.44}$$

to be contrasted, of course, with eq. (7.5.40). The reasoning can very easily be extended to the case of N objects, by considering different ways of distributing these N objects among the two halves, D′ and D″, of the detector. Clearly, now there are $N + 1$ cases to be considered: k objects in D′, $(N − k)$ in D″ with $k = 0, 1, \dots, N$. To each one of these cases corresponds a number of combinations $^NC_k = N!/k!(N − k)!$, which is the number of different states for distinct quantons. Thus, we obtain the relation

$$\mathscr{P}^{\neq}(N)S^N = \mathscr{P}^{\neq}(N)(\tfrac{1}{2}S)^N + {}^NC_1\mathscr{P}^{\neq}(N − 1)(\tfrac{1}{2}S)^{N−1}\mathscr{P}^{\neq}(1)(\tfrac{1}{2}S) + \cdots$$
$$+ {}^NC_k\mathscr{P}^{\neq}(N − k)(\tfrac{1}{2}S)^{N−k}\mathscr{P}^{\neq}(k)(\tfrac{1}{2}S)^k + \cdots$$
$$+ \mathscr{P}^{\neq}(N)(\tfrac{1}{2}S)^N, \tag{7.5.45}$$

i.e.,

$$\mathscr{P}^{\neq}(N) = \frac{1}{2^N} \sum_{k=0}^{N} {}^NC_k\mathscr{P}^{\neq}(N − k)\mathscr{P}^{\neq}(k) \tag{7.5.46}$$

where, by definition, $\mathscr{P}(0) = 1$.

As is easily verified, this recursion relation has the solution

$$\mathscr{P}^*(N) = [\mathscr{P}(1)]^N = \mathscr{P}^N. \tag{7.5.47}$$

For bosons, on the other hand, there is just one single final state for each value of k, and thus only one amplitude, obtained by adding the $^N C_k$ amplitudes corresponding to the different possible modes. This gives, in place of eq. (7.5.45),

$$|\mathscr{A}(N)|^2 S^N = |\mathscr{A}(N)|^2 (\tfrac{1}{2}S)^N + |^N C_1 \mathscr{A}(N-1).\mathscr{A}(1)|^2 (\tfrac{1}{2}S)^N + \cdots$$
$$+ |^N C_k \mathscr{A}(N-k).\mathscr{A}(k)|^2 (\tfrac{1}{2}S)^N + \cdots$$
$$+ |\mathscr{A}(N)|^2 (\tfrac{1}{2}S)^N, \tag{7.5.48}$$

or, in terms of the probabilities,

$$\mathscr{P}^b(N) = \frac{1}{2^N} \sum_{k=0}^{N} (^N C_k)^2 \mathscr{P}^b(N-k)\mathscr{P}^b(k), \tag{7.5.49}$$

which should be compared to eq. (7.5.46). The solution now is

$$\mathscr{P}^b(N) = N! \, [\mathscr{P}^b(1)]^N = N! \, \mathscr{P}^N \tag{7.5.50}$$

and we recover the bosonic factorial.

7.6. The number-phase inequality and the classical limit

Consider the light emitted from a lamp, or better, from a laser, with a well-defined pulsation ω. Suppose that this beam is emitted over a certain period of time, between the switching on at time t and switching off at time t' of the laser. Classically, the beam would be described by an electromagnetic wave, characterized by its phase. The instants of time t and t' can only be determined to a precision Δt, dependent on the very nature of the phenomena of switching on and off. From this there results an intrinsic imprecision in the phase,

$$\Delta\varphi = \omega \, \Delta t. \tag{7.6.1}$$

Furthermore, a dispersion Δt of the phenomenon in time implies, in view of the temporal Heisenberg inequality, a typically quantum dispersion ΔE, in the energy content E of the beam, so that

$$\Delta E \, \Delta t \gtrsim \hbar. \qquad (7.6.2)$$

Within the quantum framework, this beam is thought of as a system of photons. The number N of photons is obtained from the energy E, using

$$N = E/\hbar\omega, \qquad (7.6.3)$$

since all the photons, being of pulsation ω, have the energy $\hbar\omega$. It now follows, from the dispersion in energy ΔE, that the number of photons N is not defined more accurately than ΔN, where

$$\Delta N = \Delta E/\hbar\omega. \qquad (7.6.4)$$

Combining relations (7.6.1), (7.6.2) and (7.6.4), we obtain the remarkable inequality

$$\Delta N \, \Delta\varphi \gtrsim 1, \qquad (7.6.5)$$

between the dispersion in the number of photons in the beam and the dispersion in its phase. It is clear, from the ideas that have been used, that this inequality ought to hold, at least intrinsically, for any system of photons and, more generally, for any system of quantons (provided that, like the mono-chromatic photons in this case, they are all in the same state).

This important inequality is of a specifically quantum nature. This assertion might seem paradoxical, since the quantum constant $\hbar$ does not appear in inequality (7.6.5), unlike in the case of the spatial, temporal and rotational Heisenberg inequalities, where it comes in explicitly. Here, on the other hand, the 'quanticity' shows up merely through the appearance of the magnitudes N and φ: the *number* of particles and the *phase* of the wave, concepts which a priori refer to physical objects of a different nature. The reason that there exists a relation between these magnitudes is because, appropriately generalized, they now characterize another, and unique, kind of object, a quanton.

Conversely, whenever any one of the magnitudes N and φ refers exclusively to one of the classical concepts of a particle or a wave, inequality (7.6.5) plays a valuable role in characterizing the classical approximations of the quantum theory.

In particular, the case of fermions is readily treated. Inequality (7.6.5) actually deals with quantons in the same state (same energy, same momentum, etc.). However, the Pauli principle forbids the existence of more than one fermion in the same state! The dispersion in the number of fermions is thus limited, by definition, to

$$\Delta N_f \leqslant 1. \tag{7.6.6}$$

It follows that the phase of such a system is necessarily undefined, since

$$\Delta \varphi_f \gtrsim 1. \tag{7.6.7}$$

Thus, it is not possible to construct a system of fermions having a well-defined phase, which means that in so far as fermions are concerned, the classical wave approximation cannot have any validity. This is very clear in the case of electrons, for example, which under certain circumstances (of interest, among others, to engineers constructing a particle accelerator) may be described using the concepts of classical mechanics (trajectories), but not of wave theory: there is no classical 'electron field' theory. This does not exclude, of course, the existence of useful formal analogies between the quantum amplitudes of electrons and the amplitudes of a classical field (see chs. 4 and 5). But there does not exist an 'electron field' as a classical physical object which would be analogous to the electromagnetic field.

The bosonic case is quite different altogether: the number N_b of bosons in any one state being unlimited, we can think of systems in which this number is not fixed (the system is not in a proper state of N_b). The average number N_b can now be big enough for the dispersion itself to be high. If we now have

$$\Delta N_b \gg 1, \tag{7.6.8}$$

nothing prevents us from obtaining

$$\Delta \varphi_b \ll 1, \tag{7.6.9}$$

and the phase of the system becomes sufficiently well-defined, so that it can properly be described by a classical wave, which may also have a well-defined amplitude, to the extent that $\Delta N_b/N_b \ll 1$. On the other hand, one can no longer describe the system as a collection of classical particles, since their number itself is not defined.

In practice, however, not *all* bosonic systems admit such a classical undulatory description. Consider, e.g., π-mesons with kinetic energy ε. Recall that they have a mass–energy mc^2. The total energy of a system of N π-mesons would be

$$E = N(mc^2 + \varepsilon) \simeq Nmc^2, \tag{7.6.10}$$

if the kinetic energy is assumed to be small. To a dispersion ΔN in their number, corresponds an energy dispersion

$$\Delta E = \Delta N \, mc^2, \tag{7.6.11}$$

and, hence, a characteristic time

$$\Delta t \simeq \frac{\hbar}{\Delta E} \simeq \frac{1}{\Delta N} \frac{\hbar}{mc^2}. \tag{7.6.12}$$

In other words, the overall state of the system, characterized by a classical wave amplitude, would exhibit temporal variations on this Δt-scale, i.e., it would fluctuate over much too small a time interval for the classical wave approximation to be useful: for example, for π-mesons, $\hbar/mc^2 \simeq 10^{-24}$ s.

The case of photons is, therefore, unique because of their vanishing mass, which allows for a dispersion in number along with a minimal dispersion in energy and hence an enormous stability, over time, of the wave-like description. This explains why it is only in electromagnetism – the theory of photons – that the notion of a wave or a classical field has been so enormously successful.

Table 7.2
From the quantic to the classical.

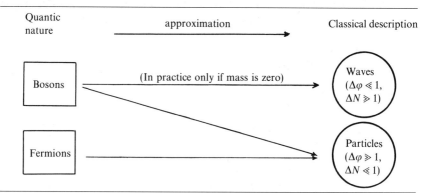

Of course, for bosons, of zero mass or otherwise, there is nothing to prevent us from having

$$\Delta N_b \ll 1 \quad \text{and} \quad \Delta\varphi_b \gg 1, \tag{7.6.13}$$

leading to the complete validity of a corpuscular description.

Table 7.2 summarizes our discussion on the conditions for the validity of the classical approximations, wave-like and corpuscular, for the two major types of quantons.

7.7. A new duality?

The discussion in the preceding sections has amply demonstrated that the fermion–boson dichotomy is of enormous importance and that it leads to two very different types of quantum behaviour. There exists another aspect of this dichotomy which could possibly be even more fundamental. Let us consider primary physical phenomena, in the course of which elementary quantons appear or disappear. We have seen in sect. 7.3.3 that a composite system behaves like a boson (and possesses an integral angular momentum) or like a fermion (and possesses a half-integral angular momentum) depending on whether it consists of an even or an odd number of fermions. It follows from this, that during an arbitrary physical process, the parity of the number of fermions cannot change. Stated otherwise, the number of fermions can only change in twos: a fermion can appear or disappear only if another fermion appears or disappears (or 3 others, etc.). Bosons, on the other hand, can appear or disappear in arbitrary numbers. The following few classic reactions between quantons, for which we have indicated the fermionic (F) or bosonic (B) nature, as appropriate, illustrate these assertions

$$
\begin{array}{cccc}
e^+ & +n & \to p & +\bar{v} \\
(F) & (F) & (F) & (F),
\end{array}
$$

$$
\begin{array}{cccc}
p & +\bar{p} & \to e^+ & +e^- \\
(F) & (F) & (F) & (F),
\end{array}
$$

$$
\begin{array}{ccccc}
p & +\bar{p} & \to \pi^+ & +\pi^- & +\pi^0 \\
(F) & (F) & (B) & (B) & (B),
\end{array}
$$

$$
\begin{array}{ccccc}
e^- & +p & \to e^- & +p & +\gamma \\
(F) & (F) & (F) & (F) & (B),
\end{array}
$$

$$\pi^- \to e^- + \bar{v}$$
(B)   (F)   (F),

$$\pi^0 \to e^+ + e^- + \gamma$$
(B)   (F)   (F)   (B),

$$\pi^0 \to \gamma + \gamma$$
(B)   (B)   (B),

$$\pi^- + p \to e^- + e^+ + n$$
(B)   (F)   (F)   (F)   (F).

Furthermore, *all* known fermions carry one or more charges, electrical or other (baryonic, leptonic), while certain bosons (such as the photon γ or the neutral pion π^0) are devoid of any charges and, so to speak, only transport dynamical magnitudes (energy, momentum, angular momentum). This experimentally observed fact is not yet completely understood, although the existence, if not the exact form, of such a connection can be given an interpretation within the quantum theory of fields (Lurçat and Michel 1965).

In any case, as a result of all this, we are very naturally led to attributing a different status to fermions – in some sense more stable – from what we attribute to bosons. Certainly, a fermion can appear or disappear – but never alone and never without leaving a trace in the form of either carrying away or transmitting certain charges. By contrast, a boson may seem to appear out of nothingness or return to nothingness. Thus, there is a certain permanence in fermions which does not exist for bosons.

It is because of this essential difference that the trend nowadays is to consider, at least implicitly, the elementary fermions (quarks, leptons,...) as being the 'true' constituents of matter – the elementary bosons (photons, gluons and intermediate bosons) playing the more ancillary role of interaction vectors, i.e., the radiations.

These remarks could also be looked at from the point of view of the ongoing search for elementarity. When an object is analyzed into its (more) elementary constituents, it is impossible to construct fermions from bosons alone, while it may seem possible to combine fermions to form a boson. It ought to be noted, nevertheless, that the said composite boson would exist as an independent particle only if the constituent fermions were bound by some interaction – and this latter would then be transmitted... by bosons. Thus, it is not possible to construct bosons without bosons: the hydrogen atom is made up of a proton, an electron *and* photons which bind the two; the pion is composed of two quarks and gluons which bind them, etc. Even though bosons may appear to be 'less real' than fermions, we can scarcely afford to overlook them.

Thus, even after having insisted on the universality of the concept of the quanton, which seemed to undermine the classical wave–particle duality (see ch. 2), we see a new duality appearing on the quantum level – related, for sure, in some complex way, to this classical duality (see preceding section and table 7.2), but yet more profound. We gladly leave open the question of determining whether this dialectic of one and of two, and its successive incarnations within physical theory, relates to the object of science or to its subject – assuming that *this* dichotomy makes sense.

Exercises

7.1. The diameters of stars, even of the nearest ones, cannot be measured directly. Michelson had shown how classical interferometry enables one to make such a measurement. In 1956, Hanbury-Brown and Twiss proposed a new method, called '*intensity interferometry*', to make such measurements. They used two parabolic mirrors, separated by an adjustable distance d. Using two photomultipliers to transform the luminous intensity into an electric current, they measured the product of the currents corresponding to the two mirrors, i.e., the correlation between the signals received. At that time, not too many people understood the idea behind the experiment, thinking that since the light would consist of photons arriving either at one or the other mirror, there could not be any correlation between the signals here. Consider the following simplified model of the Brown and Twiss experiment: two light sources A and B, separated by a distance D, are observed by the two receptors a and b, placed at a distance L. The counters, connected to a and b, measure arrival probabilities (with a fixed time lapse) of one photon inside a receptor, call it $\mathscr{P}_1$, or of two photons – one inside each receptor – call it $\mathscr{P}_2$. Let $\langle a|A \rangle$ be the probability amplitude for the arrival at a (within the time lapse being considered) of a photon originating in A. Similarly define the amplitudes $\langle a|B \rangle$, $\langle b|A \rangle$ and $\langle b|B \rangle$.

(a) Assuming a monochromatic light with undulation k, show that the amplitudes have the form

$$\langle a|A \rangle = C\,e^{ikL_1}, \qquad \langle b|A \rangle = C\,e^{-ikL_2},$$

where C is a constant, which can be taken to be real, and L_1 and L_2 are the distances aA and bA, respectively. In the same way, write $\langle a|B \rangle$ and $\langle b|B \rangle$.

(b) Compute the probabilities for the arrival of a photon at a, $\mathscr{P}_1(a)$, of a photon at b, $\mathscr{P}_1(b)$, and of a photon at a or b, $\mathscr{P}_1(a \text{ or } b)$. Beware! Can one tell whether a given photon has originated in A or B (possibly by making independent observations on A and B)?

(c) We want to compute the probability $\mathscr{P}_2$ for the arrival of two photons, one at a and another at b, during the time period considered. These photons could come, both from A, both from B, or one from A and the other from B. Should the amplitudes for the three cases be added, or the probabilities?

(d) Show that the probability of observing two photons coming from the same source

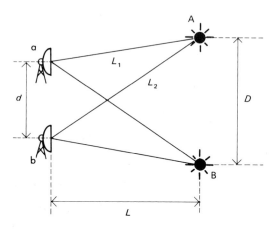

is given by

$$\mathscr{P}_2(AA) = \mathscr{P}_2(BB) = C^4.$$

(e) Show that the probability of observing two photons from different sources is

$$\mathscr{P}_2(AB) = 4C^4 \cos^2 k(L_2 - L_1).$$

Hence, deduce the total probability for observing two photons

$$\mathscr{P}_2 = 2C^4[2 + \cos 2k(L_2 - L_1)].$$

Does the bosonic nature of the photons figure in the calculations? What would have been the expression obtained for the total probability $\mathscr{P}_2^{\neq}$, had the two detected quantons been distinct? Show that $\mathscr{P}_2^{\neq}$ is also the classical limit of $\mathscr{P}_2$. In practice, one measures the *correlation* $\Gamma = (\mathscr{P}_2 - \mathscr{P}_2^{\neq})/\mathscr{P}_2^{\neq}$ between two photons. Write out Γ.

(f) If the distance L is large, compared to the distances D and d, show that approximately one gets

$$L_2 - L_1 \simeq \tfrac{1}{2} d\alpha,$$

where α is the angular separation between the sources. As d increases from zero, for what value d_m of the distance does the correlation Γ pass through a minimum?

(g) For a source having an extended surface, such as a star, and a non-monochromatic light, the computation of Γ leads to a much more complicated expression than the formula above. Nonetheless, Γ still has its maximum at $d = 0$ and decreases to a characteristic distance d_m, of the same order as the one computed above (D being taken to be the diameter of the star). In an actual experiment, photons coming from Sirius are observed. The correlation Γ, measured as a function of d, is given in the figure below. Assuming that the light is nearly monochromatic, of wavelength $\lambda = 0.5 \ \mu m$, evaluate the apparent diameter of Sirius, and then its diameter D, knowing that its distance $L \simeq 9$ light years.

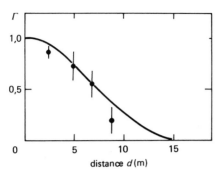

distance d(m)

[From Hanbury-Brown and Twiss, Nature *178* (1956) 1048]

7.2   Consider a one-dimensional potential well. Let $\varphi_k(x)$ be the (time-independent) wavefunction of the stationary state corresponding to the kth energy level ε_k (φ_1 is the ground state wavefunction and ε_1 its energy). Next consider a *system with two identical quantons*, mutually non-interacting, within this same potential well. Answer the following questions for:

(1) two fermions;

(2) two bosons.

(a) What is the ground state energy of the two-quanton system?

(b) What is the wavefunction $\Phi(x, y)$ of the ground state (i.e., the probability amplitude for the localization, at the two points x and y, of the two quantons)? Make sure that the wavefunction is correctly normalized – i.e., verify that $\iint dx\, dy\, \rho(x, y) = 1$, where $\rho(x, y) = |\Phi(x, y)|^2$ is the localization probability density.

(c) What is the probability density $\sigma(x)$ for localization at x of *one* of the quantons of the system?

(d) What is the conditional probability density $\tau_y(x)$ for one quanton to be at x *given that* the other is localized at y?

(e) Work out the preceding results explicitly when the potential is an infinite well with a flat bottom of width a. Obtain the expression for $\tau_y(x)$ and draw the graph of this function when $y = \frac{1}{4}a$ (quarter of the well).

(f) Repeat problems (a) to (e) for the first excited state of the system.

7.3.   The *differential effective cross-section* $\chi(\theta)$, for two spin-zero bosons is, in view of the symmetrization of the amplitude, double the value, at $\theta = \frac{1}{2}\pi$, of what it would have been had the quantons been distinct [see eq. (7.2.40)]. It is not clear if it would even pass through a maximum! For reasons of symmetry, the differential effective cross-section certainly passes through an extremum at $\theta = \frac{1}{2}\pi$. But is this a minimum or a maximum?

(a) Writing the amplitude in the form

$$f(\theta) = |f(\theta)| \exp[i\alpha(\theta)],$$

show that, in the neighbourhood of $\theta = \frac{1}{2}\pi$, we have

$$\chi(\tfrac{1}{2}\pi + \varepsilon) = \chi(\tfrac{1}{2}\pi)\left[1 + \varepsilon^2\left(\frac{|f|''}{|f|} - \alpha'^2\right)_{\theta = \frac{1}{2}\pi} + O(\varepsilon^3)\right],$$

and that the effective cross-section could go through a minimum or a maximum at $\theta = \frac{1}{2}\pi$, depending on how $|f|$ and α vary.

(b) What happens when the amplitude is real ($\alpha = 0$)?

(c) What is the situation in the Coulomb case [use expressions (5.3.12)–(5.4.14)]?

7.4. (a) Consider the *Coulomb scattering* of two beams of classical particles, having the same mass, charge Zq_e and velocity v in the centre of mass frame. A detector D picks up particles coming either from the left and deflected through an angle θ, or from the right and deflected through an angle $\pi - \theta$ (see fig. 7.3). Corresponding to these two cases, there are two types of trajectories, defined by two different impact parameters b and b'. What is the order of magnitude of the difference between the lengths of the trajectories, covered in the two cases by the detected particle from its source.

(b) Consider now the Coulomb scattering, under the same conditions as before, of two identical quantons. The total effective cross-section, as a function of the scattering angle θ, shows oscillations, due to the interference between the two quantum amplitudes corresponding to the two preceding classical cases. From the above results, deduce the order of magnitude of the phase difference between the two amplitudes, and then show that the scale of the angular oscillations can be evaluated using

$$\delta\theta \simeq (Z^2\alpha)^{-1}\frac{v}{c}.$$

Compute $\delta\theta$ and compare it with the experimental results in the following two cases:
– ^{12}C–^{12}C scattering at 5 MeV (fig. 7.6a),
– ^{28}Si–^{28}Si scattering at 20 MeV (fig. 7.6b).

(c) What can be said when the conditions for the validity of the classical approximation are fulfilled? Why are the interference terms not seen any more?

7.5. Consider the scattering of two spin-$\frac{1}{2}$ quantons by an interaction which can act on the spin. Let $f(\theta)$ be the scattering amplitude without change of spin and $g(\theta)$ the *spin-flip scattering* amplitude, for each quanton. Recall that the total spin must remain unchanged during the scattering.

(a) Write down the differential effective cross-section for an initial state ↑↑.

(b) Write down the differential effective cross-section for an initial state ↑↓.

(c) Write down the differential effective cross-section when the quantons are unpolarized.

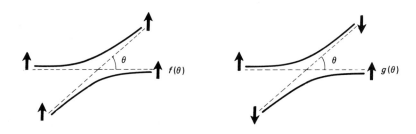

7.6. Consider the *scattering of two identical spin-s quantons*. Assume that the interaction is spin-independent (i.e., it does not change the spin state of the quantons) and denote by $f(\theta)$ the scattering amplitude when the individual spin states of the two quantons are different.

(a) What is the scattering amplitude in the direction θ, for two quantons in identical individual spin states? (Distinguish between the bosonic and fermionic cases).

(b) For spin-s quantons, how many possible spin states are there for the two-quanton *pair*? For how many of these do the two quantons have identical individual spin states?

(c) Show that the differential effective cross-section for an unpolarized beam can be written as

$$\chi(\theta) = |f(\theta)|^2 + |f(\pi - \theta)|^2 + \frac{2(-1)^{2s}}{2s+1} \operatorname{Re} \overline{f(\theta)} \, f(\pi - \theta).$$

In particular, compute the differential effective cross-section for $\theta = \tfrac{1}{2}\pi$, and compare it to the case where the two quantons are distinct.

(d) Why do we recover the classical result in the limit of $s \to \infty$?

7.7. Consider the mutual *scattering* of two identical helium atoms, in a highly simplified model, in which the interaction is represented by a one-dimensional flat potential well, of depth V_0 and width a, and where we are only interested in very low energies $E \ll V_0$.

(a) Using the results and notation of exercise 6.17, show that the effective cross-section can be written as $\sigma = |1 \pm e^{-ipa}|^2$, depending on whether we are dealing with two bosons ($^4\mathrm{He}$) or two fermions ($^3\mathrm{He}$). [We have written $E = p^2/2M$, where $M = \tfrac{1}{2}M_{\mathrm{He}}$ is the reduced mass.]

(b) For what values of the energy is the effective cross-section a minimum? Compare with the results of exercise 6.17, using the numerical values of that problem. Show that in this way the shape of the curve in fig. 6.16a can be correctly interpreted.

7.8. A system of N identical fermions is enclosed in a box of volume V. Use the *Heisenberg–Pauli inequality* to prove that, in the ground state of the system, the average energy per quanton is of the order of

$$\varepsilon \simeq (h^2/2m)\rho^{2/3},$$

where $\rho = N/V$ is the density of the system. Evaluate ε numerically
– for the valence electrons of a metal,
– for the nucleons of a nucleus.

7.9. Consider a system of N identical quantons, of mass m, interacting via their mutual *gravitational attractions* only. Show that, with the notation of sect. 7.4.3b, the energy of the system can be estimated using

$$E \simeq N \frac{\tilde{p}^2}{2m} - \frac{N^2}{2} \frac{Gm^2}{L}.$$

[Compare with eq. (7.4.18): we see that the screening effect does not occur in the gravitational case, where all forces are attractive.] Hence deduce that the energy of the ground state depends on the number of particles:

for bosons: $\quad E_N \simeq -N^3 \dfrac{Gm^4}{\hbar^2}$;

for fermions: $\quad E_N \simeq -N^{7/3} \dfrac{Gm^4}{\hbar^2}$.

7.10. In the absence of the Pauli principle, the *Coulomb forces* would not be saturated, and the energy of a system of N atoms would vary as $N^{5/3}$. Show that a system of $2N$ atoms would then have an energy lower than that of two N-atom systems. Evaluate numerically, the amount of energy that would be released under these conditions, by bringing together into a single system, two initially separated systems, of one mole of hydrogen each. Compare this with the energy released by the explosion of a powerful H-bomb, equivalent to 10 megatons of TNT (1 kiloton of TNT $\simeq 4 \times 10^{12}$ J).

7.11. *White dwarfs and the Chandrasekhar limit.* Consider the ground state of a material system, governed by gravitational forces (sect. 7.4.3c of the present chapter).

(a) Show that, in view of the considerations elaborated, the characteristic momentum of an electron is of the order of

$$p \simeq N^{2/3} \frac{GM^2 m}{\hbar}.$$

Hence deduce the existence of a critical value

$$N_c = \left(\frac{\hbar c}{GM^2} \right)^{3/2} = \alpha^{-2/3} N_t, \quad \text{with } N_t \text{ given by eq. (7.4.33)}$$

(where α is the fine structure constant), such that for $N \gtrsim N_c$, Einsteinian relativity takes over.

(b) Consider again the problem of the evaluation of the ground state energy, using the Einsteinian expression

$$E = (p^2 c^2 + m^2 c^4)^{1/2} - mc^2,$$

for the kinetic energy of the electrons (instead of $p^2/2m$). Show that for $N > N_c$, the expression for the energy does not have a minimum and that the system cannot be stable, but collapses into itself. Calculate the critical mass $\mathscr{M}_c = N_c M$ (Chandrasekhar limit) and compare it to the characteristic solar mass $M_\odot \simeq 10^{30}$ kg.

7.12. Certain *organic pigments* are made up of linear ion-molecules, consisting of a chain of n carbon atoms between two atoms of nitrogen. We are interested in the case where n is odd. The $n + 3$ electrons (n coming from the C atoms, $+4$ from the two N atoms, -1 for the ionization), which are conventionally attributed to double bonds according to the scheme

$$N^+ = C - C = C - C = C - N,$$

can in reality move along the chain, according to a more realistic scheme of the type:

$$(\text{>}N\text{---}C\text{---}C\text{---}C\text{---}C\text{---}C\text{---}N\text{<})^{+}.$$

They may be considered as being independent electrons inside an infinite flat potential well of width $L_n = (n-1)a + 2b + 2d$, where $a = 1.4$ Å is the C–C distance, $b = 1.34$ Å the N–C distance and d takes any possible edge effects into account:

$$|-N-C-C-C-C-C-N-|$$

$$|\quad d\ |\ b\ |\ a\ |$$

$$|\longleftarrow L_n \longrightarrow|$$

(a) What are the individual energy levels of the electrons?

(b) Taking the fermionic and spin-$\frac{1}{2}$ nature of the electrons into account, what is the energy of the ground state? Of the first excited state?

(c) Show that the wavelength of the light absorbed during a transition from the ground state to the first excited state is

$$\lambda_n = \frac{8mc}{h} \frac{L_n^2}{n+4}.$$

Compute λ_n numerically, for $n = 3, 5, 7$, under the assumption that $d = 0$, and compare with the experimental results shown in the figure for the pigments nos. 1, 2, 3.

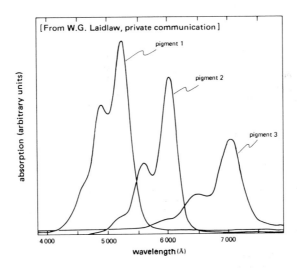

[From W.G. Laidlaw, private communication]

pigment 1

pigment 2

pigment 3

absorption (arbitrary units)

4 000   5 000   6 000   7 000

wavelength (Å)

(d) To obtain better theoretical results, evaluate d using the experimental results shown in the figure.

(e) Should one expect from this that a compound corresponding to $n = 9$ ought to be a coloured pigment?

[From W. G. Laidlaw, *Introduction to Quantum Concepts in Spectroscopy*, McGraw Hill, New York.]

7.13. Evaluate, using eq. (7.4.48), the *quantum pressure* in the ground state of an N-boson system, confined to a volume V. What happens to this pressure when we consider a system with fixed density N/V, and the volume V is allowed to grow?

7.14. Consider a simplified *model of a metal* in which the external electrons are treated as free electrons, constrained to remain inside the volume V of the crystal (free-electron gas).

(a) Express, as a function of the density of the free electrons $n = N/V$, the maximal momentum of these electrons.

(b) Estimate the order of magnitude of this quantity and hence of the corresponding velocity. How do you understand this result?

(c) Show that the product of this quantity with a_0, the Bohr radius, is of the order of unity (in a system of units in which $\hbar = 1$).

(d) Hence deduce that the Fermi level of metals is of the same order of magnitude as the Rydberg ($E_H = 13.6$ eV).

(e) Explicitly calculate ε_F [using (7.4.43)] for the following five metals, the electronic densities for which have been given.

	Metal				
	Li	Na	K	Ca	Ag
N/V (10^{22} cm^{-3})	4.70	2.65	1.40	8.47	5.86

7.15. Examine the validity of the *free-electron approximation*, on which the computation of ε_F, carried out in sect. 7.4.4 (see also exercise 7.14), is based.

(a) Show that this approximation is not justified if p_F (defined by $\varepsilon_F = p_F^2/2m$) is appreciably smaller than the distance separating the centre of the Brillouin zone from its boundaries.

(b) Incorporating now the elementary crystal mesh into a cube, show that the condition formulated in (a) is not satisfied for monovalent metals.

7.16. The *specific heat* C of a solid (and, more generally, of a material) is defined by the relation $C = dE/dT$. dE is the increase in internal energy per unit mass of the solid, when its temperature changes from T to $T + dT$. It is verified experimentally that at very low temperatures, the specific heat of solids depends linearly on the temperature, so that $C = \gamma T$. In this temperature region, the ions are 'frozen' and only the delocalized electrons participate in the thermal agitation: at the temperature T, each electron possesses an energy of the order of kT.

(a) Can all delocalized electrons experience an increase in their energies as the temperature changes from T to $T + dT$?

(b) Show that only a fraction, of the order of kT/ε_F ($\varepsilon_F =$ Fermi level of the solid), contributes to the specific heat. Now estimate the order of magnitude of this fraction.

(c) Using the table given below (in which the last line gives the measured values of γ), test the validity of the foregoing argument ($k = 10^{-23}$ J K^{-1}).

	Metal						
	Li	Na	K	Cu	Ag	Ca	Fe
ε_F (eV)	4.74	3.24	2.12	7.00	5.49	4.69	11.1
n = density of delocalized electrons $(10^{22}$ cm$^{-3})$	4.70	2.65	1.40	8.47	5.86	4.61	17.0
γ $(10^{-4}$ cal mol^{-1} K$^{-2})$	4.2	3.5	4.7	1.6	1.6	6.5	12

7.17. *Pauli paramagnetism.* The paramagnetic susceptibility of metals is about 104 times weaker than that of non-metallic solids. The paramagnetic susceptibility measures how readily the spin magnetic moments (of the ions or the electrons) align themselves in the direction of the field, when a sample of the material being examined is placed in a magnetic induction $\mathscr{B}$. It is defined through the relation $\chi = \mu_0 \mathscr{M}/\mathscr{B}$, where μ_0 is the permittivity of free space, $\mathscr{B} = |\mathscr{B}|$ and $\mathscr{M}$ is the modulus of magnetization, i.e., the magnetic moment per unit volume. Various indications lead one to believe that the low value of χ for metals is due to the relative 'freedom' of the external electrons.

(a) Let $\mu_e = q_e h/2m$ be the absolute value of the spin magnetic moment of an electron. (See exercise 1.17.) Show that

$$\mathscr{M} = \mu_e(n_+ - n_-),$$

where n_+ denotes the number of electrons per unit volume, which are in a state of spin $S_z = \frac{1}{2}$, and n_- the number of them in a state $S_z = -\frac{1}{2}$. What is the value of $\mathscr{M}$ in the absence of the magnetic induction?

(b) Study the individual energy levels of an electron, respectively in a spin state $S_z = \frac{1}{2}$ and a state $S_z = -\frac{1}{2}$, in the absence and in the presence of the field $\mathscr{B}$. Assume that the metal under consideration is an alkali (for example, sodium) and that the delocalized electrons of the metal may be treated like a free-electron 'gas'.

(c) Show that the density of states of an electron of spin $S_z = +\frac{1}{2}$ is

$$\rho_+(\varepsilon) = \tfrac{1}{2}\rho(\varepsilon + \mu_e\mathscr{B}),$$

where ρ denotes the density of states of a free electron (when $\mathscr{B} = 0$). Similarly, show that for $S_z = -\frac{1}{2}$

$$\rho_-(\varepsilon) = \tfrac{1}{2}\rho(\varepsilon - \mu_e\mathscr{B}).$$

Assume that the two types of electrons have the same Fermi level. Hence obtain the integral expressions for n_+ and n_-. The sample is assumed to be maintained at $T = 0$ K.

(d) Let n be the number of external electrons per unit volume for the sample in question. Given that the number of external electrons is not modified by the application of the field, show that to first order in $\mu_e\mathscr{B}/\varepsilon_F$, the Fermi level of the electrons is the same in the presence as in the absence of the field. Taking the order of magnitude of ε_F and μ_e into account, for what values of $\mathscr{B}$ would this cease to be true? Next determine the magnetization $\mathscr{M}$ at $T = 0$ K. Hence deduce that χ can be put into the form

$$\chi = \frac{\alpha^2}{\pi} \frac{a_0 \not{p}_F}{\hbar},$$

where α is the fine structure constant $(e^2/\hbar c)$, a_0 the Bohr radius and $\not{p}_F$ the momentum corresponding to ε_F.

(e) Calculate χ numerically for the different alkali metals. Find out from a book on the physics of solids or on statistical mechanics or from an encyclopedia, the microscopic explanation for the paramagnetism of *non*-metallic bodies (Langevin paramagnetism). Compare the numerical values of the corresponding susceptibilities with those found here. Hence deduce that the Pauli exclusion principle prevents the orderly alignment of the spins, in the direction of the applied field, much more effectively than thermal agitation.

	Metal				
	Li	Na	K	Rb	Cs
n $(10^{22}$ cm$^{-3})$	4.70	2.65	1.40	1.15	0.91

7.18. When a metal is irradiated with a light of frequency v, larger than a threshold frequency v_0, characteristic of the metal, one observes that the solid emits a current of electrons, proportional to the intensity of the incident wave. This is the photo-electric or *photoemission* effect. The quantity $W = hv_0$, called the work function, measures the amount of energy that has to be imparted to one of the electrons of the metal, in order to release it from the crystal potential.

(a) Treat the external electrons of the atoms of the metal as a free-electron gas, held inside the crystal. Set up a relation connecting W, ε_F (the Fermi level) and V_0 (depth of the potential well constraining the electrons to the interior of the crystal).

(b) Using the data in the table below, compute V_0 for the five given metals.

(c) In the free-electron model, the individual states (and their wavefunctions) are identified with those of a quanton placed in an infinitely deep well. How would you assess the validity of this approximation, in view of the numerical results of question (b)?

(d) One could try to refine the above reasoning by taking the finite value of V_0 into account. The electronic states are thus no longer completely confined to the interior of the crystal. For energies close to the Fermi level ε_F, evaluate the dimensions of this 'spill-over' region (see exercise 6.14). Find the shape of the charge distribution on either side of the surface, and show that, to a first approximation, the surface may be represented by a uniform distribution of electric dipoles.

(e) Show that, under this hypothesis, an electron outside the metal is subject to a constant potential U.

(f) Show that the work function W must now be interpreted as a measure of the difference $U - \varepsilon_F$, independently of V_0.

Metal	W(eV) (measured)	ε_F(eV) (calculated from the free-electron model)
Li	2.38	4.74
Na	2.35	3.24
K	2.22	2.12
Cu	4.4	7.00
Ag	4.3	5.49

7.19. The technique of *field emission*, which underlies a procedure in microscopy, consists of applying to a metallic sample an intense electric field $\mathscr{E}$ ($\simeq 10^9$ V cm^{-1}) so as to facilitate the release of the metallic electrons. The potential configuration at the surface of the metal is then as shown in fig. 1. Determine the polarity of $\mathscr{E}$.

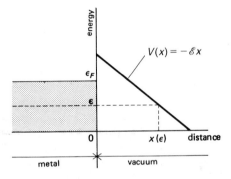

Figure 1

(a) Using the results on the tunnel effect, described in ch. 6, show that the current obtained is of the form

$$I \propto \int_0^{\varepsilon_F} \exp\left\{-\int_0^{x(\varepsilon)} [2m(V(x)-\varepsilon)]^{1/2}\,dx\right\}\rho(\varepsilon)\,d\varepsilon.$$

(b) Prove that it is mainly the electrons close to the Fermi level which contribute to the current (called the 'field emission') so produced, and that

$$I \propto \exp[-(2m)^{1/2}\,W^{3/2}/q_e\mathscr{E}].$$

Compare this result with the experimental curve (fig. 2).

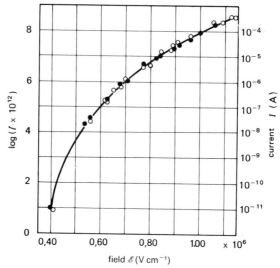

field $\mathscr{E}$ (V cm^{-1})

[From Millikan and Eyring, Phys. Rev. 27 (1976) 51]

Figure 2

(c) With the availability of perfectly monochromatic lasers, it has become possible to perform photo-assisted field emission experiments, combining field emission and photo-emission. One passes through a tungsten sample, subjected to a static field, a monochromatic electromagnetic wave, i.e., monoenergetic photons. Figure 3 gives the energy distribution of the emitted electron current. The peak to the right is obtained in the absence of illumination, that in the middle when the sample is illuminated by a light of energy 2.602 eV (visible region), and the one to the left with ultraviolet radiation of 3.531 eV. Interpret these results as well as the curves in fig. 4. How would you explain the drop in the curve towards the high intensity end of the field $\mathscr{E}$?

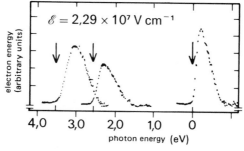

$\mathscr{E} = 2,29 \times 10^7$ V cm^{-1}

photon energy (eV)

[From M.J.G. Lee, Phys. Rev. Lett. 30 (1973) 1193]

Figure 3

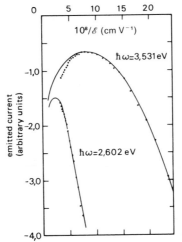

[From M.G.J. Lee, Phys. Rev. Lett. 30 (1973) 1193]

Figure 4

(d) Field emission is profitably used in the technique of microscopy which bears its name. A metallic point ($\simeq 10$ Å in diameter) is placed in a vacuum enclosure. A potential of about 10^4 V is established between this point and an anode. The electrons released from the metal follow the lines of force and are picked up by a spherical fluorescent screen (fig. 5) of about 10 cm in radius. Estimate the order of magnitude of the theoretical magnification of this microscope. Show that in this way one can hope to 'see' the atoms making up the metallic point. Actually, the resolving power is limited by the diffraction of the electrons. To remedy this, the polarity of the field is reversed and helium atoms are introduced into the tube. These latter are ionized, and following radial trajectories end up on the screen, producing an image. Why is the resolution improved in this way?

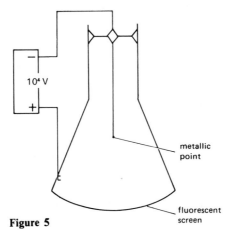

Figure 5

7.20. *Junction resistance*

(a) Consider a gas of N free fermions. In the diagram in fig. 1, the vertical axis represents the scale of the individual levels. In diagram (a), the density of states for one fermion is shown on the abscissa. (Why does the curve have a parabolic shape?) In diagram (b), the step function $Y(\varepsilon - \varepsilon_F)$ is represented (again on the abscissa), where ε_F denotes the Fermi level of the system being considered. Show that

$$N = \int_{-\infty}^{\infty} \rho(\varepsilon)\, Y(\varepsilon - \varepsilon_F)\, d\varepsilon.$$

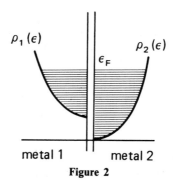

Figure 1

Now interpret diagram (c).

(b) A metal–insulator–metal junction is the interface between two metals, 1 and 2, possibly of different types and separated by a thin insulating layer (10 Å of oxide, for example). At equilibrium, the Fermi levels of the two metals (treated as two free-electron gases) are equal. This is shown in fig. 2, in which the situation in metal 1 appears to the left, and to the right the situation in the electron gas of metal 2. The interval separating the two portions of the curve denotes the insulating layer. The left side of the junction is now maintained at a potential $-V < 0$, relative to the right side. How does this change the diagram in fig. 2?

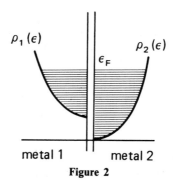

Figure 2

(c) The Fermi levels not being the same any more, the junction is crossed by a current of electrons. Determine the direction of flow of this current.

(d) The results reported in fig. 3 are for symmetrical junctions of the type $Al-Al_2O_3-Al$. These junctions, numbered 1 to 5, all have the same width (50 Å) and are in the ratios 1, 2, 3, 4, 5, the unit surface being 5×10^{-2} cm^2. It would seem that for low voltages, the junctions being considered behave as ohmic resistances: the current I crossing it is proportional to the applied voltage V. Evaluate the resistivity of the junction $Al-Al_2O_3-Al$. Compare it with the resistivities of metals, semiconductors and insulators.

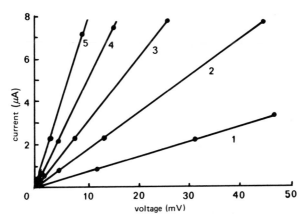

[From J.C. Fisher, I. Geaver, J. Appl. Phys. 32 (1961) 172]

Figure 3

(e) We would like to interpret the constant of proportionality between I and V. Show that the number of electrons, of energies between ε and $\varepsilon + d\varepsilon$, which cross the junction from left to right is,

$$dN(\varepsilon) = \mathscr{P}_{12}(\varepsilon)\, \rho_1(\varepsilon - q_e V)\, Y(\varepsilon - q_e V)\, \rho_2(\varepsilon)\, [1 - Y(\varepsilon)]\, d\varepsilon,$$

where $\mathscr{P}_{12}(\varepsilon)$ denotes the probability for an electron of energy ε to cross the insulating barrier by means of the tunnel effect. (The zero of the energy is chosen to be the Fermi level of the metal to the right.) Set up an analogous relationship giving the number of electrons crossing the junction from right to left.

(f) Hence deduce that the total current crossing the junction is

$$I \propto \int_0^\infty \mathscr{P}_{12}(\varepsilon)\, \rho_1(\varepsilon - q_e V)\, \rho_2(\varepsilon)\, [Y(\varepsilon - q_e V) - Y(\varepsilon)]\, d\varepsilon.$$

(g) Assume that the applied voltage is sufficiently low, so that the variation of $\mathscr{P}_{12}(\varepsilon)$ with ε may be neglected and $\rho_1(\varepsilon - q_e V)$ and $\rho_2(\varepsilon)$ identified with their values at the origin (that is, at the Fermi level of aluminium). Show that one then has

$$I = A\rho_1(0)\, \rho_2(0)\, q_e V.$$

N.B. Recall that the step function is a distribution and its derivative is the Dirac δ-distribution.

7.21. (a) Figure 1 shows the band structure of *gallium arsenide* (GaAs). Is GaAs a conductor, insulator or semiconductor?

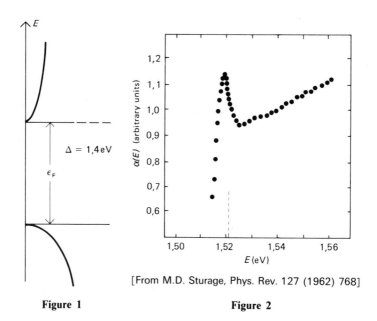

[From M.D. Sturage, Phys. Rev. 127 (1962) 768]

Figure 1 **Figure 2**

(b) Figure 2 reproduces the variation of the light absorption coefficient $\alpha(E)$, of a sample of GaAs, as a function of the energy E of the incident photons. Hence obtain an experimental value for the width Δ of the forbidden band of GaAs.

(c) Show that the curve $\alpha(E)$ must reproduce the variation of the density of the individual stationary states of the conduction band. Verify this is the case here. We shall not take into consideration the absorption line, seen in fig. 2 near the absorption threshold, and which requires a more powerful theory.

(d) The technique of epitaxy using molecular jets, developed in the seventies, enables one to stack alternate layers of different crystals, the thicknesses of which are controlled to nearly interatomic distances. Consider first, a set-up consisting of a layer of GaAs, of thickness d (between 3 Å and 4000 Å) sandwiched within much thicker layers (from 5000 Å to 2 μm) of a complex arsenic, gallium and aluminium salt (denoted GaAlAs) (fig. 3). The potential felt by the electrons in the direction Oz is represented as follows:

$V(z) = \infty$ in GaAlAs,

$V(z) = 0$ in GaAs.

The electrons are free in the two directions Ox, Oy.
Write down the stationary wavefunctions for the electrons and show that the corresponding energies are of the form

$$\varepsilon = \varepsilon_q + \frac{1}{2m}(p_x^2 + p_y^2) \quad \text{with} \quad \varepsilon_q = \frac{1}{2m}\left(\frac{q\pi}{d}\right)^2, \quad q \text{ an integer.}$$

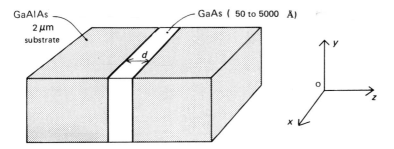

Figure 3

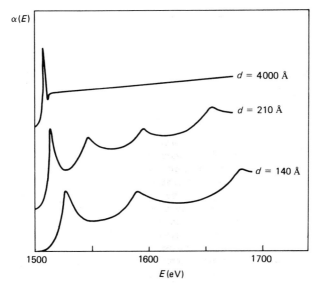

[From R. Dingle, W. Wiegmann, C. Henry, Phys. Rev. Lett. 33 (1974) 827]
Figure 4

(e) Find the density of states $\rho(\varepsilon)$ for the value $q = 1$ of the parameter q. What is it for $q = 2$? (Use the results of exercise 6.5.) Show (after tracing it) that the curve $\rho(\varepsilon)$ consists of a series of plateaus, and determine their positions in relation to the curve giving the density of states of a quanton in a three-dimensional well.

(f) Figure 4 reproduces the absorption spectra [the $\alpha(E)$ curves] as measured with the above set-up for three different thicknesses $d = 140$, 240 and 4000 Å. Compare this last result with that of fig. 2, and interpret the form of the three spectra. Compare the calculated values for the absorption thresholds, on the basis of the results in question (e), to those seen from fig. 4. Where, in your judgement, does the discrepancy between the experimental and the theoretical results come from?

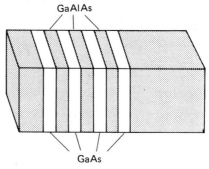

Figure 5

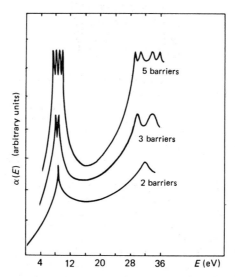

[From R. Tsu and L. Esaki, Appl. Phys. Lett. 22 (1973) 562]

Figure 6

(g) The above set-up is now made more elaborate by adjoining it to other devices of the same type (fig. 5). The distance separating two consecutive layers of GaAs (of thickness $d = 50$ Å) is of the order of a few tens of angstroms. Figure 6 reproduces the absorption spectra obtained for 1 layer of GaAs (2 barriers), 2 layers of GaAs (3 barriers) and 4 layers of GaAs (5 barriers). What is the form of the potential felt by the electrons in the direction Oz? Can you interpret the shape of the curves obtained?

7.22. In a semiconducting material, the Fermi level coincides with the top of an allowed band. In other words, a forbidden band of width Δ extends from the last occupied level to the first free level. Thus, the material can absorb a photon only when the latter can deliver to an electron an amount of energy $E > \Delta$, enough to enable the electron to pass into a state within the free band. Photons with energy $E > \Delta$ cannot be absorbed. If a semiconductor is irradiated with normal white light, it is only these photons, with energy $E < \Delta$, which are transmitted, giving the material its *apparent colour*. Thus, what is the apparent colour

    (a) of vermillion HgS, for which $\Delta = 2.1$ eV,

    (b) of cadmium yellow CdS, for which $\Delta = 2.6$ eV,

    (c) of diamond, for which $\Delta = 5.4$ eV?

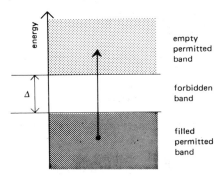

7.23. The *colour of a metallic body* is determined by the spectral composition of the light it reflects, that is, light which it does not absorb.

    (a) A metallic crystal receives a flux of photons, of visible light, having pulsation ω. Determine the order of magnitude of the energy of these photons. Compare this energy with that of the electrons of the crystal. Hence, deduce that only the external (delocalized) electrons of the metal enter into the (visible) radiation/matter interaction process.

    (b) We try to analyze the absorption of a photon by an electron, by treating the delocalized electrons of the metal in the free-electron approximation. Compute the order of magnitude of the momentum of an absorbed photon and compare it to that of the electrons. Show that it is impossible to ensure simultaneously the conservation of energy and of momentum during the process of absorption of a photon by a free electron.

    (c) However, the electrons of a metal are not free and their energy spectrum contains many allowed bands. Write down the conservation laws, which take into account the partial invariance under spatial translation of the photon + electron system, placed in the crystalline potential. For simplicity, consider a one-dimensional model.

    (d) Show that, in view of the different orders of magnitude, the absorption of a photon by an electron is possible under the condition that the electron make a transition from one allowed band to another. Prove that this transition is represented on the curve $E(\not{p})$ (dispersion pseudo-curve) by a 'vertical' displacement (at $\not{p} = $ const.) of the point representing the state of the electron.

It is also necessary, for this transition to be possible, that the final state of the electron be not already occupied. The figure shows the curve $E(\not{p})$ of an alkaline type: the first allowed band is not completely occupied. We realize that the curve $E(\not{p})$ does not differ significantly from a parabola except at the edges of the Brillouin zone.

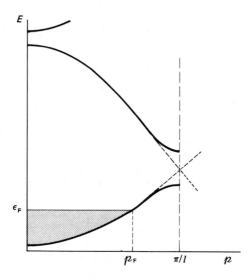

(e) Show that the spectrum of the absorption of light by the metal displays a series of bands. Prove that there exists a minimal absorption frequency and express it as a function of ε_F, the Fermi level of the metal.

(f) The table below displays the values of ε_F for different alkali metals. From this deduce that the light reflected by rubidium is richer in red than that reflected by sodium. Is this true? As a matter of fact, the absorption of light by a metal involves other processes, besides the one described here, which complicate the description.

			Metal	
	Na	K	Rb	Cs
ε_F(eV)	3.24	2.12	1.85	1.59

7.24. One of the most common modern electronic components goes by the name of the *tunnel diode*. Very schematically, a tunnel diode is obtained by bringing into contact the two halves of the same semiconducting crystal (GaAs, for example) after having 'doped' them differently (i.e., after having introduced impurities of different chemical types into them). Figure 1 summarizes the stages in the preparation from the energetic point of view.

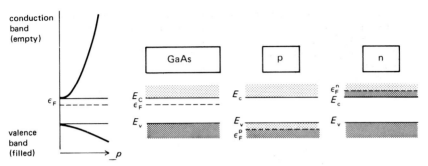

Figure 1

(a) The two halves are brought into contact. At equilibrium, the Fermi levels ε_F^p and ε_F^n are equal. Plot the energy diagram corresponding to the p–n junction and verify that no electrons can cross the junction (neither from p to n nor from n to p) since, for each electron from one half of the diode, the state of the same energy in the other half is either occupied or forbidden.

(b) A voltage V is applied to the ends of the diode (fig. 2). The p-side of the junction is brought to a potential $V > 0$ with respect to the n-side, whose potential we set equal to zero. Plot the corresponding energy diagram and show that electrons move from n to p.

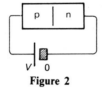

Figure 2

(c) Figure 3 displays the characteristic $I(V)$, giving the intensity I of the current crossing the diode as a function of the applied voltage V. By studying the gradual change in the energy diagrams as V is increased, explain the form of the curve $I(V)$. Trace, in particular, the diagram corresponding to the points A and B.

(d) The tunnel diode has a negative resistance in the AB portion of its characteristic. Show that by inserting the device of fig. 2 into an ordinary oscillating L, R, C circuit, one can prevent the absorption of the oscillations. Does this contradict the second law of thermodynamics?

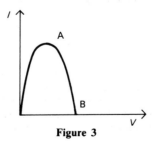

Figure 3

7.25. Consider *electrical conduction* in a one-dimensional metallic crystal, the band structure of which is shown in the figure below. Denote by ε_F the Fermi level of the metal and by $\not{p}_F$ the corresponding quasi-momentum.

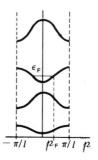

(a) Study the movement of *one* electron under the effect of an electric field $\mathscr{E}$ applied to the crystal. The electron is represented by a wave packet formed from states of momenta close to p. Inside the crystal, the electron is subject to an external force $- q_e V$. The second law of motion, which now operates on the quasi-momentum $\not{p}$, can be written as [see eq. (7.4.50)]

$$-q_e \mathscr{E} = \frac{\mathrm{d}\not{p}}{\mathrm{d}t}.$$

Establish the formula giving the group velocity v_g of the wave packet. (To do this, write down the variation in the energy, δE, of the electron while the field is applied, over an interval of time δt.) Plot the curve of the variation of v_g as a function of the average quasi-momentum $\not{p}$ of the wave packet from one end of the Brillouin zone to the other.

(b) Show that the acceleration of the electron under the effect of the field $\mathscr{E}$ is

$$\frac{\mathrm{d}v_g}{\mathrm{d}t} = -q_e \frac{\partial^2 E}{\partial \not{p}^2} \mathscr{E}.$$

(c) Compare this relation with that which governs, in an analogous manner, the motion of an electron in free space (outside the crystal). Hence deduce that inside the crystal, everything happens as though the electron had an effective (and fictitious) mass

$$m^* = \left(\frac{\partial^2 E}{\partial \not{p}^2}\right)^{-1}.$$

(d) Trace the curve of the variation of m^* as a function of $\not{p}$. Under what conditions could one replace m^* by the real mass m? How is the existence of negative values for m^* to be understood? What does $|m^*| = \infty$ signify? Can you understand, qualitatively, why the effective mass is minimal at the centre and at the edges of the Brillouin zone?

(e) The crystal does not contain 1 but rather N electrons, and the current I, generated by the application of the field $\mathscr{E}$, results from the contributions coming from all the electrons. Establish the relation

$$\frac{dI}{dt} = \sum_N \frac{d}{dt}(-q_e v_g),$$

and show that inside the crystal

$$\frac{dI}{dt} = \frac{1}{\pi} q_e^2 \mathscr{E} \int \frac{\partial^2 E}{\partial \not{k}^2} \, d\not{k},$$

where the integral is over all the occupied individual states.

(f) Comparing this result with that which would have been obtained for N free electrons, subject to the same field $\mathscr{E}$, show that the number of electrons effectively participating in the conduction of the metal is

$$N_{\text{eff}} = \frac{2m}{\pi} \left(\frac{\partial E}{\partial \not{k}} \right)_{\not{k} = \not{k}_F}.$$

Hence recover the result: the electrons of a band do not contribute to the electric current if the band is filled.

7.26. A free neutron is unstable and decays by β-radioactivity, $n \to p + e^- + \bar{v}$. How is it then that *stable nuclei* can contain neutrons? Show that the Pauli principle enables us to answer this question.

7.27. Astrophysicists acknowledge the existence of *neutron stars* (these are the so-called pulsars) – highly dense objects, resulting from the contraction of an ordinary star into a volume comparable to that of the earth, and thus having a mass $M \simeq 10^{30}$ kg and a radius $R \simeq 10^4$ km. Why is it that such an object would consist essentially of neutrons and contain no protons and electrons, while in principle neutrons are unstable and decay exactly into protons and electrons: $n \to p + e^- + \bar{v}$? Observe that, unlike in the case of ordinary nuclei (see exercise 7.26), the electrons coming from such a decay (but not the neutrinos) would be held back in the system by gravitational attraction. But we shall see that their presence is not favoured energetically.

(a) Assuming that the object is made up of hydrogen (protons and electrons), calculate the density of electrons and their Fermi level ε_F.

(b) Show that this Fermi level is much higher than the mass energy mc^2 of the electron, and hence that it is necessary to modify formula (7.4.43). Show that, in the 'ultra-relativistic' case we have

$$\varepsilon_F = \hbar c (NV)^{1/3}$$

(use the results of exercise 6.5d). Recompute ε_F under these conditions.

(c) Compare the Fermi level of the electrons ε_F with the difference between the mass energies of the neutron and the proton:

$$m_n c^2 - m_p c^2 = 0.8 \text{ MeV},$$

and hence deduce that neutron stars are indeed made up ... of neutrons.

7.28. The interaction between molecules of the same substance can be represented, in a fairly general manner, by means of a potential having a universal shape, but with the range R and the depth U depending upon the nature of the molecules, of mass m, being considered.

(a) What is the characteristic action A of the interaction between the molecules? We want to compare the *low-temperature properties of the fluids* made up of:

– diatomic molecules of the isotopes of hydrogen, namely H_2, D_2, T_2, for which $R = 3.2$ Å and $U = 2.9 \times 10^{-3}$ eV;

– monatomic helium molecules and the isotopes of helium, namely ^{3}He, ^{4}He, ^{6}He ($R = 2.9$ Å; $U = 0.8 \times 10^{-3}$ eV);

– monatomic molecules of the isotopes of hydrogen in the polarized form, namely H↑, D↑, T↑ (case in which $R = 4.1$ Å; $U = 0.5 \times 10^{-3}$ eV).

(b) For each one of these nine substances, compute the dimensionless 'quantic parameter' $\xi = A/h$. According to certain theoretical computations, the value of ξ determines the possibility for a system of bosonic molecules to exhibit a phase transition to a specifically quantum state, depending on its position relative to a critical value $\xi_c = 2.8$. What substances should one expect to show a superfluid transition?

(c) Knowing that for ^{4}He the transition (here, superfluidity) occurs at $T_\lambda = 2.18$ K, can one estimate the superfluid transition temperature for the other fluids?

[From W. C. Stwally and L. H. Nosanow, Phys. Rev. Lett. 36 (1976) 910.]

7.29. Show, using expressions (7.5.32), for the average number of photons in a state of energy $\hbar\omega$, and eq. (7.5.34), for the density of states, that the average (volume) density of photons (of all energies) in *black-body radiation* at temperature T is given by

$$d = A(kT/\hbar c)^3,$$

where A is a numerical constant. For example, how many photons are there, on the average, in a dark region of 10^3 m at normal temperature?

7.30. Generalize the Planck formula for *black-body radiation*, eq. (7.5.36), to the case of bosons of mass m. Show that if the temperature T is such that $kT \ll mc^2$, the spectral density, per unit volume, of the energy may approximately be written as

$$u_m(E) \simeq \frac{1}{2\pi^2}(2m^5c^4 E)^{1/2}\exp[-(mc^2 + E)/kT].$$

Hence deduce that the total energy

$$\rho_m = \int_0^\infty u_m(E)\, dE,$$

has the ratio

$$\rho_m/\rho_0 = C(mc^2/kT)^{5/2}\exp(-mc^2/kT),$$

with respect to the energy of a zero-mass radiation (given by Stefan's law), where C is a numerical constant. Evaluate this ratio numerically for π-mesons ($mc^2 \simeq 140$ MeV) at normal temperature ($kT \simeq \frac{1}{40}$ eV).

7.31. Consider a *fermionic black-body radiation*, i.e., an ensemble of fermions, assumed to be of mass zero (of neutrinos, for example), at thermal equilibrium, at the temperature T, with an (emitting or absorbing) material.

(a) Using a two-level atomic model, as in the text (sect. 7.5.2c), show that the Pauli principle, forbidding the emission of a fermion into an already occupied state, enables one to compute the average number n (necessarily between 0 and 1) of fermions in a state having energy $\hbar\omega$, from the equilibrium equation [homologue of eq. (7.5.30)]

$$\mathcal{N}_f\, n\mathcal{P} = \mathcal{N}_e(1-n)\mathcal{P}.$$

(b) Hence obtain the average number $n(\omega)$ of fermions in a state of energy $\hbar\omega$ and then the spectral density, per unit volume, of the radiation. Compare it with Planck's law, eq. (7.5.36). Does it admit a classical limit? Discuss (see sect. 7.6).

Subject Index